AF412728

Developments in Geotechnical Engineering, 79

Coupled Thermo-Hydro-Mechanical Processes of Fractured Media

Mathematical and Experimental Studies

Further titles in this series:
Volumes 2, 3, 5-7, 9, 10, 12, 13, 15, 16A, 22 and 26 are out of print

1.   G. SANGLERAT – THE PENETROMETER AND SOIL EXPLORATION
4.   R. SILVESTER – COASTAL ENGINEERING. 1 AND 2
8.   L.N. PERSEN – ROCK DYNAMICS AND GEOPHYSICAL EXPLORATION
      Introduction to Stress Waves in Rocks
11.   H.K. GUPTA AND B.K. RASTOGI – DAMS AND EARTHQUAKES
14.   B. VOIGHT (Editor) – ROCKSLIDES AND AVALANCHES. 1 and 2
17.   A.P.S. SELVADURAI – ELASTIC ANALYSIS OF SOIL-FOUNDATION INTERACTION
18.   J. FEDA – STRESS IN SUBSOIL AND METHODS OF FINAL SETTLEMENT CALCULATION
19.   Á. KÉZDI – STABILIZED EARTH ROADS
20.   E.W. BRAND AND R.P. BRENNER (Editors) – SOFT-CLAY ENGINEERING
21.   A. MYSLIVE AND Z. KYSELA – THE BEARING CAPACITY OF BUILDING FOUNDATIONS
23.   P. BRUUN – STABILITY OF TIDAL INLETS – Theory and Engineering
24.   Z. BAŽANT – METHODS OF FOUNDATION EGINEERING
25.   Á. KÉZDI – SOIL PHYSICS - Selected Topics
27.   D. STEPHENSON – ROCKFILL IN HYDRAULIC ENGINEERING
28.   P.E. FRIVIK, N. JANBU, R. SAETERSDAL AND L.I. FINBORUD (Editors) – GROUND FREEZING 1980
29.   P. PETER – CANAL AND RIVER LEVÉES
30.   J. FEDA – MECHANICS OF PARTICULATE MATERIALS - The Principles
31.   Q. ZÁRUBA AND V. MENCL – LANDSLIDES AND THEIR CONTROL
      Second completely revised edition
32.   I.W. FARMER (Editor) – STRATA MECHANICS
33.   L. HOBST AND J. ZAJÍC – ANCHORING IN ROCK AND SOIL
      Second completely revised edition
34.   G. SANGLERAT, G. OLIVARI AND B. CAMBOU – PRACTICAL PROBLEMS IN SOIL MECHANICS AND
      FOUNDATION ENGINEERING, 1 and 2
35.   L. RÉTHÁTI – GROUNDWATER IN CIVIL ENGINEERING
36.   S.S. VYALOV – RHEOLOGICAL FUNDAMENTALS OF SOIL MECHANICS
37.   P. BRUUN (Editor) – DESIGN AND CONSTRUCTION OF MOUNDS FOR BREAKWATER AND
      COASTAL PROTECTION
38.   W.F. CHEN AND G.Y. BALADI – SOIL PLASTICITY - Theory and Implementation
39.   E.T. HANRAHAN – THE GEOTECTONICS OF REAL MATERIALS: THE $\varepsilon_g \varepsilon_k$ METHOD
40.   J. ALDORF AND K. EXNER – MINE OPENINGS - Stability and Support
41.   J.E. GILLOT – CLAY IN ENGINEERING GEOLOGY
42.   A.S. CAKMAK (Editor) – SOIL DYNAMICS AND LIQUEFACTION
43.   A.S. CAKMAK (Editor) – SOIL-STRUCTURE INTERACTION
44.   A.S. CAKMAK (Editor) – GROUND MOTION AND ENGINEERING SEISMOLOGY
45.   A.S. CAKMAK (Editor) – STRUCTURES, UNDERGROUND STRUCTURES, DAMS, AND
      STOCHASTIC METHODS
46.   L. RÉTHÁTI – PROBABILISTIC SOLUTIONS IN GEOTECTONICS
47.   B.M. DAS – THEORETICAL FOUNDATION ENGINEERING
48.   W. DERSKI, R. IZBICKI, I. KISIEL AND Z. MROZ – ROCK AND SOIL MECHANICS
49.   T. ARIMAN, M. HAMADA, A.C. SINGHAL, M.A. HAROUN AND A.S. CAKMAK (Editors) – RECENT
      ADVANCES IN LIFELINE EARTHQUAKE ENGINEERING
50.   B.M. DAS – EARTH ANCHORS
51.   K. THIEL – ROCK MECHANICS IN HYDROENGINEERING
52.   W.F. CHEN AND X.L. LIU – LIMIT ANALYSIS IN SOIL MECHANICS
53.   W.F. CHEN AND E. MIZUNO – NONLINEAR ANALYSIS IN SOIL MECHANICS
54.   F.H. CHEN – FOUNDATIONS ON EXPANSIVE SOILS
55.   J. VERFEL – ROCK GROUTING AND DIAPHRAGM WALL CONSTRUCTION
56.   B.N. WHITTAKER AND D.J. REDDISH – SUBSIDENCE – Occurrence, Prediction and Control
57.   E. NONVEILLER – GROUTING, THEORY AND PRACTICE
58.   V. KOLÁŘ AND I. NĚMEC – MODELLING OF SOIL-STRUCTURE INTERACTION
59A. R.S. SINHA (Editor) – UNDERGROUND STRUCTURES – Design and Instrumentation
59B. R.S. SINHA (Editor) – UNDERGROUND STRUCTURES – Design and Construction
60.   R.L. HARLAN, K.E. KOLM AND E.D. GUTENTAG – WATER-WELL DESIGN AND CONSTRUCTION
61.   I. KASDA – FINITE ELEMENT TECHNIQUES IN GROUNDWATER FLOW STUDIES
62.   L. FIALOVSZKY (Editor) – SURVEYING INSTRUMENTS AND THEIR OPERATION PRINCIPLES

Developments in Geotechnical Engineering, 79

Coupled Thermo-Hydro-Mechanical Processes of Fractured Media

Mathematical and Experimental Studies

RECENT DEVELOPMENTS OF DECOVALEX PROJECT FOR RADIOACTIVE WASTE REPOSITORIES

Edited by

Ove Stephansson
Lanru Jing
Department of Civil and Environmental Engineering
Royal Institute of Technology
S-100 44 Stockholm, Sweden

and

Chin-Fu Tsang
Earth Science Division
Lawrence Berkeley National Laboratory
Berkely, California, USA

1996

ELSEVIER
Amsterdam — Lausanne — New York — Oxford — Shannon — Tokyo

ELSEVIER SCIENCE B.V.
Sara Burgerhartstraat 25
P.O. Box 211, 1000 AE Amsterdam, The Netherlands

ISBN: 0-444-82545-2

This book is printed on acid-free paper.

Printed in The Netherlands

FOREWORD

The characteristic that most distinguishes rock from other materials used in engineering is that rock is discontinuous on all scales from submillimeter microcracks and pores through joints and fractures on the scale of meters to faults on the scale of kilometers or more. The ubiquitously discontinuous nature of rock gives rise to distinctive behavior and properties such as low tensile strength despite relatively great compressive strength. However, the most significant consequence in many situations is the ability of rock to store and transport fluids. This ability is of obvious interest in geology and geophysics, hydrology, the petroleum industry and, most recently, waste disposal.

An important consequence of the storage and transport of fluids in rocks is that it gives rise to significant coupling between mechanical, hydrological, thermal and chemical processes. Such coupled processes can affect significantly the performance of structures and operations engineered in rock. Coupling between fluid pressures and mechanical stresses has long been recognized as being of primary importance in many problems and is handled using the concept of "effective stress". Thermomechanical stresses are well recognized in the field of mechanics. Simple calculations show that convective transport of heat by fluids in many situations easily overwhelms conduction of heat in many rocks. Multi-phase convective transport can result in "heat-pipe" phenomena that are even stronger in their effects on energy transport. Dissolution and precipitation of minerals is, on the one hand affected by temperature and stress and, on the other hand, can result in significant changes in permeability as well as in the "retardation" of solute transport.

The need to isolate nuclear wastes from the biosphere through geologic disposal has served to emphasize the importance of coupled processes in rocks. To demonstrate effective isolation it is necessary to have the capability of modeling and predicting the effects of coupled processes on isolation over periods of time unprecedented in engineering. Most of the field experiments that have been done in relation to the geologic isolation of nuclear wastes have revealed the dominance of fractures in determining the behavior of rock masses. Fractures result in mechanical, thermal and hydraulic behaviors that can differ by orders of magnitude from those which would be expected based on the mechanical, thermal and hydraulic properties of the rock matrix, as derived from tests on laboratory samples of intact rock. These findings have stimulated a great deal of research into the behavior and properties of fractures. Quickly it was discovered that the simplest, zero-order abstractions of fracture behavior, such as laminar flow between parallel plates to represent fluid flow in fractures, failed to capture adequately the geologic complexity. In the past decade, significant advances have been made in capturing conceptually the processes involved in the mechanical deformation of individual fractures and fluid flow through them as well as in analyzing these processes analytically and numerically. Similar progress has been made with respect to fractured rock masses.

This book constitutes a compendium of nineteen chapters authored by individuals who have made major contributions to recent progress throughout the world. The book revolves around the DECOVALEX project and is edited by the leaders of that project Professor Ove Stephansson and Dr. Lanru Jing from The Royal Institute of Technology, Stockholm in Sweden and Dr. Chin-Fu Tsang of the Lawrence Berkeley National Laboratory in the USA.

The acronym DECOVALEX stands for the **De**velopment of **Co**upled Models and their **Val**idation against **Ex**periments in nuclear waste isolation, an international project sponsored by organizations in Canada, Finland, France, Japan, Sweden, U.K. and USA, between 1992 and 1995. Likewise, the authors of the nineteen chapters in this book come from all over the world. DECOVALEX revolved around the numerical simulation of three Bench Mark Tests (BMT's) and six Test Cases (TC's). The BMT's are hypothetical two-dimensional coupled, thermal-hydraulic-mechanical problems (chemical processes and phase changes were not included) and the TC's were based on laboratory or field experiments.

Studying the results reported in the nineteen chapters of this volume leads one to conclude that most of the numerical codes are based on correct physical assumptions and lead to similar and mutually consistent predictions. It is interesting, but not surprising, to find that temperature predictions for the BMT's agree and that convective heat transport is relatively minor, given the absence of phase change. Much more significant however, are comparisons between results predicted by numerical simulation and measurements in real laboratory or field tests. Here, again, in general temperature predictions seem to be the most robust but significant disparities arise in both flow predictions and, particularly, those for displacement. Clearly, the implication of this is that the abstractions used for the mechanical and hydraulic properties of both fractures, which are discontinuous, and clay buffer materials, which are continuous, save on the smallest scale, do not represent adequately the real properties of rocks and clays.

Validation of numerical models or codes is a subtle concept that must encompass tests for a range of potential sources of error. First, the codes must be based on physically realizable abstractions. These abstractions cannot contradict basic laws, such as those of thermodynamics. Second, the actual code must implement these abstractions correctly and without excessive numerical error. Comparing codes against one another, as in the BMT's, and against known analytical solutions is effective in guarding against these kinds of errors. The results presented here suggest that most of the codes probably do not contain mistakes from these two potential sources of error. Third, and most significantly, the abstractions used in codes may fail to capture adequately the actual mechanical hydraulic and thermal behavior of real rocks and clays. The results of the TC's suggest that there is much room for improvement in capturing the mechanical and hydraulic behavior of rock fractures in particular. To achieve this, it is necessary to have a much more sophisticated understanding of the actual processes of deformation at the contacts between the two rough surfaces of rock fractures and of fluid flow through the void spaces between these surfaces than has been incorporated into the numerical models described in this volume. Some of this understanding already exists but much of it requires further experimental observation and numerical simulation of the processes of deformation and flow.

Careful thought leads to the conclusion that the validation of models or codes is valuable but not nearly to decisive or convincing as the <u>invalidation</u> of codes. Any single instance where a numerical code <u>fails</u> to replicate the results of an analytical solution, another code, or, especially sound experimental measurements raises profound questions. Usually, it can be shown that a code faithfully follows the physical abstractions upon which it is based. If such a code is invalidated by comparison with experimental results, <u>fundamental conceptual changes</u> are required before we can have any confidence in the results of numerical simulation.

In some significant degree the results from the Test Cases <u>invalidated</u> one or more aspects of every code. This demonstrates a compelling need for further work on the properties of fractures and clays and for DECOVALEX II.

The research and analysis described in this volume have indeed served a useful purpose. The overall conclusion must be that, despite substantial progress, we are not yet able to simulate all of the important fundamental processes that will occur in and around nuclear waste repositories. The authors of this work, their colleagues and sponsors are to be commended for bringing this important deficiency into such clear focus, thereby, revealing both the need for further work and the topics on which this work should be focused.

Neville G. W. Cook

Donald McLaughlin Professor of Materials Science
and Mineral Engineering
Professor of Civil Engineering
Professor of Nuclear Engineering
Chair, Energy and Resources Group
Faculty Senior Scientist, Earth Sciences Division,
Lawrence Berkeley Laboratory
University of California at Berkeley

PREFACE

Over the last ten years, substantial efforts have been made in theoretical and experimental studies of the effects of the couplings of temperature gradient, hydrologic flow and mechanical deformation in fractured rocks. Much of the impetus behind these efforts is the concern over solute transport through a rock mass hosting a heat-releasing nuclear waste repository. However, the problem is of wider interests, ranging from coupled processes associated with the construction of underground openings, geothermal energy extraction, micro-earthquakes induced by fluid injection, to secondary recovery from deep petroleum reservoirs using injection of water colder than that of *in-situ* fluids.

The coupling of thermo-hydro-mechanical (THM) processes presents a major challenge to the scientific community. From the theoretical standpoint, the three processes have widely different characteristic time constants and spatial scales. The thermal gradient has relatively large time constant and spatial scale, since it is a function of the long heating cycle, and the thermal dispersivity smoothes out the effects of "local" spatial property variations. Mechanical effects, on the other hand, propagate through the rock with the speed of sound waves and the deformation is strongly affected by faults and fractures, though much less so by medium property variations. Finally, the hydrologic flow and transport are very sensitive to smaller-scale medium heterogeneity, but with time scale corresponding to large solute transport times. Numerically these processes are commonly handled by different techniques, such as finite difference methods, finite element methods, discrete element methods and others. To combine all these into an efficient model for simulating coupled THM processes in fractured rock is no easy task.

From the point of view of laboratory experiments of coupled THM processes in rock samples, the challenge lies in providing a well-defined set of conditions and data. The effects of the deformation of the equipment set-up and the hydraulic perturbation due to monitoring holes cannot be ignored. Generalization of the results from laboratory experiments to generic relationships useful for practical applications is always a significant task. Field experiments face the usual challenge of properly defining the system, especially with regards to its boundary conditions and their time variations. Coupled THM experiments require duration of months and years and the durability of monitoring tools also becomes a matter of concern.

Due to the challenges described above, a group of scientists from eight countries came together from 1992 to 1996 under the auspices of the funding agencies of their respective countries and participated in an international cooperative research project, under the management of the Swedish Nuclear Power Inspectorate (SKI). The project is titled as Development of Coupled THM Models and Their Validation Against Experiments (DECOVALEX). With the completion of this project, a follow-up project called DECOVALEX II was proposed and agreed upon for 1996-1998 and this new project is now underway. The cooperative work so far yielded three major benefits: (1) encouraging the development of coupled THM codes by the national research teams and providing peer review and advice to each other, (2) defining both simple and realistic benchmark test (BMT) problems, so that the national research teams could study and carry out code verification studies on these problems and compare compuational results with those from other teams, and (3) collecting and documenting major laboratory and field test cases (TC), so that the national research teams can use them to perform validation studies of their models and codes.

The present book brings together the results, information and data that emerged out of all the efforts of the research teams of the DECOVALEX cooperative project. The general purpose is to present the state-of-the-art and to stimulate further study and research into coupled THM processes in rock masses. The book may be grouped into four sections. The first section, composed of the first two chapters, provides a conceptual introduction to coupled THM processes for the readers and a description of the DECOVALEX project that can act as a guide to the later parts of the book. The second section, composed of Chapters 3 through 9, gives a state-of-the-art survey of topics ranging from constitutive models for rock joints, formulation of coupled thermohydroelastic phenomena in variably saturated rocks, to a discussion of various modeling approaches and numerical issues. The third section, composed of Chapters 10 through 12, covers the description of the three BMT problems and the results from the research teams, providing a view of the state of our capability in addressing such problems. The fourth section is composed of the next seven chapters. Chapters 13 through 18 describe the laboratory and field experimental test cases (TC) and the results of the research teams in their attempts to simulate the experiments. The final chapter, Chapter 19, gives the lessons learned from all these efforts by three participants of the DECOVALEX project. The book concludes with an appendix of simple descriptions of the codes used by the research teams.

We envision that the book may be of value to three groups of readers. The students at the university graduate school level may find it useful to learn the state-of-the-art of the subject and various approaches being applied to coupled THM modeling. The readers who are researchers may want to study the various attempts given in the book by the different research teams and develop new and better methods for modeling or for experimental studies. The third and fourth sections of the book on the BMT's and TC's may be of particular value to them in testing their new methods. Finally, the readers who are concerned with practical implications of coupled THM processes with respect to safety of nuclear waste geologic repository or to other practical problems may find it useful to have a feeling of the state of capability of the scientists in providing information that may be of use to their decision-making. We shall be happy if the book promotes interest in this challenging area of science and stimulates its further study and development.

The book would not have happened without the help of many people. We would first like to acknowledge the support and guidance from the funding agencies of the different countries involved in the DECOVALEX project. The steady support and encouragement through the years by Fritz Kautsky of the Swedish Nuclear Power Inspectorate (SKI), Stockholm, Sweden, has been instrumental in the success of DECOVALEX as well as the completion of this book project. The cooperation of all the DECOVALEX research teams and authors through the painful process of writing their chapters, having them reviewed, and rewriting them is much appreciated. Finally, we are most grateful to Professor Neville G.W. Cook of the University of California at Berkeley, who wrote the Foreword for the book to provide the perspective of the subject. We have benefited from his discussions and ideas over the years and hope that the readers would be motivated by his insightful remarks to get more into the challenging field of coupled THM processes in rock masses.

Ove Stephansson and Lanru Jing
Royal Institute of Technology, Stockholm, Sweden

Chin-Fu Tsang
Ernest Orlando Lawrence Berkeley National Laboratory, Berkeley, CA, USA

LIST OF CONTRIBUTORS

R. Ababou
IMFT, Allée du Professor Camille Soulas, 31400 Toulouse, France

Mikko P. Ahola
Center for Nuclear Waste Regulatory Analyses, Southwest Research Institute, 6220 Culebra Road, San Antonio, TX 78238, USA

L. Börgesson
Clay Technology AB, Ideon, S-223 70 Lund, Sweden

A. Bougnoux
ENSMP-CGES, 35 Rue Saint Honoré, 77305 Fontainebleau Cedex, France

T. Chan
AECL Whiteshell Laboratories, Pinawa, Manitoba R0E 1LO, Canada
Current correspondence address: c/o Ontario Hydro H16 G27, 700 University Avenue, Toronto, Ontario M5G 1X6 Canada

Asadul H. Chowdhury
Center for Nuclear Waste Regulatory Analyses, Southwest Research Institute, 6220 Culebra Road, San Antonio, TX 78238, USA

L. Dewiere
Agence Nationale pour la gestion des Déchets Radioactifs (ANDRA), Parc de la Croix Blanche, 1/7 Rue Jean Monnet, 92298 Chatenay-Malabry Cedex, France

S. Follin
Department of Engineering Geology, Lund University, S-211 00 Lund, Sweden

T. Fujita
Power Reactor and Nuclear Fuel Development Corporation, Tokai-mura, Naka-gun, Ibaraki-ken, Japan

A.W. Herbert
School of Earth Sciences, University of Birmingham, Edgbaston, Birmingham B15 2TT, United Kingdom

Sui-Min Hsiung
Center for Nuclear Waste Regulatory Analyses, San Antonio, TX 78238, USA

Jan I. Israelsson
Itasca Geomekanik AB, Box 17, 781 21 Borlänge, Sweden

L. Jing
Division of Engineering Geology, Royal Institute of Technology, S-100 44 Stockholm, Sweden

J. Kajanen
Helsinki University of Technology, Laboratory of Rock Engineering, Vuorimiehentie 2, FIN-02150 VTT, Espoo, Finland

Daniel D. Kana
Southwest Research Institute, 6220 Culebra Road, San Antonio, TX 78238, USA

F. Kautsky
Swedish Nuclear Power Inspectorate, S-106 58 Stockholm, Sweden

K. Khair
Applied Geoscience Branch, Whiteshell Laboratories, Pinawa, Manitoba R0E 1L0, Canada

A. Kobayashi
Department of Agricultural Engineering, Iwata University, 3-18-1 Ueda, Morioka, Iwate, Japan

A. Kuusela-Lahtinen
Technical Research Centre of Finland, Communities and Infrastructure, PO Box 19041, FIN-02044 VTT, Espoo, Finland

Axel Makurat
Norwegian Geotechnical Institute, PO Box 3930, Ullevaal Hageby, N-0806 Oslo, Norway

A. Millard
CEA/DMT/SEMT, CEN Saclay, 91191 GIF/YVETTE Cedex, France

Sitakanta Mohanty
Center for Nuclear Waste Regulatory Analyses, San Antonio, TX 78238, USA

T.S. Nguyen
Atomic Energy Control Board, Wastes and Impact Division, 280 Slater Street, Ottawa K1P 5S9, Canada

Jahan Noorishad
Earth Sciences Division, Ernest Orlando Lawrence Berkeley National Laboratory, University of California, Berkeley, CA 94720, USA

Y. Ohnishi
School of Civil Engineering, Kyoto University, Yoshida-honmachi, Sakyo-ku, Kyoto, Japan

F. Plas
Agence Nationale pour la gestion des Déchets Radioactifs (ANDRA), Parc de la Croix Blanche, 1/7 Rue Jean Monnet, 92298 Chatenay-Malabry Cedex, France

J. Pöllä
Technical Research Centre of Finland, Communities and Infrastructure, PO Box 19041, FIN-02044 VTT, Espoo, Finland

G. Rehbinder
Dept. of Civil and Environmental Engineering, The Royal Institute of Technology, S-100 44 Stockholm, Sweden

A. Rejeb
Institut de Protection et de Sûreté Nucleaire, Départment de Protection de l'Environment et des Installations, 60-68 Avenue de Général Leclerc, B.P. 6, 92265 Fontenay-aux-Roses Cedex, France

J. Rutqvist
Division of Engineering Geology, Royal Institute of Technology, S-10044 Stockholm, Sweden

Ove Stephansson
Division of Engineering Geology, Department of Civil and Environment Engineering, Royal Institute of Technology, S-10044 Stockholm, Sweden

A. Stietel
CEA/DMT/SEMT, CEN Saclay, 91191 Gif/Yvette Cedex, France

Alain Thoraval
Laboratoire de Mecanique des Terrains, Institut National de L'Environment Industriel et des Risques, Parc de Saurupt (INERIS), 54042 Nancy, France

S-M. Tijani
Ecole Nationale Supérieure des Mines de Paris, Centre de Géotechnique et d'Exploitation du Sous-sol, 35 Rue Saint Honoré, 77305 Fontainebleau Cedex, France

E. Treille
CEA/DMT/SEMT, CEN Saclay, BP No. 2, 91191 Gif/Yvette Cedex, France

Chin-Fu Tsang
Earth Sciences Division, Ernest Orlando Lawrence Berkeley National Laboratory, One Cyclotron Road, Berkeley, CA 94720, USA

G. Vouille
Ecole Nationale Supérieure des Mines de Paris, Centre de Géotechnique et d'Exploitation du Sous-sol, 35 Rue Saint Honoré, 77305 Fontainebleau Cedex, France

E. Vuillod
INERIS, Ecole des Mines, Parc de Saurupt, 54042 Nancy Cedex, France

P. Wilcock
AEA Technology, 424.4 Harwell, Didcot, Oxfordshire OX11 0RA, United Kingdom

CONTENTS

O. Stephansson, L. Jing and C.-F. Tsang (Editors)
Coupled Thermo-Hydro-Mechanical Processes of Fractured Media
Developments in Geotechnical Engineering, vol. 79
© 1996 Elsevier Science B.V. All rights reserved.

A Conceptual Introduction to Coupled Thermo-Hydro-Mechanical Processes in Fractured Rocks

Chin-Fu Tsang[a] and Ove Stephansson[b]

[a]Earth Sciences Division, Ernest Orlando Lawrence Berkeley National Laboratory, University of California, Berkeley, CA 94720, USA

[b]Division of Engineering Geology, Department of Civil and Environment Engineering, Royal Institute of Technology, S-10044 Stockholm, Sweden

Abstract

The chapter presents an introduction to the conceptualization of coupled hydro-mechanical and thermo-hydro-mechanical processes in fractured rocks and points out promising directions for future research.

1. INTRODUCTION

The last decade has seen substantial progress in experimental and theoretical studies of the effects of coupling temperature gradient, hydrologic flow and mechanical deformation in fractured rocks. Much of the impetus behind these efforts is the concern of solute transport through a rock mass hosting a heat-releasing nuclear waste repository (Tsang, 1987). However, the problem is of wider interests, ranging from coupled hydromechanical processes associated with the construction of underground openings, geothermal energy extraction, earthquakes induced by fluid injection, to guidelines on injection pressures for stimulation of deep petroleum reservoirs with water colder than in-situ fluids. In this contribution the emphasis will be mainly on problems related to isolation and storage of radioactive waste.

Figure 1 gives an illustration of the various coupled processes involving thermal (T), hydrologic (H) and mechanical (M) effects for a nuclear waste repository in fractured rocks. This figure shows examples of coupling (1) T: heat flow due to radioactive waste heat release; (2) H: ground water flow through the rock fractures and matrix; and (3) M: deformation of rock matrix and rock fractures. The couplings indicated in the figure between T and H are: (a) water buoyancy flow and (b) heat transfer by convection. The couplings between T and M are: (c) thermal stress and (d) mechanical energy conversion. Finally, the couplings between H and M are: (e) changes in fracture apertures and rock porosity and (f) water pressure influence on effective stress. These and additional examples of coupled processes may be found in Tsang (1991), Stephansson (1995) and elsewhere in this book.

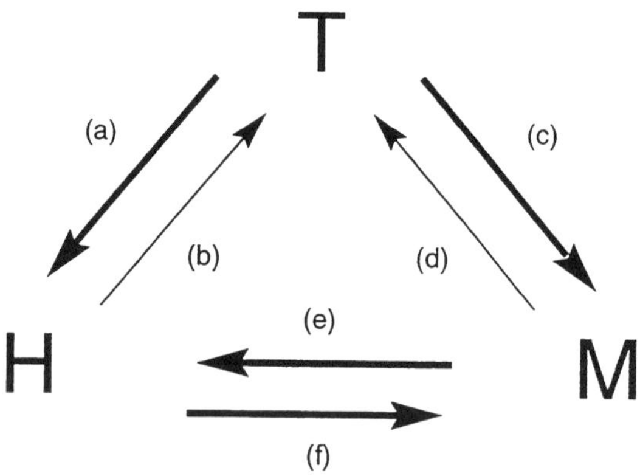

Figure 1. Coupled thermal (T), hydro (H), and mechanical (M) processes around a potential nuclear waste repository in fractured rock.

Additional processes and couplings exist, e.g., those with chemical effect, but they are not covered in this book.

Considerable progress has been made in the study of these coupled processes by many research groups and through the international cooperative DECOVALEX project (Jing et al., 1993, 1995, 1996; Stephansson et al., 1994). It is of particular note that such studies have encouraged hydrogeologists and rock mechanics researchers to learn from and cooperate in each other's area of interest. Through this, new methods and results have been developed to provide a better understanding of behavior of rock masses under complex T, H, M conditions.

The purpose of the present chapter is not to give a detailed overview of coupled processes or technical details of recent developments, which can be found elsewhere (e.g., Tsang, 1991 and Stephansson, 1995) and from the other contributions in this book. Rather we shall endeavor to present a conceptual model for the study of these coupled processes and to point out some directions where further research should be carried out.

2. A CONCEPTUAL MODEL OF COUPLED PROCESSES ASSOCIATED WITH A FRACTURED ROCK MASS

A typical rock mass consists of intact rock and discontinuities or fractures imbedded in it. These fractures have a large range of lengths, from mm to more than tens of meters. The gaps or apertures of fractures can also range from almost zero to mm size, and in some cases as much as one or two cm. Generally they vary over the fracture plane, and sometimes the gaps are filled with clay or other materials.

For intact rock without any fractures, the flow of fluid would be through the small pores randomly distributed in its matrix. The rock's thermal, hydrologic and mechanical behaviors can be described by continuum mechanics and thermohydraulic flow through a homogeneous medium. The modeling of such a system is relatively straightforward, except for non-linear effects causing significant numerical difficulties.

On the other hand, the presence of fractures qualitatively changes the system's behavior. Fluid flow would be mainly through these fractures which commonly form a network in the rock mass. Thus, fractured the rock is sometimes referred to as double-permeability system, one permeability (larger) associated with the fracture network and the other (smaller) associated with the rock matrix, that is, the medium between fractures. If the fractured rock is unsaturated with water (i.e., both the gas and water phases are present in the rock), gas will prefer to be in the fractures and water would tend to stay in the finer pores of the matrix because of capillary suction effects. Therefore, water flow will be mainly through the rock matrix and the small aperture areas of the fractures (Wang and Narasimhan, 1985). On the other hand, for water-saturated fractured rock, water flow will be mainly through the fractures because of their large permeability. For crystalline rocks, often the fracture permeability is several orders of magnitude larger than that of the rock matrix, so that the latter can be neglected and one needs only to consider the fracture network in flow calculations.

For mechanical stress and strain in the rock, fractures will control the stress-strain characteristics since the stress gradients tend to concentrate near the fractures. These gradients are particularly large at fracture tips or intersections between fractures. On a local scale how the aperture of a fracture changes depends on the presence of neighboring fractures and the regional stress field or the mechanical boundary conditions around the domain containing the fractures. Furthermore it depends on the variations of water pressure in the fracture and matrix pores, since they change the effective stress across the fracture. The interference of the different effects is complex and it is entirely possible that the aperture change is not uniform over the single fracture. On a larger scale to study the stress-strain behavior of a system of many fractures in detail is a very computational intensive effort, unless the fractures are arranged in a regular pattern, whose symmetries may allow calculational simplifications. Attempts on representing the multiple-fracture system as an equivalent anistropic continuum have been made (see for example Oda, 1985; Kawamoto, 1988; Pariseau, 1993; Cai and Horii, 1993; and also Stietel et al., this book), but the results are not yet satisfactory, except for the simplest applications. Further complications arise due to potential fracture propagations and bridging of adjacent fractures. Some aspects of these phenomena are discussed in section 4 below.

2.1. Terms on Hydromechanical Properties of a Single Fracture

Let us consider a single rock fracture in more detail. We may imagine it to be composed of two rock surfaces separated by a small space or aperture. Typically, these surfaces are not flat and smooth; rather they are rough, so that the aperture is not a constant, but variable over the fracture plane. It is useful to define eight terms often used in the context of the fracture hydromechanical behavior. These are illustrated in Fig. 2 (Hakami, 1995).

Roughness is the unevenness of the two rock surfaces forming the fracture. If the fracture is created by cracking the rock under tension, the roughness of the upper and lower fracture surfaces should be the same.

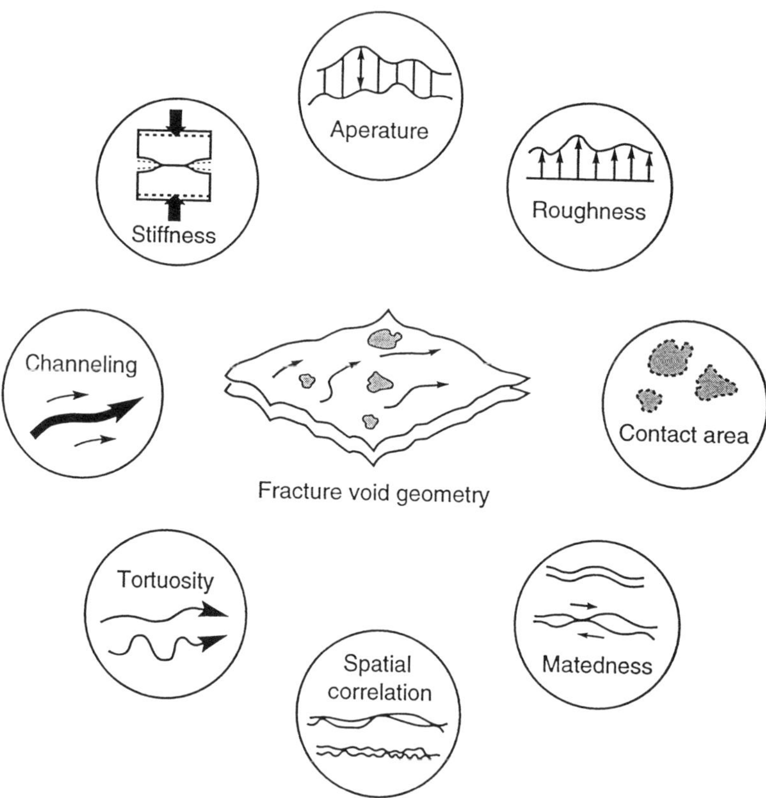

Figure 2. Various terms and concepts related to hydro-mechanical behavior of a rock fracture (Hakami, 1995).

The aperture is the gap between the upper and lower surfaces of a fracture. It is in general variable over the fracture plane (Bourke, 1987; Hakami and Barton, 1990).

Matedness is a description how well the upper and lower rock surfaces of a fracture match each other. If the upper and lower surfaces have the same roughness characteristics, then a well-mated case corresponds to one where the upper and lower surfaces have not been shifted relative to each other. In this case the aperture should be approximately constant and small. On the other hand, if there was a significant relative lateral shift between the upper and lower surfaces, it is a poorly-mated case and the aperture will be highly variable, being large in some areas and zero in others.

Contact areas are areas on the fracture surface where the upper and lower rock surfaces touch, i.e., where the aperture is zero.

Stiffness is a measure of how the rock fracture deforms under normal and shear loads.

Spatial correlation is a measure in length over which the fracture aperture value at one location is correlated with its neighboring points. Very roughly the spatial correlation length is directly related to the length over which the aperture may have similar values.

Tortuosity relates to the fact that the flow through a fracture with variable aperture does not follow a straight path, but a tortuous and longer one. Tortuosity is then the ratio of the length of the actual tortuous path, from one point to another, to the length of the straight line between the two points.

Channeling describes the phenomenon where because of the variable aperture over the fracture surface, fluid will tend to flow along a few major paths of least resistance. These paths, by which most of the flow goes, typically occupy only 20% of the fracture surface (Bourke 1987). Thus the flow velocity is much higher than that obtained by assuming the flow to be spread uniformly over the fracture surface.

2.2. Flow Through Single Fracture Under Normal Stress

Having explained the often used terms above, let us characterize a single fracture by an aperture distribution b(x,y), where (x,y) are points on the fracture surface (see Fig. 3). At each location, if the flow is laminar and steady, the permeability to flow is assumed to be given by the parallel-plate flow solution:

$$k(x,y) = b^2(x,y)/12$$

so that flow

$$q(x,y) \propto b.L.k(x,y)$$
$$\propto b^3(x,y)$$

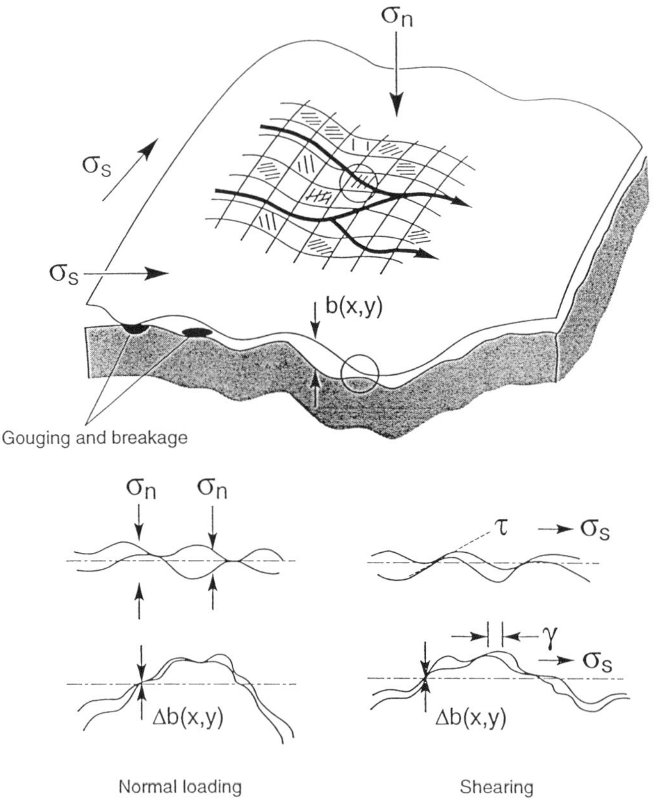

Figure 3.   Conceptual model of coupled HM processes around a single fracture.

where L is the fracture dimension and (b.L) is the cross-sectional area normal to the flow.   Early work in this field assumed a constant b value over the entire fracture so that

$$q \propto b^3$$

which is the so-called cubic law for flow through fractures.   The conditions under which this law is valid were studied by and Witherspoon et al. (1980) and others.

Under normal or shear stress, the function b(x,y) will be modified:

$$b(x,y) \rightarrow b'(x,y)$$

For relatively small normal stresses, b′(x,y) may be approximated:

$$b'(x,y) = b(x,y) - \Delta b \quad \text{or} \quad 0 \quad \text{if} \quad \Delta b > b(x,y)$$

where Δb is a displacement of constant value.   This is the approach taken by Tsang and Witherspoon (1981) and followed by several researchers, e.g.,

Hakami (1995). For large normal stresses, one would need to take care of the geometry near contact points between the upper and lower fracture surfaces, to account for local deformation near these points and to ensure no solid disappears (Hopkins et al., 1990). One may represent this refinement as

$$b'(x,y) = b(x,y) - Db$$

where Db is a displacement of constant value, except for all points (x_n, y_n) where $b(x,y) < \Delta b$.

The values $b'(x_n \pm \delta, y_n \pm \delta)$ are then estimated as having a larger contact area around (x_n, y_n) to ensure no net loss of materials. On the other hand, there are cases where the rock material is relatively weak so that the large normal stresses at contact points causes local gouging of material. In these cases $b'(x_n \pm \delta, y_n \pm \delta)$ has to be suitably modified to approximate the gouging result.

Flow through the system is calculated based on $b(x,y)$ or $b'(x,y)$, corresponding to initial flow condition or flow condition after the application of normal stress. Flow through such a system has been extensively studied by Tsang and co-worker (1987, 1988, 1989), Moreno and co-workers (1988, 1990, 1991, 1994), and Nordqvist et al. (1992). These authors assume that $b(x,y)$ can be represented by three parameters, i.e., mean aperture b_m, standard deviation of logarithm of b, σ_{lnb} and spatial correlation length λ. From these, flow permeability of the fracture can be calculated. The calculation of the normal stress-displacement relationship is usually based on contact areas. Gangi (1978) used a "bed of nails" or asperity model, whereas Tsang and Witherspoon (1981) assumed the stress stiffness to be inversely proportional to distances between contact points.

Two limiting approximations may deserve further comments. The first assumes that $b(x,y)$ can be represented by a number of asperities with $b = 0$, with the rest of fracture area having a constant aperture b_0 (Fig. 4). The flow through such a system can be calculated (Zimmerman et al., 1992), while the stress-strain relationship can be obtained using the Gangi method. Under normal stress, b_0 will decrease and each of the contact areas will increase. In some cases, two adjacent contact areas may merge (Fig. 4). This has been experimentally demonstrated using transparent replicas of natural joint surfaces (Hakami and Barton, 1990). This is a simple model on which the coupled HM effect on the fracture can be evaluated. The other limiting approximation assumes that the upper and lower surfaces of the fracture are mostly in contact except for certain pore bodies and pore throats, the latter being also referred to as aperture constrictions (Fig. 5). Here flow permeability is controlled by the sizes of the constrictions, whereas tracer travel time is dominated by the size of the pore bodies. Under normal stress, the size of the pore bodies will decrease as well as the constriction areas. In some locations the constrictions will be closed out to zero, causing a major perturbation on the flow.

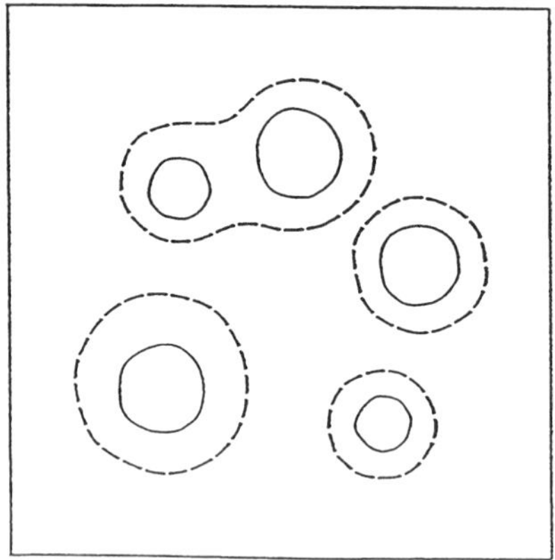

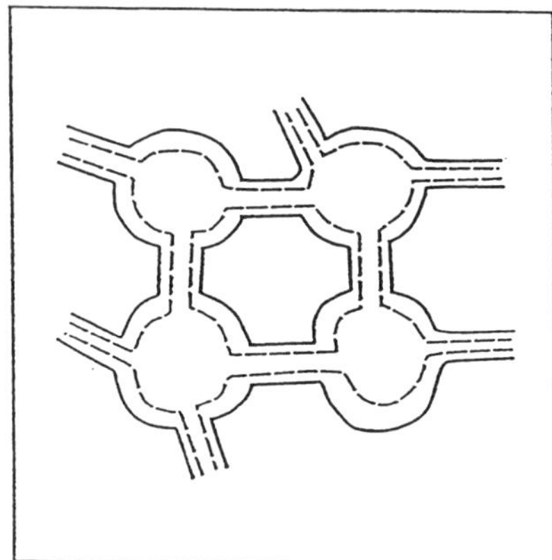

Figure 4. Asperity model of a rock fracture. Areas bounded by solid lines indicate contact areas and broken lines indicate increased contact areas under normal stress.

Figure 5. Void model of a rock fracture. Solid lines indicate the pores connected by pore throats (i.e., aperture constrictions). Broken lines indicated reduced pores under normal stress.

Concerning stress-strain relationship there exists a large number of test data, most of which fit the so-called hyperbolic model of closure versus normal stress (Goodman, 1976, Bandis et al., 1983). It is to be noted that at a very high normal stress, complete closing of apertures is not to be accomplished because the stiffness at the apertures will approach that of the intact rock as the remaining apertures become more circular in shape. This is why one finds fracture flow even at high stresses. In the discussion here, it is obvious that for a specified flow rate the pressure field in the rock fracture is controlled by pore throats or small aperture constrictions, whereas the tracer travel time depends on the pore volume the tracer has to cover and is therefore controlled by the arithmetic mean of apertures. This contrast leads to different calculated effective aperture values based on analysis of pressure flow or tracer transport experiments. The different "types" of apertures are explained and discussed by Tsang (1992).

2.3. Flow Through Single Fracture Under Shear Stress

The behavior under shear stress is quite different and is dependent on the magnitude of the normal stress. In this case we have to consider the roughness of the upper and lower fracture surfaces separately. If the upper and lower surfaces are described by

$s_1(x,y)$ and $s_2(x,y)$

then

$b(x,y) = s_1(x,y) — s_2(x,y) + \Delta$

where Δ is a separation distance between the two surfaces to ensure that the two surfaces are touching with no overlap of materials. Typically a fracture is formed with

$s_1(x,y) = s_2(x,y) = s(x,y)$

and the upper and lower surfaces may be displaced by a distance γ with respect to each other (Fig. 3). Then

$b(x,y) = s(x,y) - s(x,y - \gamma) + \Delta$.

For a well-mated fracture, $\gamma = 0$ and $\Delta = 0$ and then $b(x,y) = 0$ and there is no flow. It is the degree of mismatch, γ, that gives the fracture its aperture for water flow.

Under shear stresses, γ will change, which will require a larger value for Δ. Since flow through parallel fracture surfaces is proportional to the cube of the fracture aperture, this implies a large increase of hydraulic conductivity. The relationship between shear stress and displacement γ depends on frictional properties of the fracture surfaces and the normal stress. Vouille and Bougnoux (1994) proposed to develop this relationship by assuming a representative case where $s(x,y)$ can be represented by a two-dimensional, saw-tooth structure (see Fig. 6a), while Ohnishi (1994) proposed to analyze it by assigning a more general form to $s(x,y)$ (Fig. 6b).

At high normal stresses, the situation will be more complex. The asperities on each fracture surface (i.e., nails in the bed-of-nails model) will deform elastically or plastically with shear stress. In softer rock the asperities may completely break off. The water flow will then carry away the broken rock materials. Such "muddy" water flow was found in the laboratory experiment reported by Ahola et al. (1996, this book). The effect on water flow can be represented as a change in viscosity and density of the flowing fluid. On the other hand, a much more drastic effect can be expected when these broken rock materials clog up some of the flow channels or aperture constrictions (see Fig. 3) causing a substantial local flow decrease and forcing the fluid to find new flow paths. Thus under shear stress, dependent on the magnitude of normal stress and rock material, flow permeability can dramatically increase or decrease.

So far we have considered only a conceptual model of the mechanisms of coupled hydromechanical processes associated with a rock fracture. Imposition of a temperature gradient across the system is usually not expected to produce drastic effects. The thermal process will result in rock expansion which will modify $s(x,y)$ and $b(x,y)$, and could also cause convective

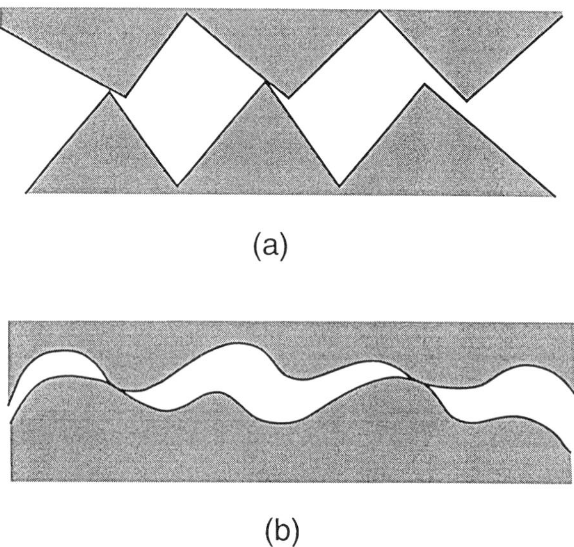

(a)

(b)

Figure 6.   Relationship between shear stress and displacement:
(a) Conceptual model of Couille and Bougnoux (1994); (b) conceptual model of
Ohnishi (1994).

flow of the water in the fracture. However, if the temperature change is
large, the temperature dependence of some material properties might have to
be considered. For high temperature gradients, thermomechanical effects
may be so large that multiple cracking may occur at borehole or tunnel walls.
Furthermore, if we were to study this system for a long time span we also
have to consider the possibilities of mineral alterations and the development
of new mineral assemblies on the fracture surfaces.

3. COUPLED PROCESSES NEAR UNDERGROUND OPENINGS

Underground openings represent a region of stress release causing
mechanical deformations and of pressure drop inducing fluid flow.
Dependent on the regional stress fields, portions of the rock around the
opening may experience compression while other portions are under tension.
For hard rocks and in the presence of pre-existing fractures, mechanical
effects will be concentrated at the fractures. Redistribution of the local stress
field around the opening may result in instability and release of rock blocks
from walls into the opening. The rate of such compression and tension
changes is fast, controlled by the sound velocity in the rock. On the other

hand, the rate of water drainage through the fracture controls the rate of pressure transients in the water. Since the sound velocity in rock is higher than fracture water flow, during the initial short time after an opening is formed in the rock, interesting strong transient effects may occur. One expects that in the interval of time when the rock responds to the stress release and before water responds, there will be strong variation in water pressure in the rock fractures. For those fractures under compression, there will be a strong transient pressure rise, which will then decay. Similarly a sharp transient water pressure drop will occur in fractures in a region of tension.

Noorishad et al. (1992) carried out a numerical modeling calculation for a generic scoping study of such a phenomenon. They evaluated the behavior of transient water pressure in a vertical fracture in the roof of a drift right after the drift had been excavated. A two-dimensional model was used in which the drift was represented by a circular opening. The opening was considered to be constructed in ten steps by removing ten horizontal benches of equal thickness from top to bottom. An example of their results is shown in Fig. 7. The water pressure in the fracture is given on the vertical axis, and the horizontal axis gives the distance along the fracture from the wall of the underground opening which has just been constructed. The different curves represent the pressure profiles at different times. The first ten time steps correspond to the removal of ten benches to make the opening. In this case the fracture is in a region of compression upon opening of the tunnel. A strong transient effect is observed. When the top two or three benches are removed, the opening and the corresponding mechanical changes are relatively small, and water drains from the vertical fracture and water

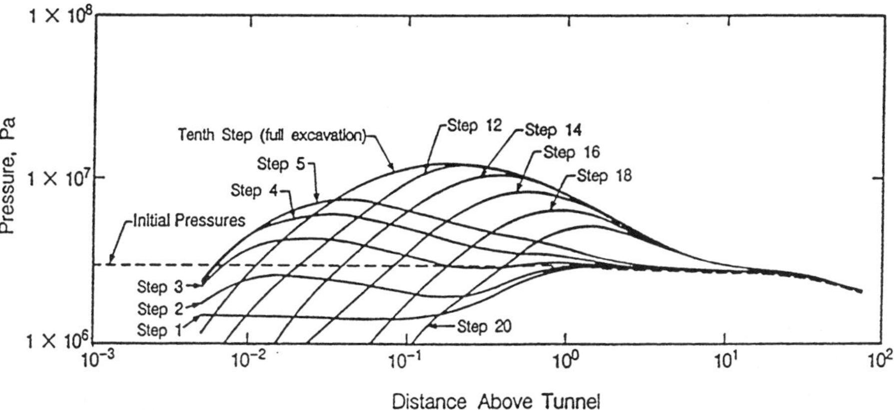

Figure 7. Evolution of the pressure (vertical axis in Pa) inside the vertical fracture (horizontal axis in m) during and after excavation. Steps 1 to 20 indicate time steps after the initiation of excavation (Noorishad et al., 1992).

pressure in the fracture decreases. As more benches are removed to form the opening, rock responds mechanically and the fracture is compressed to a smaller aperture value. Since the mechanical compression rate is higher than the water drainage rate, water pressure builds up to as much as four times the initial value. This then decays after the completion of the opening, when the mechanical compression is stabilized and the water is allowed to drain from the fracture.

The effects of such hydraulic pressure transients can be speculated, including the possibility of local hydrofracturing in certain cases. If such pressure transients can be measured, it opens up the interesting possibility of deducing compression-tension spatial distributions, and the general in-situ stress distribution. Transducers for measuring transient water pressures at short time intervals are well developed and have high accuracies. Strategic emplacements of these sensors before construction of a major underground opening may make such a study feasible.

This type of study can also include an evaluation of the problems associated with the slow advance of the tunnel face using a tunnel-boring machine versus the shock waves generated by a conventional drilling and blasting operation. Recent studies of rock failure initiation in the vicinity of tunnels by microseismic arrays at the Underground Research Laboratory (URL) in Canada have revealed severe fracturing over a distance more than a tunnel diameter ahead of the face (Martin et al., 1995). Similar results have been obtained in the tunnel work at the Hard Rock Laboratory at Åspö in Sweden. The failure process will allow water to penetrate the microfractures in the rock and enhance the transient pressure effects.

Other coupled processes near underground openings including the impact of water drainage on the stability of tunnel walls, and, inversely, the effect of mechanical deformation near tunnel walls on the permeability of the tunnel wall surface layer (Wilson et al., 1981). For the latter case, it was suggested that a release of water pressure at the opening may result in water degassing (Olsson, 1992). How the presence of both gas and water in a deformable fracture affects its hydromechanical behavior may be an interesting topic for further study.

4. COUPLED PROCESSES IN MULTIPLE FRACTURE SYSTEM UNDER CHANGING STRESS FIELDS

The rock mass near underground openings consists of intact rock and discontinuities or fractures embedded in it. Various sets of fractures in the rock mass have different persistence or sizes. The intact rock between the tips of non-persistent fractures is named a rock bridge (Fig. 8). The coexistence of rock fractures and rock bridges is the main reason for the complicated mechanical response of brittle rocks to stress loading.

Shen (1993) studied the simple system of two fractures and an interjacent bridge subjected to compressive loading. Depending on the magnitude of the boundary stress, the fracture responses can be divided into three different phases as illustrated in Fig. 8. At low stress level the rock fracture responds by sliding deformation (Fig. 8b). At intermediate level, fractures undergo sliding and tensile (mode I) failure propagation and typical wing cracks are formed at the tip of the fractures (Fig. 8c). When the ultimate stress level is reached, the bridge fails and rock fractures coalesce (Fig. 8d).

Fracture coalescence can develop by tensile failure, shear failure and mixed tensile and shear failure. The mode of failure depends upon relative position and orientation of two or more fractures in relation to the direction of loading. Shen (1993) used a fracture mechanics approach to predict the tensile failure of bridges, and Shen et al. (1995) made a series of uniaxial compression tests using gypsum specimens with pre-existing cracks to study the mechanism of fracture coalescence. Experimental results showed that both open and closed cracks can coalesce by tensile or shear failure and that coalescence load of closed cracks is about 25% higher than that of open cracks.

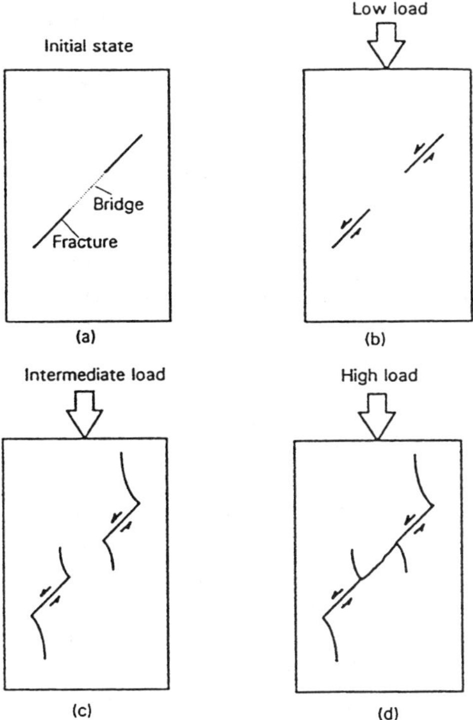

Figure 8.   Models of fracture response at different levels of compressive loading. (a) initial fracture-bridge system; (b) fracture deformation by shear at low load; (c) tensile failure and wing crack propagation at intermediate load; and (d) fracture coalescence at high load (Shen, 1993).

The model of fractures and intervening bridges was implemented in a boundary element code and applied to a KBS 3 concept for storage of spent nuclear fuel (Shen and Stephansson, 1995). A number of models with different fracture geometries and loading conditions were analysed. A low stress ratio applied ($\sigma_3/\sigma_1 \leq 0.05$) at the boundary of the models resulted in tensile fracture propagation and the failure occurred in the direction of the maximum principal stress. Intermediate stress ratio ($0.10 \leq \sigma_3/\sigma_1 \leq 0.15$) caused fracture propagation in both tension and shear mode. A stress ratio of $0.20 \leq \sigma_3/\sigma_1 \leq 0.25$ only produced shear failure, while a stress ratio greater than 0.25 is unlikely to cause fracture propagation. Hence, from the results of this study it is obvious that location of a repository in an area of large deviatoric horizontal *in-situ* stress will lead to shearing and fracturing at the tip of existing joints. This might lead to enhanced groundwater flow in the near-field of tunnels and deposition holes.

On an even larger scale, multi-fracture blocks under stress field variations such as advancing or retreat of glacier sheets have been considered by Stephansson and Shen (1991), Hansson et al. (1995) and Shen and Stephansson (1995). The far-field study conducted by Hansson et al. (1995) treated the rock mass as an assemble of discrete blocks defined by a number of major faults and fracture zones.

During the life time of the waste four major events are considered as loading to the rock mass in a repository. They are: (i) excavation of tunnel and deposition holes, (ii) swelling pressure from the highly compacted backfill, (iii) thermal loading from the waste and (iv) loading from an ice sheet in glaciated terrains. The different events will appear at different time and may influence the repository separately or simultaneously. The listed loading sequences (i) to (iv) were simulated by the 3DEC code (Shen and Stephansson, 1995). From the numerical results it was found that a maximum temperature of 48°C would be reached 200 years after the emplacement of the waste canisters. The average increase of maximum principal stress due to thermal loading is 9.5 MPa horizontally and 20.2 MPa vertically due to glaciation. These changes in stress magnitude and orientation might affect ground water flow in and around the repository.

To illustrate the main impact of different events on the repository, principal stresses at two points, one in the wall of the deposition hole and the other in the roof of the tunnel are presented in Fig. 9. Thermal loading after 200 years of waste deposition causes almost twice as high stresses as under virgin condition. High stresses can cause plastic yielding of intact rock and displacement of joints and fracture zones. The large variations in stress magnitude are also followed by a major shift in the orientation of the principal stresses, so that minor becomes major and visa versa. Rock masses are not particularly good at sustaining repeated loading and the fatigue might enhance ground water flow and reduce their stability.

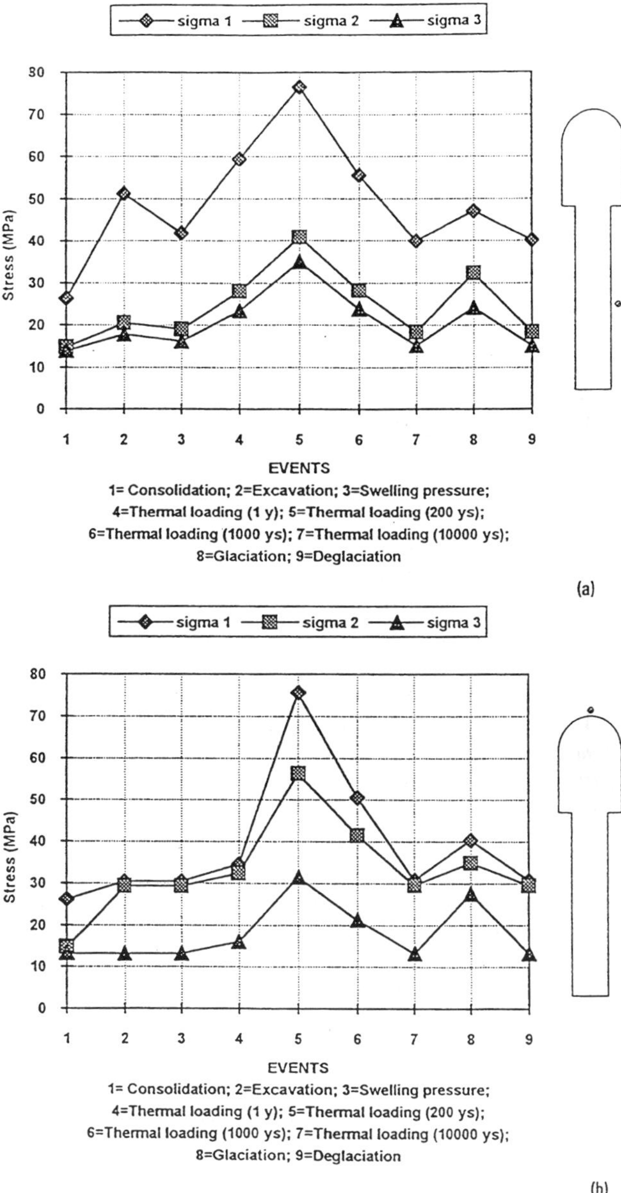

Figure 9. Result of three-dimensional discrete element modelling of a waste repository subjected to nine different loading events; a) principal stresses in the wall at the mid-height of the deposition hole; b) principal stresses in the roof of the tunnel (Shen and Stephansson, 1995).

5. FIELD TEST METHODS UTILIZING COUPLED PROCESSES

This is relatively a new area to be developed. The question is, since thermal, hydraulic and mechanical effects are coupled, can one take advantage of this coupling to make one type of measurement and obtain the properties or rock characteristics that control another effect? If we can do this, it would open up a new area of investigation. An example was indicated earlier in the section on coupled transient effects near underground openings. Another one is given below based on the work of Rutqvist and co-workers (1992, 1995).

Conventional methods to study in-situ stress fields at a given location include well hydrofracturing. Such methods are drastic, causing permanent local change in the system. The method proposed by Rutqvist's work is based on coupled hydromechanical processes and is not destructive. Through careful theoretical and experimental studies these researchers were able to show that for injection into a well intercepted by a fracture, the fracture undergoes elastic aperture dilation, even at an injection pressure well below the lithostatic value. Such an aperture dilation causes a flow increase, reflecting a larger permeability, during the course of the testing. The particular experimental set-up is shown in Fig. 10, where packers are emplaced in a well to bracket an intercepted fracture. For this particular case, the well intersected four fractures which were studied one by one. Typical flow variations measured under constant injection pressure are shown in Fig. 11. Without mechanical deformation, one expects from conventional hydrology that the flow will have an early sharp rise, corresponding to the initial imposition of injection pressure, and then decay to a constant value. However, due to the elastic behavior of the fracture opening, which is controlled by the stiffness of the fracture and surrounding rock mass, the flow rate actually rises after the early decrease. Rutqvist et al. demonstrated this phenomenon experimentally. They also showed that if the injection pressure is applied as a pulse, i.e., for a finite duration, the fracture does not respond mechanically and the aperture remain the same. Thus, in a field experimental if one is to perform a step-injection test as well as a pulse-injection test, one would be able to measure the mechanical dilation properties of the fracture through a careful comparative analysis of the results of the two tests. Often a well intersects several fractures which may have different orientations. Investigations of coupled hydromechanical processes of these fractures may yield mechanical conditions reflecting the different orientations.

Imposition of a temperature gradient by emplacing a heater in the well provide an additional means of perturbing the system to evaluate its properties and condition.

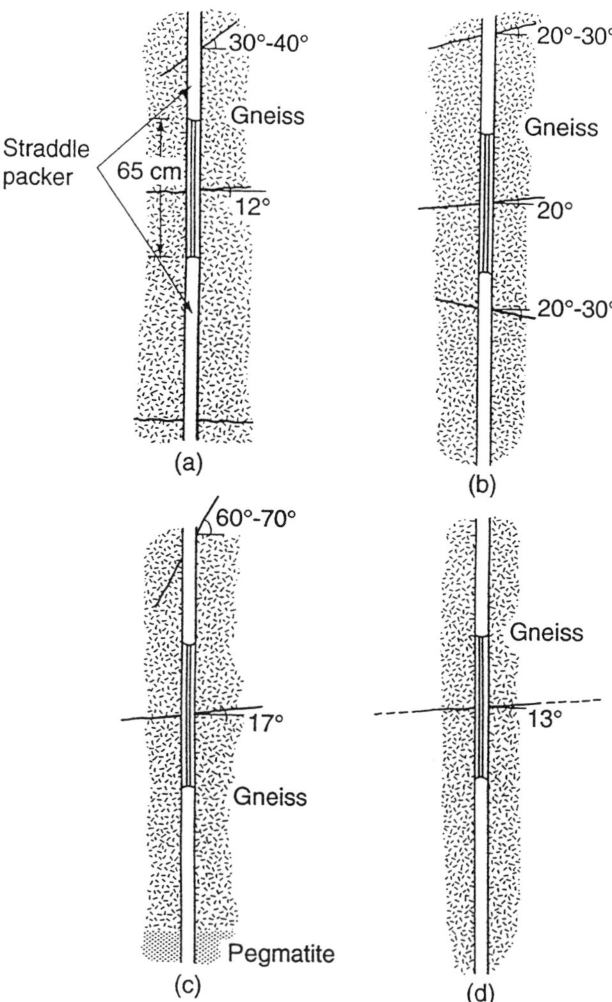

Figure 10. A coupled HM well test technique, using packers applied across fractures at four different depths, (a) 81 m, (b) 131 m, (c) 257 m and (d) 356 m (Rutqvist et al., 1992).

                                            C.-F. Tsang and O. Stephansson

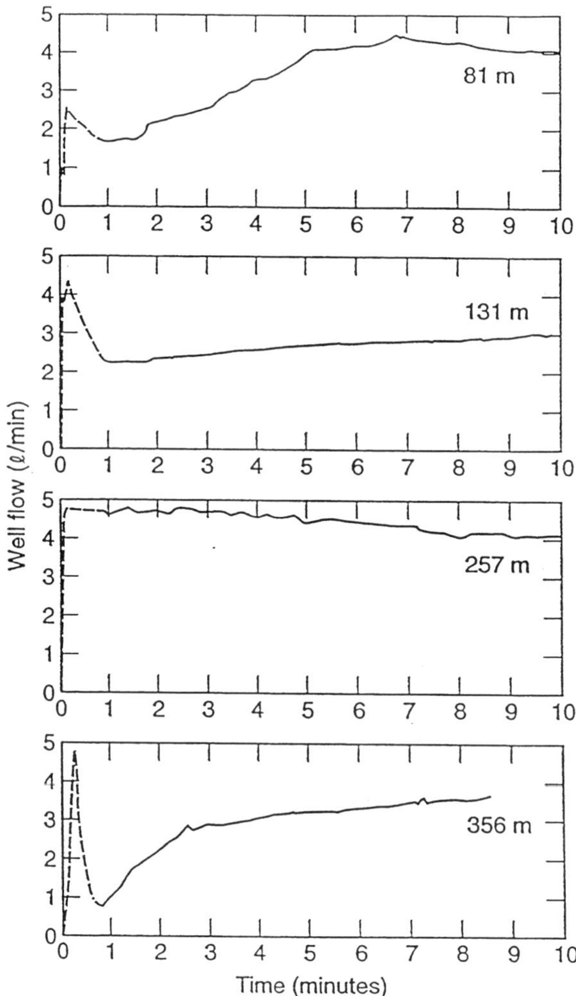

Figure 11. Flow rate variations under constant injection pressures applied to the four fractures shown in Fig. 10 (Rutqvist et al. 1992).

6. ROCK ENGINEERING SYSTEM APPROACH

In 1992, a new methodology for approaching rock engineering problems was presented and named the "Rock Engineering System (RES; Hudson, 1992). The principle of the RES approach is that, for all rock engineering projects, one should start with a top-down approach in order to ensure that all aspects and processes of the problem are considered. One analyses the different variables and the interactions between them. The method is intrinsically objective-oriented, and proceeds always with the objective in mind. The basic device used in the RES methodology is the interaction matrix. The main variables or parameters are identified and listed along the leading diagonal of a square matrix, and the interactions between the parameters are then given by the off-diagonal terms. The interactions between each one of the variables are then studied with a clockwise convention for the influence direction.

The RES methodology was used to structure and complete the long list of Features, Events and Processes (FEP) existing within SKB and SKI in Sweden for the final disposal of spent nuclear fuel (Eng et al., 1994). Based on the list of FEPs, various scenarios of a repository were studied. Figure 12 illustrates the basic interaction matrix for a typical waste repository in hard rocks. Water flow at the centre of the diagram links the near-field with the far-field and the loops inside the matrix are of different type. This so-called 'soft' RES approach indicates the structure of the problem and the main mechanism pathways. The RES methodology has proved to work satisfactory in structuring the difficult couplings of variables in radioactive waste isolation and it has been demonstrated to various disciplines and scales of the waste isolation problem.

The RES methodology has been further developed to a system called the fully-coupled model (FCM; Jiao and Hudson, 1995). It considers the interaction matrix as a mechanism network. Graph theory is used to assess the contributions of all the mechanisms in all the pathways. The binary matrix is transformed into a fully-coupled global interaction matrix and each step in the coupling process can be studied. FCM has shown to be a very powerful tool to study couplings of a complicated system once the theory and governing equations for each interaction is known. Thus, the FCM should be applicable to the THM couplings presented in this book.

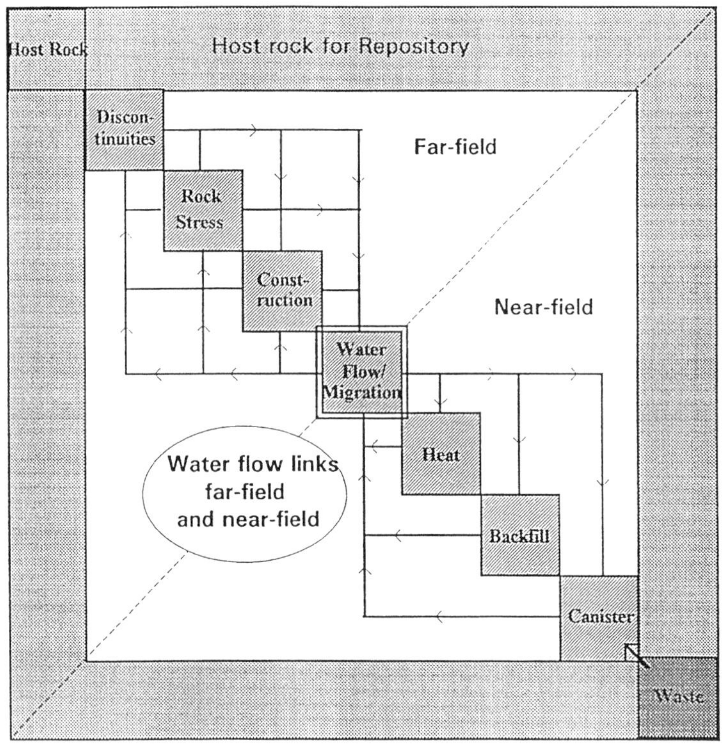

Figure 12. A 9 × 9 interaction matrix of the RES methodology to indicate the most important variables and main variables for a waste repository in hard rocks.

7. CONCLUDING REMARKS

The present chapter gives an introductory discussion and a conceptual model of coupled THM processes in fractured rocks. It is a rich field of research with many new possible directions, some of which have been outlined and discussed. Of particular promise is the development of new field testing techniques using coupled processes to evaluate in-situ rock properties and conditions.

Practically, THM processes are expected to be present around an underground radioactive waste repository. An important part of the safety analysis of a potential repository is the evaluation of the effects of THM couplings, and in order to help meet this objective we need to develop mathematical models capable of describing these processes. The models are the prime tools because we have to study processes active over a very long time span. For this we need a systematic approach, such as the Rock Engineering System (RES) methodology, to define the problem and the relevant couplings. Thus, the RES might help us to identify the most important variables, processes and their couplings, which will then be incorporated into a fully-coupled model to evaluate repository safety.

8. ACKNOWLEDGMENTS

We appreciate discussions with members of the international DECOVALEX project, which have helped to shape many of the ideas in this chapter. Careful review and helpful comments from Lionel Dewiere of ANDRA, France, and Marcelo Lippman of Berkeley Laboratory are gratefully acknowledged. The work is jointly supported by grants from the Swedish Nuclear Power Inspectorate, Stockholm, and by the Office of Basic Energy Sciences, Engineering and Geosciences Division, Office of Energy Research of the U.S. Department of Energy under Contract No. DE-AC03-76SF00098.

9. REFERENCES

Ahola, M.P, Mohanty, S. and Makurat A., Coupled mechanicl shear and hydraulic flow behaviour of natural rock joints (contribution to this book) (1996).

Bandis, S.C., Lumsden, A.C. and Barton, N., Fundamentals of rock joint deformation, Int. J. Rock Mech. Min. Sci. & Gemech. Abstr. 20 (1983) 249.

Bourke, P.T., Channeling of flow through fractures in rock, Proceedings of the GEOVAL87 Symposium, pp. 167–177, Swed. Nucl. Power Insp., Stockholm, 1987.

Cai, M. and Horii, H., A. Constitutive model and FEM analysis of jointed rock mass, Int. J. Rock Mech., Min. Sci. & Geomech. Abstr., 30, No. 4 (1993) 351–359.

Eng, T., Hudson, J.A., Stephansson, O., Skagius, K. and Wiborgh, M., Scenario development methodologies. SKB Technical Report 94-28. Swedish Nuclear Fuel and Waste Management Co, Stockholm (1994).

Gangi, A.F. Variation of whole and fractured porous rock permeability with confining pressure, Int. J. Rock Mech. Min. Sci., 15 (1978) 249.

Goodman, R.E., Method of Geologic Engineering in Discontinuous Rocks, West Publishing, New York, 1976.

Hakami E., Aperture distribution of rock fractures. Doctoral Thesis, Division of Engineering Geology, Royal Institute of Technology, Stockholm (1995)

Hakami E. and Barton, N., Aperture measurements and flow experiments using transparent replicas of rock joints, in Proc. Int. Symp. Rock Joints, Loen, Norway, Balkema, Rotterdam (1990) 383.

Hansson, H., Jing, L. and Stephansson, O., 3-D DEM modelling of coupled thermo-mechanical response for a hypothetical nuclear waste repository, in Proc. 5th Int. Symp. Numerical Models in Geomechanics - NUMOG V, Davos, Switzerland (1995) 257.

Hopkins, D.C., Cook, N.G.W. and Myer, L.R., Normal joint stiffness as a function of spatial geometry and surface roughness, Proc. Int. Symp. Rock Joints, Loen, Norway, Balkema, Rotterdam (1990) 203.

Hudson, J.A., Rock Engineering Systems- theory and Practice, Ellis Horwood, London (1992).

Jiao, Y. and Hudson, J.A., The fully-coupled model for rock engineering systems, Int. J. Rock Mech. Min. Sci. & Geomech. Abstr., 32 (1995) 491.

Jing, L., Stephansson, O., Rutqvist, J., Tsang, C.-F. and Kautsky, F. DECOVALEX-Mathematical models of coupled T-H-M processes for nuclear waste repositories, Phase I report, SKI Technical Report 93, No. 31, Stockholm (1993).

Jing, L., Stephansson, O., Rutqvist, J., Tsang, C.-F. and Kautsky, F. DECOVALEX-Mathematical models of coupled T-H-M processes for nuclear waste repositories, Phase II Report, SKI Technical Report, Stockholm (1995).

Jing, L., Stephansson, O., Rutqvist, J., Tsang, C.-F. and Kautsky, F. DECOVALEX-Mathematical models of coupled T-H-M processes for nuclear waste repositories, Phase III report, SKI Technical Report, Stockholm (1996), in press.

Kawamoto, T., A Modelling of Jointed Rock Mass, Numerical methods in geomechanics, G. Swoboda, Ed., A.A. Balkerma Publishers, rotterdam, Netherlands (1988).

Martin, C.D., Martino, J.B. and Dzik, E.J. Comparison of borehole breakouts from laboratory and field tests, in Proc. Eurock´94, Delft, Balkema, Rotterdam (1995) 183.

Moreno, L. and Tsang, C.-F., Multiple-peak response to tracer injection tests in single fractures: A numerical Study, Water Resour. Res., 27 (1991) 2143.

Moreno, M. and Tsang, C.-F., Flow channeling in strongly heterogeneous porous media: A numerical study, Water Resour. Res., 30 (1994) 1421.

Moreno, L., Tsang, Y.W. and Tsang, C.-F., Some anomalous features of flow and solute transport arising from fracture aperture variability, Water Resour. Res., 26 (1990) 2377.

Moreno, L., Tsang, Y.W., Tsang, C.-F., Hale, F.V. and Neretnieks, I. Flow and tracer transport in a single fracture: A stochastic model and its relation to some field observations, Water Resour. Res., 24 (1988) 2033.

Noorishad, J., Tsang, C.-F. and Witherspoon, P.A., Theoretical and field studies of coupled hydromechanical behavior of fractured rocks– 1. Development and verification of a number simulator, Int. J. Rock Mech. Min. Sci. & Geomech. Abstr., 29 (1992) 401.

Nordqvist, A.W., Tsang, Y.W., Tsang, C.-F., Dverstorp, B. and Andersson, J., A variable aperture fracture network model for flow and transport in fractured rocks, Water Resour. Res., 28 (1992) 1703.

Oda, M., Permeability Tensor for discontinuous rock mass, Geotechnique, 35(4) (1985) 483–495.

Ohnishi, Y. Private communication at the DECOVALEX meeting, May, Oxford, UK (1994)

Olsson, O. (ed.), Stripa Project, Site characterization and validation–Final report, SKB Technical Report 92-22, Swedish Nuclear Fuel and Waste Management Company (SKB), Stockholm, Sweden, April (1992).

Pariseau, W.G., Equivalent Properties of a Jointed Biot Material, Int. J. Rock Mech., Min. Sci. & Geomech. Abstr., 30(7) (1993) 222–232.

Rutqvist, J., Determination of hydraulic normal stiffness of fractures in hard rock from well testing, Int. J. Rock Mech. Min. Sci. & Geomech. Abstr., 32 (1995) 513.

Rutqvist, J., Noorishad, J., Stephansson, O., Tsang, C.F., Theoretical and field studies of coupled hydromechanical behaviour of fractured rocks–2, Field experiment and modeling, Int. J. Rock Mech. Min. Sci. and Geomech. Abstr., 29 (1992) 411.

Shen, B., Mechanics of fractures and intervening bridges in hard rocks, Doctoral thesis, Division of Engineering Geology, Royal Institute of Technology, Stockholm (1993).

Shen, B. and Stephansson, O., Modelling of rock fracture propagation for nuclear waste disposal, SKI report, Stockholm (1995).

Shen, B., Stephansson, O., Einstein, H., Ghahreman, B., Coalescence of fractures under shear stress in experiments, J. Geophys. Res., 100 (1995) 5975.

Stephansson, O. (ed.), Special issue on Thermo-Hydro-Mechanical coupling in rock mechanics, Int. J. Rock Mech. Min. Sci. & Geomech. Abstr., 32 (1995).

Stephansson, O. and Shen, B., Modelling of rock masses for site location of a nuclear waste repository, in Proc. 7 Congress Int. Soc. Rock Mech., Vol 3, Aachen, Germany (1991) 157.

Stephansson, O., Jing, L., Tsang, C.-F. and Kautsky, F. Development of coupled models and their validation against experiments–DECOVALEX Project, Proc. GEOVAL '94 Validation Through Model Testing, in Proc. NEA/SKI Symp., Paris, France (1994).

Tsang, C.-F. (ed.), Coupled Processes Associated with Nuclear Waste Repositories, Academic Press (1987).

Tsang, C.-F., Coupled thermomechanical hydrochemical processes in rock fractures. Review of Geophysics, 29 (1991) 537.

Tsang, Y.W., Usage of "Equivalent Apertures" for rock fractures as derived from hydraulic and tracer tests, Water Resour. Res., 28 (1992) 1451.

Tsang, Y.W. and Tsang, C.-F., Channel model of flow through fractured media, Water Resour. Res., 23 (1987) 467.

Tsang, Y.W. and Tsang, C.-F., Flow channeling in a single fracture as a two-dimensional strongly heterogeneous permeable medium, Water Resour. Res. (1989) 2076.

Tsang, Y.W., Tsang, C.-F., Neretnieks, I. and Moreno, L. Flow and tracer transport in fractured media: A variable aperture channel model and its properties, Water Resour. Res., 24 (1988) 2049.

Tsang, Y.W. and Witherspoon, P.A., Hydromechanical behavior of a deformable rock fracture subjected to normal stress, J. Geophys. Res., 86 (1981) 9287.

Vouille, G. and Bougnoux, A., Private Communication at the DECOVALEX meeting, May, Oxford, UK (1994).

Wang, J.S.Y. and Narasimhan, T.N., Hydrologic mechanisms governing fluid flow in a partially saturated, fractured, porous medium, Water Resour. Res., 21 (1985) 1861.

Wilson, C.R., Long, J.C.S., Galbraith, R.M.K., Karasaki, H.K., Endo, A.O. DuBois, M.J. McPherson, and G. Ramqvist, Geohydrological data from the macro-permeability experiment at Stripa, Sweden, Lawrence Berkeley Laboratory report, LBL-12520, March (1981).

Witherspoon, P.A., Wang, J.S.Y., Iwai, K., Gale, J.E., Validity of cubic law for fluid flow in a deformable rock fracture, Water Resour. Res., 16(6) (1980) 1016-1024.

Zimmerman, R.W., Chen, D.-W. and Cook, N.G.W., The effect of contact area on the permeability of fractures, J. Hydrology, 139 (1992) 79.

O. Stephansson, L. Jing and C.-F. Tsang (Editors)
Coupled Thermo-Hydro-Mechanical Processes of Fractured Media
Developments in Geotechnical Engineering, vol. 79
© 1996 Elsevier Science B.V. All rights reserved.

Validation Of Mathematical Models Against Experiments For Radioactive Waste Repositories - DECOVALEX Experience

L. Jing[a], C.F. Tsang[b], O. Stephansson[a] and F. Kautsky[c]

[a] Division of Engineering Geology, Royal Institute of Technology, S-100 44 Stockholm, Sweden

[b] Lawrence Berkeley Laboratory, 1 Cyclotron Road, Berkeley, CA94720, USA

[c] Swedish Nuclear Power Inspectorate, S-106 58 Stockholm, Sweden

Abstract

The international co-operative research project - DECOVALEX- is presented. The project was initiated as a collective research effort to further our understanding of the coupled thermo-hydro-mechanical processes in fractured media and engineered buffer materials, and validate mathematical models and computer codes against experiments for radioactive waste isolation. The presentation summarizes the organization, structure, approaches and management of the project briefly, followed by an outline of the Bench-Mark-Test (BMT) and Test Case (TC) problems studied in the project. The scientific achievements and lessons learned are briefly discussed at the end.

1. INTRODUCTION

The **DECOVALEX** (acronym for **DE**velopment of **CO**upled models and their **VAL**idation against **EX**periments in nuclear waste isolation) project is an international effort to develop mathematical models, numerical methods and computer codes for coupled thermo-hydro-mechanical (T-H-M) processes in fractured rocks and buffer materials for geological isolation of spent nuclear fuel and other radioactive wastes, and validate them against laboratory or field experiments [1]. The project was jointly sponsored by nine Funding Organizations from Canada, Finland, France, Japan, Sweden, UK, and USA. The project consisted of three phases, began in 1992 and ended in 1995.

1.1. Coupled T-H-M Processes for Radioactive Waste Repositories in Rocks

The rock mass response to storage of radioactive waste and spent nuclear fuel is a coupled phenomenon involving thermal (T), hydrological (H), mechanical (M) and chemical (C) processes [2]. The term "Coupled Processes" implies that one process affects the initiation and progress of others and vice versa. Therefore, the rock mass response to waste storage cannot be properly predicted by considering each process independently. Traditional disciplines in geoscience and geoengineering (e.g. structural geology, geophysics, geohydrology, rock mechanics and rock engineering, mechanics of porous media, etc.) have to be integrated interactively together to gain a proper understanding of coupled T-H-M behaviour of rock masses in which radioactive waste repositories are built. Mathematical models and computational methods are specially important to performance and safety assessment of radioactive waste repositories in rocks since the processes involved are very complex, of long-term, and of large scales. The chemical processes (e.g. nuclide migration and absorption, etc.) were excluded in the DECOVALEX project.

1.2. The Objectives of the DECOVALEX Project

Though the importance of coupled thermo-hydro-mechanical (T-H-M) processes has been recognized for some years, the current capability of modelling such processes is still very limited, partially because of lack of applicable and realistic test cases (experimental studies) for establishing conceptual models and to validating mathematical models and computer codes. The performance and safety assessment of repositories need robust and reliable computer codes capable of modelling fully coupled T-H-M processes in geologic systems. It is to this end that the DECOVALEX project was established. The overall goal of DECOVALEX project is to increase our understanding of the various aspects of coupled T-H-M processes of importance in the release and transport of radionuclides from a repository to the biosphere and to develop and validate mathematical models and computer codes simulating these processes. The objectives of the DECOVALEX project can be summarized below:

- to support the development of mathematical models and computer algorithms and codes for T-H-M modelling;
- to validate and verify different mathematical models and computer by comparing theory and model calculations with results from laboratory or field experiments;
- to design new experiments of coupled T-H-M processes for further code development and validation/verification.

2   PROJECT STRUCTURE

Members of the DECOVALEX project consisted of nine Funding Organizations who paid a yearly fee to the project and also provided funding to their own research teams working on the problems defined in the project (Fig. 1 and Table 1). The representatives from these nine Funding Organizations formed a Steering Committee to direct the project activities. The Steering Committee was assisted by a Managing Organization (Swedish Nuclear Power Inspectorate) for managing the project economy and a Secretariat (Division of Engineering Geology, Royal Institute of Technology, Stockholm, Sweden) for all administration and technical matters. The Commission of European Communities (CEC) participated in the project as a Funding Party by providing research funds to a number of research teams within the EC countries, but not paying the yearly fee to the project. The Atomic Energy

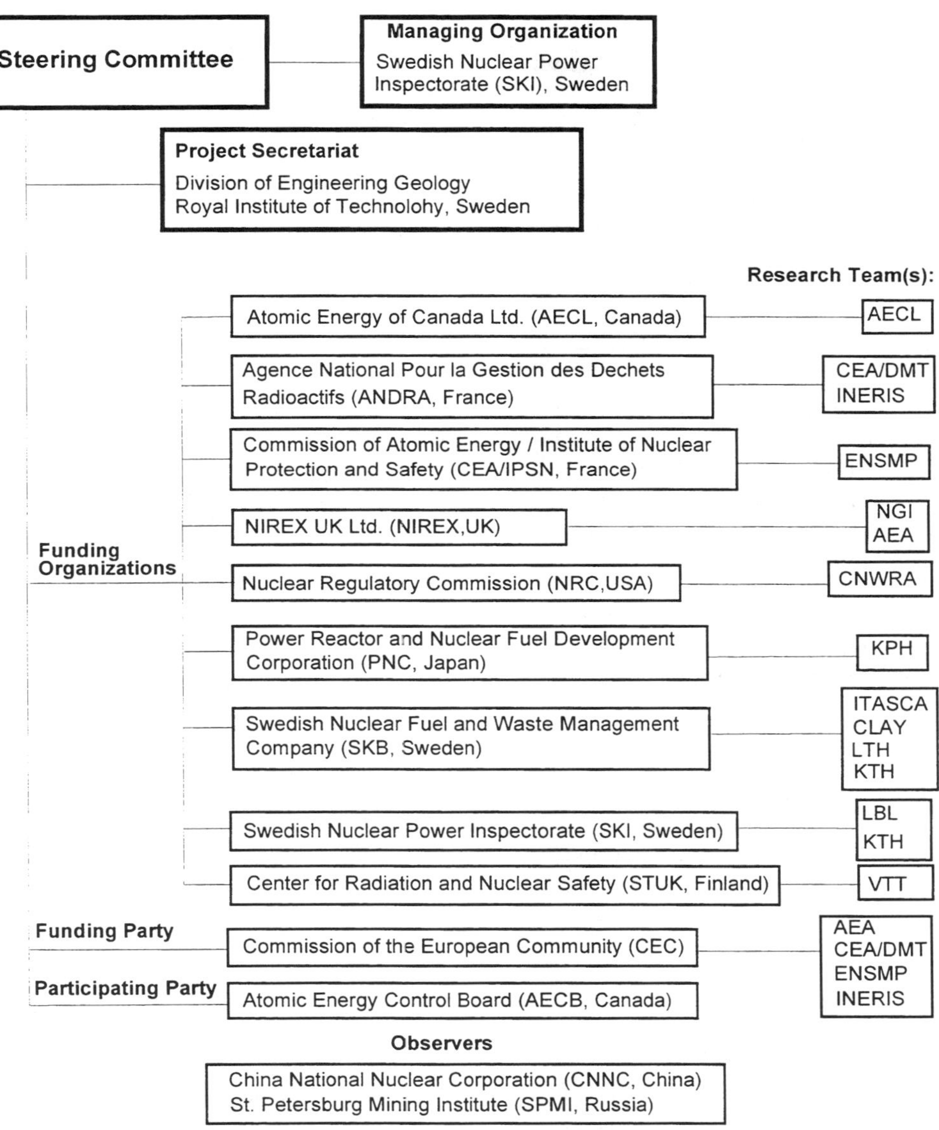

Figure 1 Members and Organization of the DECOVALEX project, Phase I, II and III.

Table 1 Research teams in DECOVALEX Project, Phase I, II and III

Full name	Acronym
Applied Geoscience Branch, Whiteshell Laboratories, AECL, Canada	AECL
Commissariat à l'Energie Atomique, Department de Mecanique et de Technologie, France	CEA/DMT
INERIS/Laboratoire de Mécanique des Terrains Parc de Saurupt, France	INERIS
Ecole Nationale Supérieure des Mines de Paris, France	ENSMP
AEA Technology, UK	AEA
1) Kyoto University, Kyoto, Japan 2) Power Reactor and Nuclear Fuel Development Corporation, Japan 3) Hazama Corporation, Japan	KPH
Norwegian Geotechnical Institute, Norway	NGI
Center for Nuclear Waste Regulatory Analysis, Southwest Research Institute, Texas, USA	CNWRA
Road, Traffic and Geotechnical Laboratory, Finland	VTT
Lawrence Berkeley Laboratory, University of California, Berkeley, USA	LBL
Division of Engineering Geology, Royal Institute of Technology, Stockholm, Sweden	KTH(g)
Division of Hydraulics, Royal Institute of Technology, Stockholm, Sweden	KTH(h)
Clay Technology AB, Lund, Sweden	CLAY
Lund University of Technology, Lund, Sweden	LTH
ITASCA Geomekanik AB, Sweden	ITASCA

Control Board (AECB) of Canada participated in the project as a party, working on the problems defined in the project on their own initiatives. The China National Nuclear Corporation (CNNC) of China and St. Petersburg Mining Institute (SMPI) of Russia were received by the project as observers. The acronyms of organizations, parties and research teams will be used, instead of their full names, in the rest of this chapter.

3. APPROACHES

3.1. General Approach

The activities of the DECOVALEX project was organized around the numerical simulations of three Bench-Mark Tests (BMT1, BMT2, and BMT3) and six Test Cases (TC1, TC1:2, TC2, TC3, TC4, TC5, TC6), by international research teams using different mathematical models and computer. The BMTs are hypothetical problems defined specifically to represent the complete or part of coupled T-H-M processes of a repository in fractured rocks as either near or far-field problems. The TCs are previous or ongoing laboratory or field experiments. Both were defined as initial-boundary value problems with proper thermal, hydraulic and mechanical initial-boundary conditions and loading sequences. The number of the BMT and TC problems were limited so that each of them was studied by several research teams in parallel and at successive phases. The results were presented and compared at regularly held workshops, leading to in-depth discussion and exchange of information and scientific knowledge from different disciplines and approaches. Based on these studies, new experiments were proposed to provide more rational tests of concepts and models, and to advance the state-of-the-art of mathematical modelling of the coupled T-H-M processes in fractured rocks and buffer materials. Analytical and semi-analytical solutions to the coupled problems were also developed whenever possible. The representatives from the national radioactive waste management organizations participating in the project also took an active part in the whole process that project was conducted equally from scientific, engineering and managing points of view. The physical processes studied in the BMTs and TCs are listed in Table 2.

Table 2   Physical phenomena studied in DECOVALEX

Components	Phenomena
Physico-mechanical processes	Thermal expansion, diffusion and convection in fractured rocks and buffer materials
	Fluid flow in fractured rocks and buffer materials
	Deformation of fractured rocks and buffer materials
	Constitutive laws for rock fractures, fractured rock masses and buffer materials
Geometrical factors and properties	Rock fracture networks and their characterization and representation
	Rock fracture properties (aperture, roughness, gouge production, filling, conductivity, storativity)
	Variability and representability of network connectivity
	Swelling pressure and suction potential of the buffer materials

3.2. Numerical Methods and Computer Codes

The different numerical methods applied in the DECOVALEX project include the Finite Element Method (FEM), the Finite Difference Method (FDM), the Distinct Element Method (DEM) and Discrete Fracture Network Method (DFN). The FEM and FDM simulate the media as continua, the DEM treats the rock mass as an assemblage of deformable blocks, and the FDN considers the connected fracture spaces for flow analysis only (Fig. 2). Table 3 lists the computer codes and their characteristics applied in the DECOVALEX Project.

4. Bench-Mark-Test Problems

Three BMT problems were defined and studied in the DECOVALEX project: BMT1, BMT2 and BMT3. The they were defined as genetic studies without support of experimental data, but contained the most important aspects of coupled thermo-hydro-mechanical processes in fractured rocks [3, 4, 6 - 10]. Their sizes vary from a very small scale (0.75 x 0.5 m in BMT2), to an intermediate scale (50 m x 50 m in BMT3), then to a very larger scale (3000 m x 1000 m in BMT1). Conversion of mechanical work to thermal energy, and phase

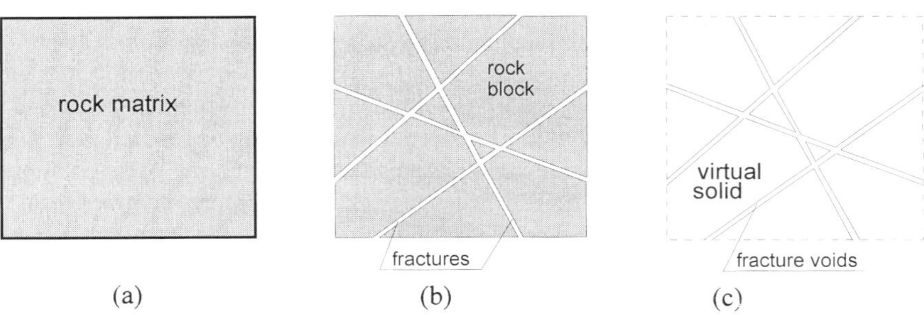

Figure 2 Geometrical representations of a fractured medium by a) a continuum (FEM, FDM, b) a discontinuum (DEM), and c) a discrete fracture network (DFN).

Table 3 Computer codes and their characteristics used in DECOVALEX

Code	User	Major Characteristics
MOTIF	AECL	3-D FEM solution for problems of transient and steady-state groundwater flow, heat transfer and mechanical deformation in porous and fractured media
THAMES	KPH	3-D FEM solution of coupled T-H-M processes for porous and fractured media
ADINA-T & JRTEMP	VTT	2-D FEM solutions of heat transfer (ADINA-T) and deformation analysis (JRTEMP) without fluid flow
ROCMAS	LBL, KTH(g)	2-D FEM solution of coupled T-H-M problems of continua with joint elements
CHEF, VIPLEF & HYDREF	ENSMP	2-D and 3-D FEM solution of coupled T-H-M problems for porous and fractured media with joint elements
TRIO-EF & CASTEM- 2000	CEA/DMT	2-D FEM solution of coupled T-H-M 2-D & 3-D problems for porous media
NAPSAC	AEA	3-D Discrete fracture network solution for water flow
FRACON	AECB	2-D FEM solution with joint elements for coupled H-M processes of continua
ABAQUS	CNWRA, CLAY	3-D FEM solution of coupled T-H-M problems for continua
FLAC	ITASCA	2-D FDM solution of coupled T-H-M problems of continua
UDEC	CNWRA, NGI, VTT, INERIS ITASCA	2-D DEM solution of coupled T-H-M problems of discrete, deformable block assemblages
3DEC	INERIS	3-D DEM solution for coupled T-M problems of discrete, deformable block assemblages

change of water were not considered. All three BMTs were defined as 2-D problems with initial and boundary conditions for thermal, hydraulic and mechanical loads.

4.1   BMT1- A Far-Field Model

This 2-D BMT is designed to simulate the T-H-M processes in a large volume of jointed rock mass with a repository located at a depth of about 500 meters [6, 7]. The objective is to examine the thermal, hydraulic and mechanical effects of the repository on its far-field environment. The model measures 3000 m × 1000 m and contains two perpendicularly intersecting sets of parallel discontinuities, see Fig. 3. Three different spacing (100m, 50m, and 25m, respectively) of the discontinuity sets are specified as different cases. The mechanical, hydraulic and thermal boundary conditions for the model are given as:

a) Mechanical conditions: Free top surface, zero normal displacement at bottom
   surface, and varying stresses (with gravity) along two vertical boundaries. The
   initial mechanical condition for the model is the in-situ stresses due to gravity.

b) Hydraulic conditions: Zero flux along two vertical boundaries and the bottom
   boundary. A variable hydraulic head, H, at the top surface, given by

$$H = 100 + 25\cos(\pi(x - 1250)/3000) \tag{1}$$

c) Thermal conditions: Zero thermal flux on two vertical boundaries, a constant
   temperature ($T^0 = 283^o K$) at the top surface and natural geothermal flux for the
   bottom boundary. The initial temperature distribution was obtained as a result of
   solving the thermal problem under specified thermal boundary conditions. The heat
   flux from the repository was given by

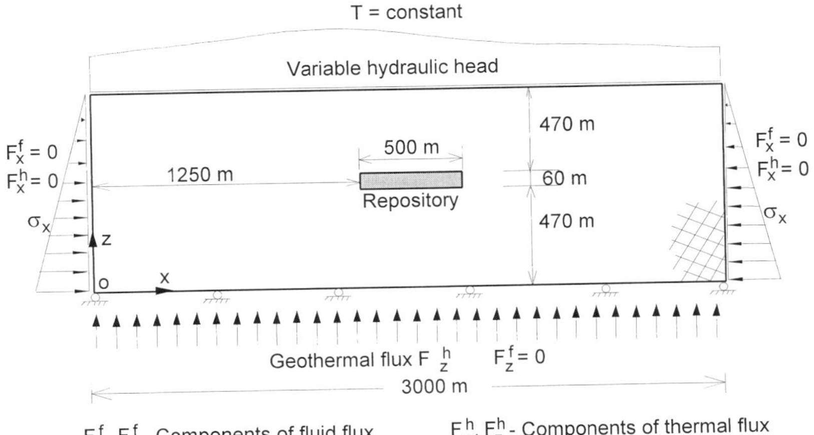

Figure 3 Thermal, hydraulic and mechanical boundary conditions of BMT1.

$$F_s^h = F_0^h e^{-\beta t} \tag{2}$$

where $F_0^h = 0.5$ W/m^3, $\beta = 0.02$ 1/year and t is time. The thermal loading time is 500 years. The research teams studied this problem are listed in Table 4.

Table 4   Research teams and codes applied to study BMT2

Research Team (Sponsor)	Code	Method
CEA/DMT(ANDRA, CEC))	CASTEM 2000	FEM with plastic failure analysis, no heat convection
ENSMP (CEC, IPSN)	VIPLEF, CHEF, HYDREF	FEM with joint elements, elastic analysis, no heat convection
INERIS (ANDRA, CEC)	UDEC	DEM, elastic analysis, no heat convection
KPH (PNC)	THAMES	FEM, elastic analysis, with heat convection

The major findings from this BMT study can be summarized as:

- Temperature results agrees well among the results obtained by difference teams and are mesh-dependent;
- Thermal convection (considered only by KPH model) does not have significant effect on temperature results;
- Negligible effect of different spacing of discontinuity sets on temperature;
- Results of Hydraulic head, velocity and flow rate at selected points in the computational models differ significantly among different teams;
- Different spacing of discontinuity sets have insignificant effect on hydraulic head and fluid velocity;
- Mechanical results (stresses and displacements) at selected points in the computational models agree reasonably well among different models;
- Different spacing of discontinuity sets have significant impact on DEM results of stresses and displacements, but not on FEM results.
- Difference in mechanical results between FEM and DEM models is significant, and is due to intrinsic differences in model formulation.

4.2. BMT2 - A Near-Field Model of Multiple Fractures

This two-dimensional model measures 0.75 m × 0.50 m and consists of nine rock blocks, separated by two vertical and two horizontal discontinuities (Fig. 4) [8, 9]. The initial conditions of the model include a constant temperature of $15^o C$, constant compressive normal stresses of 4 MPa in both directions, a constant pressure of 10 KPa, and a constant aperture of 300 μm. The initial normal and shear displacements are set to zero. The normal displacements along all boundaries are set to zero. Zero flux (both thermal and hydraulic) is

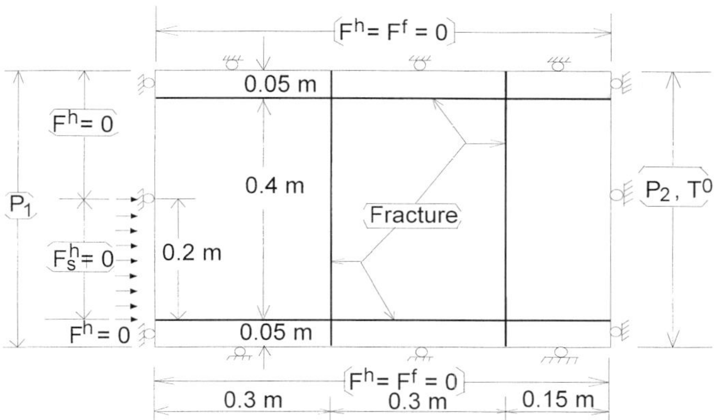

Figure 4. Thermal, hydraulic and mechanical boundary conditions of BMT2.

imposed on the top and bottom boundaries. Hydraulic pressures of $P = 10$ KPa and $P = 11$ KPa are imposed on the left and right vertical boundaries, respectively, to stimulate water flow. For thermal boundary conditions, zero thermal flux is imposed on the top and bottom boundaries of the model. A constant temperature ($T = 15°C$) is imposed on the right vertical boundary. For the left vertical boundary, a heat source with flux $F_s^h = 60 W/m^2$ is prescribed for portion 0.05 m $\leq$ y ≤ 0.25 m. The rest of the left vertical boundary has a zero thermal flux boundary condition. The forced heat convection along the fractures was an option for modelling. The thermal loading time is 10^7 seconds. See Table 5 for the research teams and codes applied for this study .

Table 5   Research teams and codes applied to study BMT2.

Research Team (Sponsor)	Code	Method
AECL (AECL)	MOTIF	FEM with heat convection, elastic analysis
CNWRA (NRC)	UDEC	DEM without heat convection, elastic analysis
INERIS (ANDRA, CEC)	UDEC	DEM without heat convection, elastic analysis
LBL (SKI)	ROCMAS	FEM with convection, elastic analysis
VTT (STUK)	ADINA-T, JRTEMP	FEM without fluid flow and heat convection, elastic analysis

The major findings from this study are:

- BMT2 is a suitable near-field T-H-M model of fractured rocks for genetic studies;
- Although the general trend of temperature field produced by different teams is consistent, significant difference can be detected by considering or not considering the effect of heat

convection by fluid flow;
- Thermal expansion is the major cause for rock deformation since no excavation is considered;
- Heat convection has also major impact on the model deformation, including fracture closure;
- Consistent results of stresses and displacements (including fracture aperture changes) between continuum and discrete approaches, probably due to fixed boundaries and a low heat power output of the heater;
- Closure is the major mode of deformation of rock fractures under heating. Inclusion or exclusion of heat convection in calculations may change the fracture closure by almost one order of magnitude;
- Hydraulic results (including both head and velocity) predicted by different models are also consistent in general, and heat convection affects the fluid velocity to some extent.

4.3. BMT3 - A Near-Field Model with a Realistic Fracture Network

BMT3 is a near field repository model set up as a two-dimensional plane-strain problem in which a tunnel with a deposition hole is located in a fractured rock mass. The model is 50 x 50 m in dimension and 500 meters below the ground level (Fig. 5) [10]. The fracture network is a two-dimensional realization of 6580 fractures from a realistic three-dimensional fracture network model from the Stripa Mine, Sweden (Fig. 6). The problem is set up as a fully coupled Thermo-Hydro-Mechanical near-field problem of a repository. The thermal effect is caused by the heat release of the radioactive waste in the deposition hole (the heater). The heat output decays exponentially with time. The rock matrix is assumed to be isotropic and linearly elastic, and its mechanical properties do not change with temperature variations. The thermal conductivity and expansion of the rock matrix are also assumed to be isotropic. The fractures are assumed to consist of parallel, planar, smooth surfaces at the macroscopic level with an effective hydraulic aperture. The initial and boundary conditions for the mechanical, thermal and hydraulic effects are shown in Fig. 5. The heating is maintained for 100 years. See Table 5 for the research teams and codes applied to this study and Fig. 7 for the representations and simplifications used in the models. The CEA/DMT and KPH teams used Oda's crack tensor approach to determine the equivalent hydro-mechanical properties of the fractured rock with different homogenization scales (REV size) [11].

Table 5 Research teams and codes applied to study BMT3

Research team (Sponsor)	Code	Method
AEA (NIREX)	NAPSAC	DFN - fracture flow only
CEA/DMT (ANDRA, CEC)	CASTEM 2000	FEM - homogenization scale 25 m
CNWRA (NRC)	UDEC	DEM with simplified fracture network
INERIS (ANDRA, CEC)	UDEC	DEM with simplified fracture network
ITASCA (SKB)	FLAC	FDM without homogenization
KPH (PNC)	THAMES	FEM - homogenization scale 10 m
NGI (NIREX)	UDEC	DEM with simplified fracture network
VTT (STUK)	UDEC	DEM with simplified fracture network

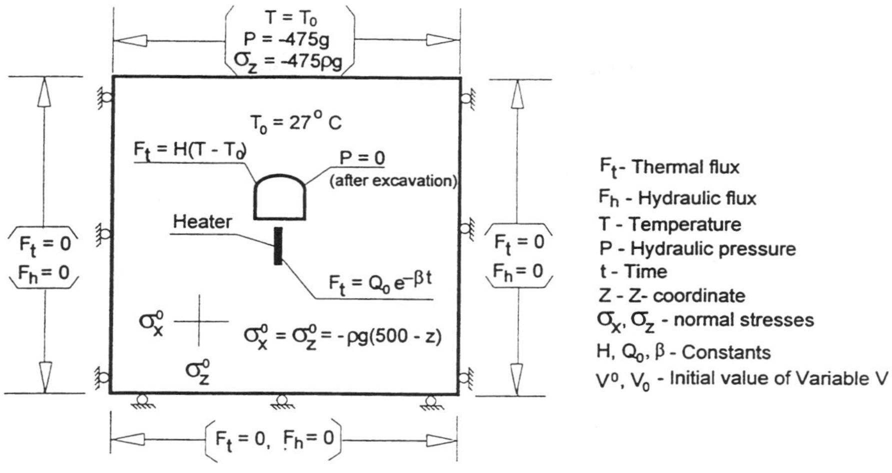

Figure 5 Definition of BMT3: model geometry and boundary conditions.

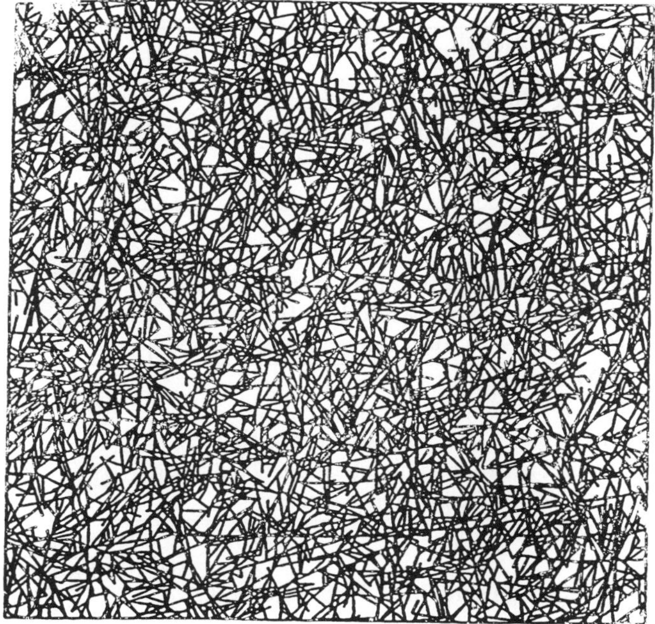

Figure 6 Reference fracture network for BMT3.

AEA (NAPSAC code, DFN model) Complete and explicit representation of 6580 fractures by DFN method, without heat and rock deformation, but with analytical model for stress approximation	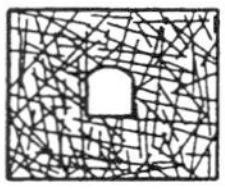
CEA/DMT (CASTEM 2000 code, FEM model) Equivalent continuum approach with FEM method (1344 elements, 4216 nodes), homogenization (scale 25 x 25 meters) by Crack Tensor theory (Oda, 1986)	
CNWRA (UDEC code, DEM model) Vertical symmetry and simplified fracture network with 295 fractures, 337 blocks, 1540 finite difference elements, and 10853 nodes. A small inner region with random fractures and lager outer region with regular artificial fractures	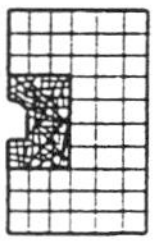
ITASCA (FLAC code, FDM model) Equivalent continuum approach with FDM (735 finite ifference elements and 792 nodes)without formal homogenization. Vertical symmetry and inhomogeneous permeability	
INERIS (UDEC code, DEM model) Simplified fracture network with inner region of random fractures and outer regions of regular artificial fractures (564 fractures, 677 blocks, 24391 finite difference elements, and 16791 nodes)	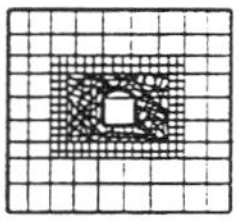
KPH (THAMES code, FEM model) Equivalent continuum approach with FEM method (674 elements, 2127 nodes). Homogenization (scale 10 x 10 meters) by Crack Tensor theory (Oda, 1986). Heat convection considered	
NGI (UDEC code, DEM model) Two models of vertical symmetry (one left half and one right half) with simplified fracture network without heating, about 512 fractures, 510 blocks, and 1580 finite difference elements	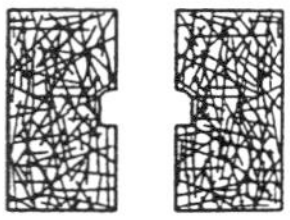
VTT (UDEC code, DEM model) Vertical symmetry (left half) with simplified fracture network, 814 fractures, 496 blocks, 1308 finite difference elements, and 2222 nodes.	

Figure 7 Different model representations and simplifications for BMT3.

The major findings from this BMT study include:

- This BMT is a well defined near-field problem with both a realistic fracture network which may be most possibly encountered in practice and complete aspects of coupled T-H-M processes. This BMT was regarded as a excellent model for both testing capabilities of mathematical models and computer codes, and provoked a number of important scientific issues to be addressed in future research;
- Agreement in temperature results from all research teams is remarkable;
- Heat convection has negligible effects on the temperature results;
- Displacement results agree reasonably well among the research teams and between discrete and continuum approaches. Large discrepancies occur only when monitoring points are located on loose blocks of large movement, close to the boundary of the tunnel;
- Stress results agree less well than the displacement results, especially at points closer to the heat source and tunnel boundary, although a consistent general trend in the stress results can be observed;
- The flow rate results are very divergent both between the discrete and continuum approaches, and among models in the same approach group;
- Two homogenization scales were obtained using the same Crack Tensor theory, by two teams, with the validity of REV and its relationship to the size of finite element unresolved; Satisfying all physical laws when simplifying fracture network by discrete approach was not thoroughly considered.

5. TEST CASE PROBLEMS

5.1. TC1 - A Coupled Shear-Flow experiment of A Single Rock Joint

The TC1 is a numerical simulation of an laboratory test of a rock joint contained in a large diameter core, under controlled boundary stresses with multiple loading-unloading cycles while fluid is conducted through the joint [3, 12, 13]. The deformations (closure, dilatancy and shear displacements) and stresses (normal and tangential) of the joint, and the flowrates were recorded simultaneously during the test, and were compared with numerical predictions. The proposed model of TC1 included a rock core containing a single joint, grouted in epoxy to form a rectangular shape and tested under horizontal and vertical boundary stresses (supplied by flat jacks during tests). The model measures 260 mm × 260 mm with its bottom and right lateral boundaries fixed. Water is injected at one end of the joint. The options of linear and non-linear joint deformability were given and different loading conditions were specified. The purpose of this test case is to compare performances of different modelling methods and constitutive laws of rock joints, including coupled fluid-rock responses, in different computer codes applied to study this problem. See Table 6 for the research teams and codes for this problem. The test set up and the specification of the computational model are illustrated in Fig. 8. The steel plate for loading with hydraulic jacks were considered by most of the research teams. The major findings of this study are:

- TC1 was a difficult test case for numerical simulation. The problem definitions did not correctly reflect the experimental conditions.

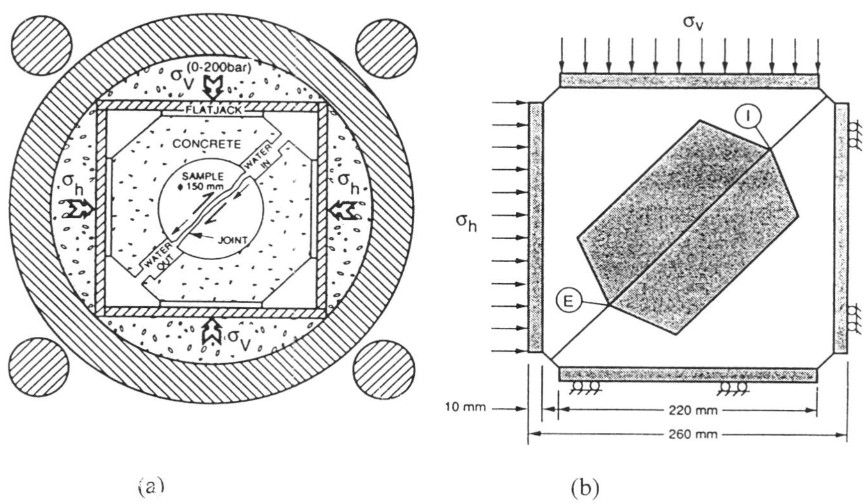

(a)   (b)

Figure 8 Test set up (a), model configuration and boundary conditions (b) of TC1. I – Water injection point; E - Water outlet point.

Table 6 Research teams and their computer codes for study of TC1.

Research team (Sponsor)	Code	Method
CNWRA (NRC)	UDEC	DEM model including the steel plate for loading
ITASCA (SKB)	UDEC	DEM model including the steel plate for loading
LBL (SKI)	ROCMAS	FEM model without the steel plate for loading
NGI (NIREX)	UDEC	DEM model including the steel plate for loading

- The numerical results diverge significantly from the experimental data for normal stress, normal displacement (dilatancy) and effective hydraulic aperture of the rock joints.
- Considerable differences exist also among the DEM models using the same computer code (UDEC), due partially to the differences in mesh sizes, values of constitutive properties for the joint and the interface between the rock and epoxy materials used to grout the test sample.
- The difference between the FEM model and the DEM models is significant for the calculated mechanical aperture of the joint, but becomes less for calculated normal displacement (dilatancy) and normal stress.
- An improvement to the definition of TC1 was recommended with more detailed descriptions of geometry and test equipment, higher joint roughness, detailed loading procedure and using only non-linear joint models. These recommendations led to a new test case, TC1:2.

5.2. TC1:2 - A Coupled Shear-Flow Experiment of A Single Rock Joint, An improvement of TC1

TC1:2 is an improved shear-flow test of single rock joint to the original TC1 [4, 14]. Besides the increased joint roughness, the model geometry is completely redefined to reflect more closely the test conditions and shape of the core sample (Fig. 9). The research teams studying this problem are listed in Table 7. All computational models considered the steel plate as the outer boundary.

Table 7 Research teams and their computer codes for study of TC1:2.

Research team (Sponsor)	Code	Method
AECL (AECL)	MOTIF	FEM with modified BB joint model
LBL (SKI)	ROCMAS	FEM model with Goodman's joint model
NGI (NIREX)	UDEC	DEM with BB joint model

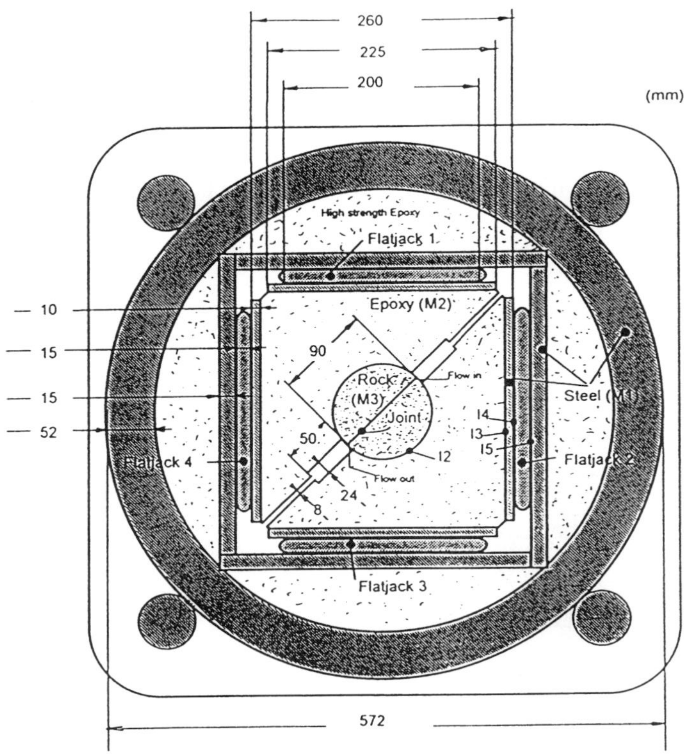

Figure 9 Test set up and definition of TC1:2. The inner steel plate contacting epoxy was taken as the outer boundary of the computational model.

The major findings from this study are:

- Better agreements were achieved for the normal stress-normal displacement curves between the numerical and experimental results, but only for loading path. The unloading path reveals still considerable discrepancies among the research teams.

- The numerical results differs significantly from the experimental results about the normal stress -hydraulic aperture relationship. The numerical results are much smaller than the test results.

- Numerical results about normal stress versus mechanical aperture differs considerable among the research teams, possibly due to different constitutive laws used.

- The numerical results about normal dilatancy and hydraulic aperture during shear differs significantly from the experimental results. Good agreement is limited to compressive loading path (without shear) only.

5.3. TC2 - Fanay-Augères In Situ Thermo-Mechanical Experiment

TC2 is a numerical simulation of an in situ experiment of coupled H-M processes of fractured granite rocks, carried out at Fanay-Augères mine in France [5, 15]. The objective of TC2 is to validate, although partially, the numerical models for fractured rocks in three-dimensions, against results from an in situ T-M experiment. The hydraulic aspect of the experiment was not considered in TC2 because of the unsaturated conditions at the test site and very limited fracture permeability measurements. Figure 10 illustrates the test room geometry. The experimental site, consisting of a test room, an access gallery, and a side gallery, is located in a granite rock mass at a depth of 100 m and can be accessed through a drift. The rock at the Fanay-Augères test site is moderately fractured granite. The test volume measures 10 m × 10 m× 5 m, was fractured by six major and a great deal of minor

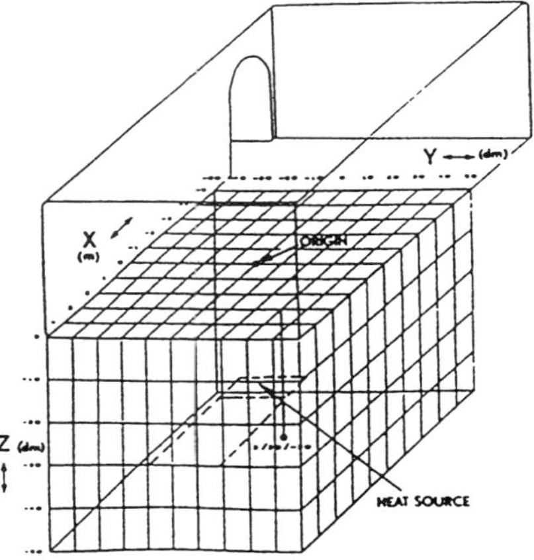

Figure 10   Spatial diagram of the Fanay-Augères test site.

fractures, and was heated with five cylindrical heat sources buried 3 m below the floor (top surface of the test volume), with total power of 1 KW. A total of 153,185 measurements for temperature, vertical displacements, heaving and sagging of the test room floor, longitudinal and transversal strains in the test room floor, and opening, closing and vertical shear displacements of fractures 1 and 2, were carried out, using thermal sensors, borehole and electric extensometers and photo-electric cells.

A uniform initial temperature of 13°C was used for the rock mass. The initial stress state was determined by hydraulic fracturing as $\sigma_v = 0.001\ \gamma\ z$ (vertical stress), $\sigma_h = 0.92\ \sigma_v$ (minor horizontal stress, 60° from North), $\sigma H = 1.35\ \sigma_v$ (major horizontal stress, 150° from North), where $z = 100$ m is the depth of the test volume below the ground surface.

The research teams studied this test case are listed in Table 8. The ENSMP team used two models, $ENSMP_c$ with continuum approach, and $ENSMP_d$ with the discrete approach.

Table 8 Summary of TC2 models.

Research Teams (Sponsor)	Code	Method
INERIS (ANDRA)	3DEC	3-D DEM with 4-noded tetrahedral elements
$ENSMP_d$ (CEC, IPSN)	VIPLEF	3-D FEM with 4--noded tetrahedral elements
CLAY (SKB)	ABAQUS	3-D FEM with 8 - noded hexahedral elements
$ENSMP_c$ (CEC, IPSN)	VIPLEF	3-D FEM with 10 - noded tetrahedral elements

The thermal and mechanical boundary conditions adopted by the various teams for the thermal calculations are presented in Tables 9 and 10, respectively. Γ_{outer} means the outer boundary of the surrounding rock, Γ_{inner} the boundary consisting of the roof, walls, and floor of the excavated test room, and Γ_{heater} the boundary of the heat sources. The thermal conductivity, λ, used in the CLAY model varies linearly with temperature. All the other teams used a constant λ at room temperature. The constitutive model for fractures is a Drucker-Prager type for CLAY model and Mohr-Coulomb type for the others. The loading steps are: i) applying the in situ stresses; ii) excavating the test room; iii) applying heating.

The major findings from this test case are:

- The TC2 is a very useful test case for simulating three-dimensional problems of a heated and fractured rock mass and identifying the possibilities and limitations of the three-dimensional computer codes used.
- Both continuum and discontinuum approaches are able to model the coupled thermo-mechanical behaviour of a volume of fractured rocks. The main difficulty is the adequate representation of the three-dimensional geometry and a reasonable constitutive law for fractures. The computer codes used all showed drawbacks in mesh generation process.
- Comparisons can only be made in a general sense since the assumptions, parameters and the geometry of the models, especially the extension of fractures, are quite different among the research teams.
- The calculated temperature fields agree well with the experimental results and can be further improved if more representative thermal parameters of the fractured rock mass are used.

Table 9   Thermal boundary conditions

Team	CLAY	$ENSMP_c$, $ENSMP_d$	INERIS
Γ_{outer}	Adiabatic	Adiabatic	Adiabatic
Γ_{inner}	Heat convection $K = 10W/m^2\,K$ $\theta_i = 13°C$, $\theta_f = 9°C$	Linear decrease of temperature during the test (13°C to 10°C)	Adiabatic
Γ_{heater}	8 nodes (1.2 m²) No electric breakdown	5 cylinders with electric breakdowns	5 x (50 points) with electric breakdowns

Table 10   Mechanical boundary conditions

Boundary	CLAY	$ENSMP_c$, $ENSMP_d$	INERIS
Γ_{outer}	Fixed normal displacements	Weight of the overburden on the roof; normal displacement is fixed on lateral surfaces	$\sigma = \sigma_i$
Γ_{inner}	Free	Free	8 corners of the chamber are fixed

- The prediction of floor heaving by the discrete models is closer, qualitatively, to the measured results than that obtained by the continuum models. However, numerical models, in general, did not produce reasonable predictions for the floor surface deformation, compared with measured data.
- The calculated openings and vertical shear displacements of the fractures by the discrete models did not agree well with the experimental data, due probably to the lack of a proper constitutive law for rock fractures.
- Further developments are needed to improve the three-dimensional modelling of fractured rock masses, especially where coupled hydro-mechanical processes are expected.

5.4. TC3 - A Laboratory Experiment of Engineered Barrier System - Big-Ben Experiment

The TC3 is a numerical simulation of a large scale laboratory experiment of the coupled T-H-M behaviour of the engineered barrier system, using the Big-Ben (Big-Bentonite) Facility at Tokai Works, PNC, Japan [5, 16]. The aim of the TC3 is to test and compare the capabilities of different numerical models and codes for simulating coupled T-H-M behaviour of the buffer materials in one of the disposal pits. The test facility is composed of an electric heater, a carbon steel overpack, buffer material and cylindrical man-made rock with a borehole (Fig. 11a). The mixture of 70% bentonite and 30% quartz sand was packed with a tamper directly in the borehole. The heater was operated at 0.8 kW. Water was injected from the circumference of the borehole at the pressure of 0.05 MPa. For the comparison of the results, the temperature, the water content or volumetric water content, the vertical and horizontal stress were calculated at 9 points in the buffer material after 1 month and 5 months (Fig. 11b).

The simulations were performed by three research teams in the DECOVALEX project: i) CLAY (sponsored by SKB, Sweden), ii) CNWRA (sponsored by NRC, USA), and iii) KPH (sponsored by PNC, Japan). The CLAY and CNWRA teams used the general purpose finite element code ABAQUS and KPH team used the finite element code THAMES. A two dimensional axisymmetric model was used by all three teams. The initial and the boundary conditions are summarized in Tables 11 and 12, respectively.

The major findings from this test case are:

- The calculated temperature gradient is in fairly good agreement with the measured data, but differences between the calculated and measured temperature values calculated by the three teams, were not negligible. This is in clear contrast with temperature calculations in fractured rocks (e.g. TC2 and BMT3) in which calculated and measured temperatures usually agree very well.

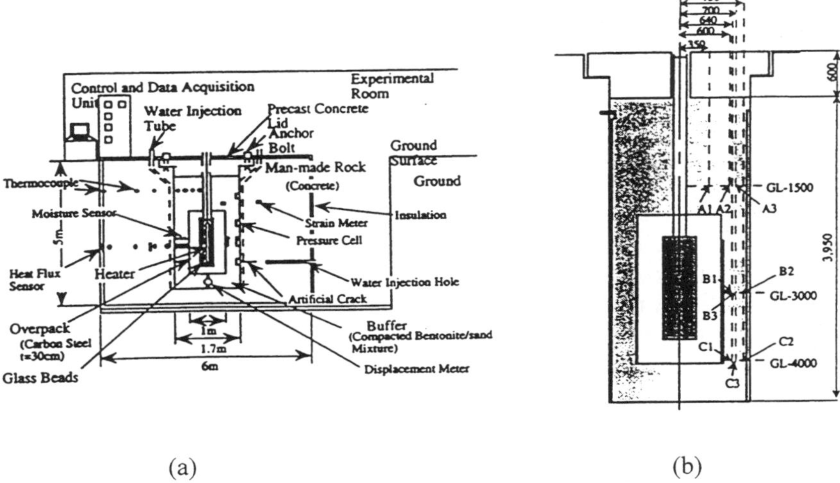

(a)   (b)

Figure 11   a) Schematic view of BIG-BEN experiment equipment: b) Locations of monitoring lines.

Table 11 Initial Conditions

Parameter	CLAY	CNWRA	KPH
Temperature (°C)	15	15	15
Water Content (%)	16.6	16.5	16.5
Void Ratio	0.7		0.69
Effective Average Stress (MPa)	0.8		0.02 (radial); 0.05(Vertical)
Pore Water Pressure (MPa)	-1.231		-14.31

Table 12 Boundary conditions (B.C.)

Boundary conditions (B. C)	CLAY	CNWRA	KPH
Thermal B. C. with given values of heat transfer coefficient (W/m^2 K)	0.1 (upper) 0.5 (right) 0.1 (lower	0.5	1.16
Hydraulic B. C. with pressure	Hydrostatic pressure	0.05 (MPa)	0.05 (MPa)
Mechanical B. C. with displacement	stiff (fixed)	Stiff (fixed)	slider (flexible)

- The calculated and measured water content distribution agree quite well, except at the bottom of the test pit, caused probably by the convective water flow caused by greater thermal gradient.
- Considerable disagreement was found among the three numerical models for stresses. This may be caused by different numerical methods, material assumptions, boundary conditions, and especially the uncertainty about the law for swelling pressure of the buffer material, density change and the effect of phase change of water due to heating. It appears that classical solid elasticity or elasto-plasticity may not be adequate enough to describe the mechanical behaviour of buffer material.

5.5 TC4 - A Laboratory Test of Coupled Hydro-Mechanical Behaviour of Rock Joints

The TC4 is a laboratory investigation on coupled stress-flow behaviour of rock fractures. The aim of TC4 is to provide an experimental data base for developing and validating more comprehensive and realistic constitutive models of rock fractures. The TC4 covers only the first part of the study, i.e. the tests on relationship between normal stress-hydraulic conductivity in a single rock fracture without shear displacements. Temperature effect was not considered. The basic idea is to find out the changes in hydraulic conductivity under variable normal stress across the fracture.

The tests were conducted using a MTS 815 Rock Mechanics Test System at the Laboratory of Rock Engineering at Helsinki University of Technology, using a TestStar digital controller (Fig. 12). Data collection and recording was done using TestWare-SX application programme. The rock fracture samples were either natural or man-made by splitting rock cores of 102 mm in diameter and 100 mm in length into two equal parts along the centreline. Profiles of the fracture surfaces were measured using optical laser-profilometer. During the tests, the fracture opening/closure due to varying water pressure, the circumferential deformation of the sample due to the axial and confining stress, and the deformations of the neoprene jacketing sleeve around the sample are recorded. During the tests the sample was subjected to 20 MPa axial loading, at a rate of 0.8 MPa/sec, and to 2 - 16 MPa of normal stress across the fractures by increasing the confining pressure at a rate of 0.12 MPa/sec. The water flow was kept steady at the rate of 0.05 cm^3/sec. To end the test, the water pressure, the confining pressure and the axial loading were decreased to the zero level at a rate of 0.25 MPa/sec, 0.25 MPa/sec, and 0.8 MPa/sec, respectively. The parameters measured as histories versus time during the test are listed in Table 13.

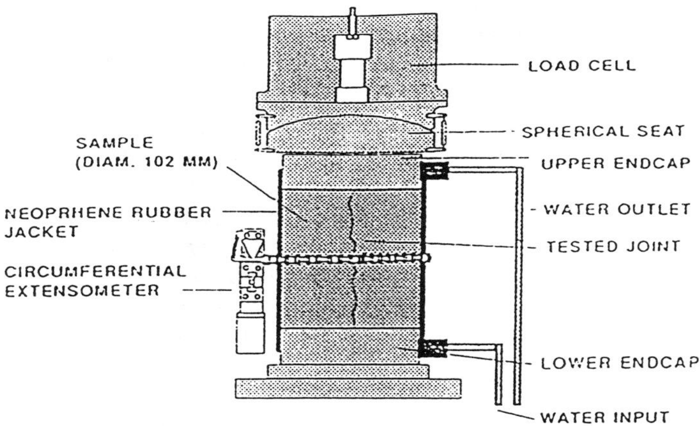

Figure 12   Sample instrumentation.

Table 13   Measured parameters in water conductivity tests.

Parameter	Measuring device
Time, t (sec)	computer clock
Axial displacement, u_a (mm)	actuator LVDT
Axial force, F_a (KN)	2500 KN load cell
Circumferential displacement, u_c (mm)	circumferential extensometer
Confining pressure, P_c (MPa)	confining pressure transducer
Confining volume, V_c (cm^3)	confining intensifier piston LVDT
Water pressure, P_w (MPa)	water pressure transducer
Water flow rate, V_w (cm^3)	water intensifier piston LVDT

   The Reynolds number of the flow during preliminary test was found to be around 0.26, ensuring that the flow was laminar. The cubic law for flow through parallel plates was adopted to interpret the flow results, calculate hydraulic aperture and estimate initial mechanical aperture. Both Goodman's hyperbolic [17] and Zhao and Brown's [18] logarithmic models for normal-stress - normal closure relationships of rock fractures were used to fit the data of normal stress -normal closure. The models developed by Gangi [19], Swan [20] and Gale [21] were used to fit the data for hydraulic aperture and normal stresses. The test data and predictions by these three models, using material parameters identified from the test data, agreed very well.

   Although this test case was not simulated by any research team during the DECOVALEX project, it provided a valuable data base for further developing and validating constitutive models for coupled hydro-mechanical behaviour of rock fractures.

5.6 TC5 - A Laboratory Experiment of Coupled Hydro-Mechanical Behaviour of Rock Joints

The TC5 is a laboratory experiment about the coupled fluid flow and stresses of rough natural rock joints, carried out at the Center for Nuclear Waste Regulatory Analyses (CNWRA), using the CNWRA's direct shear testing apparatus, Fig. 13 [5, 22]. The test was also simulated using ABAQUS code. The test equipment was modified to allow linear fluid flow experiments to be conducted within the joint under constant normal stresses of 2.0, 4.0, 5.0 and 8.0 MPa and shear displacements up to 2.54 cm. The fluid was injected at a constant flowrate of 4.0 cc/min over the entire width of the fracture on the left edge of the top specimen, and collected over the entire width of the fracture from the right edge of the top specimen. The flowrate was chosen such that laminar flow is maintained within the fracture for all normal mechanical loading. Absolute and differential pressure transducers were used to measure the inlet and outlet fluid pressures. A rectangular rubber gasket was placed around the specimen to prevent water leakage.

The rock fracture specimens were prepared from 45.7-cm-diameter core of a welded tuff of the Apache Leap formation. Statistical analysis on 113 uniaxial Apache Leap welded tuff samples and 72 triaxial specimens were carried out to determine mechanical properties of both rock matrix and joints, including parameters required by Barton-Bandis (BB)-model for rock joints.

Surface profile measurements were taken of the joint surfaces on both the top and bottom specimens prior to and after completion of the direct shear tests. The scanning interval in both the x and y directions was set at 0.64 mm. Attempts were made to try to numerically mate the two upper and lower joint surfaces to facilitate calculation of the approximate initial aperture distribution over the joint area under zero applied normal stress. Based on this exercise, the mean aperture was determined to be approximately 0.032 in. (0.81 mm).

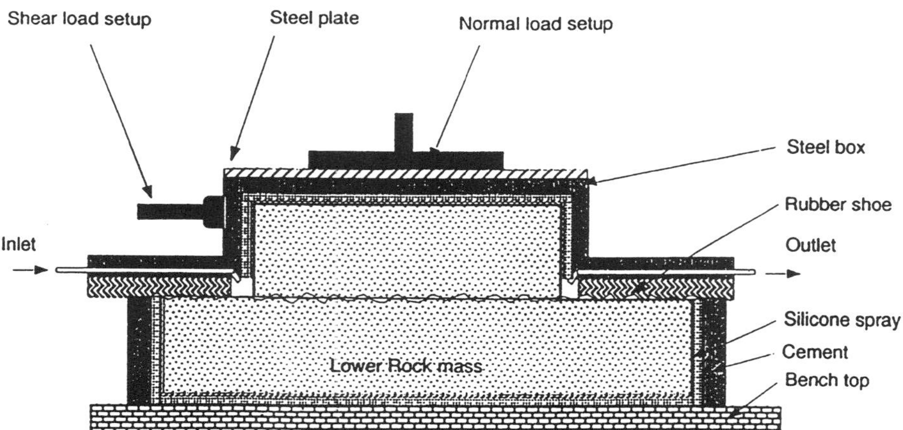

Figure 13 Shear test arrangement for TC5.

The normal loading was conducted up to a maximum load of 8.0 MPa while monitoring the joint normal displacement (i.e., joint closure) and fluid pressure changes across the sample for a fixed flowrate. The shear loading were performed with cyclic shear cycles, up to four shear cycles with 2.54 cm of the maximum shear displacement. However, it was only for the fourth cycle that corresponding hydraulic data was obtained. Numerical simulation was carried out only for the mechanical response of the joint, using the 3D finite element code ABAQUS with both a Mohr-Coloumb model and a modified Drucker-Prager/Cap plasticity model. The latter used a 13 mm thin layer to simulate the fracture.

The major findings from this test case are:

- The TC5 experimental results show that the hydraulic properties of the rock fracture can change by up to a factor of 3 under shear deformation. However, additional tests and improvements to the experimental apparatus are necessary in order to make more quantitative assessments of changes in such properties and gouge production during shear.
- The two constitutive models could not simulate the initial closing of rock joints occurred during the shear test. They could not reproduce with a reasonable closeness to the measured shear stress - shear displacement curve. However, the prediction to the peak shear strength of the fracture agreed well with the measured one. The maximum amount of dilation obtained by the Drucker-Prager/Cap model was significantly lower than that measured experimentally for the first shear cycle.
- The numerical results obtained using ABAQUS show that, although the fractures can be represented by thin zones having plastic or jointed material behaviour, there appear to be several limitations in both the type of shear stress-shear displacement and dilation behaviours obtained. Future work should look into enhancing explicit modelling of fractures within code ABAQUS to better handle fracture behaviour under shear.

5.7. TC6 - An In Situ Experiment of Borehole Injection

TC6 is about the modelling of hydraulic field experiments of a single sub-horizontal fracture which were conducted at the Luleå University of Technology, Sweden [5, 23]. The test case addresses the role of hydro-mechanical response to the interpretation of hydraulic conductivity field tests. The tests were conducted in a vertical borehole with a diameter of 56 mm and located in hard crystalline rock. A natural fracture, that has been tested before by hydro-fracturing stress measurements, intersects the borehole at a depth of 356.7 m (Fig. 14). The borehole section containing the fracture was isolated by pressurised straddle-packers and three types of injection tests were performed: (1) pressure pulse test; (2) step pressure test (hydraulic jacking); and (3) constant pressure injection test with a pressure exceeding the overburden pressure. The hydraulic jacking test was carried out by increasing the pressure stepwise in the isolated borehole section and monitoring the rate of water injection into the section. At each pressure step an apparent stationary flow is obtained within 3 minutes before the pressure is increased for the next step. The test result is presented as pressure versus flow rate.

The pulse test was conducted by first instantaneously increasing the pressure in the isolated borehole section and thereafter closing a valve immediate above it. The borehole pressure then declines to the pre-pulse pressure as water flows into the fracture. The rate of pressure decay depends on the fracture properties, and the volume and compliance of the sealed-off section. The constant pressure injection test was conducted by increasing the downhole pressure to 11.2 MPa and then maintaining it constant during the entire experiment.

Laboratory experiments were conducted on samples of a drill core (41 mm in diameter) from the borehole to determine the density, Young's modulus, Poisson's ratio and uniaxial compressive strength of the rock material. A compression test was conducted on the drill core containing the fracture to obtain a curve that relates the fracture normal stress to the normal displacement. The in-situ normal stress across the fracture has been estimated by hydraulic stress measurement to be 10.2 MPa.

The TC6 is a calibration of the numerical models against the measured results of the pulse, hydraulic jacking and constant head injection tests. The calibration should result in an estimation of the equivalent Young's modulus for the surrounding rock mass, initial hydraulic aperture at initial stress, type and location of the outer boundary of the fracture, and the effective normal stress as a function of normal displacement of the fracture.

Five Research Teams listed in Table 14 modelled TC6, in which 'Jacking' means the hydraulic jacking test, 'Pulse' means the pulse test and

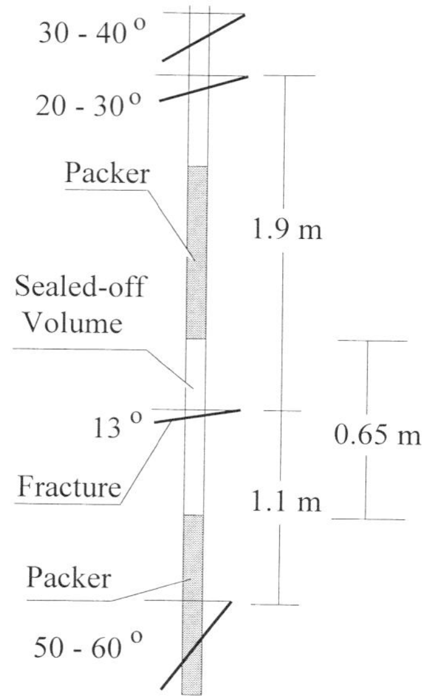

Figure 14 Down-hole geometry of the fractures for the injection test, TC6.

Table 14 Research Teams and models for modelling of TC6.

Research Team (Sponsor)	Code	Method	Simulation Task		
KTH (SKI)	ROCMAS	FEM	Pulse	Jacking	Injection
AECB	FRACON	FEM	Pulse	Jacking	Injection
AECL (AECL)	MOTIF	FEM	Pulse	Jacking	Injection
AEA (NIREX, CEC)	NAPSAC	DFN	Pulse	Jacking	
LTH (SKB)		FEM			Injection

'Injection' means the constant pressure borehole injection test, respectively. KTH and AECB discretized domain of Test Case 6 to an axisymmetric model with the vertical borehole in its centre. The sub-horizontal fracture was assumed to be penny-shaped and centred at the borehole intersection. AECL used a three-dimensional square model with the borehole in one of the corners. AEA used a fracture network model. The hydro-mechanical coupling in code NAPSAC is provided by three different empirical relations between the effective normal stress and fracture's transmissivity [4, 24], but the code cannot model fluid injection with pressures greater than the in situ normal stress. LTH used the diffusion equation together with the "cubic law" for flow through fractures and an empirical relationship between rock porosity and rock hydraulic conductivity [25].

The major findings from this study can be summarized as:

- The numerical modelling of TC6 shows that for injection pressures below the virgin stress normal to the fracture plane, the hydro-mechanical properties of the joint are dictating the pressure versus flow response. The key parameters are the hydraulic aperture, joint normal stiffness and effective normal stress. In a calibration of the models it is therefore recommended to start with the initial hydraulic aperture from the pulse test and from the lowest pressure steps in the hydraulic jacking test. This can be conducted without attention to the effects of joint opening. Thereafter, the hydro-mechanical coupling may be considered and the joint normal stiffness can be calibrated from the pressure versus flow response of the hydraulic jacking test.

- The modelling of the TC6 showed a fairly good agreement to the field test data. The discrepancies between different teams are in most cases due to slightly different choice of initial hydraulic aperture and joint normal stiffness. However, all the teams obtained an initial joint stiffness that was lower than the stiffness determined from laboratory testing of a core sample containing the joint. This indicates that the joint stiffness is dependent of the sample size.

- TC6 demonstrates that coupled hydro-mechanical modelling can be a tool for analysis of hydraulic field testing and can also be used for determination of in situ hydro-mechanical properties of single joints.

6. ACHIEVEMENTS AND LESSONS LEARNED

6.1 Scientific Basis of the BMT and TC Problems

Table 16 summarizes the scales and processes studied in all three BMT and six TC problems studied in the DECOVALEX project, Phase I, II and III. In general, the TC and BMT have addressed, to various degrees, following important aspects of coupled thermo-hydro-mechanical processes:

- Constitutive laws for single rock fractures (TC1, TC1:2, TC3, TC4 and TC5);
- Significance of heat convection due to fluid flow in fracture networks (BMT2);
- Conceptualization of the THM behavior of a fracture rock with realistic problem size and fracture network geometry for underground nuclear waste repositories (BMT1 and BMT3) with approaches of:
    - Equivalent continuous porous media
    - Discrete media
    - Explicit, discrete fracture networks
- Validation (verification) of predictive modelling capabilities of computer algorithms and codes (all TC problems).

Table 15 Summary of BMTs and TCs studied in DECOVALEX.

Problem	Brief Description	Scale	Processes	Remarks
BMT1	Fractured rock with 2 orthogonal sets of persistent fractures and a heat source.	3000 × 1000 m	THM	2D far field problem
BMT2	Fractured rock with 4 discrete fractures and a finite-length heat source.	0.75 × 0.5 m	THM	2D near-field problem
BMT3	Fractured rock with a realistic fracture network of 6580 fractures from Stripa mine data.	50 × 50 m	THM	2D near-field problem
TC1	Laboratory shear-flow test on a rock core sample with a single joint.	8 cm (length)	HM	Coupled H-M behaviour of an rock fracture.
TC2	Field experiment in fractured rock at Fanay-Augères, France	12 × 10 × 5 m	THM	3D problem; unsaturated rock; mainly T-M coupling
TC3	Large-scale laboratory experiment of engineered buffer material (Big - Ben).	5 (high) × 6 (diameter) m	THM	3D T-H-M problem, unsaturated bentonite blocks.
TC4	Laboratory coupled stress-flow tests on rock fractures.	10 (length) × 8.6 (diameter) cm	(T)HM	Coupled H-M behaviour of rock fractures.
TC5	Laboratory shear-flow experiment of a rock block with a single joint.	20 × 10 cm	HM	Coupled H-M behaviour of rock fractures.
TC6	Field experiment of hydraulic injection test on fractures at 356 m depth .	15 m (radius of influence)	HM	Comparative study of flow results with constant and with pulse injection using packers.

All three BMT problems were proved to be suitable generic models for a relatively complete spectrum of the coupled thermo-hydro-mechanical processes in fractured rocks. Both far-field (BMT1) and near-field (BMT2 and BMT3) problems of a repository were considered. They served as a scientific base for development and validation of mathematical models, computer algorithms and their numerical implementation.

The six TC problems are laboratory or in situ experiments representing different aspects of coupled T-H-M processes in fractured rocks (TC2 and TC6), buffer materials (TC3) or single rock fractures (TC1, TC1:2, TC4 and TC5). Except for TC2 which was completed before the start of the DECOVALEX project, all other experiments were carried out in parallel with predictive numerical modelling so that experiments data could be compared with numerical predictions. The combination of properly defined BMT and TC problems, predicative numerical modelling and experimental data base has proven to be a successful strategy for validation of mathematical models and computer codes for coupled processes.

6.2. Major Scientific Findings from Studies of BMT and TC Problems

6.2.1 Temperature

The temperature predictions made by the different approaches and computer codes for all TC or BMT problems are very consistent among themselves and in good agreement with the experimental data. The agreement is better for fractured rocks than that for buffer materials. We may conclude that the mathematical model for the heat transfer equations used in the computer codes in the DECOVALEX project is a realistic representation of the thermal process in fractured rocks and buffer materials and can be trusted to produce realistic thermal results for performance assessments of nuclear waste repositories. The minor differences observed are generally due to either differences in mesh resolution or whether the thermal convection is taken into account. The latter is not expected to be significant for fractured rock systems except in the very close vicinity of the heat source.

6.2.2 Stress-deformation (for both isothermal and thermal loading cases)

Stress-deformation (strain or displacement) predictions for the rock masses, including thermal stress induction and volume expression, are quite consistent among the numerical approaches and codes, and in fairly good qualitative agreement with experimental data. The major source of discrepancy is the thermo-mechanical disturbances caused by heat sources or excavations. The agreement between the numerical predictions and experimental measurements improves as the distance from the thermo-mechanical sources disturbance increases. Near the thermo-mechanical disturbance sources, the discrete approach tends to produce larger displacements than the continuum approach, due to its ability to allow larger movement along discontinuities. We feel confident to use current models and codes for predictions of stress-deformation behaviour of rock matrix surrounding a nuclear waste repository and for design of excavation and support measures.

The major drawback in the stress-deformation predictions by the current numerical models used in the DECOVALEX project is the fact that predictions for the aperture and stress distributions across and along rock fractures during complex deformation paths are not in good agreement with the experimental results. This is caused by the lack of progress in developing comprehensive and reliable constitutive models for rock fractures. This lack of progress affects the prediction capabilities of numerical codes for hydraulic processes to a significant extent.

6.2.3 Hydraulic Field (for both isothermal and thermal loading cases)

Unlike thermal and mechanical results, there are significant discrepancies in hydraulic results, from one conceptual approach to another and between models (codes) using the same conceptual approach. The discrepancies in pressure heads and flow rates are quantitative if not qualitative, both in time and in space, especially when large number of fractures with a complex geometry is concerned.

The reasons for these discrepancies in predicting fluid flow in fractured rocks are multifolds:

- Complete information about the fracture population and geometry for a fractured rock can never be obtained in detail.
- Some fractures must be removed to built up a simplified discrete computational model which can be run within a reasonable time frame, but with a changed fracture connectivity. The conservation laws must be kept before and after the fracture network simplification. This has not been checked in the DECOVALEX project.
- Capability to predict aperture changes of rock fractures by current constitutive laws is too limited.
- Equivalent material properties of the fractured rocks established through a homogenization process, for continuum approaches, is valid only over a certain REV (Representative Elementary Volume). The size of this REV may conflict with the size of excavation or the finite element mesh resolution for a reasonable smoothness of stress-deformation analysis. For fractured rocks, a REV may not exist at all. The homogenization must also be made separately for hydraulic and mechanical processes. The REV for a coupled hydro-mechanical processes should then be the larger REV produced between the hydraulic homogenization and mechanical homogenization. This has not been checked in the DECOVALEX project.

In general, the geometric simplification procedure used in discrete approaches and homogenization method used for continuum approaches need to be re-evaluated. The contribution of the DECOVALEX project is that it has brought this into focus for further research efforts.

6.3. Key Scientific Uncertainties

A number of scientific uncertainty was identified during the project and requires further research. These key scientific uncertainties are summarized as:

- Constitutive models for rock fractures considering especially roughness representation, gauge production, aperture change and time effects (e.g. creep of fractures);
- Constitute laws of large scale faults and fracture zones which cannot be tested in laboratories;
- Swelling pressure laws for buffer materials;

- Change of mechanical and hydraulic properties of rocks, fractures and buffer materials under high temperature and over a long duration of time;
- Dynamic behaviour and properties of rocks fractures;
- Coupled fluid flow and stress-deformation processes in three-dimensional fractured rocks for discrete element methods;
- Homogenization methods for fractured media with new constitutive theories;
- Data base for heat convection coefficient for fluid flow through rock fractures;
- Large scale computation techniques for both continuum and discontinuum approaches for simulating problems of field scales.

7 CONCLUDING REMARKS

The first stage of project DECOVALEX, called DECOVALEX I, was complete. Good progress was made and the project will continue for a second stage, DECOVALEX II. From the studies of DECOVALEX I, we may draw following conclusions:

1) The DECOVALEX project has been successful in its effort to bring together a number of both scientifically sounded and well defined Bench-Mark Tests and Test Cases for numerical analyses. Broad aspects in repository design and performance assessment are covered in these problems. The Bench-Mark tests are carefully defined to represent the important aspects of a repository in situ. The Test Cases are those either completed recently or ongoing experiments which provide a sound basis to validate the mathematical models.

2) The results and experiences obtained in these studies have improved our understanding of the coupled T-H-M processes in fractured rocks and buffer materials. They demonstrated our abilities and limitations to simulate complex coupled T-H-M processes and pointed out future research directions. The identified scientific uncertainties will become forefront research subjects.

3) Integration of different geoscience and geoengineering disciplines has been implemented. The disciplines include: physics, applied mathematics, geohydrology, hydraulics, engineering geology, soil and rock mechanics, and civil and mining engineering. The integration of these disciplines are manifested in the formulation of BMTs and TCs, evaluation of models and results, and estimation of uncertainties. National waste managing agencies are also actively involved in every aspect of the project.

4) Integration of different numerical methods based on both continuum and discontinuum approaches has been a particular aspect of DECOVALEX I and proves to be very fruitful and educational. Convergence rather than divergence appears to be the general trend to different approaches. This integration is proved to be a very useful approach to improve the understanding of capabilities, limitations, and uncertainties in their respective formulations.

5) Some uncertainties and difficulties remain in the project, especially regarding the constitutive laws for rock fractures, swelling pressure in buffer materials, large scale computations for discrete approaches and homogenisation methods for continuum

representation of fractured media. Our present experience shows that the greatest difficulty lie in the interpretation of physical processes and obtaining the material properties/parameters involved.

6) Recognizing that a ultimate verification against nature is impossible to achieve, our aim is the confidence building in our models and codes with respect to the rationality of the basic assumptions and practical needs. Recognizing also the importance of models and codes for the design and performance assessments of repositories, the validation of the models and codes will remain necessary for the radioactive waste management enterprises and regulators until a relatively high level of confidence is achieved.

The DECOVALEX II project will continue for another three years until 1998 with four tasks: i) a shaft sinking project undertaken by NIREX, UK, at Sellafield involving coupled H-M process in fractured rocks; ii) an in situ T-H-M experiment of buffer material surrounded by fractured rocks at Kamaishi Mine, Japan, undertaken by PNC, Japan; iii) review of currently available constitutive models for rock joints; and iv) preparation of a state-of-science on coupled thermo-hydro-mechanical processes for safety and performance assessments of nuclear waste repositories. The objective of this continuous effort is to apply the mathematical models and computer codes developed and validated during the DECOVALEX I to large scale field experiments of coupled T-H-M processes for fractured rocks and engineered buffer materials, and draw up some concluding statements about the state-of-the-science of the coupled T-H-M processes for safety and performance assessments of nuclear waste repositories.

ACKNOWLEDGEMENTS

The authors of this contribution, on behalf of the Steering Committee of the DECOVALEX project, would like to thank all research teams and individuals for their scientific contributions to the project, and Funding Organizations and Funding Parties for their financial support.

REFERENCES

1. Jing, L., Stephansson, O. and Tsang, C-F., DECOVALEX - an international co-operative research project on mathematical models of coupled THM processes for safety analysis of radioactive waste repositories, Int. J. Rock. Mech. Min. Sci. Geomech. Abstr., 32(5), 389-398, 1995.

2. Tsang, C-F. (ed.), *Coupled processes associated with nuclear waste repositories*, Academic Press Inc., 1987.

3. Jing, L., Stephansson, O., Rutqvist, J., Tsang, C-F., and Kautsky, F., DECOVALEX - mathematical models of coupled T-H-M processes for nuclear waste repositories, Phase I report, SKI Technical Report 93:31, June, 1993.

4. Jing, L., Stephansson, O., Rutqvist, J., Tsang, C-F., and Kautsky, F., DECOVALEX - mathematical models of coupled T-H-M processes for nuclear waste repositories, Phase II report, SKI Technical Report 94:16, 1994.

5. Jing, L., Stephansson, O., Rutqvist, J., Tsang, C-F., and Kautsky, F., DECOVALEX - mathematical models of coupled T-H-M processes for nuclear waste repositories, Phase III report, 1996 (in press).

6. DECOVALEX Secretariat, Definition of Bench-Mark-Test 1, Far - Field Model, DECOVALEX Series Document Doc91/103, Division of Engineering Geology, Royal Institute of Technology, Stockholm, Sweden, 1991.

7. Millard, A., Durin, M., Stietel, A., Thoraval, A., Vuillod, E., Baroudi, H., Plas, F., Bougnoux, A., Vuoille, G., Kobayashi, A., Hara, K., Fujita, T. and Ohnishi, Y., Discrete and continuum approaches to simulate the thermo-hydro-mechanical couplings in a large, fractured rock mass, Int. J. Rock. Mech. Min. Sci. Geomech. Abstr., 32(5), 409-434, 1995.

8. DECOVALEX Secretariat, Definition of Bench-Mark-Test 2, Multiple Fracture Model, DECOVALEX Series Document Doc91/104, Division of Engineering Geology, Royal Institute of Technology, Stockholm, Sweden, 1991.

9. Chan, T., Khair, K., Jing, L., Ahola, M., Noorishad, J. and Vuillod, E., International comparison of coupled thermo-hydro-mechanical models of a multiple-fracture bench mark problem: DECOVALEX Phase I, bench mark test 2, Int. J. Rock. Mech. Min. Sci. Geomech. Abstr., 32(5), 435-452, 1995.

10. DECOVALEX Secretariat, Definition of Bench-Mark-Test 3, Near - Field Repository Model, DECOVALEX Series Document Doc92/112, Division of Engineering Geology, Royal Institute of Technology, Stockholm, Sweden, 1992.

11. Oda, M., An equivalent continuum model for coupled stress and fluid flow analysis in jointed rock massses. Water Resoc. Res., 22(31), pp. 1845-1856, 1986.

12. DECOVALEX Secretariat, Definition of Test Case 1, Stress - Flow Model, DECOVALEX Series Document Doc91/105, Division of Engineering Geology, Royal Institute of Technology, Stockholm, Sweden, 1991.

13. Makurat, A., Ahola, M., Khair, K., Noorishad, J., Rosengren, L. and Rutqvist, J., The DECOVALEX Test-Case one, Int. J. Rock. Mech. Min. Sci. Geomech. Abstr., 32(5), 399-408, 1995.

14. DECOVALEX Secretariat, Definition of Test Case 1:2, Coupled Stress - Flow Model, Phase II, DECOVALEX Series Document Doc92/111, Division of Engineering Geology, Royal Institute of Technology, Stockholm, Sweden, 1992.

15. Gros, J. C., DECOVALEX - Phase 3,- Test Case 2 - Fanny-Augéres THM Test, Suggested test case for DEVOVALEX, Commissariat à l'Energie Atomique (CEA), Institut de Protection et de Sûreté Nucléalre (IPSN), and Commission of the European Communities (CEC), DPEI/SERGD Report No. 92/43, 1993.

16. Fujita, T,. Moro, Y., Hara, K., Kobayashi, A. and Ohnishi, Y,. DECOVALEX - Specifications for test case 3 - Big-Ben Experiment, Power Reactor and Nuclear Fuel Development Corporation, Tokyo, Japan., 1993.

17. Goodman, R. E., The mechanical properties of joints, Proc. 3rd Cong. ISRM, Denver, U.S. National Academy of Sciences, pp.127-140., 1974.

18. Zhao, J. and Brown, E. T., Hydro-thermo-mechanical properties of fractures in the Carnmenellis granite, Quart. J. Eng. Geol., Vol. 25, 279-290, 1992.
19. Gangi, A. F., Variation of whole and fractured porous rock permeability with confining pressure, Int. J. Rock Mech. Min. Sci. & Geomech. Abstr. 15(5), 249-257, 1978.
20. Swan, G., Determination of stiffness and other joint properties from roughness measurements. Rock Mech. & Rock Engng., 16, 19-38, 1983.
21. Gale, J. E-, The effects of fracture type (induced versus natural) on the stress-fracture closure-fracture permeability relationships, Proc. 23rd US Symp. on Rock Mechanics, University of California, Berkeley, Aug. 25-27, pp. 290-298, 1982.
22. Hsiung, S. M., Kana, D. D., Ahola, M. P., Chowdhury, A. H. and Ghosh, A., Laboratory Characterization of rock joints. Research Report of Centre for Nuclear Waste Regulatory Analyses, TX, USA, NUREG/CR-6178, CNWRA 93-013, 1993.
23. DECOVALEX Secretariat, Definition of Test Case 6, Borehole Injection Test, DECOVALEX Series Document Doc93/139, Division of Engineering Geology, Royal Institute of Technology, Stockholm, Sweden, 1993.
24. Wilcock, P. M., The results of applying the NAPSAC fracture network code to model BMT3: the near field test case, AEA Report, D&W 0640, NSS/R365 (Draft), AEA Technology, Harwell, UK, 1993.
25. Claesson, J., Follin, S. and Hellström, G., High pressure testing on fractured crystalline rock - a note on the diffusion equation regarding Test Case 6 of DECOVALEX, SKB Arbetsrapport 94-48, Swedish Nuclear Waste Management Co. (SKB), May, 1994.

O. Stephansson, L. Jing and C.-F. Tsang (Editors)
Coupled Thermo-Hydro-Mechanical Processes of Fractured Media
Developments in Geotechnical Engineering, vol. 79
© 1996 Elsevier Science B.V. All rights reserved.

Constitutive Models For Rock Joints

Y. Ohnishi[a], T. Chan[b] and L. Jing[c]

[a] School of Civil Engineering, Kyoto University, Kyoto 606, Japan

[b] AECL (Atomic Energy of Canada Ltd.), Pinawa, Manitoba, Canada

[c] Division of Engineering Geology, Royal Institute of Technology, S-100 44 Stockholm, Sweden

Abstract

Constitutive models for rock joints are reviewed in this contribution. Emphasis is given to the different approaches for theoretical and empirical formulations of constitutive models, their validation against experimental results, new developments for roughness representations, dilation laws and coupled hydro-mechanical behaviour of rock joints.

1. INTRODUCTION

The experimental results of rock joints, observed in laboratory tests or field experiments, provide raw data under specific test conditions. These data provide the basis for determining the joint behaviour. A general mathematical model must be developed to represent the mechanical behaviour of the rock joints under general loading paths and histories, such that the universal laws of classical solid physics are always satisfied. Such models serve as constitutive models for rock joints and they are utilized as predictors of complete histories of stresses and displacements of rock joints in various numerical methods and computer programs under general loading conditions. In view of the extreme complexity of joint behaviour, usually only part of the most important aspects of the rock joint behaviour, most often resulted from laboratory tests, can be approximately represented in a constitutive model. These are usually relations between stresses and displacements (or their increments). Constitutive models are often conceived intuitively from conceptual models obtained as typical behaviour of rock joints under laboratory test conditions (see, for example, Fig. 1).

The formulations of constitutive models follow basically two approaches: the empirical approach and the theoretical approach. In the empirical approach, a constitutive model is obtained either by curve fitting to laboratory results or from conceptual mathematical functions. The constants appearing in these models may or may not have physical meanings and can only represent rock joint behaviour under special test conditions. The empirical

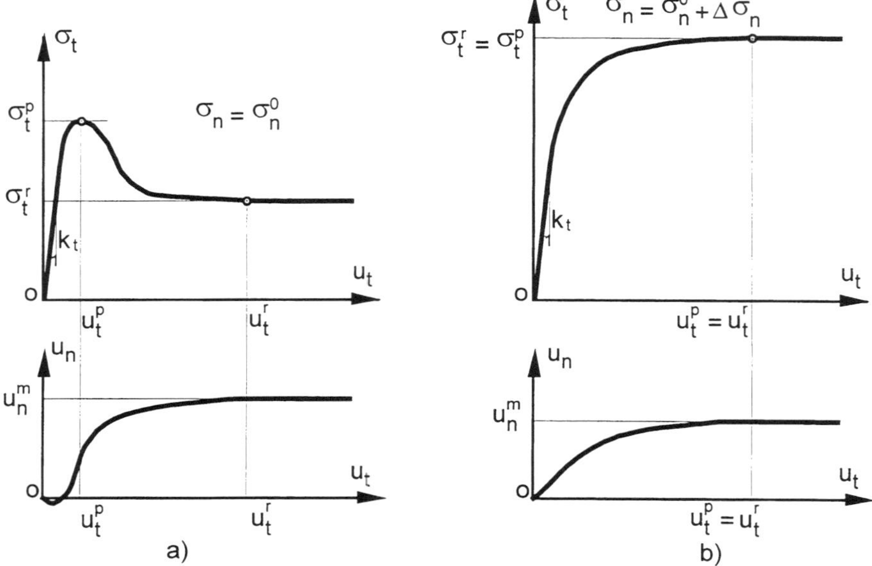

Figure 1. Conceptual behaviour of rock joints under constant normal stress (a) and constant system stiffness (b), where σ_t is the shear stress, σ_n is the normal stress, u_t is the shear displacement, σ_t^p is the peak shear stress at shear displacement u_t^p, σ_t^r is the residual shear stress at shear displacement u_t^r, and k_t is the shear stiffness of the joint.

constitutive models obtained in this manner can provide practical solutions to engineering problems so long as the parameters lie within the range covered by laboratory tests or their reasonable extrapolations. The shortcomings of this approach are: i) the model developed might not conform with physical laws commonly accepted as axioms in all branches of classical solid physics. ii) the range of validity of the parameters in these models cannot be completely covered by laboratory experiments or field measurements. Necessary extrapolations must often be made, and these may introduce some errors whose significance may become unpredictable. In such circumstances, the model will lose both its physical basis and practical utility. iii) the empirical models thus obtained are usually relations between total stresses and total displacements, not in an incremental form which is often better suited for numerical modelling.

The theoretical approach, on the other hand, is based on a formal description of joint behaviour with the understanding that basic laws of classical solid physics must be obeyed as the first priority. The models are usually formulated using theories of different branches of solid mechanics, for example, the theory of plasticity, damage mechanics, and Hertz's contact theory of elasticity, etc.. The fundamental laws of these branches of mechanics are then obeyed in the formulation. However, models formulated by this approach may suffer from the limitation brought about by that fact that today no existing mathematical theory of classical solid physics or mechanics can conveniently represent, even approximately, all aspects of rock joint behaviour. Some less important aspects of the joint behaviour must be ignored and the developed model must be validated against laboratory test data. The behaviour of rock joints under certain experimental conditions, therefore, is the common foundation for both approaches.

The simplest constitutive model for rock joints is perhaps the Coulomb friction law in which the joint behaviour is simply characterized by a single value of friction angle. Some more complicated joint models appeared later accompanying the development of numerical methods. Notable among them are Goodman's empirical model [1], Barton-Bandis' empirical model [2, 3, 4], Amadei-Saeb's analytical model [5] and Plesha's theoretical model [6]. All of them are two-dimensional models. These joint models are briefly reviewed, followed by a detailed derivation of a 2-D constitutive model for rock joints, as an extension of the Plesha's original model [7,8]. New developments of joint models during the DECOVALEX project [9] by Nguyen and Selvadurai [10] and Ohnishi et al. [11] are then presented in detail. These models are selected for presentation because they are representative in their approach to formulation, or have been widely applied in practice or especially in the DECOVALEX project, or represent new directions of current research.

2. EMPIRICAL CONSTITUTIVE MODELS FOR FROCK JOINTS

2.1. Goodman's Empirical Model

Goodman [1] formulated the first comprehensive constitutive model, in an empirical approach, for two-dimensional rock joints. The model is formulated in total stresses and displacements, based on the observed joint behaviour under constant normal stresses observed in the laboratory tests.

The normal stress (σ_n) - normal displacement (u_n) relation is given by:

$$\frac{\sigma_n - \xi}{\xi} = A\left(\frac{u_n}{u_n^m - u_n}\right)^t \tag{1}$$

where ξ is the "seating pressure", u_n^m is the maximum closure (Fig. 2a), and A and t are material constants. The normal stiffness of the joint is then written

$$k_n = \frac{\partial \sigma_n}{\partial u_n} = \frac{t}{u_n^m - u_n}(\sigma_n - \xi) \tag{2}$$

The complete curve of shear stress (σ_t) versus shear displacement (u_t) is composed of five connected parts for monotonic shear paths (Fig. 2b). The shear stress - shear displacement relations in each part of the curve are given by

                                                                          Y. Ohnishi et al.

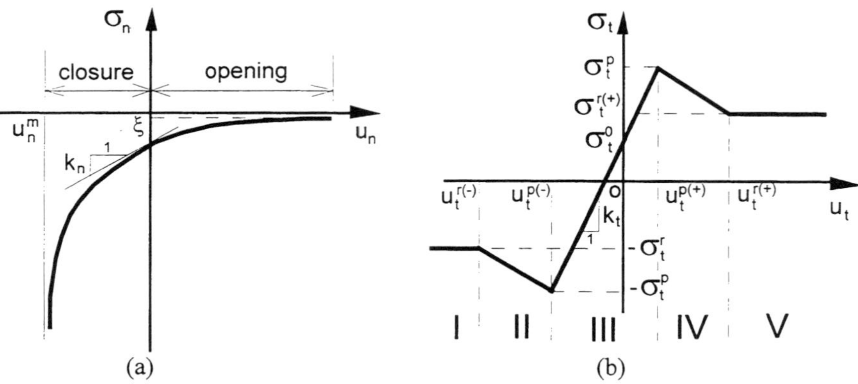

Figure 2. Stress - displacement curves for Goodman's constitutive model for rock joints [1].
a) Normal stress - normal displacement curve; b) Shear stress - shear displacement curve.

$$\text{I)}\quad \sigma_t = -\sigma_t^r \qquad\qquad (u_t \le u_t^{r(-)}) \qquad (3)$$

$$\text{II)}\quad \sigma_t = -\sigma_t^P + \frac{\sigma_t^P - \sigma_t^r}{u_t^P - u_t^r}(u_t - u_t^{P(-)}) \qquad (u_t^{r(-)} \le u_t \le u_t^{P(-)}) \qquad (4)$$

$$\text{III)}\quad \sigma_t = k_t u_t + \sigma_t^0 \qquad\qquad (u_t^{P(-)} \le u_t \le u_t^{P(+)}) \qquad (5)$$

$$\text{IV)}\quad \sigma_t = \sigma_t^P + \frac{\sigma_t^P - \sigma_t^r}{u_t^P - u_t^r}(u_t - u_t^{P(+)}) \qquad (u_t^{P(+)} \le u_t \le u_t^{r(+)}) \qquad (6)$$

$$\text{V)}\quad \sigma_t = \sigma_t^r \qquad\qquad (u_t \ge u_t^{r(+)}) \qquad (7)$$

where σ_t^P is the peak shear stress (shear strength) and σ_t^r is the residual shear stress. k_t is the shear stiffness and σ_t^0 is the initial shear stress. $u_t^{P(+)}$ and $u_t^{P(-)}$ are the shear displacements corresponding to the occurrence of peak shear stress in positive and negative shear directions, respectively, and $u_t^{r(+)}$ and $u_t^{r(-)}$ are the shear displacements corresponding to the occurrence of residual shear stress in positive and negative shear directions, respectively. They are given by

$$\begin{cases} u_t^{P(+)} = u_t^P = \dfrac{\sigma_t^P - \sigma_t^0}{k_t}, & u_t^{P(-)} = -u_t^P - \dfrac{2\sigma_t^0}{k_t} \\[2em] u_t^{r(+)} = u_t^r = M\dfrac{\sigma_t^P - \sigma_t^0}{k_t}, & u_t^{r(-)} = -u_t^r - \dfrac{2\sigma_t^0}{k_t} \end{cases} \qquad (8)$$

where M is a material constant. The normal displacement increment caused by dilatancy over surface asperities is given by

$$\Delta u_n = -\frac{d_p}{u_t^p}\left(|u_t| + \left|\frac{\sigma_t^0}{k_t}\right|\right) \qquad (u_t^{r(-)} \leq u_t \leq u_t^{r(+)}) \qquad (9)$$

and

$$\Delta u_n = -\frac{d_p}{u_t^p}\left(u_t^{r(+)} + \left|\frac{\sigma_t^0}{k_t}\right|\right) \qquad (u_t \geq u_t^r) \qquad (10)$$

where d_p is the dilatancy of the joint when the peak shear stress occurs.

Besides Coulomb friction model, this model is among the earliest constitutive model for rock joints implemented into computer programs for rock mechanics analysis and enjoyed wide engineering applications. However, considerable discrepancy often exists between the model predictions and experimental data. The surface roughness degradation with accumulated shear deformation is not considered and the model works only for monotonic shear.

2.2. Barton-Bandis' Empirical Model

Barton - Bandis' model (often called BB-model) is an empirical model for rock joints, based on numerous experimental results [2, 3, 4]. This model is formulated in total stresses and displacements. The normal stress (σ_n) - normal displacement (u_n) relation is given by:

$$\sigma_n' = \sigma_n - p = k_{n0}\left(\frac{u_n}{1 - u_n / u_n^m}\right) \qquad (11)$$

where k_{n0} is the initial normal stiffness of the joint, u_n^m is the maximum closure and σ_n' is the effective normal stress, defined as the total stress (σ_n) minus the fluid pressure p. The mobilized shear stress, σ_t^{mob}, at any given shear displacement is given by

$$\sigma_t^{mob} = \sigma_n' \tan\left[\left(JRC_{mob}\log\left(\frac{JCS_{mob}}{\sigma_n'}\right) + \phi_r\right)\right] \qquad (12)$$

where ϕ_r is the residual friction angle, JRC_{mob} is the mobilized Joint Roughness Coefficient of the joint at the corresponding shear displacement, JCS_{mob} is the mobilized Uniaxial Compressive Strength of the rock material. Under very high normal stress, the term

JCS_{mob} is replaced by $(\sigma_1 - \sigma_3)$ which stands for the confined shear strength of the asperity.

A simplified BB-model developed by Guvanasen and Chan [12] was utilized by Chan and Khair [13] to simulate BMT2 of the DECOVALEX Project [14, 15]. In this simplified model, the normal stress - normal displacement relation of a joint is the same as that of equation (11) and the normal stiffness is given by

$$k_n = \frac{\partial \sigma_n}{\partial u_n} = k_{ni}\left(\frac{u_n^m}{u_n^m - u_n}\right)^2 = \frac{\sigma_n u_n^m}{(u_n^m - u_n)u_n} \tag{13}$$

The shear stress - shear displacement relation is given by

$$\sigma_t = \frac{\sigma_t^p}{u_p} u_t \ , \qquad\qquad |u_t| \le |u_t^p| \tag{14}$$

$$\sigma_t = \sigma_t^p + \frac{\sigma_t^r - \sigma_t^p}{u_t^r - u_t^p}(u_t - u_t^p) \ , \quad |u_t^p| \le |u_t| \le |u_t^r| \tag{15}$$

$$\sigma_t = \sigma_t^r \ , \qquad\qquad |u_t| \ge |u_t^r| \tag{16}$$

where

$$\sigma_t^p = \sigma_n \tan(\phi_r + \alpha) = \sigma_n \tan\left[\phi_r + JRCLog_{10}\left(\frac{JCS}{\sigma_n}\right)\right] \tag{17}$$

$$u_t^p = A(JRC)^B \tag{18}$$

$$u_t^r = m(u_t^p) \tag{19}$$

$$\sigma_t^r = \left(\frac{0.5 + \phi_r/\alpha}{1.0 + \phi_r/\alpha}\right)\sigma_t^p \tag{20}$$

where the term α is the effective dilatancy (or roughness) angle of the joint surface. The parameters A, B, and m are empirical constants. From this simplified BB-model, the shear stiffness with scale effect of the joint can be identified as

$$k_t = \frac{|\sigma_t^p|}{u_t^p} = \sigma_n \tan\left[JRC\log_{10}\left(\frac{JCS}{\sigma_n}\right) + \phi_r\right]\bigg/ \frac{L}{500}\left(\frac{JRC}{L}\right)^{0.33} \tag{21}$$

where L is the length of the joint under consideration.

This model is similar, in both formulation approach and functional form, to the Goodman's model. It works for cyclic normal compression, but only for monotonic shear loading. Both models were applied to investigate coupled hydro-mechanical behaviour of single rock joints, defined in BMT2 and TC1 of the DECOVALEX project [4, 14, 15]. Considerable difference still exists between predictions by these two models or between either model and experimental results. Although some of the discrepancies might have been caused by unreliable experimental measurements. It is clear that these empirical models cannot provide a reasonable simulation of joint behaviour under general laboratory test conditions. It may also be difficult for them to be used for more general and complex shear paths and loading histories under *in-situ* conditions.

3. THEORETICAL CONSTITUTIVE MODELS FOR ROCK JOINTS

3.1. Amadei-Saeb's Theoretical Model

Amadei and Saeb [5] developed a two-dimensional constitutive model for rock joints with consideration for the different normal deformabilities of rock joints with mated and unmated initial positions, and the effects of the deformability of the rock mass surrounding the joints on the joint behaviour. The model is formulated in an incremental form, given by

$$\begin{Bmatrix} d\sigma_n \\ d\sigma_t \end{Bmatrix} = \begin{bmatrix} k_{nn}, & k_{nt} \\ k_{tn}, & k_{tt} \end{bmatrix} \begin{Bmatrix} du_n \\ du_t \end{Bmatrix} \tag{22}$$

where $[k_{ij}]$ $(i, j = n, t)$ denotes the stiffness tensor of the joint. The elements of this stiffness tensor can be written

$$k_{nn} = \left[\frac{-u_n k_2}{\sigma_c} \left(1 - \frac{\sigma_n}{\sigma_c} \right)^{k_2-1} \tan(i_0) + \frac{(u_n^m)^2 k_{n0}}{k_{n0} u_n^m - \sigma_n} \right]^{-1} \tag{23}$$

$$k_{nt} = -\left(1 - \frac{\sigma_n}{\sigma_c} \right)^{k_2} \tan(i_0) \left[\frac{-u_n k_2}{\sigma_c} \left(1 - \frac{\sigma_n}{\sigma_c} \right)^{k_2-1} \tan(i_0) + \frac{(u_n^m)^2 k_{n0}}{k_{n0} u_n^m - \sigma_n} \right]^{-1} \tag{24}$$

$$k_{tn} = \frac{u_t}{u_t^p} k_{nn} \frac{\partial \sigma_t^p}{\partial \sigma_n}, \qquad (u_t < u_t^p) \tag{25}$$

$$k_{tn} = \frac{k_{nn}}{u_t^p - u_t^r} \left\{ \frac{\partial \sigma_t^p}{\partial \sigma_n} (u_t - u_t^r) + (u_t^p - u_t)C \right\}, \qquad (u_t^p \leq u_t \leq u_t^r; \ \sigma_n < \sigma_T) \tag{26a}$$

$$\text{with} \quad C = \left[\frac{\partial \sigma_t^P}{\partial \sigma_n} \left(B_0 + \frac{1 - B_0}{\sigma_c} \sigma_n \right) + \frac{\sigma_t^P}{\sigma_n} (1 - B_0) \right] \tag{26b}$$

$$k_{tt} = \frac{u_t}{u_t^P} k_{nt} \frac{\partial \sigma_t^P}{\partial \sigma_n} + \frac{\sigma_t^P}{u_t^P}, \qquad (u_t < u_t^P) \tag{27}$$

$$k_{tt} = \frac{\sigma_t^P - \sigma_t^r}{u_t^P - u_t^r} + \frac{k_{nt}}{u_t^P - u_t^r} \left[\frac{\partial \sigma_t^P}{\partial \sigma_n} (u_t - u_t^r) + (u_t^P - u_t)C \right], \; (u_t^P \le u_t \le u_t^r ; \sigma_n < \sigma_T) \tag{28}$$

where σ_c is the uniaxial compressive strength of the rock material, i_0 is the initial dilatancy angle, $B_0 \; (0 \le B_0 \le 1)$ is the ratio of residual shear stress over peak shear stress under zero normal (or very low) stress, and k_2 is a material constant. The expressions for σ_t^P, σ_t^r and $\partial \sigma_t^P / \partial \sigma_n$ depends on the selected criterion for the peak shear stress. A revised Ladanyi and Archambault criterion for peak shear stress is used here, given by

$$\sigma_t^P = \sigma_n \tan(\phi_\mu + i)(1 - a_s) + a_s s_r \tag{29}$$

where

$$i = \tan^{-1} \left[\left(1 - \frac{\sigma_n}{\sigma_c} \right)^{k_2} \tan(i_0) \right] \tag{30}$$

$$s_r = s_0^i + \sigma_n \tan(\varphi_0^i) \tag{31}$$

and

$$a_s = 1 - \left(1 - \frac{\sigma_n}{\sigma_c} \right)^{k_1} \tag{32}$$

where ϕ_μ is the friction angle when sliding along the asperities (equivalent to ϕ_0 in the Ladanyi & Archambault criterion [16]), s_0^i, φ_0^i denote the cohesion and internal friction angle of the intact rock matrix, respectively, and k_1 is a material constant. The term s_r stands for the shear strength of the intact rock matrix assuming a Mohr-Coulomb criterion. With this peak shear stress criterion, then

$$\frac{\partial \sigma_t^p}{\partial \sigma_n} = a_s \tan(\varphi_0) + \frac{S_r}{\sigma_c} k_1 \left(1 - \sigma_n/\sigma_c\right)^{k_1-1} + (1 - a_s)F_1 + F_2 \tag{33a}$$

$$F_1 = \tan(\phi_\mu + i) \tag{33b}$$

$$F_2 = -\frac{\sigma_n}{\sigma_c} \frac{(1-a_s)k_2}{\cos^2(\phi_\mu + i)} \frac{\tan(i_0)\left(1 - \dfrac{\sigma_n}{\sigma_c}\right)^{k_2-1}}{1 + \left(1 - \dfrac{\sigma_n}{\sigma_c}\right)^{2k_2} \tan^2(i_0)} - \frac{\sigma_n k_1 F_1}{\sigma_c}\left(1 - \frac{\sigma_n}{\sigma_c}\right)^{k_1-1} \tag{33c}$$

When $u_t > u_t^r$, k_{tt} vanishes. The same applies when $\sigma_n \geq \sigma_c$ and $u_t \geq u_t^p = u_t^r$.

When $\sigma_n \geq \sigma_c$ and $u_t < u_t^p$, $k_{tt} = \sigma_t^p / u_t^p$.

There two distinct advantages associated with Amadei-Saeb model: i) the model can predict different deformabilities in the normal direction of joints for both mated (fully interlocked asperities as the initial position of joints) and unmated (the asperities are mismatched initially) joints; and ii) the constraint due to deformability of surrounding rocks to the joints can be considered. The major shortcoming is that dilatancy due to asperities may become reversible because the assumption that the dilatancy rate is given by

$$d = \partial u_n / \partial u_t = \tan(i) = \left(1 - \sigma_n/\sigma_c\right)^{k_2} \tan(i_0) \tag{34}$$

When the normal stress σ_n decrease s, the dilatancy rate d increases rather than decreases.

3.2. Plesha's Theoretical Approach For Model Formulation

Plesha [6] developed a theoretical 2-D constitutive model for rock joints with consideration of surface damage and capacity of cyclic shear paths. The joint surface is assumed to be macroscopically planar. The model was based on the theory of plasticity and an assumption of uniform tooth-shaped asperities with the stress transformation given by

$$\begin{cases} \sigma_t^c = \eta(\sigma_t \cos\alpha_k + \sigma_n \sin\alpha_k) \\ \sigma_n^c = \eta(-\sigma_t \sin\alpha_k + \sigma_n \cos\alpha_k) \end{cases} \tag{35}$$

where η is a constant representing the aerial ratio active asperity surface over base surface of the representative asperity ($\eta = L_k / L^0$, $k = L, R$) and α_k is the current active asperity angle (Fig. 3). Assuming that for a solid interface composed of two contacting surfaces in a two-dimensional space, the total displacement increment, du_i can be divided into a reversible part, du_i^e and an irreversible part du_i^p, respectively, i.e.

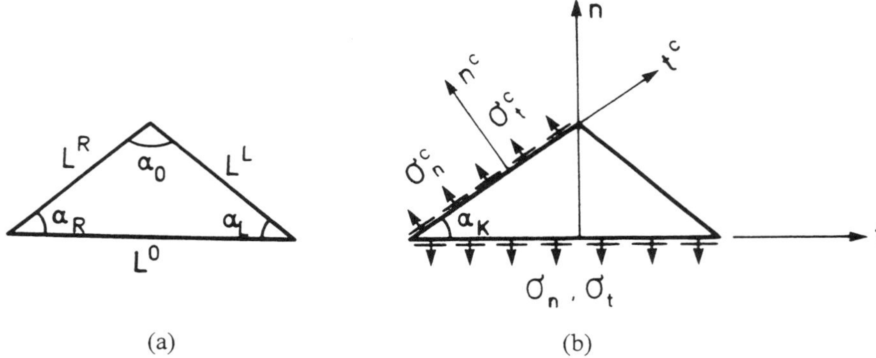

Figure 3. Stress transformation on a representative asperity [6, 7]. a) different active asperity surfaces for shearing forward (α_R) and backward (α_L); b) transformation of macroscopic (σ_t , σ_n) and microscopic (σ^c_t , σ^c_n) stress components on the active asperity surface.

$$du_i = du_i^e + du_i^p \tag{36}$$

where $i = n,\ t$ are the normal and tangential directions defined on the joint plane. The reversible part of the displacement can be represented by an elastic response through a stiffness tensor, k_{ij}, given by

$$d\sigma_i = k_{ij} du_j^e \tag{37}$$

where $j = n,\ t$. The irreversible part is described by a sliding rule given by

$$du_i^p = \begin{cases} 0, & F < 0 \\ \lambda\, \partial Q / \partial \sigma_i , & F \geq 0 \end{cases} \tag{38}$$

where F and Q are the slip function and sliding potential, respectively, defined for the joint, similar to the yield function and flow potential in the theory of plasticity and λ is a positive scalar quantity. The dissipated energy during the sliding, dW^p, is therefore, given by

$$dW^P = \sigma_i du_i^p = \lambda \sigma_i (\partial Q / \partial \sigma_i) \tag{39}$$

Further assuming that the loading function $f(\sigma_i , W^P)$ has a linear form, given by

$$f(\sigma_i, W^P) = F(\sigma_i) - mW^P \tag{40}$$

where $m = h > 0$ for shear-strengthening before peak and $m = s < 0$ for shear -weakening after peak, and using the consistency condition $df = 0$, following expression can be derived

$$\partial F/\partial \sigma_i = k_{ij}\left(du_j - \lambda \,\partial Q/\partial \sigma_i\right) - m\sigma_i \lambda(\partial Q/\partial \sigma_i) = 0 \tag{41}$$

where the non-negative parameter λ is given by

$$\lambda = \frac{(\partial F/\partial \sigma_i)k_{ij}du_j}{(\partial F/\partial \sigma_r)k_{rp}(\partial Q/\partial \sigma_p) + m\sigma_r(\partial Q/\partial \sigma_r)} \tag{42}$$

where the range $r, p = n, t$ represents the normal and tangential directions. Substitution of equations (36), (38) and (42) into (37) leads to an constitutive model for solid interfaces

$$d\sigma_i = \left(k_{ij} - \frac{k_{ir}(\partial Q/\partial \sigma_r)(\partial F/\partial \sigma_p)k_{pj}}{(\partial F/\partial \sigma_r)k_{rp}(\partial Q/\partial \sigma_p) + m\sigma_r(\partial Q/\partial \sigma_r)}\right)du_j \tag{43}$$

Equation (43) represents the general constitutive model of a solid interface, using Plesha's theoretical approach. The constitutive model developed by Plesha in [6], however, did not consider the post-peak shear stress behaviour, i.e. the shear -weakening part of the shear stress - shear displacement curve, and the shear stiffness of the interface was taken as constant. To take into account of the shear - weakening effects, it is necessary to introduce a restriction to the values of parameters so that the second law of thermodynamics of the solids are not violated when the shear- weakening part of the joint behaviour is simulated.

3.3. Jing's Constitutive Models For Rough Rock Joints

Considering the limitations in the Plesha's original model, a new 2-D theoretical constitutive model for rock joints with considerations for post-peak shear-weakening, changing normal stiffness with normal displacement, changing shear stiffness with normal stresses and surface damage accumulated over shear processes was developed by Jing [7, 8]. The basic approach is that a general theoretical basic model is developed first to form a mathematical frame, then specially developed functions representing special stress-deformation-damage mechanisms are developed separately according to experimental results, which are introduced into the general basic model to produce a final constitutive model for rock joints. The second law of thermodynamics is then used to restrict the limit of parameter values.

The second law of thermodynamics requires that the entropy production of the sliding system, $d\mathrm{E}$, be positive, i.e.

$$d\mathrm{E} = (d\mathrm{H} + dW^P) / \mathrm{T} \geq 0 \tag{44}$$

where H is the heat source, T is the temperature, respectively. Assuming an isothermal condition (i.e. without considering the heat source and convection between the mechanical energy and thermal energy), equation (44) is simplified into

$$dW^P = \lambda \sigma_i \frac{\partial Q}{\partial \sigma_i} \geq 0 \tag{45}$$

Equation (43) plus inequality (45) are the general constitutive model for a contact-friction problem of solid interfaces. Different functional forms of F and Q will determine different interface behaviour.

For formulation of constitutive models for rock joints, attention is paid to following characteristics special to rough surfaces:

i) representation of slip function F and sliding potential Q, given by [6]

$$F = \left| \sigma_t \cos \alpha_k + \sigma_n \sin \alpha_k \right| + \tan \phi_r (-\sigma_t \sin \alpha_k + \sigma_n \cos \alpha_k) - C \tag{46}$$

$$Q = \left| \sigma_t \cos \alpha_k + \sigma_n \sin \alpha_k \right| \tag{47}$$

where ϕ_r and C represent the residual friction angle and cohesion of joints, respectively;

ii) representation of surface damage, given by [17]

$$\alpha_k = \alpha_k^0 \exp(-D_m W^P) \tag{48}$$

where term D_m is the damage coefficient of asperities, a material constant and α_k^0 is the initial asperity angle.

iii) representation of shear-strengthening and shear-weakening functions h and s, in the place of modulus m in equation (43), given by [7, 8]

$$h = k_t (u_t^P - u_t) / (u_t^P - u_t^0) \qquad\qquad (u_t^0 \leq u_t \leq u_t^P) \tag{49}$$

$$s = -s_c \frac{(u_t^r - u_t)(u_t - u_t^P) \, k_t}{(u_t^r - u_t^P)^2 \, Q} \sin^2 \alpha_k \tag{50}$$

where s_c is a material constant.

iv) representation of hyperbolic variation of normal stress-normal displacement behaviour, given by [7, 8]

$$k_n = \frac{k_n^0}{(1 - u_n / u_n^m)^2} \qquad \text{(for closure)} \qquad (51)$$

$$k_n = 0 \qquad \text{(for opening)} \qquad (52)$$

where k_n^0 is the initial normal stiffness.

v) representation of stress dependency of shear stiffness, written [7, 8]

$$k_t = \frac{\sigma_n}{\sigma_c}\left(A - \frac{\sigma_n}{\sigma_c}\right)k_t^m \qquad \text{for } (0 \le |\sigma_n| \le |\sigma_c|) \qquad (53)$$

$$k_t = 0 \qquad \text{for } (|\sigma_n| > |\sigma_c|) \qquad (54)$$

where k_t^m is the maximum shear stiffness when $\sigma_n = \sigma_c$, and A and C are the material constants.

These representations are determined according to either experimental results and/or conceptual behaviours from literature.

Substitution of the above representations into equation (43) leads to an explicit, incremental constitutive model for rough rock joints, given by following equation,

$$\begin{cases} d\sigma_t = \left(k_t - \dfrac{b_1 k_t^{\,2}}{b_1 k_t + b_2 k_n + mQ}\right)du_t - \dfrac{b_3 k_t k_n}{b_1 k_t + b_2 k_n + mQ}du_n \\[2mm] d\sigma_n = -\dfrac{b_4 k_t k_n}{b_1 k_t + b_2 k_n + mQ}du_t + \left(k_n - \dfrac{b_2 k_n^{\,2}}{b_1 k_t + b_2 k_n + mQ}\right)du_n \end{cases} \qquad (55)$$

where $b_1 = \cos^2 \alpha_k - \tan \phi_r \, \mathrm{sgn}(\sigma_t^c)\sin \alpha_k \cos \alpha_k$,

$\qquad b_2 = \sin^2 \alpha_k + \tan \phi_r \, \mathrm{sgn}(\sigma_t^c)\sin \alpha_k \cos \alpha_k$

$\qquad b_3 = \sin \alpha_k \cos \alpha_k + \tan \phi_r \, \mathrm{sgn}(\sigma_t^c)\cos^2 \alpha_k$

$\qquad b_4 = \sin \alpha_k \cos \alpha_k - \tan \phi_r \, \mathrm{sgn}(\sigma_t^c)\sin^2 \alpha_k$

and $\mathrm{sgn}(\sigma_t^c)$ is the plus or minus sign of the microscopic shear stress component σ_t^c on the active asperity surface.

Applying the thermodynamic restriction represented by the inequality (45), the ranges of validity of parameters are: $\alpha_k \le \pi / 4$, $\phi_r \le \pi / 4$, $k_n \ge k_t$ and $0 \le s_c \le 4.0$. These restrictions imply that the model should not be applied for very rough joints with asperity

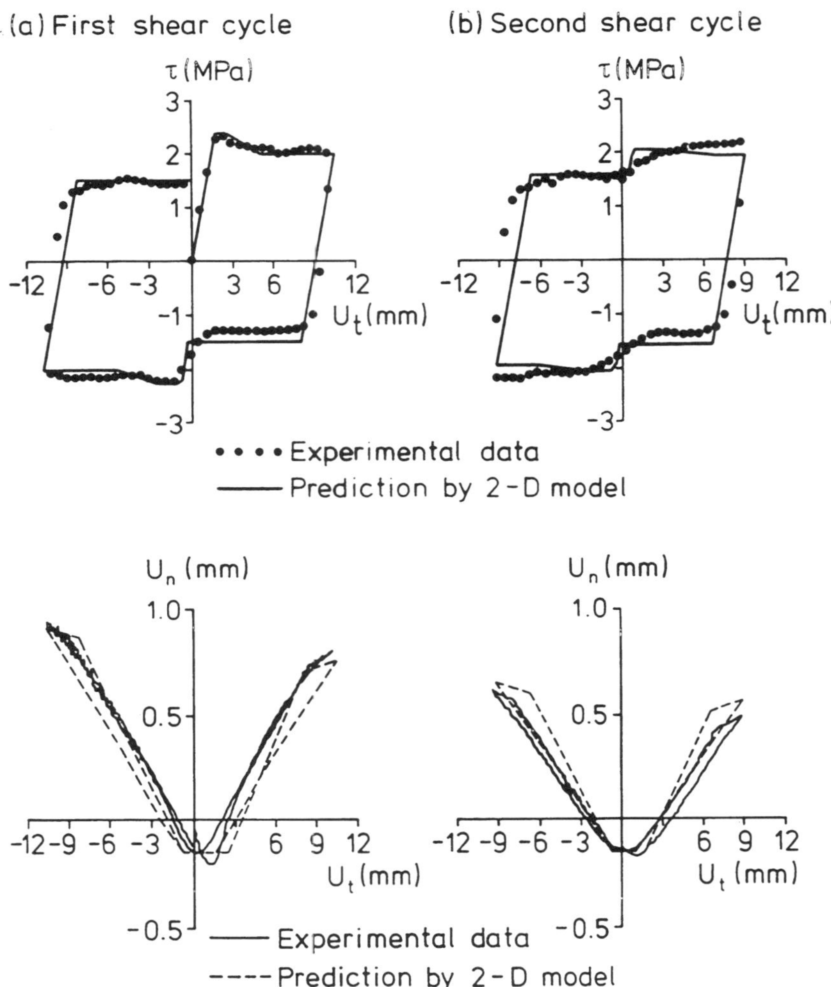

Figure 4. Comparison of experimental and numerical results for two cycles of shear tests under constant normal stress, showing shear stress and normal displacement versus shear displacement for a) the first cycle and b) the second cycle [7, 8].

angles larger than 45 degrees and the magnitude of the normal stiffness should not be less that of the shear stiffness, which is the general case.

The model was implemented into an early version of a distinct element code UDEC. Figure 4 illustrates the verification of the model against experimental results. A 3-D constitutive model for rock joints using the same approach was also developed [7, 18].

4. COUPLED HYDRO-MECHANICAL MODELS

A joint in rock is bounded by two rough surfaces which are subject to some normal and shear stresses. When the joint deforms under stress changes, the asperities on the joint surfaces deform and the contact area (or, equivalently, the void space) between the two opposing surfaces of the joint changes. These changes will, in turn, affect the hydraulic behaviour of the joint when water flow is present in the joint.

Considerable effort [19 - 28] has been devoted to investigating the relationship between stress, deformation and fluid flow in rock joints. Based on laboratory, field and theoretical studies, several investigators have developed constitutive models to predict the hydraulic properties of a joint as a function of the surface roughness characteristics and the applied stress. This section surveys the various coupled hydro-mechancal constitutive models for rock joints and discusses how they compared with experimental results. The present survey draws heavily on previous reviews and original works by Barton and Bakhtar [29] and Gale et al. [30]. The constitutive models reviewed include those proposed by Gangi [31], Walsh [32], Swan [33, 34], Tsang and Witherspoon [35], Barton et al. [3] and Gale et al. [30].

Among these models the first four can be classified as phenomenological models based on the theories of elastic deformation and fluid flow along with a mathematical description of joint roughness characteristics through a statistical distribution function of the joint asperities or local apertures. In these models it has been assumed that the roughness of the joint surface is uncorrelated and the two surfaces of the joint are mismatched. They all contain adjustable parameters which must be determined by calibration using the elastic modulus of the rock along with either characterization of the morphology of the joint surfaces or the stress-deformation curve of the joint.

The models of Barton et al. [3] and Gale et al. [30] are empirical models. These two empirical models consider the variation of joint hydraulic properties with both normal and shear stresses, as well as loading-unloading cycles, hysteresis and permanent deformations. Both matched and mismatched joints can be simulated. By contrast, the four phenomenological models alluded to above deal with normal stress only.

4.1. The Gangi Model

Gangi [31] developed a model for the variation of the permeability of a single rock joint using the Hertzian approach. In his model the rough surface of a joint was visualized as a "bed of nails" lying on a planar surface. He described this "bed of nails" by means of a cumulative distribution function $N(x)$ of the (rod) asperity shortness, which he defined as the height, w, of the highest asperity minus the height of the asperity in question, given by a power law,

$$N(x) = I\left(\frac{x}{w}\right)^{n-1} \qquad (1 \le n < \infty) \qquad (56)$$

where $N(x)$ is the number of rods with shortness less than or equal to x and I is the total number of rod asperities. The variation of joint permeability, k, was then derived as a function of effective normal stress,

$$k(\sigma'_n) = k_0 \left[1 - (\sigma'_n / Y_n)^m \right]^3 \tag{57}$$

where σ'_n is the effective stress, k_0 is the permeability under zero applied stress or self-weight of the test specimen, $m = 1/n$ and $Y_n = YA_n / A$, Y is the Young's modulus of the intact rock, A_n is the contact area, i.e., the area covered by the rods and A is the total nominal planar area of the joint.

Gangi found his models to fit the laboratory test data for artificial joints in sandstone by Nelson and Handin [36] and in carbonate rocks by Jones [37]. Other workers such as Gale et al. [30], Brar and Stesky [38], Tsang and Witherspoon [35] and Elliott et al. [24] were unable to achieve a reasonable fit between experimental data and Gangi's model.

4.2. The Walsh Model

Walsh and Grosenbaugh [39] combined the theory of compressibility for jointed rock with the tribological model of Greenwood and Williamson [40] for the elastic deformation of rough contacting surfaces to derive a normal stress - closure relationship for a rock joint. Their model related changes in both joint aperture and contact area to changes in normal stress. Assuming the tips of the aspirates to be spheres with the same radius of curvature and assuming an exponential distribution function for asperity height, they showed that the normal stiffness of a joint should vary linearly with the effective normal stress with the standard deviation of asperity height as the constant of proportionality. They found that *in-situ* joint stiffness measured by Pratt et al. [25] in a large-block experiment did vary linearly with applied normal stress as predicted by their model.

Using the above closure model and taking the advantage of the analogy between heat flow in a sheet containing non-conducting cylindrical inclusions and fluid flow in a planar joint with contacting asperities to show how contact area affects the hydraulic conductivity of a joint, Walsh [32] deduced the following relationship between the joint hydraulic conductivity, K, and the effective normal stress,

$$K/K_0 = \left(1 - 2\sqrt{2}(b/E_0)\ln(\sigma'_n / \sigma'_0) \right)^3 \tag{58}$$

where K_0 and E_0 are the joint conductivity and aperture at some reference effective stress, σ'_0, and b is the standard deviation of the asperity height distribution. Walsh found good agreement between equation (58) and experimental results for artificial joints by Jones [37], Kranz et al. [41] and Brar and Stesky [38].

Gale et al. [30] fitted their experimental normal stress versus normalized flow rate data to Equation (58) and found moderately strong to weak correlation between pairs of best-fit parameters and lack of correlation between any of the model parameters and the flow rate. Furthermore, the asperity height distribution measured by profilometer did not agree with the best-fit values of b and E_0. They concluded, however, that since the Walsh model was formulated from basic mechanics, it had considerable merit and perhaps required only some redefinition of the roughness parameters for it to be viable.

4.3. The Swan Model

Swan [33, 34] presented a constitutive model for the relationship between normal stress and closure of a rough joint. His model is basically the same as that of Walsh and Grosenbaugh [39] with a few more details. Assuming an exponential distribution of asperity heights and peak to peak contacts for the asperities, Swan showed that the contact area was a function of b / E_0, where the symbols has the same meanings as given in the previous subsection. The normal stiffness of the joint turned out to be a function of the same ratio and was directly proportional to the normal stress.

Swan [33, 34] made topographic profiling measurements on several slate cleavage plane surfaces and performed normal loading tests to confirm the validity of his and Walsh and Gresenbaugh's normal stress-closure relationship for normal stress less than 30 MPa. He then combined his joint closure model with the assumed validity of the parallel-plate joint flow model to express joint hydraulic conductivity as a function of normal stress and the roughness properties as follows

$$\sqrt{K / K_0} = (1 - a_b / E_0) - (b / E_0) \ln(\sigma_n') \tag{59}$$

where K is the joint hydraulic conductivity, K_0 is the hydraulic conductivity at zero stress or self-weight, a_b is a constant, and E_0, b, σ_n' have been defined for equation (58).

Swan further suggested that Equation (59) can be simplified as

$$\sqrt{K / K_0} = C_1 - C_2 \ln(\sigma_n') \tag{60}$$

where C_1 and C_2 are constants. This would rely on the existence of an empirical relationship between b and E_0.

Swan did not measure the contact area or flow rate in his experiments but used a numerical model to simulate the closure of rough surfaces in contact and found the model results to agree qualitatively with Iwai's [19] measurements of contact area as a function of normal loading. His model indicated that contact area increased linearly with stress, decreased with increasing initial aperture and was less than 5% at 20 MPa normal stress. The small contact area implies peak to peak contacts at a few taller asperities only, which is probably valid at low stress only. Some of the profiling data by Gale et al. [30] generally confirmed this while a small percentage of their data indicated mated surfaces. On the other hand, experimental results obtained by Pyrak-Nolte et al. [28] and Gentier [42] suggested that contact area varies nonlinearly with increasing stress.

Gale et al. [30] compared their normal stress-normalized flow rate experimental results with the Swan model for different loading cycles and found that, as with the Gangi model, the best-fit model parameters varied from cycle to cycle and furthermore, there was hardly any correlation between any of the best-fit parameters and the normalized flow rate.

4.4. The Tsang and Witherspoon Model

Tsang and Witherspoon [35] developed a dual-concept model which encompassed a void deformation model for stress-joint closure behavior and a distributed aperture (distributed asperity height) model for the stress-flow relationship.

In the Tsang and Witherspoon model a single joint is represented by a collection of flat voids between contacting asperities and the closure of the joint under stresses due to deformation of the voids. Following an earlier formulation of Walsh [43], they obtained an expression for the ratio between the effective modulus of the rock with voids and the intrinsic modulus of the intact rock in terms of the average length of the void, d. The effective/intrinsic moduli ratio also varies as $1/d$. Thus the experimental stress-displacement curve for the rock sample containing a joint can be used to calculate d and the number of contacts, N_c, and hence also the asperity height distribution $n(h)$ by numerical differentiation. Once these roughness characteristics of the joint have been determined, the statistical average joint aperture $\langle e \rangle$ can be calculated, as a function of joint deformation, Δu_n, and effective normal stress, σ'_n, without profiling the surface and without fitting parameters to stress-flow data, given as follows

$$\langle e^3(\Delta u_n, \sigma'_n) \rangle = \int_0^{e_0 - \Delta u_n} (e_0 - \Delta u_n - h)^3 n(h)dh \bigg/ \int_0^{e_0} n(h)dh \tag{61}$$

where e_0 is the maximum aperture of the joint at zero stress and can be obtained from an estimate of the joint contact area for the joint walls at a specified stress. The joint transmissivity is given by the cubic law where the statistical average $\langle e^3 \rangle$ is used in place of the smoothed-wall aperture for the usual parallel plate model.

The theory showed good agreement with the experimental data of Iwai [19] for tensile joints in granite and basalt with contact area ratios between 10% and 20%. Gale et al. [30] found poor fit of the Tsang and Witherspoon model to their laboratory data for the URL samples. They concluded that since this model treats both the loading and unloading parts of the stress-joint flow curve, further efforts may be warranted to investigate the proper approach to extracting the key model parameters from the experimental data.

4.5. The Barton-Bandis Model

Barton et al. [3] compared published experimental data for mechanical aperture, E, with the hydraulic aperture, e, that was derived from fluid flow measurements by means of the cubic law and ascribed the difference between E and e to the head losses due to tortuosity and surface roughness. They proposed an empirical equation between E, e (both in μm) and the Joint Roughness Coefficient, JRC, of the joint, given by

$$e = \min\left\{E; E^2 / (JRC)^{2.5}\right\} \tag{62}$$

The intrinsic permeability, k, of the joint is given by the modified cubic law

$$k = e^2 / 12 \tag{63}$$

To determine the complete normal stress-joint permeability curve, it is necessary to know the initial mechanical aperture, E_0, or the initial hydraulic aperture, e_0. Based on the values of *JRC* and *JCS* recorded by Bandis [44], Barton and Bakhtar [29] suggested the following empirical equation for estimating E_0

$$E_0 \approx 0.2(JRC)\left(0.2\sigma_c / JCS\right) - 0.1)\tag{64}$$

where the symbols have been defined above in this and previous sections. Alternatively, the initial in situ hydraulic aperture can be back calculated from borehole hydraulic test data. At any particular normal stress the mechanical aperture is calculated by

$$E = E_0 - \Delta E\tag{65}$$

where ΔE is given by the normal stress-joint closure relationship discussed in Section 2 above. Equations (62 - 64), together with the normal closure relationship, provide the empirical model for coupled hydro-mechanical behaviour of a rough joint subjected to normal stress.

As discussed in Section 2.2 above, changes of mechanical aperture caused by dilation can be calculated from the tangent of the dilation angle and the increment of shear displacement. In conjunction with the empirical equation relating E and e, this provides an empirical model for the shear dilatancy - joint permeability relationship.

Very limited experimental data existed where all the required parameters had been measured or can be estimated. Barton et al. [3 , 29] showed that the normal stress - flow and shear dilation - flow responses calculated by their empirical models were consistent with the laboratory experiments of Maini and Hocking [26] and with in situ flat-jack loaded block tests in welded tuff in the G-tunnel in the Nevada Test Site and the heated block test conducted by Hardin et al. [27].

Gale et al. [30] compared their extensive laboratory testing data on the URL samples with the Barton-Bandis (BB)model. They found some agreement in the slope of the log-log plot of the flow rate versus normal stress curve between the BB model and the measurements. However, since the BB model was based on the parallel plate concept, the hydraulic aperture under normal stress given by the BB model, as well as the derived fluid velocities, did not agree with the flow experiments. The model also produced much higher shear displacements and lower peak shear stresses than the experiments. They suggested that either the friction factor and the roughness were not properly characterized by the measurement of the index parameters or they were not properly simulated by the BB model.

4.6 Gale's Modified Barton-Bandis Model

The original BB model [3] was formulated primarily for large shear displacements (larger than the shear displacement at peak shear stress). It does not allow for the initial contraction (negative dilation) during shearing observed in some tests by Gale et al. [30]. Joint surface damage and the effects of gouge production are not included. Only mated joints are simulated. Normal loading is assumed to be at constant shear stress and shear loading is assumed to be at constant normal stress. Cyclic normal loading is permitted while only one shear loading is permitted. Complete unloading is allowed but assumed to simply reverse the same shear stress - shear displacement curve as loading.

Gale et al. [30] proposed a modified BB model (MBB) based on the paper by Bandis et al. [2]. The detailed equations are too lengthy for presentation in this paper. For these equations the reader is referred to Gale et al. [30]. Finite element implementation of these equations is discussed in Guvanasen and Chan [45]. Briefly the changes include (1) shear displacement behaviour modeled in terms of mobilized joint roughness with the pre-peak shear displacement modeled by a hyperbolic shear stress - shear displacement curve; (2) mobilized dilatancy angle as a function of the ratio of shear stress/normal stress so that upon shearing the joint will first contract and then dilate; (3) normal stiffness variation due to shear displacement; (4) shear displacement during normal loading of unmated joints and (5) effects of loading history on shear displacement. The coupled hydro-mechanical responses predicted by the MBB model agreed much better with the laboratory experiments conducted by Gale et al. [30] for some of their samples but did not perform quite well for the other samples. Apparently, the stress - permeability relationship has to be refined to reflect flow in rough joints in order to predict a compatible flux and velocity relationship as a function of stress.

In AECL's participation in BMT2, TC1, and TC6 of the DECOVALEX project [13, 14], only the early version of MOTIF with the BB model was utilized. For simplicity, shear dilatancy was not activated in the simulations. The latest version of MOTIF code with the MBB model is currently undergoing detailed testing.

4.7 Cook's Empirical Model for Flow through Rough Rock Joints

Cook [46] proposed an empirical model for fluid flow through a single, rough, natural rock joint, based on the observed discrepancy between the results from laboratory experiments and predictions by the cubic law. Let the initial mean aperture of the discontinuity be e_0 and the mean closure be d, under a normal stress, the increment of the discontinuity's mean effective aperture can be written as $\Delta(e_0 - d) = \Delta e_c$ and is distributed over the whole area, A, of the discontinuity. This increment of mean aperture must be accompanied by a decrement in the mean thickness of the void space adjacent to the contact area, Δe_a, between two rough surfaces of the discontinuity. Δe_a is, however, distributed over an area $(A - a)$ where a is the actual contact area. Under an uniform strain in the direction normal to the mean plane of the discontinuity, the following relation must hold

$$\Delta e_a = \Delta e_c / (1 - a/A) \tag{66}$$

Assuming that the fraction of the discontinuity area in contact at any stress can also be approximately given by the ratio of the mechanical closure under that stress over the mean initial aperture under zero stress, i.e., d/e_0, then

$$a / A = d/e_0 = 1 - e_c/e_0 \tag{67}$$

Substituting equation (67) into equation (66) leads to

$$\Delta e_a = e_0(\Delta e_c / e_c) \tag{68}$$

or in differential form

$$de_a = e_0(de_c / e_c) \tag{69}$$

where

$$e_a = e_0\left[1 + \ell n(e_c / e_0)\right] \tag{70}$$

since $e_a = e_c = e_0$ at zero normal stress.

Defining the out-of-plane tortuosity, ς, at any stress, as the ratio of the mean discontinuity aperture at zero normal stress to the mean discontinuity aperture at the current normal stress, i.e. $\varsigma = e_0 / e_c$, the empirical flow law is then given in the form

$$\overline{q}_x = -L_y \frac{g}{12v}\left\{e_0[1 - \ell n(\varsigma)\varsigma]\right\}^3 \left(\frac{\varsigma}{2\varsigma - 1}\right) + q_r \tag{71}$$

or

$$\ell n\left[\frac{-12(\overline{q}_x - q_r)}{L_y(e_0)^3 \, dh/dx}\right] = \ell n\left\{[1 - \ell n(\varsigma)\varsigma]^3\left(\frac{\varsigma}{2\varsigma - 1}\right)\right\} \tag{72}$$

where $\overline{q}_x$ is the mean flow rate in the x-direction, $\overline{q}_r$ is the residual flow rate independent of applied normal stress and variation of aperture, L_y is the dimension of discontinuities perpendicular to the flow direction (x-direction), i.e. the width of the flow, v is the viscosity of fluid and h is the hydraulic head. The slope of the line relating the logarithms of the specific flow $q/(\Delta h) = \overline{q}_x/(dh/dx)$ and the aperture by this relation is lager than 6, not 3 as given by the cubic law. This empirical flow law fits well with many features of the experimental data, including both the high exponents at low stresses and the aperture-independent flow at high stresses, but also requires more material parameters to be determined by laboratory tests or extrapolations, which is by no means an easy or reliable task. The flow property of the rock joints may also be scale-dependent, which is little understood today.

5. NEW DEVELOPMENT OF JOINT MODELS

It was demonstrated in the study of TC1, BMT2, TC5 and TC6 of the DECOVALEX Project [13, 14] that the existing constitutive models, especially the Goodman's and Barton-

Bandis', are not able to provide a reasonable prediction of rock joint behaviour under even very simple laboratory test conditions. It is doubtful that they can be applied with confidence for practical problems under realistic in-situ conditions. Efforts were devoted during the project to develop new constitutive models which may shed more light on understanding and simulating the rough rock joints. The focus of the research was on the characterization of roughness of joint surfaces and coupled hydro-mechanical aspects.

5.1 The Nguyen and Selvadurai Model

Nguyen and Selvadurai [10] extended the constitutive model for rock joints by Plesha [6] to include hydraulic behaviour, with the parameters estimated from the Barton-Bandis empirical coefficients *JRC* and *JCS*.

During shearing, the joint can experience dilation, leading to an initial increase in its permeability. Experiments have shown that the rate of increase of the permeability decreases as shearing proceeds to a very late stage. This behaviour is attributed to gouge production. It is assumed that gouge production is related to the plastic work of the shear stress, which enables the derivation of a relationship between the permeability of the joint and its mechanical aperture.

The theory of plasticity is used to formulate the stress-displacement relationship and the approach is very similar to that of Plesha's and Jing's [6, 7, 8]. The only difference is that the plastic work , W^P , includes only works of shear stress over shear displacements.

Nguyen and Selvadurai [47] proposed that parameters used in the model could be estimated from the Barton's empirical coefficients *JRC* and *JCS*, using the Barton-Bandis empirical expression for the peak shear envelope, similar to that developed by Guvanasen and Chan (cf. equations (22) and (23)).

The parallel plate model, see equation (61), is used to calculate the permeability k of the joint, using the hydraulic aperture of the joint, e. Witherspoon et al. [48] and Elliot et al. [24] proposed a linear relationship between the hydraulic and mechanical apertures

$$e = e_0 + f\Delta E \tag{73}$$

where e_0 is the initial hydraulic aperture, ΔE is the variation in mechanical aperture due to the combined effects of compression and shear, and f is a proportionality factor.

The effect of gouge production on the joint permeability is assumed to be related to the total plastic work due to shear. If the exponential law for surface damage (equation (48) is adopted, the factor f in equation (73) may be related to the plastic work produced by the shear forces according to the following equation:

$$f = f_0 e^{-c_f W^P} \tag{74}$$

where c_f is a gouge production factor. It is very likely that the additional parameters f_0 and c_f could be empirically related to *JRC*, *JCS* and σ_n. A detailed experimental program will be needed in order to arrive at specific correlation.

The above joint model was implemented in the finite element code FRACON [10, 47] and used to simulate typical laboratory experiments on rock joints with fairly good agreements between the model predictions and experimental results for shear stress, dilatancy and hydraulic aperture versus shear displacement of a single joint test. The predictions for joint dilatancy and permeability changes with respect to shear displacement, made by code FRACON against experimental results are showing Figures 5 and 6, respectively.

5.2. Ohnishi Approach For Roughness Characterization and Model Formulation

Ohnishi et al. [50] proposed to use directly digitized joint surface profile data to simulate the dilatancy-shear displacement behaviour of rock joints. The proposed model uses the digitized profile data to produce the load-deformation (stress-displacement) curves with both dilatancy and shear stress relationships, without using *JRC* or fractal dimension for a mathematical characterization of surface roughness. The technique is still in experimental stage and can be extended to wider range of joint conditions.

5.2.1 Roughness Measurement

Many researchers have proposed different ways to characterize the roughness of rock joint surfaces, especially the fractal dimension, but the measures they proposed very often relates to the *JRC* index, which has to be experimentally determined and may not be representative at all. The method proposed by Ohnishi et al. [50] uses the direct data of digitized surface profiles. A joint surface is set on the measuring table which has a laser beam displacement apparatus on the top. The position of the surface is measured incrementally in the x and y-directions on the surface of the joint sample. The amount of increment is minimum 0.5 mm

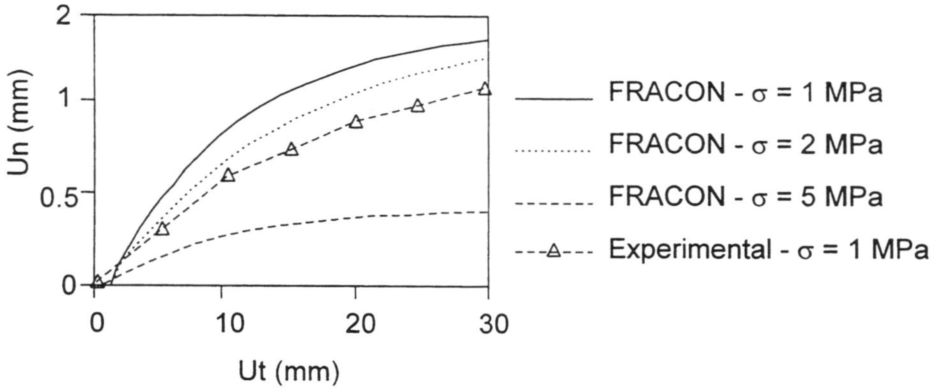

Figure 5. Experimental results and computational prediction by code FRACON for dilatancy versus shear displacement of a rock joint during a direct shear test [47].

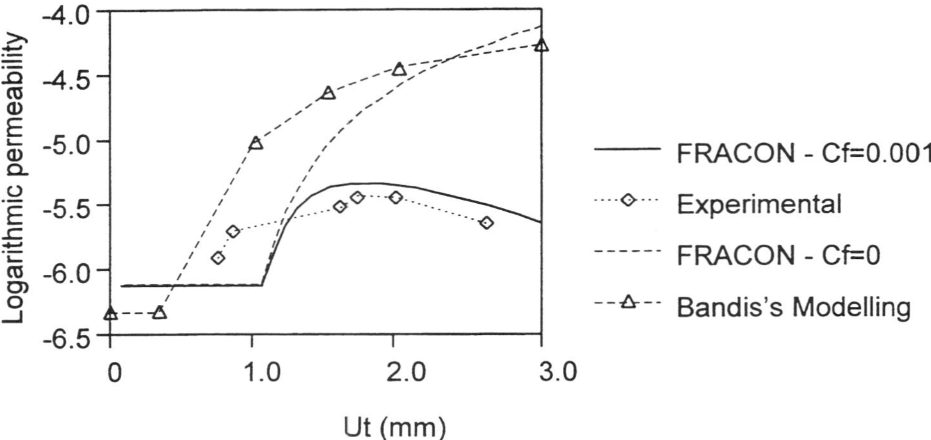

Figure 6. Experimental results and computational prediction by code FRACON for permeability versus shear displacement of a rock joint during a direct shear test [47].

in both directions (practically 1 mm pitch). Asperity height can be measured in the range of accuracy of 1/100 - 1/1000 mm depending on the resolution of the apparatus (Figure 7). The measured data is fed into computer and stored.

5.2.2. Modeling Procedure

For simplicity the explanation is given only to the direct shear test under constant normal stresses. During shear the upper half of the joint overrides on the asperities of the lower half of the sample, causing normal displacement of the upper half (dilatancy). The amount of the normal displacement changes with shear displacement.

It is assumed that the asperities are rigid without deformation. In the calculations the measured joint surface forms an imaginary joint block and its upper half is subjected to a small increment of shear displacement, Δu_t. A new position of the joint profile for the upper half is then calculated and checked whether any overlapping with lower half surface exists. If there is overlapping, the upper block is lifted upward by the amount computed from the maximum slope angle of the asperities which was determined initially from the measured data. If there is no contact between two half blocks, the shear displacement will be increased until the block contact is created. In this way the shear and normal displacement relationship (dilatancy curve) is obtained. The procedure for a 3-D profile rock joint is the same as in the 2-D case. The maximum slope angle is taken only in the shear direction.

In reality the sharp and weak asperities can be destroyed due to crushing, wearing and shearing. It is not easy to simulate all of these phenomena accurately. At present shearing of the asperity tip is taken into account in the proposed model. In the computer simulation, if there are overlapping, the upper half block is lifted in the way explained above and at the same time the tips of the overlapped asperities are cut off in a certain amount (for example

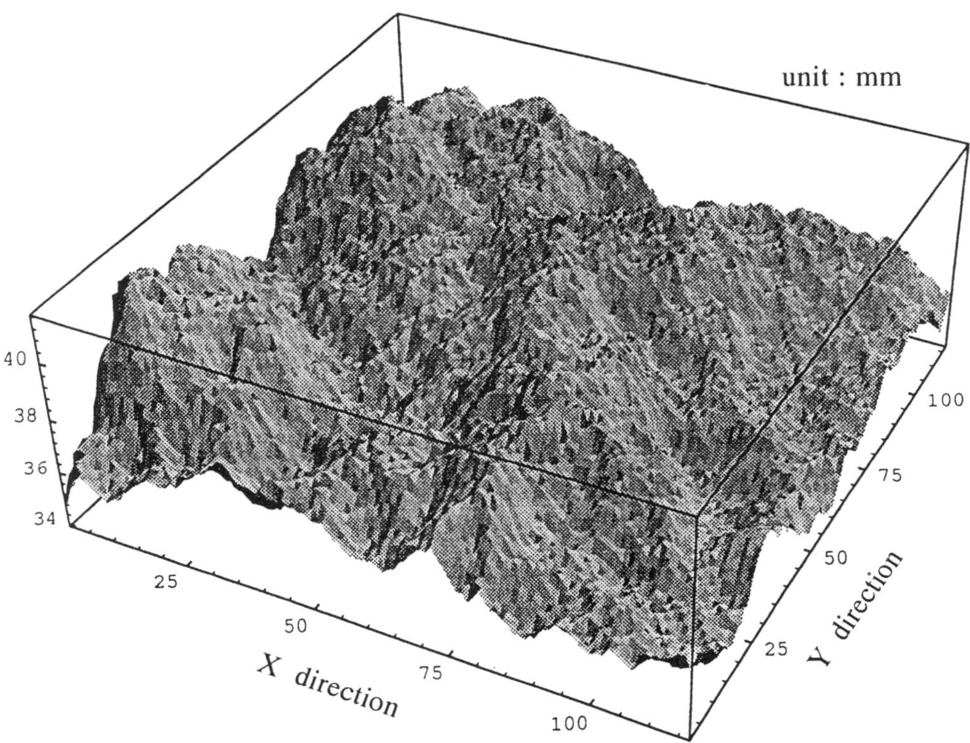

Figure 7. Digitized surface of a rock joint (lower half) [50].

0.5 mm in the normal direction). Once the asperity tips is cut, the overlapping is rechecked and the process is iterated until no contact is detected, then the shear displacement is increased for the next step. The number of iteration n and the cut off length k are arbitrary. In order to determine the realistic values of n and k the calibration test is recommended to match the dilatancy curve.

The actual direct shear test was done using the specimen of Figure 7 to evaluate the feasibility of the proposed method. The specimen was opened at the time of 10.0 mm shear displacement and the crushed area (rock wall color changed) was identified by using a picture analyzer. The dark area in Figure 8 are the area of cut off after 10.0 mm of shearing by simulation with $k= 0.1$ and $n= 400$, using data taken at the specimen shown in Figure 7.

At the initial stage the contact area is very small and increases with development of crushing due to shearing. Most of the crushed area is around the steep slope zone. This shear process and aperture change can be visualized on the computer screen and the change of flow path may be analyzed sequentially.

At this stage the physical background of parameters k and n has not been investigated yet. Some theoretical consideration such as plastic work and degradation must be incorporated.

                                      Y. Ohnishi et al.

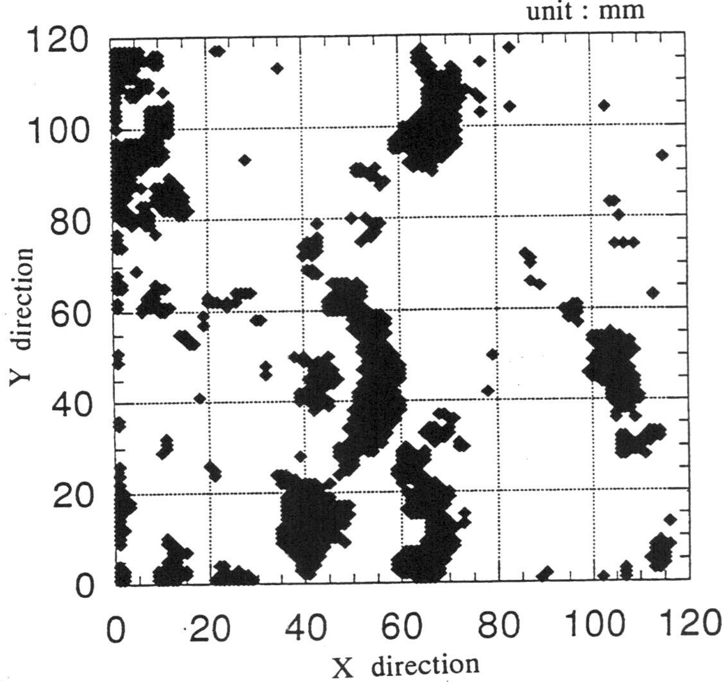

Figure 8. Contact areas (dark area) on the joint surface shown in Figure 7 during shear [50].

The influence of these parameters on the dilatancy and stress-strain behavior of the joint must also be quantitatively determined. Figure 9 shows the dilatancy curve obtained before the shear test for a joint specimen by using surface profile data. It can be seen that the calculated dilatancy curve depends on the amount of iteration n. The best number of iteration could be determined from a reference test.

5.2.3. Shear Load-Deformation Curve

The shear stress σ_t corresponding to the shear displacement u_t^j can be calculated by using either one of the following two equations

$$\sigma_t\big|_{u_t=u_t^j} = \sigma_n \tan\phi_b + \sigma_n (\partial u_n/\partial u_t)_{u_t=u_t^j} \tag{75}$$

$$\sigma_t\big|_{u_t=u_t^j} = \sigma_n (\tan i + \tan\phi_b)\Big/\left(1 - \tan\phi_b (\partial u_n/\partial u_t)_{u_t=u_t^j}\right) \tag{76}$$

where ϕ_b is the basic friction angle of the rock surface and i is the slope angle of the profile.

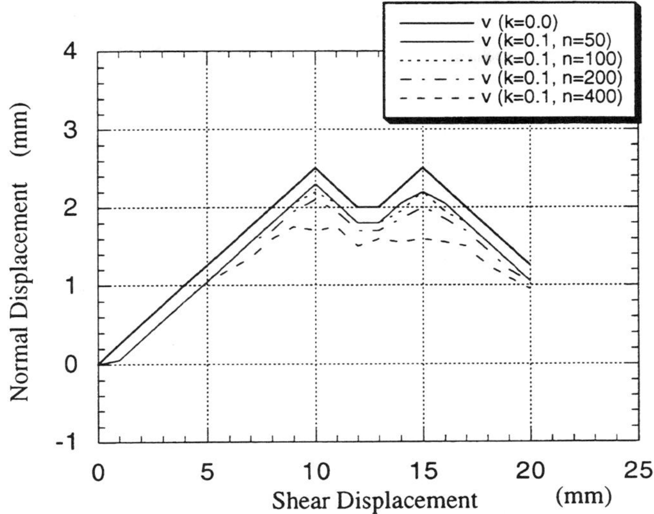

Figure 9 Measured and predicted (before shear test) dilatancy curves [50].

Equation (75) is derived only considering the geometric relation of the joint surface profile in which the degradation of the asperities are not taken into account. Equation (76) is derived from energy balance equations considering wearing and degradation by Seidel and Haberfield [51]. The term $\left(\partial u_n / \partial u_t\right)\big|_{u_t=u_t^j}$ in these equations can be determined in the process of shearing in the digital shear test simulation using the following relation

$$\left(\partial u_n / \partial u_t\right)\big|_{u_t=u_t^j} = (u_n^j - u_n^{j-1})/\Delta u_t \tag{77}$$

where u_n^j is the normal displacement at $u_t = u_t^j$ and Δu_t is the increment of shear displacement which is used in the calculation step. The maximum slope angle of asperities at $u_t = u_t^j$, denoted as $i_{\max,j}$ can be obtained from the joint surface profile at the beginning of each shear displacement increment step.

5.2.4. Model Simulation of Experiments

Two types of test specimens were prepared for comparison with the digital shear test. The direct shear test specimen of Figure 10(a) has a repeated regular saw-tooth and has a same cross section perpendicular to the shear direction. It is an artificial joint of concrete and is named "Type R (Regular)" joint. Another joint specimen was made by casting concrete on a natural rock joint which is shown in Figure 10(b). The profile data was taken in 1 mm pitch in x and y directions. This joint is called "Type IR (Irregular)" joint just for identification. Figure 11(a) shows the dilatancy curve obtained by the "Type R" shear experiment with the

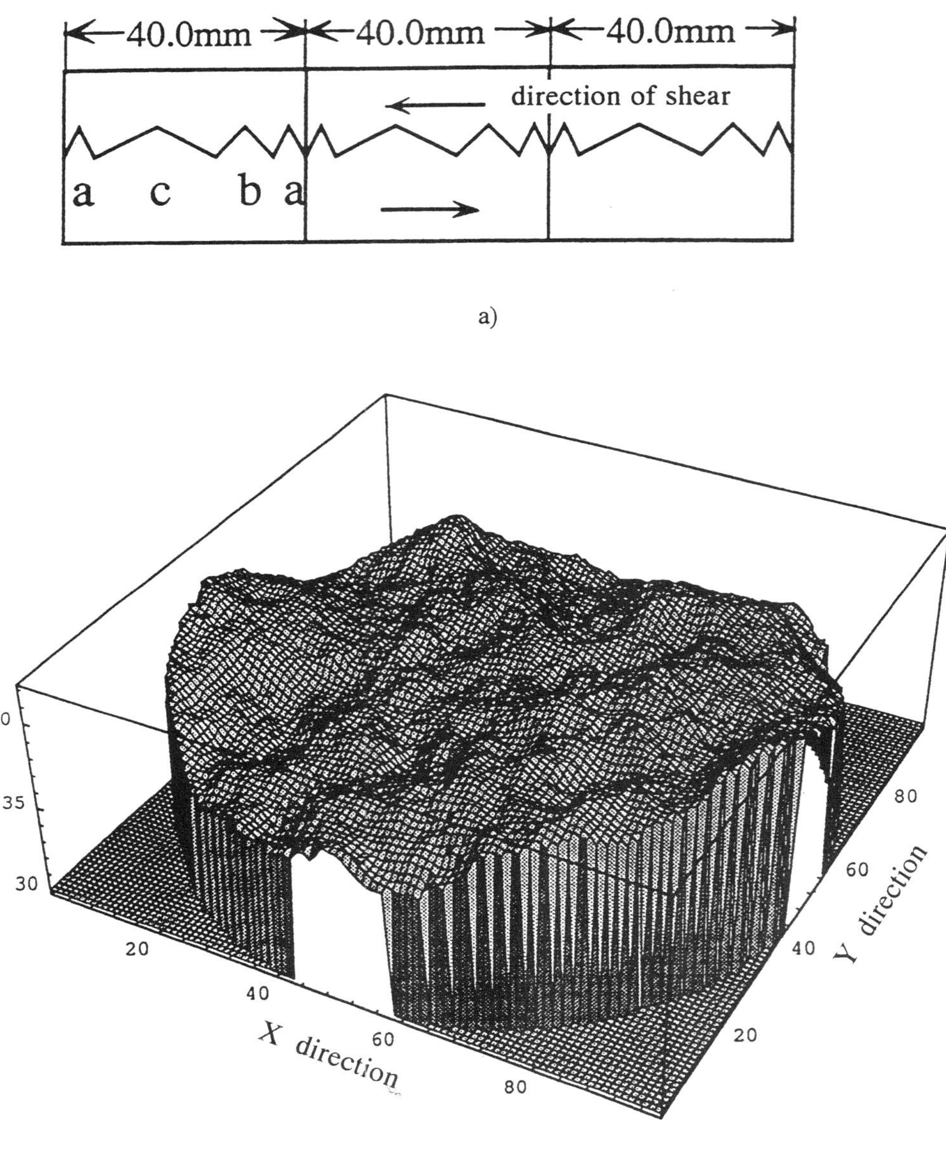

a)

(b)

Figure 10. Regular saw-tooth profile of an artificial joint (a) and an concrete cast of lower half of a rock joint with irregular surface asperities (b) [50].

normal stress of 0.5 MPa and by the computer simulation in which the rock wall is assumed to be rigid (k=0). Although there is a small gap between the experiment and the simulation after the first peak, the initial volumetric increase of the joint is well represented by the simulation. Figure 11(b) shows the measured and the predicted shear stress-shear displacement relationship. Simulation 1 in the figure means that the stress-displacement was calculated by using equation (75) and Simulation 2 by equation (76). The results of Simulation 2 fit better with the test data. The degradation of the asperities were not considered because small value of the normal stress.

In Figure 12(a) the measured "Type IR" dilatancy data was successfully fitted by numerical simulation results using the parameters of $k=0.07$ and $n=80$. The necessity to use the k value greater than 0 indicates effect of asperity degradation. This dilatancy curve was used to predict the shear stress-displacement curve as shown in Figure 12(b). The predictions made in Simulation 1 became better. Simulation 2 shows a sharp peak stress which is not indicated by the experiment. The experimental result fall in between both predictions, which suggest that these equations (75) and (76) should be modified to make closer predictions.

The proposed technique to predict the shear stress-shear displacement of a joint using digitized roughness measurement data may be useful for investigating the microscopic joint behavior. The joint aperture distribution, contact area, and degradation of the asperities can be analyzed. However there are several points that need to be discussed in more detail:

1) determination of k and n: In order to determine the parameters k and n, at least one dilatancy curve of a real direct shear test of a joint is necessary. The most appropriate k and n may be fixed by fitting dilatancy curve of an experiment. The normal stress dependency of k and n must carefully be explored. The increase of normal stress will activate the degradation of joint asperities, then k and n will increase. Here we used a simple relationship of k and n with the normal stress and the dilatancy curve at the different normal stress was estimated. Then the peak stresses were calculated using equation (75) or (76). The Mohr-Coulomb failure curves by the experiment and the prediction by the Simulation are shown in Figure 13. The strength calculated by the computer simulation (Simulation 1) is slightly lower than the actual one. However, the rough estimate of the strength parameters, such as cohesion and friction angle, can be obtained by using the roughness data only.

2) Shear stress-deformation relationship: This relationship is fundamental to the proposed method. In Figure 13 the results of Simulation 1 are only depicted. In order to consider the wearing, crushing, production of gouge and its effect in the stress-strain relation, more detailed examination of rock joint is urgently needed.

6. CONCLUSION

The investigation of rock joint behaviour and development of constitutive models for rock joints have been conducted for many years. New theories and models have been proposed, formulated, partially verified and implemented into various computer codes for use in numerical analyses. The validity of these proposed constitutive models and their associated confidence level, however, has not been seriously challenged against carefully designed laboratory experiments under combined hydro-mechanical loading conditions.

The DECOVALEX project tried to evaluate the performance of available constitutive laws

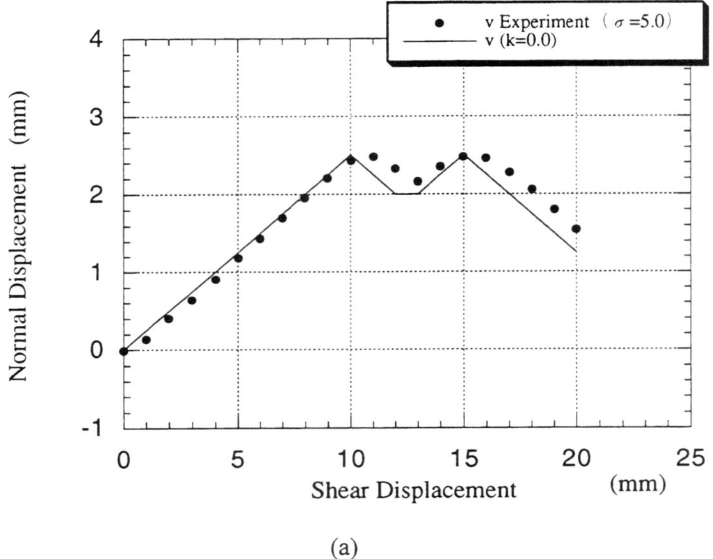

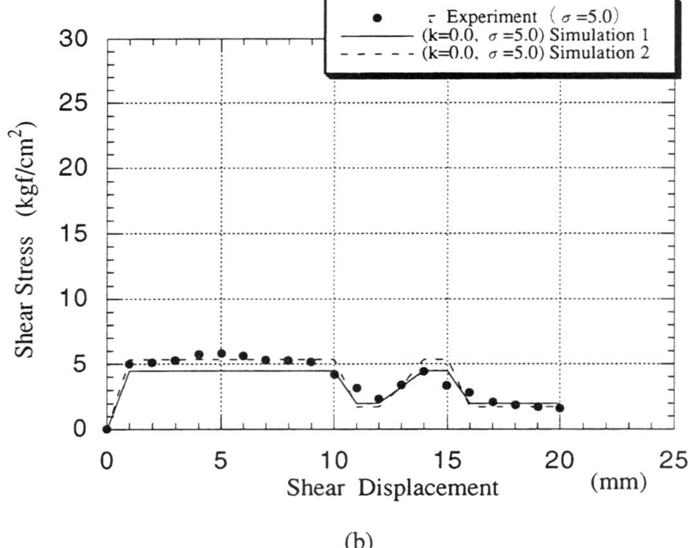

Figure 11. Measured results and model predictions for dilatancy (a) and shear stress (b) versus shear displacement of Type R joint [50].

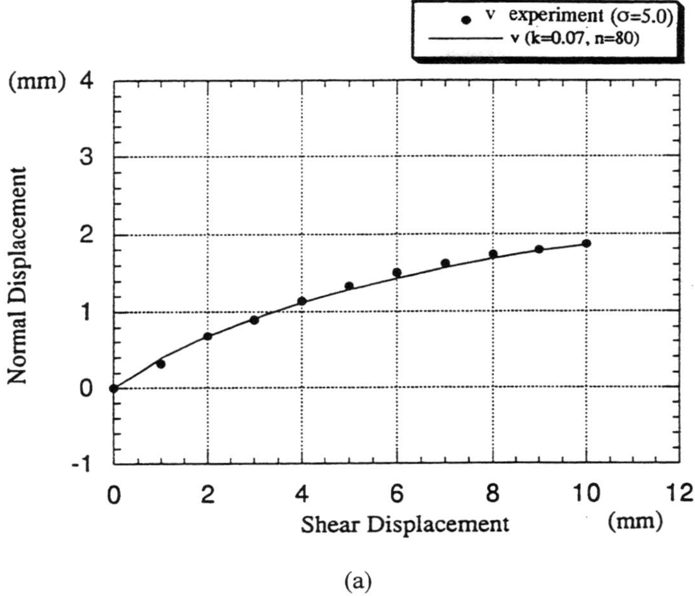

(a)

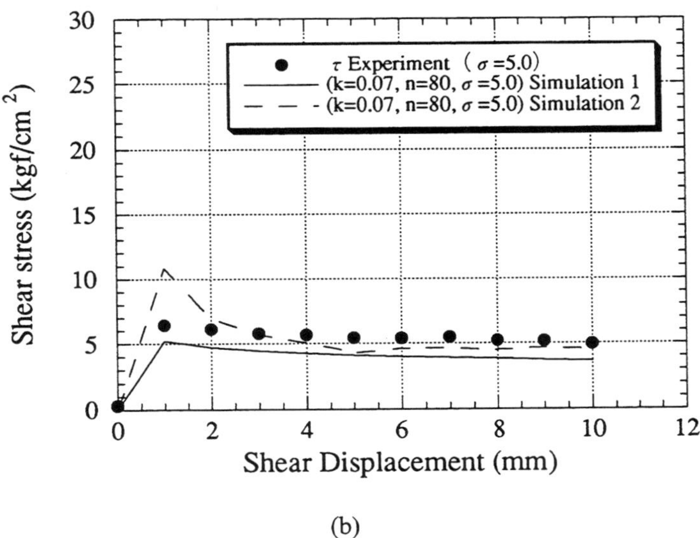

(b)

Figure 12. Measured results and model predictions for dilatancy (a) and shear stress (b) versus shear displacement of Type IR Joint [50].

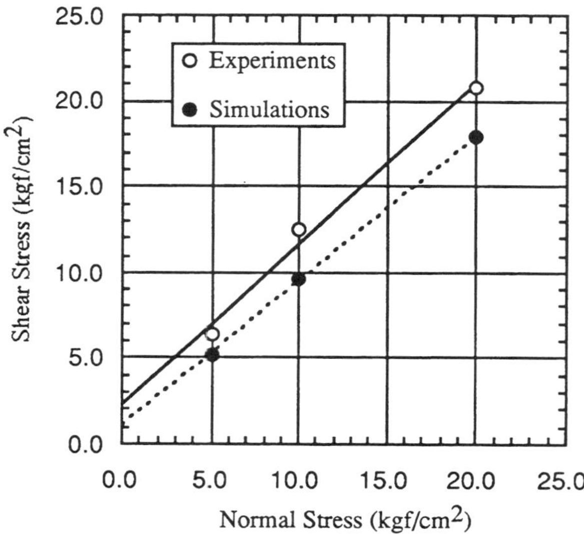

Figure 13. Measured and predicted Mohr-Coulomb Failure curves for the Type IR joint [50].

for single joints and for rock masses with a large number of joints in relation to the complex coupled T-H-M problems. The results from the several TC and BMT studies indicated that the existing constitutive laws are not satisfactory and can not represent the actual behavior of a rock joint which may appear in the nuclear repository site. New efforts must be made in better understanding the physical behaviour and properties of rock joints and development more representative constitutive models.

The points of discussion to be continued in the development of the joint constitutive laws are as follows:

1) Empirical relationships available now show relatively good performance when they are used independently in the normal and shear directions. There is no adequate consistent relationship to correlate them. Empirical dilatancy curves proposed do not fully explain the non-linear behavior of the joint.

2) In the continuum mechanics the elasto-plastic theory can rigorously derive the dilatancy relation based on the flow rule and plastic potentials. However in this category the roughness information is not fully utilized and the complex joint behavior is not yet digested into the plastic theory.

3) The shear-weakening of a joint is the phenomenon which is not fully understood. The numerical difficulty is often encountered to represent the shear-weakening behavior of joints.

4) The influence of joint roughness to the strength of the joint was almost established by the effort of Barton and his coworkers and many others who proposed the empirical shear strength criteria using different roughness measures like JRC. However nonlinear stress-strain relationship and deformability of a joint is still subject to the discussion.

5) Asperities on the joint surfaces are subjected to damage during shear deformation. The degradation of the joint surface and effect of gouge production have found only very limited representations in constitutive laws explicitly, and their validity needs to be established through carefully designed experiments.

6) Flow in a single joint has been thought to be rather simple until many laboratory tests were done. The flow path, tortuosity, initial contact area of a joint makes flow in the joint complex. No qualitatively better model than the parallel plate model has yet been proposed.

7) It is known that mechanical and hydraulic properties interactively influence each other as a stress-flow coupling effect. The joint dilates or contracts during shearing and it changes aperture distribution which strongly affect the state of flow.

8) The degradation of asperities will produce gouge material which may block the flow paths in the joint. The amount of degradation and subsequent gouge production depends on the mechanical and chemical properties of rock joint and the applied load. Microscopic investigation inside the joint during shearing will be necessary.

ACKNOWLEDGMENT

The authors are grateful to the researchers who provided very knowledgeable discussions about constitutive laws of rock joints during a number of workshops of the DECOVALEX project, which motivated much of the new developments in this contribution. The work of the second author (Tin Chan) has been supported by Canadian Nuclear Fuel Waste Management Program, which is jointly funded by Atomic Energy of Canada Ltd. (AECL) and Ontario Hydro under the auspices of the CANDN Owners Group.

REFERENCES

1. R. E. Goodman, Methods of geological engineering in discontinuous rocks. West Publishing Company, San Francisco, 1976.
2. S. Bandis, A. C. Lunsden and N. R. Barton, Fundamentals of rock joint deformation. Int. J. Rock Mech. Min. Sci. & Geomech. Abstr., 20(6), 249-268, 1983.
3. N.R. Barton, S. Bandis and K. Bakhtar, Strength, deformation and conductivity coupling of rock joints. Int. J. Rock Mech. Min. Sci. & Geomech. Abstr., 22(3), 121-140, 1985.
4. A. Makurat, M. Ahola, K. Khair, J. Noorishad, L. Rosengren and J. Rutqvist, The DECOVALEX Test Case 1. Int. J. Rock Mech. Min. Sci. & Geomech. Abstr., 32(5), 399-408, 1995.
5. B. Amadei and S. Saeb, Constitutive models of rock joints. In Rock Joints (Barton & Stephansson, eds.), Balkema, Rotterdam, 1990.
6. M. E. Plesha, Constitutive models for rock discontinuities with dilatancy and surface degradation, Int. J. Num. Analy. Methods in Geomech., 11, 345-362, 1987.
7. L. Jing, Numerical modelling of jointed rock masses by distinct element method for two and three-dimensional problems, Ph.D. Thesis, Luleå University of Technology, 1990.
8. L. Jing, O. Stephansson and E. Nordlund, Study of rock joints under cyclic loading conditions. Rock Mech. Rock Engng., 26(3), 215-232, 1993.

                                                              Y. Ohnishi et al.

9. L. Jing, C-F. Tsang and O. Stephansson, DECOVALEX - an international co-operative research project on mathematical models of coupled THM processes for safety analysis of radioactive waste repositories. J. Rock Mech. Min. Sci. & Geomech. Abstr., 32(5), 389-398, 1995.

10. S. T. Nguyen and A. P. S. Selvadurai, A model for coupled mechanical and hydraulic behavior of a rock joint, Presentations at DECOVALEX workshop in Washington, 1995.

11. Y. Ohnishi, M. Kato and Y. Yano, Method of Quantifying the roughness of a rock joint, 26th Japan Rock Mechanics Symposium, JSCE, 1995 (in Japanese).

12. V. Guvanasen and T. Chan, A three-dimensional finite-element solution for heat and fluid transport in deformable rock masses with discrete fractures. In Proc. Int. Conf. of the Int. Assoc. Computer Methods and Advances on Geomech., May 6 - 10, Cairns, Australia, 1547-1552, Bakema, Rotterdam, 1991.

13. T. Chan and K. Khair, Simulation and results of the multiple fracture model, DECOVALEX Project - Phase I, Bench-Mark-test 2. AECL Research Report AECL-10780, 1993.

14. L. Jing, J. Rutqvist, O. Stephansson, C-F. Tsang and F. Kautsky, DECOVALEX - Mathematical models of coupled T-H-M processes for nuclear waste repositories, Report of Phase I, SKI Technical report 93:21, 1993.

15. T. Chan, K. Khair, L. Jing, M. Ahola, J. Noorishad and E. Vuillod, International comparison of coupled thermo-hydro-mechanical models of a multiple fracture Bench-Mark problem: DECOVALEX Phase I, Bench-Mark Test 2. Int. J. Rock Mech. Min. Sci. & Geomech. Abstr., 32(5), 435-452, 1995.

16. B. Ladanyi and G. Archambault, Simulation of shear behaviour of a jointed rock mass. Proc. 11th US Symp. Rock Mechanics, Berkeley, 1969.

17. R. W. Hutson, Preparation of duplicate rock joints and their changing dilatancy under cyclic shear. Ph.D. Thesis, Northwestern University, Evanston. Il., 1987.

18. L. Jing, E. Nordlund and O. Stephansson, A 3-D constitutive model for rock joints with anisotropic friction and stress dependency in shear stiffness. Int. J. Rock Mech. Min. Sci. & Geomech. Abstr., 31(2), 173-178, 1994.

19. K. Iwai, Fluid flow in simulated fractures. American Institute of Chemical Engineering Journal, Vol. 2, 259-263, 1976.

20. P. A. Witherspoon, J. S. Y. Wang, K. Iwai, and J. E. Gale, Validity of the cubic law for fluid flow in a deformable rock fracture. Water Resour. Res., 16(6), 1012-1024, 1980.

21. T. Engelder and G. H. Sholz, Fluid flow along very smooth joints at effective pressures up to 200 Megapascals. In Mechanical behaviour of crustal rocks, Am. Geophys. Union, Monograph 24, 147-152, 1981.

22. J. E. Gale, The effects of fracture type (induced versus natural) on the stress-fracture closure-fracture permeability relationships, Proc. 23rd US Symp. on Rock Mechanics, University of California, Berkeley, Aug. 25-27, 290-298, 1982.

23. K. Raven and J. E. gale, Water flow in a natural rock fracture as a function of stress and sample size. Int. J. Rock Mech. Min. Sci. & Geomech. Abstr, 22(4), 251-261, 1985.

24. G. M. Elliot, E. T. Brown, P. I. Boodt and J. A. Hudson, Hydromechanical behaviour of joints in the Carnmenellis granite, S. W. England, Proc. Int. Symp. on Fundamentals of Rock Joints, Bjorkliden, Sweden (Stephansson, ed.), CENTEK Publ. 249-258, 1985.

25. H. R. Pratt, H. S. Awolfs, W. F. Brace, A. D. Black and J. W. Hardin, Elastic and transport properties of in situ jointed granite, Int. J. Rock Mech. Min. Sci. & Geomech. Abstr., Vol. 14, 35-45, 1977.
26. T. Maini and G. Hocking, An examination of the feasibility of hydrologic isolation of a high level repository in crystalline rock, Proc. Geologic Disposal of High-Level Radioactive Waste Session, Annual Meeting of the Geological Society of America, Seattle, Washington, 1977.
27. E. L. Hardin, N. Barton, R. Lingle, P. M. Board and M. D. Voegele, A heated flatjack test series to measure the thermo-mechanical and transport properties of in situ rock masses, Office of Nuclear Waste Isolation, Columbus, Ohio, ONWI-260, 193 pp, 1982.
28. L. T. Pyrak-Nolte, L. R. Myer, N. G. W. Cook and P. A. Witherspoon, Hydraulic and mechanical properties of natural fractures in low permeability rock. Proc. 6th ISRM Congress, Montreal, Vol. 1, 225-231, 1987.
29. N. Barton and K. Bakhtar, Description and modelling of rock joints for the hydro-thermo-mechanical design of nuclear waste vaults, AECL TR-418, 1987.
30. J. E. Gale, R. Mcleod, M. Gutierrez, L. Dacker and A. Makurat, Integration and analysis of coupled stress-flow: laboratory tests data on natural fractures . MUN and NGI Tests. Report submitted to Atomic Energy of Canada Limited by Fracflow Consultants, Inc. and Norwegian Geiotechnical Institute, 1993.
31. A. F. Gangi, Variation of whole and fractured porous rock permeability with confining pressure. Int. J. Rock mech. Min. Sci. & Geomech. Abstr., Vol. 15, 249-257, 1978.
32. J. B. Walsh, Effect of pore pressure and confining pressure on fracture permeability. Int. J. Rock Mech. Min. Sci. & Geomech. Abstr., Vol. 18, 429-435, 1981.
33. G. Swan, Stiffness and associated joint properties of rock. Proc. Conf. on Applications of Rock Mechanics to Cut-and-Fill Mining, University of Luleå, Sweden, June 1-3, 1980, published by Institution of Mining and Metallurgy, London, 169-178,1980.
34. G. Swan, Determination of stiffness and other joint properties from roughness measurements. Rock Mech. Rock Engng., Vol. 16, 1983.
35. Y. W. Tsang and P. A. Witherspoon, Hydromechanical behaviour of a deformable rock fracture subject to normal stress. J. Geophys. Res., Vol. 86, 1981.
36. R. A. Nelson and J. W. Handin, Experimental study of fracture permeability in porous rock. American Association of Petroleum Geology Bulletin, 61(2), 227-236, 1977.
37. F. O. Jones, A laboratory study of the effects of confining pressure on fracture flow and storage capacity in carbonate rocks. J. Petroleum Technology, January, 21-27, 1975.
38. N. S. Brar and R. M. Stesky, Permeability of intact and jointed rock, EOS, 61(46), 1980.
39. J. B. Walsh and M. A. Grosenbaugh, A new model for analyzing the effect of fractures on compressibility. J. Geophys. Res., Vol. 86, 1979.
40. J. A. Greenwood and J. Williamson, Contact of nominally flat surfaces. Proc. Royal Society , London, Serial A 295, 1966.
41. R. L. Kranz, A. D. Frankel, T. Engelder and C. H. Scholz, The permeability of whole and jointed Barry Granite. Int. J. Rock Mech. Min. Sci. & Geomech. Abstr., Vol. 16, 225-235, 1979.
42. S. Gentier, Morphological analysis of a natural fracture. In Selected papers on Hydrogeology from the 28th Int. Geological Congress, Washington D. C., USA, July, 1989, Int. Association of Hydrogeologists, (Simpson and Sharp, eds.), Vol. 1, 315-326,

1990.
43. J. B. Walsh, The effect of cracks on the uniaxial elastic compression of rocks. J. Geophys. Res., Vol. 70, 399-411, 1965.
44. S. Bandis, Experimental studies of scale effects on shear strength and deformation of rock joints. Ph.D. Thesis, Univ. of Leeds, Dept. of Earth Sciences, UK, 1980.
45. V. Guvanasen and T. Chan, A new three-dimensional analysis of hysteretic thermohydromechanical deformation of fractured rock mass with dilatancy in fractures. Proc. of 2nd Int. Symp. on Mechanics of Jointed and Faulted Rock (MJFR-2) (Rossmanith ed.), Balkema, Rotterdam, 437-442, 1995.
46. N. G. W. Cook, Natural joints in rock: mechanical, hydraulic and seismic behaviour and properties under normal stress. First Jaeger Memorial Lecture, 29th U.S. Symp. on Rock Mech. Minneapolis, 1988.
47. A. P. S. Selvadurai and T. S. Nguyen, Finite element modeling of consolidation of jointed porous media, Proc. 1993 Canadian Geotechnical Conf., 79-88, 1993
48. P. A. Witherspoon, C. H. Amick, J. E. Gale and K. Iwai, Observations of a potential size effect in experimental determination of the hydraulic properties of joints, Water Resour. Res., 15(5), 5, 1979.
49. S. Bandis, A. Makurat and G. Vik, Predicted and measured hydraulic conductivity of rock joints, Symp. on Fundamentals of Rock Joints, Bjorkliden, Sweden (Stephansson, ed.), CENTEK Publ. 249-258, 1985.
50. Y. Ohnishi, M. Kato and T. Yano, Method of quantifying the roughness of a rock joint. Proc. 26th Japanese Symp. on Rock Mechanics, JSCE, 1995 (in Japanese).
51. J. P. Seidel and C. M. Harberfield, The application of energy principles to the determination of the sliding resistance of rock joints, Rock Mech. and Rock Eng., 28(4), 211-226, 1995.

O. Stephansson, L. Jing and C.-F. Tsang (Editors)
Coupled Thermo-Hydro-Mechanical Processes of Fractured Media
Developments in Geotechnical Engineering, vol. 79
© 1996 Elsevier Science B.V. All rights reserved.

Coupled Thermohydroelasticity Phenomena in Variably Saturated Fractured Porous Rocks -- Formulation and Numerical Solution

Jahan Noorishad and Chin-Fu Tsang

Earth Sciences Division, Ernest Orlando Lawrence Berkeley National Laboratory, One Cyclotron Road, Berkeley, California 94720

Abstract

The physicochemical environment of geologic systems is host to various coupled thermal, hydraulic, mechanical and chemical processes that take place continually at varying rates, dependent on the nature and strength of the sources of these processes in the systems. Scientific interest in these coupled physicochemical processes in the earth's crust, in general, and economic and environmental concerns related to waste geologic disposal, in particular, have resulted in many research efforts aimed at understanding the coupled thermal, hydrologic, chemical, and mechanical (THMC) behavior of geologic systems subject to complex natural or man-made perturbations. In this chapter, the physical aspects of the coupled thermo-hydro-mechanical behavior (THM) are investigated. The work begins with a presentation of background development in the theoretical aspects of the THM phenomena followed by a search in the literature for the THM solution methodologies developed up to the present time. After this review of the state of the art, we first derive the (macroscopic) governing equations for simultaneous occurrences of coupled processes in fully saturated fractured porous media. Next, we propose general field equations for the coupled thermohydroelastic response of variably saturated rocks. This section is complemented by the development of an alternative formulation for the specific cases of weakly non-isothermal conditions. Following these fundamental developments, we offer a finite-element solution methodology and the related algorithms for the solution of the coupled THM problems in variably saturated porous fractured rocks subject to the condition of weak nonisothermal conditions. Finally, solutions of a number of THM sample problems addressing thermoelastic consolidation, flow to a heater test hole, thermohydraulic fracturing, post closure far-field effect in a hypothetical High Level Nuclear Waste Repository (HNLW) and the effects of placement of a HNLW canister and bentonite overpack, are discussed. The solutions to these problems were obtained using the ROCMAS code developed at the Berkeley National Laboratory, which embodies the formulation for low-temperature, coupled thermohydroelasticity phenomena.

1. INTRODUCTION

The physicochemical environment of geologic systems is host to natural chains of coupled thermal, hydraulic, mechanical and chemical processes that take place continually at varying rates dependent on the nature and strength of the sources of driving energies available to the systems. Man-made perturbations of the systems can affect these processes and sometimes induce undesirable events, such as groundwater pollution or induced seismicity. Implementation of a geotechnical project, such as a high-level nuclear waste repository, is considered to be a large-scale perturbation, the effects of which depend on proper understanding and assessment of the coupled phenomena that are being altered or induced. The first step toward such understanding is formulation of each occurring phenomenon with correct consideration of all the major contributing elements among the quadruple processes. We will try to achieve this task in this paper. However, only physical phenomena, namely, thermal (T), hydraulic (H), and mechanical (M), will be addressed.

Theoretical background. The individual processes, in the absence of full consideration of coupling to others, form the basis of very well-known disciplines. Text books in hydrogeology, heat transfer, and elasticity cover these areas thoroughly. Likewise doubly coupled phenomena, formed by pairs of individual processes, are the basis of well-developed theories of (coupled) thermoelasticity, hydrothermal flow, and consolidation (hydroelasticity = poroelasticity). The coupled phenomena of heat and fluid flow in rocks, i.e., hydrothermal flow, have been of great interest in the past few decades due to their influence on geothermal resources recovery. As a result, extensive understanding of the phenomena along with powerful numerical analysis capabilities have been developed. Recent reviews by O'Sullivan (1985) and Bodvarsson (1986) discuss the progress in the context of the development of geothermal reservoir simulators. As in other areas, more realistic understanding of hydrothermal behavior of geologic systems must await further improvement in characterization technology, constitutive relations, and simulation capability.

Coupled thermoelasticity is the phenomenon that involves processes of energy transport and mechanical deformation. Because of the rather insignificant role of mechanical energy coupling in the thermal energy transport process (Carter and Booker, 1988), the phenomenon is generally referred to as thermoelasticity. It is a well-understood and extensively investigated subject in engineering and physics disciplines. From the rock mechanics perspective, more realistic assessment of the thermomechanical behavior of rock masses requires further progress, mainly in characterization and simulation technologies. Such needs have been discussed at length in reviews by Hocking (1979), Baca (1980), and Tsang (1980).

Coupled hydroelasticity (poroelasticity), the phenomenon of coupled fluid flow and deformation of porous material, has its roots in the field of classical hydrogeology and soil mechanics. In the usual formulation and solution of fluid-flow problems in saturated porous media, deformation of the medium is considered by introducing a coefficient of specific storage (Theis, 1938). This approach, although by no means precise, is adequate to represent most fluid-flow problems. However, the well-known theory of consolidation, introduced by Terzaghi (1925), which is mainly used for settlement analysis, was the first rigorous means of analyzing the coupled effects of deformability and flow of fluid in a saturated porous medium. His concept of the effective stress law was the basis of the analysis. Biot (1941) introduced a more general theory of consolidation that makes possible a more realistic treatment of the hydromechanical behavior of saturated porous rocks. The early work of Biot (1941) appeared to be physically motivated but was later supplemented by a thermodynamic base (Biot, 1956).

Other approaches that correctly address the problem are the theories of mixtures (e.g., Green and Naghdi, 1965; Crochet and Naghdi, 1966; Aifantis, 1977), which have a sound thermodynamic base and a general associated constitutive theory. They provide a basis for developing, as a special case, a theory for flow of fluids through porous elastic solids that is equivalent to Biot's result.

The underlying assumption in these developments is the full saturation of the pore spaces in the media. In spite of the significant role of the vadose zone, the zone of partial saturation in both hydrologic and mechanical behavior of porous media, complete extension of the coupled theory has been attempted only in the past few years. Earlier, Bishop (1959) proposed the first extension of Biot's (1941) effective stress law to a partially saturated flow domain. This development made the extensions of Terzaghi's and Biot's theories to variably saturated soils easily possible. Works of Narasimhan and Witherspoon (1977), Safai and Pinder (1980), and Noorishad, *et al.* (1981) are some of the early developments. The recent publication of Schrefler and Simoni (1988) represents a unified approach to the formulation of the problem, with due consideration of gas-phase flows, based on Bishop's (1959) extension of the effective stress law using the concept of average pore space pressure. However, the effective stress law is not applicable generally and most likely is not fundamentally valid (Chang and Duncan, 1983). Findings of Colman (1962), Blight (1965), and Matayas and Radhakrishna (1968) point to the above fact. Using the concept of the "state surface for void ratio," by the latter authors, an alternative expression was introduced by Lloret and Alonso (1980). Later, based on this development, a general formulation for the coupled stress-strain and fluid-flow behavior of partially saturated soils was put forward by Alonso, *et al.* (1988). Included in this formulation is the equation for flow of air, which resulted in the development of a set of triply coupled equations to explain the phenomenon. Chang and Duncan (1983) proposed the concept of a "homogenized pore fluid" that bypasses the complexity of the above formulation for a useful range of conditions above a critical saturation value. This "fluid," a mixture of gas (air) and liquid (water), fully saturates the pore space, thus rendering Biot's effective stress law valid. The compressibility of the mixture of air and water determines the fluid compressibility used in Biot's theory. For a fundamental and thorough discussion of the mechanical behavior of unsaturated soils, the reader is referred to the recent book by Fredlund and Rahardjo (1993).

Solution background. In spite of the existence of Terzaghi's (1926) and Biot's (1941) theories, solution techniques for saturated-hydroelasticity problems were limited to simple conditions for a long period. The reasons for this limitation were the complexity of the solution of Biot's equation and the nonphysical form of the equation parameters. The reformulated form of Biot's equations by Rice and Cleary (1976), using explicit measurable physical parameters, was a fundamental step in making it possible to obtain solutions for limiting (fully drained and undrained conditions) hydroelastic behaviors of porous media. During the past few decades, a number of analytic and semi-analytic solutions of hydroelastic problems have appeared in the literature. Works of McNamee and Gibson (1960), Booker (1974), Cleary (1977), Rudnicki (1981, 1987), Cheng and Ligget (1984), Detournay and Cheng (1987), and Detournay and Cheng (1988) are among the major studies. Numerical solutions of hydroelastic phenomena first appeared around 1970 and have continued with the advent of computers. Some efforts provided a means of analyzing the problem of fluid flow in consolidating soils using Terzaghi's consolidation theory (Helm, 1974; Narasimhan and Witherspoon, 1977). In an attempt to develop a method for the solution of general consolidation problems, Sandhu and Wilson (1969) applied the variational finite-element method to the problem of fluid flow through saturated, porous elastic solids based on Biot's theory. This method was extended by Ghaboussi and Wilson (1973a and 1973b), who considered the effects of fluid compressibility and inertial effects. Later works by a number of investigators, such as Small, *et al.* (1976), Booker and Small (1977), Carter (1977), Runessen and Booker (1982), and Zienkiewicz and Shiomi (1984), developed solutions for various cases with material and geometric nonlinearities and used improved algorithms. Further developments in the solution of various hydroelastic problems, mostly using numerical schemes, are continuing at a rapid pace as new applications of the phenomenon are emerging.

Unlike soils (porous media), which exhibit the dramatic settlement behavior that motivated the development of consolidation theories, fractures and the effects of their deformation on various aspects of rock behavior went unnoticed until recently. Davis and Moore (1965) presented one of the first direct measurements of fracture deformation (of the order of microns) caused by earth tides. Snow (1968) reported strains of 10^{-7} to 10^{-8} at a distance of about 91 m from a water well in metamorphic rocks subjected to 9 m of drawdown. Further evidence of fracture deformation was obtained from (1) the difference between pumping into and pumping out of a well (Evans, 1966); (2) the nonlinear relationship between fluid pressure and flow rate during injection tests (Snow, 1965; Louis and Maini, 1970); and (3) the pressure dependence of transmissivity in the Precambrian reservoir beneath the Denver Basin (Van Poollen, 1969). For some time, hydraulic and hydromechanical analyses of fractures were achieved using an equivalent porous medium approach (Snow, 1969). Noorishad (1971) conducted theoretical and numerical studies of fluid flow in a rock mass, taking into account the deformable nature of fractures in a discrete manner. This work was based on earlier studies of discrete fracture behavior from a load-deformation point of view by Goodman, *et al.* (1968) and a fluid flow point of view by Wilson and Witherspoon (1970). This work obviously was based on what may be called a total stress approach, although a nearly exact Biot coupled behavior was simulated implicitly. This was in the face of the fact that computer solutions of Biot's theory for continuum were just being worked out (e.g., Sandhu and Wilson, 1969). Further laboratory and field tests by Gale (1975) provided strong evidence of nonlinear fracture deformability induced by fluid pressure changes and also verified the capability of earlier numerical solution techniques by Noorishad (1971).

A deterministic solution for transient flow of fluids in deformable fractured porous rocks was not achieved until later. Ayatollahi (1978) used an enumerative approach that is based upon a generalization of Biot's (1941) constitutive stress-strain law and incorporates a Gurtin-type (1964) variational principle. An extension of this work by Noorishad, *et al.* (1982) provided a general two-dimensional finite element solution technique for the study of the deformation, stress distribution, fluid storage, and hydraulic properties of a fractured porous medium under the influence of fluid flow and structural boundary conditions.

Increased interest regarding the emplacement of nuclear waste has led to a number of developments in hydroelasticity as part of the broader field of the coupled thermohydroelasticity (THM). They will be referred to in the introduction of the THM methodologies. Most of the efforts toward development of numerical tools for the study of hydroelastic phenomena in rock masses utilize a discrete representation of the fractures. In spite of the soundness of these approaches, practical considerations limit their applicability. The concept of representation of fractured media by an equivalent continuum offers promising possibilities. The main body of work is in deformation theory (e.g., Gerard, 1982; ODA, 1982; Kyoya, *et al.*, 1985; Kawamoto, *et al.*, 1988; Kaneko and Shiba, 1990; Swoboda, *et al.*, 1990; Pariseu, 1993; Cai and Hori, 1993). Works of Snow (1969), Long (1983), and ODA (1984) mark such efforts in the field of hydrogeology. Developments by ODA (1984) and Pariseu (1993) put forward the possibility of defining equivalent hydroelastic material models. A separate chapter in this book discusses very recent developments in this area.

It is essential to notice that most of the equivalent-continuum theories rely on the concept of representative elementary volume (REV), which may easily break down at a local scale in modeling applications. Such considerations may require a combination of discrete and equivalent modeling as a practical approach.

For unsaturated hydroelasticity, the solution techniques developed at an even slower pace. Perhaps the limited applicability of Bishop's effective stress law may have served as a hindrance. Safai and Pinder (1979) were the first investigators who used the Bishop-type effective stress law to do "Biot-type" coupled hydroelastic analysis in partially saturated media. Lloret and Alonso (1980) presented the first numerical solution of consolidation of unsaturated soils using the concept of the "state surface for volume change," which accommodates soil behavior in all saturation ranges. A very recent work by Alonso, *et al.* (1988) complements the earlier works of the authors and extends the application further. The work of Chang and Duncan (1983), who developed a finite element model for consolidation analysis of partially saturated elasto-plastic soils based on the "homogenized" fluid concept, offered a more practical alternative. With their method, a host of problems could be solved using a Bishop-type effective-stress law. Assumptions of an above-critical value of saturation and occluded air bubbles are central to this approach. Using a Bishop effective-stress law, Schrefler and Simoni (1988) present a classical finite-element approach for analysis of coupled deformation behavior of partially saturated media. Disregarding solid grain compressibility simplifies the latter solution scheme to that of Safai and Pinder (1979). More recently Schrefler and Xiaoyong (1993), removing the pore static air assumption from the earlier work by Schrefler and Simoni (1988), put forward a fully coupled numerical solution for consideration of air and water movement in a deforming porous medium.

The studies cited above have each focused on only two aspects of a much broader question that needs investigation. This is the fully coupled problem of how to analyze the behavior of a fractured rock mass that is subject to the effects of thermal, hydraulic, and mechanical perturbations. We shall refer to the combined effects as THM.

Efforts at formulating the coupled equations of THM seem to have culminated around the late seventies (see Wang, *et al.*, 1980). Work of Derski and Kowalski (1979) presents one of the early attempts in which complete coupled equations for flow of heat and fluid in a saturated deforming porous medium were offered. Seemingly, the first attempt at solving the triply coupled equations of the thermohydroelastic phenomenon was made by Aktan and Faroug Ali (1978). In this work, the "implicitly coupled" numerical algorithms were only amenable to iterative solution. The method was used to investigate the temperature and thermal stresses induced by hot-water injection into an aquifer. The work of Lewis and Kaharonoglo (1981) and Aboustit, *et al.* (1982) are among the first to provide "explicitly coupled" numerical algorithms for solving the coupled thermohydroelastic problems in saturated porous media. A comprehensive development of thermohydroelasticity formulation for saturated fractured rocks, along with a finite element solution scheme, was presented by Noorishad, *et al.* (1984). Continued interest in the phenomenon has resulted in the development of many computer codes capable of thermohydroelastic analysis at varied levels of sophistication. Presently two international forums, namely DECOVALEX and INTERCLAY, are providing opportunities for development, verification, and validation of such capabilities. So far, the main emphasis in the above-mentioned developments has been placed on the coupled thermohydroelastic behavior of a saturated porous medium. However, some problems of interest do not involve full saturation. Examples include the behavior of partially saturated buffer material in a high-level nuclear-waste deposition hole and consolidation of partially saturated soils in the presence of thermal gradients. For this reason, solutions for the partially saturated zone are appropriate. In this chapter, we first develop the formulation of coupled thermohydroelasticity for saturated porous rocks. Then, we treat partial saturation. A general numerical algorithm will be discussed, followed by example applications of a computer code, ROCMAS, which we developed.

2. FIELD EQUATIONS—Fully saturated case

Equations for thermohydroelastic phenomena are derived by writing conservation equations for energy, mass, and momentum. The developments will be at macroscopic levels (all variables are averaged over the REV of the medium). First, an isotropic linear elastic skeleton and full saturation with liquid is assumed. We start with the developments leading to the flow equation.

The displacement field within a saturated, porous elastic medium is defined by two field variable quantities, $\mathbf{U}_l$ and $\mathbf{U}_s$, which represent the displacement vectors of the fluid and the solid, respectively. The velocity field vectors are correspondingly defined as $\mathbf{V}_l$ and $\mathbf{V}_s$, leading to the representation of the fluid relative velocity as

$$\mathbf{V}_r = \mathbf{V}_l - \mathbf{V}_s \ . \tag{1}$$

Using (1) in the Eulerian form of the mass balance equation, given in the absence of a source term as (Mercer, 1973)

$$\frac{\partial \hat{\rho}_l}{\partial t} + \nabla \cdot \hat{\rho}_l \mathbf{V}_l = 0 \ , \tag{2}$$

yields

$$\frac{\partial \hat{\rho}_l}{\partial t} + \hat{\rho}_l \nabla \cdot \mathbf{V}_s + \mathbf{V}_s \cdot \nabla \hat{\rho}_l + \nabla \cdot \hat{\rho}_l \mathbf{V}_r = 0 \ , \tag{3}$$

where $\hat{\rho}_l = \phi \rho_l$ and ϕ and ρ_l are the porosity of the medium and the fluid mass density, respectively. Employing the comoving time derivative for following the motion of a solid particle along its trajectory, expressed as

$$\frac{D}{D^s t} = \frac{\partial}{\partial t} + V_s \cdot \nabla \qquad (4)$$

in (3), results in the following form of the continuity equation:

$$\frac{D\hat{\rho}_l}{D^s t} + \hat{\rho}_l \nabla \cdot \mathbf{V}_s + \nabla \cdot \hat{\rho}_l \mathbf{V}_r = 0 \; . \qquad (5)$$

Similarly, the Lagrangian form of the differential mass balance for the solid is obtained as

$$\frac{D\hat{\rho}_s}{D^s t} + \hat{\rho}_s \nabla \cdot \mathbf{V}_s = 0 \; . \qquad (6)$$

This equation provides an important relationship for use in (5) as follows. Substitution of $\hat{\rho}_s = (1-\phi)\rho_s$, in (6), with ρ_s, the solid grain mass density, being constant, results in

$$\frac{1}{1-\phi} \frac{D\phi}{D^s t} = \nabla \cdot \mathbf{V}_s \; . \qquad (7)$$

However, $\nabla \cdot \mathbf{V}_s$ can be expressed in terms of skeleton volume strain e as

$$\nabla \cdot \mathbf{V}_s = \frac{\partial}{\partial t} \nabla \cdot \mathbf{U}_s = \frac{\partial}{\partial t}\left(\delta_{ij} e_{ij}\right) = \frac{\partial e}{\partial t} \qquad (8)$$

($\mathbf{V}_s \cdot \nabla e$ is assumed negligible). Accordingly, (7) can be written as

$$\frac{D\phi}{D^s t} = (1-\phi)\frac{\partial e}{\partial t} \; . \qquad (9)$$

Substitution of (8) and (9) in (5), along with simple manipulations, results in a new form for the relative mass balance equation for the fluid,

$$\phi \frac{D\rho_l}{D^s t} + \hat{\rho}_l \frac{\partial e}{\partial t} + \nabla \cdot \hat{\rho}_l \mathbf{V}_r = 0 \; . \qquad (10)$$

Development of the momentum balance for the fluid flow in porous media under usual assumptions (Mercer, 1973) yields the generalized Darcy equation of motion in the Cartesian coordinate system:

$$\mathbf{q} = -\frac{\mathbf{k}_l}{\eta_l} \cdot \left(\nabla P + \rho_l g \nabla z\right) \qquad (11)$$

where:

$$\eta_l = \text{liquid dynamic viscosity}$$
$$\mathbf{q} = \phi \mathbf{V}_r$$
$$\mathbf{k}_l = \text{local intrinsic permeability tensor}$$

$$\nabla z = \begin{Bmatrix} 0 \\ 0 \\ 1 \end{Bmatrix}$$

P = liquid pressure.

Substitution of (11) in (10) yields the following equation:

$$\phi \frac{D\rho_l}{D^s t} + \hat{\rho}_l \frac{\partial e}{\partial t} = \nabla \cdot \frac{\rho_l \mathbf{k}_l}{\eta_l} \cdot (\nabla P + \rho_l g \nabla z) . \tag{12}$$

Further development of (12) requires the fluid state equation relating fluid density to temperature and pressure. Assuming small changes in density, the truncated Taylor series expansion of ρ_l about the reference density ρ_0 provides the required relationship:

$$\rho_l = \rho_0 + \left(\frac{\partial \rho_l}{\partial T}\right)_0 (T - T_0) + \left(\frac{\partial \rho_l}{\partial P}\right)_0 (P - P_0) . \tag{13}$$

The coefficients in the above equation can be defined as

$$\beta_T = \frac{(\partial \rho_l / \partial T)_0}{\rho_0} \quad \text{and} \quad \beta_P = \frac{(\partial \rho_l / \partial P)_0}{\rho_0} ,$$

where β_T is the coefficient of thermal volume expansion for the liquid and β_P is the compressibility coefficient of the liquid. Substituting β_T and β_P in (13) and taking the comoving time derivative of the fluid density, with respect to solid velocity, gives

$$\frac{D\rho_l}{D^s t} = \rho_0 \beta_T \frac{DT}{D^s t} + \rho_0 \beta_P \frac{DP}{D^s t} . \tag{14}$$

Using (14) in (12) and neglecting the $\mathbf{V}_s \cdot \nabla P$ and $\mathbf{V}_s \cdot \nabla T$ terms yields

$$\frac{\rho_l}{\rho_0} \frac{\partial e}{\partial t} + \phi \beta_P \frac{\partial P}{\partial t} + \phi \beta_T \frac{\partial T}{\partial t} = \nabla \cdot \left[\frac{\rho_l \mathbf{k}_l}{\rho_0 \eta_l} \cdot (\nabla P + \rho_l g \nabla z) \right] . \tag{15}$$

Equation (15) is the final form of the combined mass and momentum balance law for the moving fluid in thermohydroelasticity. In isothermal conditions, (15) takes the familiar form of Biot's equation, under the assumption of solid grain incompressibility. The same assumption is used in this study in the derivation of (7).

Considering the negligibility of $(e/\rho_0)(\partial \rho_l / \partial t)$ — a second order term — even in the non-isothermal case, equation (15) can be recast into exact Biot form. By introducing a variable ξ, representing the equivalent fluid volume strain, and generalizing for solid grain compressibility through Biot's coupling parameter α, we can write

$$\frac{\partial \xi}{\partial t} - \frac{e}{\rho_0} \frac{\partial \rho_l}{\partial t} = \nabla \cdot \left[\frac{\rho_l \mathbf{k}_l}{\rho_0 \eta_l} (\nabla P + \rho_l g \nabla z) \right] , \tag{16}$$

where

$$\xi = \frac{\rho_l}{\rho_0} \alpha \delta_{ij} e_{ij} + \frac{1}{M} P + \frac{1}{M_T} T \qquad (17)$$

and

$$M = 1/\phi\beta_P \quad \text{and} \quad M_T = 1/\phi\beta_T \quad .$$

Consideration of solid grain compressibility leads to additional terms in equation (16) which may be required for special cases (see Lewis, *et al.,* 1986, for such consideration). Equation (17) provides the first constitutive relationship for the analysis of thermohydroelastic phenomena. (For a complete definition of Biot's parameters see the Footnote at the end.)

The next important constitutive equation, which is complementary to (17), relates the total bulk stress components τ_{ij} to strain components e_{ij}, fluid pressure P, and the thermal stress τ_{Tij}. Biot (1941, 1956) provides such a relationship in the absence of thermal stresses. It is known that any rise or fall of temperature acts in the form of initial strains (Zienkiewicz, 1976) that are independent of other mechanical stresses. Therefore, Biot's (1941) stress-strain law for the mechanically isotropic case needs to be modified in the following way:

$$\tau_{ij} = 2\mu\left(e_{ij} - e_{Tij}\right) + \lambda\delta_{ij}\delta_{kl}\left(e_{kl} - e_{Tkl}\right) - \alpha\delta_{ij}P \quad . \qquad (18)$$

Assuming a thermal expansivity tensor β_{ij}, the above equation can be written as

$$\tau_{ij} = 2\mu e_{ij} + \lambda\delta_{ij}\delta_{kl}e_{kl} - \tau_{Tij} - \alpha\delta_{ij}P \quad , \qquad (19)$$

where $e_{ij} = \frac{1}{2}\left(U_{i,j} + U_{j,i}\right)$, λ and μ are Lame's constants, and

$$\tau_{Tij} = \left(2\mu\beta_{ij} + \lambda\delta_{ij}\delta_{kl}\beta_{kl}\right)T \quad . \qquad (20)$$

For thermal isotropy conditions, (19) can be written as

$$\tau_{ij} = 2\mu e_{ij} + \lambda\delta_{ij}\delta_{kl}e_{kl} - \gamma\delta_{ij}T - \alpha\delta_{ij}P \quad , \qquad (21)$$

where $\gamma = (2\mu + 3\lambda)\beta$ and β is the isotropic linear solid thermal expansion coefficient.

The dependent variables of the above equations (i.e., e, P, and T) are all regarded as incremental in value and represent deviation from the zero (stress-free) state. Also in the above development, thermal contact equilibrium between fluid and solid is assumed. Equations (17) and (21) are the constitutive equations relating stresses to strains in thermohydroelasticity phenomena.

The law of conservation of momentum, in the absence of inertial term, provides the equation for static equilibrium as

$$\frac{\partial\tau_{ij}}{\partial x_j} + \bar{\rho}_s f_i = 0 \quad , \qquad (22)$$

where f_i is the component of the body force, which is gravity in this case, and $\bar{\rho}_s$ is the average specific mass of the porous medium.

Finally, the expression for conservation of energy is considered. This equation is fairly well known in many scientific disciplines. The development is based on writing the energy conservation law over a REV (i.e., in a macroscopic sense), using the assumption of thermal equilibrium between the solid and the fluid. It is written as

$$(\rho C)_M \frac{\partial T}{\partial t} + T\gamma \frac{\partial}{\partial t}\left(\delta_{ij}e_{ij}\right) + (1-\phi)\gamma T \frac{\partial e}{\partial t} + \rho_l C_{Vl}\mathbf{q}\cdot\nabla T = \nabla\cdot\mathbf{K}_M\cdot\nabla T \ , \tag{23}$$

where $(\rho C)_M$ and $\mathbf{K}_M$ are the specific heat capacity and thermal conductivity of the fluid-filled media, defined as

$$(\rho C)_M = \phi\rho_l C_{Vl} + (1-\phi)\rho_s C_{Vs}$$
$$\mathbf{K}_M = \phi\mathbf{K}_l + (1-\phi)\mathbf{K}_s$$

in which C_{Vl} and C_{Vs} are the fluid and solid specific heat constants at constant volume and $\mathbf{K}_l$ and $\mathbf{K}_s$ are the total fluid thermal dispersion tensor and the solid thermal diffusion tensor.

Comparison of equation (23) with a rigorous microscopically-based development, such as that of Bear and Corapcioglo (1981), shows few missing terms. They relate to contribution from fluid mechanical energy conversion rate and are seemingly insignificant. A further simplification can also be made. If the temperature changes (T) are small compared to the absolute ambient temperature, then we can write (23) as (Carter and Booker, 1988)

$$(\rho c)_M \frac{\partial T}{\partial t} + T_\circ\,\gamma \frac{\partial}{\partial t}\left(\delta_{ij}e_{ij}\right) + \rho_l C_{Vl}\frac{\mathbf{k}_l}{\eta_l}\left(\nabla P + \rho_l g\nabla z\right)\cdot\nabla T = \nabla\cdot\mathbf{K}_M\cdot\nabla T \ , \tag{24}$$

where $T_\circ$ designates the absolute temperature in the stress-free state.

The deformation energy conversion term $T_\circ\beta\,\partial\!\left(\delta_{ij}e_{ij}\right)\!/\partial t$, not being a major contributing term to the energy balance, at least in the range of deformations of concern here, can be dropped from the equation if desired. However, we will retain this term for the sake of completeness.

Besides neglecting the energy associated with fluid dilation, thermal contact equilibrium between fluid and solid and absence of fluid shearing stresses in the macroscopic sense are assumed. An energy source term in (24) and a mass source term in (15) are neglected for the sake of brevity. Both source terms can be added to the fluid flow and heat equation with no difficulty if so desired.

As may be seen, the fundamental laws governing the static equilibrium, flow of fluid, and flow of heat are coupled through the dependent variables of solid displacement, fluid pressure, and the macroscopic medium temperature. These governing laws, expressed in (15), (22), and (24), in conjunction with constitutive equations (17), (21), and the equation of state, (13), provide the complete mathematical formulation of linear thermohydroelastic phenomena in saturated porous media. The foregoing equations, along with the following general initial and boundary conditions, define the mixed initial boundary value problem of thermohydroelasticity for which a numerical solution approach is suggested.

3. INITIAL AND BOUNDARY CONDITIONS—Fully saturated case

The general boundary and initial conditions for the saturated porous elastic medium are

$$
\begin{aligned}
\mathbf{U}(\mathbf{x}, t) &= \hat{\mathbf{U}}(\mathbf{x}, t) &\text{on}&\quad A_1 \times [0, \infty) \\
\tau(\mathbf{x}, t) \cdot \mathbf{n}(\mathbf{x}) &= \hat{\mathbf{G}}(\mathbf{x}, t) &\text{on}&\quad A_2 \times [0, \infty) \\
P(\mathbf{x}, t) &= \hat{P}(\mathbf{x}, t) &\text{on}&\quad B_1 \times [0, \infty) \\
\frac{\mathbf{k}_l}{\eta_l} \cdot (\nabla P + \rho_l g \nabla z) \cdot \mathbf{n}(\mathbf{x}) &= \hat{Q}_l(\mathbf{x}, t) &\text{on}&\quad B_2 \times [0, \infty) \\
T(\mathbf{x}, t) &= \hat{T}(\mathbf{x}, t) &\text{on}&\quad C_1 \times [0, \infty) \\
\mathbf{K}_M \cdot \nabla T \cdot \mathbf{n}(\mathbf{x}) &= \hat{Q}_h(\mathbf{x}, t) &\text{on}&\quad C_2 \times [0, \infty) \\
\mathbf{U}(\mathbf{x}, 0) &= 0 &\text{on}&\quad V \\
\tau(\mathbf{x}, 0) &= 0 &\text{on}&\quad V \\
P(\mathbf{x}, 0) &= 0 &\text{on}&\quad V \\
T(\mathbf{x}, 0) &= 0 &\text{on}&\quad V
\end{aligned}
\tag{25}
$$

A, B, and C represent parts of the boundary for stress-displacement, pressure-fluid flow, and temperature-heat flux considerations. The quantity V represents the volume under consideration. As mentioned earlier, the dependent variables $\mathbf{U}$, P, and T represent incremental deviations from the strain-free state assumed by the above choice of initial conditions.

4. CONSIDERATIONS FOR FRACTURES AND MECHANICAL NONLINEARITY IN GENERAL

As mentioned at the outset of this section, two major assumptions were made. First, the skeleton was isotropic linear elastic; and second, the medium was fully saturated. In this section we will try to relax the first assumption.

Most realistic problems defy, at least in part, the isotropicity and/or linearity assumptions. However, replacement of the isotropic linear elasticity matrix with a general anisotropic stress-deformation matrix C_{ijkl} is sufficient to meet most realistic demands. The values C_{ijkl} may be time, position, and stress dependent. As such, the stress-strain law will take the following form:

$$
\tau_{ij} = C_{ijkl} e_{kl} - C_{ijkl} \beta_{kl} T - \alpha_{ij} \delta_{ij} P \quad .
\tag{26}
$$

Notice that anisotropic thermal and hydroelastic parameters have also been introduced. Deployment of a multilinear behavior approach allows consideration of various material behaviors. Elastoplasticity (in general), creep, and other macroscopic constitutive laws can be implemented. In the following, implementation of a discrete fracture material is discussed.

Fractures, when their peculiar behavior is taken into account, can be treated similarly from the standpoint of the coupled stress and fluid-flow analyses. This approach requires that the fracture material properties be defined on a parallel basis to porous elastic media. Of particular importance are Biot's constants α, M, and M_T, which must also be defined for fractures. In general, one can expect that both α and M for a fracture should depend upon parameters such as roughness of the fracture surfaces, the filling material, and the state of stress. However, the assumptions of α equal to unity, M equal to inverse fluid compressibility, and M_T equal to inverse thermal expansivity for clean fractures seem to be reasonable. In the case of fractures with filling materials, consideration of filling-material porosity leads to a meaningful estimate of the value of M and M_T. In this case, the correct estimate of the value of α could possibly be obtained from tests similar to those described by Biot and Willis (3) for solid materials.

Having defined α and M, we can formulate constitutive and governing equations for fractures parallel to those for solids, as follows:

$$\tau_i = C_{ij}U_j^r + \alpha\delta_{i3}P \qquad\qquad i, j = 1, 3 \qquad\qquad (27\text{a})$$

$$\xi = \alpha\frac{U_3^r}{2b} + \frac{1}{M}P + \frac{1}{M_T}T \qquad\qquad (27\text{b})$$

$$\frac{\partial\xi}{\partial t} = \left[\frac{\rho_l k_i^f}{\rho_0 \eta_l}\cdot(\nabla P + \rho_l g\nabla z)\right] . \qquad \text{(Planar flow)} \qquad (27\text{c})$$

The first and second equations are nonlinear constitutive relationships in terms of the local value of normal and shear stresses and relative local deformation. Equation (27c) is the governing equation of fluid flow in the fracture's planar domain. The parameter k^f is the average planar permeability. The matrix C_{ij} represents the deformation parameters of the fracture, and they are defined in the existing joint material models such as Goodman, *et al.* (1968) and Barton and Brandis (1992). U_j^r is the relative displacement of fracture faces in direction j, and 2b represents fracture aperture. As for the flow of energy in the fracture, equation (23) applies similarly. However, as in the fluid flow formulation, the flow of heat is considered to take place only along the fracture plane. Therefore, because of the relative thinness of the fracture, no thermal gradient is assumed in the direction perpendicular to the fracture. By providing a faster fluid flow path in the rocks, fractures can take on a significant role in the development of the temperature field. This possibility complicates the analysis further. We will address this issue in a later section when solution methodology is discussed.

5. FIELD EQUATIONS—Variably saturated (general) case

In the developments that follow, the medium is assumed to take variable values of saturation. Generally, any saturated region neighbors a partially saturated one. However, there are many circumstances in which regions of phenomenological interest lie either in fully saturated regions or in partially saturated ones. Vadose zones are examples of the latter case in which many geomechanical projects are implemented. In such a zone, the phenomenon of coupled thermohydroelasticity becomes more complicated on several accounts. One needs to address at least two flowing phases; a combination of air and vapor forms a gas-phase flow, while an interstitial liquid made up of water and dissolved air forms the other mass flow phase. Of course existing gradients in air mass, moisture content, and heat content cause an interchange of mass and energy between the three phases. Also, the proper choice of stress-strain law, as discussed at length in the introduction, in the face of yet not fully finalized propositions, is a major concern. In what follows we discuss a rather complete formulation of the problem. In the next section, the formulations for the special case of the weakly non-isothermal systems are developed.

Adapting a procedure used by Pruess (1991), we approach this problem first by considering conservation laws for three components, namely, air (in the gas and in liquid phases), water (vapor in the gas and water in the liquid phase), and heat. We need to write the mass flux equations for each component within each phase and the heat flux equation, namely:

$$F_l^{(air)} = -K\frac{k_{rl}}{\eta_l}\overline{\rho}_l^{\,a}\left(\nabla P_l - \gamma_l \nabla z\right) \tag{28}$$

$$F_g^{(air)} = -K\frac{k_{rg}}{\eta_a}\overline{\rho}_g^{\,a}\left(\nabla P_g - \gamma_g \nabla z\right) - D_{va}\rho_g \nabla\left(\frac{\overline{\rho}_g^{\,a}}{\rho_g}\right) \tag{29}$$

$$F_l^{(water)} = -K\frac{k_{rl}}{\eta_l}\overline{\rho}_l^{\,w}\left(\nabla P_l - \gamma_l \nabla z\right) \tag{30}$$

$$F_g^{(water\ vapor)} = -K\frac{k_{rg}}{\eta_g}\overline{\rho}_g^{\,w}\left(\nabla P_g - \gamma_g \nabla z\right) - D_{va}\rho_g \nabla\left(\frac{\overline{\rho}_g^{\,w}}{\rho_g}\right) \tag{31}$$

$$F^{heat} = -\mathbf{K}_M \cdot \nabla T + h_l^{(air)}F_l^{(air)} + h_g^{(air)}F_g^{(air)} + h_l^{(water)}F_l^{(water)} + h_g^{(water)}F_g^{(water)} \;. \tag{32}$$

The parameters used in the above equations are:

$$F_p^q \quad = \quad \text{mass flux for component } q \text{ in phase } p$$

$$F^{heat} \quad = \quad \text{heat flux}$$

$$K \quad = \quad \text{saturated media permeability}$$

$$k_{rp} \quad = \quad \text{relative permeability of phase } p = f\!\left(S_g, S_w\right)$$

$$S_p \quad = \quad \text{saturation of phase } p$$

$$\eta_p \quad = \quad \text{viscosity of phase } p = f(P, T)$$

$$\rho_p \quad = \quad \text{mass density of phase } p = f(P, T)$$

$$\gamma_p \quad = \quad \left(\rho_p g\right) \text{ weight density of phase } p$$

$$z \quad = \quad \text{upward positive Cartesian coordinate}$$

$$
\begin{aligned}
P_p &= \text{pressure of component p or phase p} \\
\overline{\rho}_p^{\,q} &= \text{mass of component } q \text{ in unit volume of phase } p = f(P,T) \\
D_{va} &= \text{diffusion coefficient for vapor air mixture} = f(P,T,S_p) \\
h_p^{\,q} &= \text{enthalpy (energy per unit mass) of component } q \text{ in phase } p = f(P,T) \\
\mathbf{K}_M &= \text{heat conductivity of tensor rock-fluid mixture} = f(S_g, S_w, T) \\
\phi &= \text{porosity .}
\end{aligned}
$$

The conservation equations for each phase will therefore be:

$$
\phi \frac{\partial}{\partial t}\left(S_l \overline{\rho}_l^{\,a} + S_g \overline{\rho}_g^{\,a}\right) + \alpha \left(S_l \overline{\rho}_l^{\,a} + S_g \overline{\rho}_g^{\,g}\right)\frac{\partial e}{\partial t} = \nabla \cdot \left(F_l^{(air)} + F_g^{(air)}\right) \tag{33}
$$

$$
\phi \frac{\partial}{\partial t}\left(S_l \overline{\rho}_l^{\,w} + S_g \overline{\rho}_g^{\,w}\right) + \alpha \left(S_l \overline{\rho}_l^{\,w} + S_g \overline{\rho}_g^{\,w}\right)\frac{\partial e}{\partial t} = \nabla \cdot \left(F_l^{(water)} + F_g^{(water)}\right) \tag{34}
$$

$$
\frac{\partial}{\partial t}\left[(1-\phi)\rho_s C_{Vs} T + \phi \left(S_l \rho_l u_l + S_g \rho_g u_g\right)\right] + T_\circ \gamma \frac{\partial e}{\partial t} = \nabla \cdot \left(-\mathbf{K}_M \nabla T + F^{heat}\right) . \tag{35}
$$

The added parameters are defined as:

$$
\begin{aligned}
\alpha &= \text{Biot's coupling parameter} \\
e &= \left(\delta_{ij} e_{ij}\right) \text{ rock volume strain} \\
\rho_s &= \text{solid mass density} \\
C_{Vs} &= \text{solid specific heat} \\
u_p &= \text{specific internal energy of phase } p \\
\gamma &= \text{defined earlier in terms of solid thermal expansion units and elastic moduli} \\
T_\circ &= \text{absolute temperature in the stress-free state .}
\end{aligned}
$$

Notice that the mass flux equations (29) and (31) account for the convection of the air and vapor masses by the flowing gas. In the derivation of the above equations, the following assumptions are made: (1) local equilibrium is assumed, (2) air is treated as an ideal gas, with $P_g = P_a + P_{vapor}$, and (3) mass and heat source terms have not been included. One should note that since the liquid mass equation includes both water and vapor, the phase change does not appear and becomes implicit in the treatment. Equations (33) to (35) involve a large number of apparent unknowns. However, the following relationships will make it possible to finalize these equations in choices of three unknowns, such as P_g, P_l, and T :

$$
\begin{array}{llll}
\overline{\rho}_l^{\,a} + \overline{\rho}_l^{\,w} = \rho_l & , & P_l - P_g = P_c & , \\[4pt]
\overline{\rho}_g^{\,a} + \overline{\rho}_g^{\,w} = \rho_g & , & S_g + S_l = 1 & , \\[4pt]
\overline{\rho}_g^{\,a}/\rho_g = P_a / P_g & , & P_g = P_a + P_{vapor} & \\[4pt]
P_a = K_H \overline{\rho}_l^{\,a} / \rho_l & & \text{(Henry's law)} & \\[4pt]
P_c = f(S_l, S_{lr}) & & S_{lr} \text{ is the liquid residual saturation .} &
\end{array}
$$

All of the thermophysical properties of water (vapor and liquid) are given by steam table equations such as the International Formulation Committee (1967). Other constitutive relationships are needed to define relative permeabilities, air-vapor mixture diffusivities, heat conductivity and capacity, and enthalpies. A host of proper expressions for each parameter is available in many references. The length of discussions needed to explain these expressions puts such effort beyond the scope of this work. Therefore, we refer the reader to the geothermal literature such as Pruess (1983); Garg, *et al.* (1977); Prichet, *et al.* (1975); Faust and Mercer (1979a,b); and Falta, *et al.* (1992).

As discussed very early in this section, the selection of the stress-strain law, in the face of the very serious questions regarding the validity of the Bishop-type effective stress concept (see Lloret and Alonso, 1980; Duncan, 1983; Alonso, *et al.*, 1988), is not an easy one. However, the stress-path dependency of the earlier alternative "surface state" concept (Lloret and Alonso, 1980) and lack of generality in such a concept (see also Josa, *et al.*, 1992) for applicability to the saturated part of porous media (Schrefler and Xiaoyong, 1993) makes it undesirable. Under the circumstances, we resort to using the effective stress law, knowing that the formalism may not be correct below a critical saturation level. However, for pore pressure, an average pressure defined as

$$\overline{P} = S_g P_g + S_l P_l$$

is used. So in this case equation (26) is written as

$$\tau_{ij} = C_{ijkl}e_{kl} - C_{ijkl}\beta_{kl}T - \alpha\delta_{ij}\overline{P} \; . \tag{36}$$

Applying equilibrium law, we get the force-deformation equation as before:

$$\frac{\partial \tau_{ij}}{\partial x_j} + \overline{\rho}_s f_i = 0 \; . \tag{22}$$

Again, sets of governing equations (33), (34), (35), along with equation (22) in conjunction with the related constitutive laws, define the phenomenon of thermohydroelasticity in partially saturated media.

6. FIELD EQUATIONS—Variably saturated weak non-isothermal case

In situations where the temperature of the geologic environment stays well below 100°C, referred to as weakly non-isothermal, it is possible to use simpler formulations. Philip and DeVries (1957) developed the macroscopic engineering water mass (vapor mass and liquid mass) flow and heat flow equations for a nondeformable porous medium in terms of liquid content and temperature as main variables. In these developments, convection of vapor is neglected and transport of latent heat of vaporization (L) due to thermal gradient is added to sensible heat transport by inclusion of L in the thermal conductivity of the medium. Of course, the transport of the latent heat of vaporization under a pressure gradient (isothermal vapor transfer) is considered as a separate term. The reader is referred to Childs and Matstaf (1982) for a clear review of this work. Cameron (1984) worked out these equations in terms of liquid water pressure and temperature. In this development, he assumed negligibility of capillary pressure and equality of the vapor and liquid pressures. These assumptions, which are often used for strongly non-isothermal flow of liquid and vapor, certainly do not apply to the conditions here. The proper formulation of the moisture and heat transport in hysteretic, inhomogeneous porous media (near the earth's surface) is given by Milly (1982). This development uses matric–potential and temperature as the dependent variables. In the light of works by Philip and DeVries (1957) and Milly (1982), we will develop the governing equations for thermohydroelastic phenomena for weakly non-isothermal (temperature well below 100°C), variably saturated porous media.

We start with expressions of mass balance for both water and vapor, namely:

$$\nabla \cdot \theta \rho_w \mathbf{V}_w + \frac{\partial \theta \rho_w}{\partial t} - Q_v^m = 0 \tag{37}$$

and

$$\nabla \cdot \theta_a \rho_v V_v + \frac{\partial \theta_a \rho_v}{\partial t} - Q_w^m = 0 \ , \tag{38}$$

where

$$
\begin{aligned}
\theta &= \text{water content} \\
\theta_a &= \text{air content} = \phi - \theta \\
\phi &= \text{porosity} \\
\mathbf{V}_w &= \text{water velocity} = \mathbf{V}_{rw} + \mathbf{V}_s \\
\mathbf{V}_v &= \mathbf{V}_{rv} + \mathbf{V}_s \\
\mathbf{V}_s &= \text{solid velocity} \\
\mathbf{V}_{rw} &= \text{water relative velocity} \\
\mathbf{V}_{rv} &= \text{vapor relative velocity} \\
\mathbf{q}_{rw} &= \theta \mathbf{V}_{rw} \\
\mathbf{q}_{rv} &= \theta_a \mathbf{V}_{rv} \\
Q_v^m &= Q_w^m = \text{mass exchange between vapor phase and liquid phase.}
\end{aligned}
$$

The mass flow equation for each phase is written as:

$$\mathbf{q}_{rw}^m = -\rho_w \mathbf{q}_{rw} = -\frac{\rho_w k_w}{\eta_w}\left(\nabla P + \gamma \nabla z\right) \tag{39}$$

$$\mathbf{q}_{rv}^m = -\rho_w \mathbf{q}_{rv} = -D_v \nabla \rho_v \ , \tag{40}$$

where D_v is an effective molecular diffusivity of water vapor in air and k_w is the unsaturated water conductivity.

Notice that if we assume $\rho_g = \rho_\alpha$, i.e., $\overline{\rho}_g^v = \rho_v$ (these are the basic assumptions in the present case) in equation (31) and neglect convection, we get equation (40). Expanding equations (37) and (38) results in

$$\nabla \cdot \mathbf{q}_{rw}^m + \theta \rho_w \nabla \cdot \mathbf{V}_s + \frac{D(\phi S \rho_w)}{D^s t} - Q_v^m = 0 \tag{41}$$

$$\nabla \cdot \mathbf{q}_{rv}^{m} + (\phi - \theta)\rho_v \nabla \cdot \mathbf{V}_s + \frac{D(\phi - \theta)\rho_v}{D^s t} - Q_w^m = 0 \ . \tag{42}$$

Using the assumption of local equilibrium between the liquid and vapor phases, the vapor density equation, given as

$$\rho_v = \rho_{vs} \exp\left(\frac{P}{\rho_w RT}\right) = \rho_{vs}(T)h(P, T) \ , \tag{43}$$

where

$$\rho_{vs} = \text{saturated vapor density}$$
$$R = \text{universal gas constant,}$$

can be used to obtain

$$\frac{D\rho_v}{D^s t} = \left[h\frac{d\rho_{vs}}{dT} - \frac{\rho_v P}{RT^2}\right]\frac{\partial T}{\partial t} + \left[\frac{\rho_v}{\rho_w RT}\right]\frac{\partial P}{\partial t} \ . \tag{44}$$

Expanding the third terms in equations (41) and (42) and makng similar assumptions to those made in the development of the saturated flow equation, (16), result in the following equations:

$$S\rho_w \frac{\partial e}{\partial t} + \rho_w C_s \frac{\partial P}{\partial t} + \phi S\rho_0 \beta_P \frac{\partial P}{\partial t} + \phi S\rho_0 \beta_T \frac{\partial T}{\partial t} - Q_v^m = -\nabla \cdot \mathbf{q}_{rw}^m \tag{45}$$

$$(1-S)\rho_v \frac{\partial e}{\partial t} - \rho_v C_s \frac{\partial P}{\partial t} + \left[h\frac{d\rho_{vs}}{dT} + \frac{\rho_v P}{RT^2}\right]\frac{\partial T}{\partial t} + \left[\frac{\rho_v}{\rho_w RT}\right]\frac{\partial P}{\partial t} + Q_w^m = -\nabla \cdot \mathbf{q}_{rv}^m \ , \tag{46}$$

where S is the saturation and $C_s = \phi\, \partial S/\partial P$ is the moisture capacity. Equation (40) can also be expanded in terms of pressure and temperature gradient (Milly, 1982) as

$$\mathbf{q}_{rv}^m = -D_{Pv}\nabla P - D_{Tv}\nabla T \ , \tag{47}$$

where

$$D_{Pv} = \frac{D_v \rho_v}{\rho_w RT}$$

$$D_{Tv} = D_v\left[h\frac{d\rho_{vs}}{dT} - \frac{\rho_v P}{\rho_w RT^2}\right] \ .$$

Adding the two flow equations (45) and (46) and using equations (39), (40), and (47), we obtain the final total water mass flow equation as

$$\left[S\rho_w + (1-S)\rho_v\right]\frac{\partial e}{\partial t} + \left[\phi S\rho_w^0\beta_P + (\rho_w - \rho_v)C_s + \frac{\rho_v}{\rho_w RT}\right]\frac{\partial P}{\partial t} + \left[\phi S\rho_w^0\beta_T + h\frac{d\rho_{vs}}{dT} + \frac{\rho_v P}{RT^2}\right]\frac{\partial T}{\partial t}$$

$$= \nabla \cdot \left[\left(\frac{\rho_w k_w}{\eta_w} + D_{Pv}\right)\nabla P + \frac{\rho_w k_w}{\eta_w}\gamma_w \nabla z\right] - \nabla \cdot (D_{Tv}\nabla T) \ . \tag{48}$$

The above flow equation is in terms of pressure and includes the effects of water compressibility and thermal expansivity as well as skeletal deformation of the medium. In this development, the role of solid grain compressibility and the flow of the absorbed liquid due to thermal gradients are ignored for simplification. The latter effect is considered in the development of Milly (1982) for nondeforming media, and the former effect is dealt with explicitly in the isothermal water and air flow model of Schrefler and Xiaoyong (1993). We will consider some measure of the grain compressibility effect implicitly by multiplying the first terms of equation (48) by α, Biot's coupling coefficient. Regarding the question of handling hysteretic effects and other constitutive model refinement modifications, the reader is referred to Milly (1982).

For development of the heat equation, we use the same approach as is used for the saturated case. Actually with that background, we can write the energy balance equation, for this case, as:

$$(\rho C)_M \frac{\partial T}{\partial t} + T_\circ \gamma \frac{\partial}{\partial t}(\delta_{ij} e_{ij}) + \rho_w C_{Vw}(\mathbf{q}_v^m + \mathbf{q}_w^m) \cdot \nabla T = \nabla \cdot [(\mathbf{K}_M + \rho_w L D_{Tv})\nabla T + L D_{Pv} \nabla P],$$

(49)

where

$$(\rho C)_M = (1 - \phi)\rho_s C_{Vs} + \rho_w C_{Vw}\theta + \rho_v C_{Pv}(\phi - \theta)$$

$$\mathbf{K}_M = (1 - \phi)\mathbf{K}_s + \phi\mathbf{K}_w$$

$$C_{Vs},\ C_{Vw} = \text{volumetric heat capacity of solid and water}$$

$$C_{Pv} = \text{specific heat of water vapor at constant pressure.}$$

Notice that the right-hand side of the equation (49) includes transport of distillation heat due to both thermal and moisture gradients. The final equation to complete the formalism for the case under consideration is the equilibrium equation. Considering the implicit assumption of the prevalence of atmospheric conditions in the unsaturated part of the pore space, we write the effective stress law (with due consideration for grain compressibility) as

$$\tau_{ij} = C_{ijkl} e_{kl} - C_{ijkl} \beta_{kl} T - \alpha S \delta_{ij} P \ , \tag{50}$$

where S is the water saturation and other terms have been defined earlier. An improvement in the formulation would be to regard P as the homogenized pressure and apply the mixture compressibility (Chang and Duncan, 1983) in place of β_P in equation (48). The expression of the total stress tensor in equation (50), employed in the equilibrium law of equation (22), finalizes the developments in this section.

Comparison of equations (48), (49) and (22), of the partially saturated weakly non-isothermal flow and deformation phenomena, with equations (16), (23) and (22), of the saturated case, reveals the minimum complexity brought about by the simplest treatment needed for analysis of the coupled thermohydroelasticity phenomena in variable ranges of saturation. Of course, more precise treatment of the phenomena calls for usage of the complete formulation laid out in equations (33) to (35) and equation (22).

7. SOLUTION METHODOLOGIES

In the preceding sections, field equations were developed for coupled thermohydroelasticity in geologic systems, for fully saturated as well as partially saturated conditions, and for both strong and weak thermal fields. Complete solution methodologies for the fully saturated cases have been available since the early 1980s (see Noorishad, *et al.*, 1984). Regarding weak non-isothermal conditions, two very recent attempts by Borgesson (1994) and Fujita, *et al.* (1994) can be mentioned. For consideration of vapor flow, both of the solution methodologies were created by modifying single, liquid-phase (full or partial saturation), coupled, thermohydroelastic, finite-element codes. However, both approaches have assumed the isothermal vapor mass flow to be negligible, and the transport of distillation heat due to a moisture gradient is disregarded in Borgesson (1994). These assumptions, though they may have been well-justified in case of the specific applications in these works, have to be tested in general. We are not aware of any solution technique presently available for strongly non-isothermal situations. However, benefiting from the long history of model development for geothermal resource recovery at Lawrence Berkeley Laboratory (LBL), effort is currently under way to expand the capabilities of the coupled thermohydroelastic code ROCMAS for multiphase flow. In this work, though, we will only address the solution approach for the weak non-isothermal case within the context of the formulation developed here. This will be done first in non-incremental form and later will be generalized in a rate form suitable for incremental solution. We use the Galerkin finite element solution approach (Hwang, *et al.*, 1971; Zienkiewicz, 1976) on equations (22), (48), and (49) and obtain the following sets of matrix equations:

$$\mathbf{KU} + \mathbf{C}_{UP}\mathbf{P} + \mathbf{C}_{UT}\mathbf{T} = \mathbf{F} \tag{51a}$$

$$\mathbf{C}_{PU}\dot{\mathbf{U}} + \mathbf{E}_P\dot{\mathbf{P}} + \mathbf{H}_P\mathbf{P} + \mathbf{C}_{PT_1}\dot{\mathbf{T}} + \mathbf{C}_{PT_2}\mathbf{T} = \mathbf{Q}_{PT} \tag{51b}$$

$$\mathbf{C}_{TU}\dot{\mathbf{U}} + \mathbf{C}_{TP}\mathbf{P} + \mathbf{E}_T\dot{\mathbf{T}} + (\mathbf{H}_T + \mathbf{H}_{TP})\mathbf{T} = \mathbf{Q}_{TP} , \tag{51c}$$

where:

$$\mathbf{K} = \sum_{n=1}^{N} \int_{v^e} \Phi_e \mathbf{C} \Phi_e^{Tr} \, dv \tag{52a}$$

$$\mathbf{C}_{UP} = -\sum_{n=1}^{N} \int_{v^e} \Phi_e \alpha \hat{\mathbf{1}} \Phi_\pi^{Tr} \, dv \tag{52b}$$

$$\mathbf{C}_{UT} = -\sum_{n=1}^{N} \int_{v^e} \Phi_e \gamma \hat{\mathbf{1}} \Phi_\pi^{Tr} \, dv \tag{52c}$$

$$\mathbf{F} = -\sum_{n=1}^{N} \int_{v^e} \Phi_u \bar{\rho}_s \hat{\mathbf{1}} f_i \, dv + \sum_{n=1}^{N} \int_{A_2^e} \Phi_u \hat{\mathbf{G}} \, ds \tag{52d}$$

$$\mathbf{C}_{PU} = -\sum_{n=1}^{N} \int_{v^e} \Phi_\pi \alpha [S\rho_w + (1-S)\rho_v] \hat{\mathbf{1}}^{Tr} \Phi_e^{Tr} \, dv \tag{52e}$$

$$\mathbf{E}_P = -\sum_{n=1}^{N}\int_{v^e}\Phi_\pi\left[\phi S\rho_w^0\beta_P + (\rho_w - \rho_v)\mathbf{C}_s + \frac{\rho_v}{\rho_w RT}\right]\Phi_\pi^{Tr}\,dv \tag{52f}$$

$$\mathbf{H}_P = \sum_{n=1}^{N}\int_{v^e}\Phi_\theta\left(\frac{\rho_w k_w}{\mu} + \mathbf{D}_{Pv}\right)\Phi_\theta^{Tr}\,dv \tag{52g}$$

$$\mathbf{C}_{PT_1} = \sum_{n=1}^{N}\int_{v^e}\Phi_\pi\left[\phi S\rho^0\beta_T + \mathbf{h}\frac{d\rho_{vs}}{dT} + \frac{\rho_v P}{RT^2}\right]\Phi_\pi^{Tr}\,dv \tag{52h}$$

$$\mathbf{C}_{PT_2} = \sum_{n=1}^{N}\int_{v^e}\Phi_\theta\mathbf{D}_{Tv}\,\Phi_\theta^{Tr}\,dv \tag{52i}$$

$$\mathbf{Q}_{PT} = \sum_{n=1}^{N}\int_{B_2}\Phi_\pi\hat{\mathbf{Q}}_f\,ds + \sum_{n=1}^{N}\int_{v^e}\Phi_\theta\frac{\rho_w k_w}{\mu}\gamma_w\hat{\mathbf{i}}_z\,dv$$
$$+ \sum_{n=1}^{N}\int_{B_2}\Phi_\pi(D_{Pv}\nabla P)\cdot n\,ds + \sum_{n=1}^{N}\int_{B_2}\Phi_\pi(D_{Tv}\nabla T)\cdot n\,ds \tag{52j}$$

$$\mathbf{C}_{TU} = \frac{1}{T_o}\mathbf{C}_{UT} \tag{52k}$$

$$\mathbf{C}_{TP} = \sum_{n=1}^{N}\int_{v^e}\Phi_\theta\mathbf{LD}_{Pv}\Phi_\theta^{Tr}\,dv \tag{52l}$$

$$\mathbf{E}_T = \sum_{n=1}^{N}\int_{v^e}\Phi_\pi(\rho C)_M\Phi_\pi^{Tr}\,dv \tag{52m}$$

$$\mathbf{H}_T = \sum_{n=1}^{N}\int_{v^e}\Phi_\theta(\mathbf{K}_M + \rho_w\mathbf{LD}_{Tv})\Phi_\theta^{Tr}\,dv \tag{52n}$$

$$\mathbf{H}_{TP} = \sum_{n=1}^{N}\int_{v^e}\Phi_\pi\left[\rho_w\mathbf{C}_{Vw}\left(\mathbf{q}_w^m + \mathbf{q}_v^m\right)\right]\Phi_\theta^{Tr}\,dv \tag{52o}$$

$$\mathbf{Q}_{TP} = \sum_{n=1}^{N}\int_{C_2}\Phi_\pi\hat{\mathbf{Q}}_h\,ds + \sum_{n=1}^{N}\int_{C_2}\Phi_\pi(\mathbf{LD}_{PV}\nabla P)\cdot n\,ds \tag{52p}$$

$$\hat{\mathbf{i}} = \begin{Bmatrix}1\\1\\1\\0\\0\\0\end{Bmatrix} \quad\text{and}\quad \hat{\mathbf{i}}_z = \begin{Bmatrix}0\\0\\1\end{Bmatrix}\,.$$

In equations (52), $\mathbf{C}$ is the deformation compliance matrix, Φ_π is the shape function for both pressure and temperature; Φ_u is the shape function for displacement; Φ_θ is the transformation matrix for pressure and temperature gradient; Φ_e is the transformation matrix for strains; and n is the outward normal direction unit vector. The rest of the parameters have been defined earlier.

In these algorithms, we have not addressed discrete fractures explicitly for the sake of brevity. However, using appropriate constitutive laws, nearly the same expressions with reduced integration dimension by one order hold for fractures as well.

Equations (51) are coupled nonlinear nonsymmetric equations that need to be integrated in time, thus making them ready for linearization and final solution by an equation solver. However, the matrix equations can only be used to obtain one increment of change from equilibrium. Therefore, the setup is not suitable for strongly nonlinear problems requiring incremental solution. To obtain the proper incremental matrix equations, one needs to start with all variables in rate form (field variables in the field equations and in the initial and boundary conditions). This means the variables $\dot{e}$, $\dot{\tau}$, $\dot{P}$, $\dot{T}$, $\dot{q}$, $\dot{U}$, $\dot{G}$, and $\dot{Q}$ will replace their undotted counterparts in the process of formulating the problem. Application of the solution methodology, in this case Galerkin finite element, will lead to a matrix formulation that resembles equations (51) with each dependent variable replaced by its "dotted" version. Knowing this property, to obtain the incremental matrix equations we can simply proceed (Small, *et al.*, 1976) by writing equation (51a) in its incremental form and keeping equations (51b) and (51c) in their original form. Had we obtained rate matrix equations from the beginning by having started from the "rate" variables, we would need to integrate the fluid and heat flow equations once, to obtain (51b) and (51c). Thus, the rate matrix formulation will be the following:

$$\mathbf{K}\dot{\mathbf{U}} + \mathbf{C}_{UP}\dot{\mathbf{P}} + \mathbf{C}_{UT}\dot{\mathbf{T}} = \dot{\mathbf{F}} \tag{53a}$$

$$\mathbf{C}_{PU}\dot{\mathbf{U}} + \mathbf{E}_P\dot{\mathbf{P}} + \mathbf{H}_P\mathbf{P} + \mathbf{C}_{PT_1}\dot{\mathbf{T}} + \mathbf{C}_{PT_2}\mathbf{T} = \mathbf{Q}_{PT} \tag{53b}$$

$$\mathbf{C}_{TU}\dot{\mathbf{U}} + \mathbf{C}_{TP}\mathbf{P} + \mathbf{E}_T\dot{\mathbf{T}} + (\mathbf{H}_T + \mathbf{H}_{TP})\mathbf{T} = \mathbf{Q}_{TP} \ . \tag{53c}$$

Integrating equations (53) between t_1 and $t_2 = t_1 + \Delta t$ we obtain

$$\tilde{\mathbf{K}}(\mathbf{U}_2 - \mathbf{U}_1) + \tilde{\mathbf{C}}_{UP}(\mathbf{P}_2 - \mathbf{P}_1) + \tilde{\mathbf{C}}_{UT}(\mathbf{T}_2 - \mathbf{T}_1) = \mathbf{F}_2 - \mathbf{F}_1 \tag{54a}$$

$$\tilde{\mathbf{C}}_{PU}(\mathbf{U}_2 - \mathbf{U}_1) + \tilde{\mathbf{E}}(\mathbf{P}_2 - \mathbf{P}_1) + \Delta t\,\tilde{\mathbf{H}}_P\overline{\mathbf{P}} + \tilde{\mathbf{C}}_{PT_1}(\mathbf{T}_2 - \mathbf{T}_1) + \tilde{\mathbf{C}}_{PT_2}\Delta t\,\overline{\mathbf{T}} = \Delta t\,\tilde{\mathbf{Q}}_{PT} \tag{54b}$$

$$\tilde{\mathbf{C}}_{TU}(\mathbf{U}_2 - \mathbf{U}_1) + \tilde{\mathbf{C}}_{TP}\Delta t\overline{\mathbf{P}} + \tilde{\mathbf{E}}_T(\mathbf{T}_2 - \mathbf{T}_1) + (\tilde{\mathbf{H}}_T + \tilde{\mathbf{H}}_{TP})\Delta t\overline{\mathbf{T}} = \Delta t\,\tilde{\mathbf{Q}}_{TP} \ , \tag{54c}$$

where $\quad \overline{\mathbf{P}} = \zeta\mathbf{P}_2 + (1-\zeta)\mathbf{P}_1 \text{ and } \overline{\mathbf{T}} = \zeta\mathbf{T}_2 + (1-\zeta)\mathbf{T}_1 , \quad 0 \le \zeta \le 1$

and all the coefficient matrices marked by $\tilde{\ }$ are average values over the integration time. In final form, equations (54) are represented as

$$\tilde{\mathbf{K}}\Delta\mathbf{U} + \tilde{\mathbf{C}}_{UP}\Delta\mathbf{P} + \tilde{\mathbf{C}}_{UT}\Delta\mathbf{T} = \Delta\mathbf{F} \tag{55a}$$

$$\tilde{\mathbf{C}}_{PU}\Delta\mathbf{U} + (\tilde{\mathbf{E}} + \zeta\Delta t\tilde{\mathbf{H}}_P)\Delta\mathbf{P} + (\tilde{\mathbf{C}}_{PT_1} + \zeta\Delta t\,\tilde{\mathbf{C}}_{PT_2})\Delta\mathbf{T} = \Delta t\tilde{\mathbf{Q}}_{PT} - \Delta t\tilde{\mathbf{H}}_P\mathbf{P}_1 - \Delta t\tilde{\mathbf{C}}_{PT_2}\mathbf{T}_1 \tag{55b}$$

$$\tilde{\mathbf{C}}_{TU}\Delta\mathbf{U} + \Delta t\tilde{\mathbf{C}}_{TP}\Delta\mathbf{P} + [\tilde{\mathbf{E}}_T + \zeta\Delta t(\tilde{\mathbf{H}}_T + \tilde{\mathbf{H}}_{TP})]\Delta\mathbf{T}$$
$$= \Delta t\tilde{\mathbf{Q}}_{TP} - \Delta t\tilde{\mathbf{C}}_{TP}\mathbf{P}_1 - \Delta t(\tilde{\mathbf{H}}_T + \tilde{\mathbf{H}}_{TP})\mathbf{T}_1 \ . \tag{55c}$$

Using a variety of linearization schemes, we can solve the above equations directly in terms of $\Delta\mathbf{U}$, $\Delta\mathbf{P}$, and $\Delta\mathbf{T}$, as variations from the values of the dependent variables $\mathbf{U}$, $\mathbf{P}$, and $\mathbf{T}$ at time t_1. However, considering the time-constant differences between the fluid-flow equation (55b) and the heat equation (55c), an interlaced solution scheme is the optimal approach. In this method, the hydromechanical equations (55a) and (55b) are solved directly first. Depending on the interest in the evolution of the hydromechanical phenomenon and the ratios of the time constants, this solution may be transient or steady-state. For the latter, the steady state versions of (55a) and (55b) need to be solved. After fluid-flow field results are obtained, the heat equation is solved for an appropriate time step, and the results are used for the next round of solving of hydromechanical equations (55a) and (55b). Strong material nonlinearity may necessitate iterative solution of (55a), (55b), and (55c) at each time-marching step of the latter equation. However, this additional calculation is rarely needed for saturated conditions. Also, under saturated flow conditions, the steady-state solution of hydromechanical equations is usually sufficient and provides the "snapshot" of the stress-flow conditions as the overall coupled thermohydroelastic phenomenon evolves. In the interlaced approach, one may symmetricize (55a) and (55b) for the dependent variables $\Delta\mathbf{U}$ and $\Delta\mathbf{P}$ by using $\mathbf{C}_{PU} = \mathbf{C}_{UP}^{tr}$. Obviously, the approximation has to be justifiable. However, this choice allows easy conversion of the existing single-phase coupled thermohydroelastic codes for solution of problems involving two-phase weakly non-isothermal coupled phenomena.

An important consideration, brought up in the preceding chapter, is the proper solution of the heat-transport equation under the likely condition of dominance of convective transport. The classic Galerkin finite-element method cannot generally handle the movement of a sharp thermal front that may exist in the solution domain. There is an abundance of methodologies in the literature, with a promise of delivering the correct solution under challenging conditions. However, some require more sophisticated algorithms and some have a limited range of applicability. In a recent publication, Noorishad, *et al.* (1992) put forward a simple technique to augment the classic methods, enabling them to work in a wide range of Peclet and Courant numbers. The method simply requires systematic implementation of a longitudinal thermal-dispersivity coefficient equal to $\alpha_\ell = (V \cdot V\Delta T)/|V|$ throughout the domain. The parameter lends itself to the calculation of an apparent thermal-diffusion tensor. Elements of this tensor replace the corresponding elements of the physical-diffusion tensor only if the latter is greater than the former. This process introduces a measure of needed artificial diffusion into the problem, albeit at the cost of some smearing of the front.

Special consideration for unsaturated soil boundary conditions. At this boundary with the atmosphere, mass flow in the form of water vapor and heat flow by radiation and convection may become the dominant modes of mass and energy transport. The mathematical form of these fluxes (see Childs, 1982, for details) necessitates formulation of additional vectors and matrices. Since the development and its numerical implementation are straightforward, we do not elaborate on this issue and refer the reader to works such as Polinka and Wilson (1976).

8. PRELIMINARY MODELING APPLICATIONS

The complexity of the coupled thermohydroelastic phenomenon poses major obstacles for any analysis attempt. More sophisticated material models must be implemented, and any realistic THM modeling attempt demands extensive data characterization. As a result, only simple and often conceptually compromised problems have been simulated. In the following we will discuss some of the simulations we have performed with LBL's ROCMAS code over the past several years. Other chapters of the book include some of the most recent attempts at THM modeling as part of the DECOVALEX efforts.

8.1 Example 1: Thermoelastic consolidation of a sand column

Aboustit, *et al.* (1982) used a nonconvective, coupled, variational, finite-element approach to analyze the thermoelastic consolidation of a sand column. Table 1 gives the data used in this problem. Results of their analysis have been reproduced in Figure 1 along with our solution of the same problem. There is almost perfect agreement between the two solutions. The slight discrepancy is due to the interlaced solution scheme of the present work, in which temperature solution lags one step behind the hydromechanical calculations. Reduction of the time-step size will decrease the difference.

In solving this problem, the advection of energy along with the effects of fluid compressibility, thermal expansivity, and viscosity variations were assumed to be negligible. All of these assumptions, in addition to linear material behavior, can of course be relaxed, if desired, in the present work.

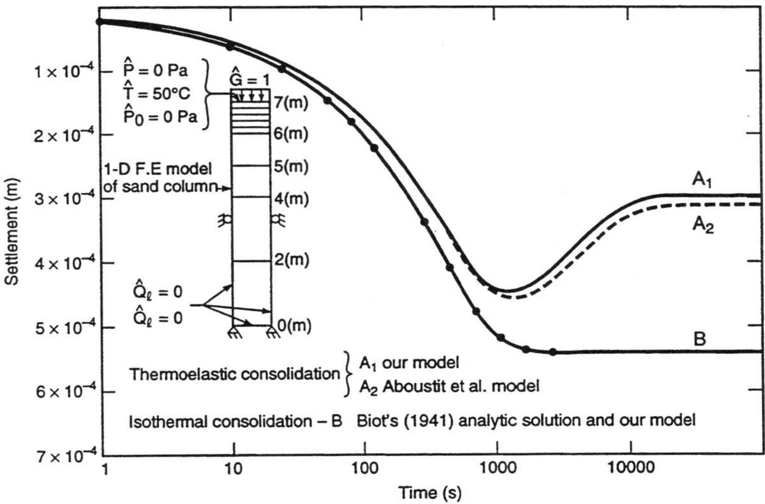

Fig. 1 Consolidated history for the one-dimensional finite-element model.

8.2 Example 2: Variation of flow to the Stripa heater test hole

In this case, the thermohydromechanical environment around a waste canister or heater and the ensuing changes of fluid inflow to the heater borehole induced by the temperature rise are studied. A 5-kW heater is located at a depth of 350 m in granite. A horizontal fracture is assumed to lie 3 m below the heater midplane and extend from the heater borehole to a hydrostatic boundary at a radial distance of 20 m from the borehole. The properties of rock and fracture are given in Table 1, and the two-dimensional axisymmetric (r,z) finite-element grid and some data are shown in Figure 2.

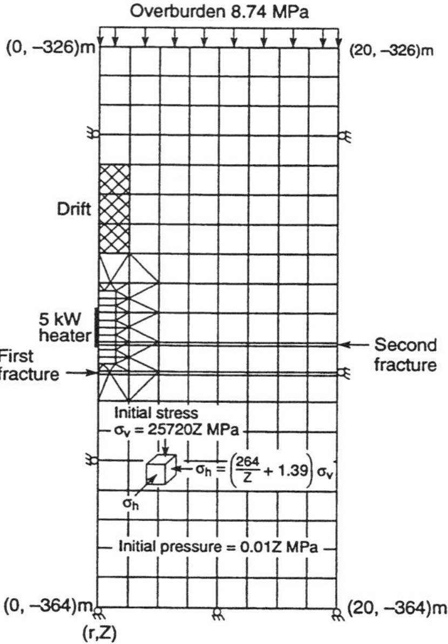

Fig. 2. Finite-element model of the heat-source environment.

The heater drift, approximated by the cylindrical hatched area in Figure 2, is simulated by assigning a very low value of Young's modulus to the elements. Before the heater raises the temperature of a large volume of the rock, the flow from the hydrostatic outer boundary to the atmospheric ("zero" hydraulic pressure) borehole is high. Later in time, with the heated rock above the fracture expanding and the fracture aperture near the heater borehole closing, the flow decreases sharply, as shown in Figure 3.

The evolution of the fracture aperture profile, together with the variations of the pressure and temperature distributions, are shown in Figure 4. As may be seen in the pressure-distance graph of this figure, at 0 day, before the tapping of the fracture, full hydrostatic pressure prevails in the fracture.

This pressure diminishes rapidly at 0.25 day before major development of the thermal front. However, as thermal stresses are established, the fracture starts closing. As a result, the pressure inside the fracture starts rising, thus leading to the establishment of full pressure in the fracture at 14 day, similar to the 0 day case. These results may provide a better understanding of some of the observations made in the *in situ* heater experiments in the

TABLE 1. Material Properties used for the example applications

Material	Parameter	Example 1	Example 2	Example 3	Example 4	Example 5
Fluid	Mass density, ρ_s	NA	997 kg m^{-3}	997 kg m^{-3}	997 kg m^{-3}	1000 kg m^{-3}
	Compressibility, β_P	NA	5.13 $\times$ 10^{-1} GPa^{-1}	5.13 $\times$ 10^{-1} GPa^{-1}	5.13 $\times$ 10^{-1} GPa^{-1}	5.13 $\times$ 10^{-1} GPa^{-1}
	Dynamic viscosity at 20°C, η	NA	10^{-3} N s m^{-2}	10^{-3} N s m^{-2}	10^{-3} N s m^{-2}	10^{-3} N s m^{-2}
	Thermal expansion coefficient, β_T	NA	3.17 $\times$ 10^{-4} °C^{-1}	3.17 $\times$ 10^{-4} °C^{-1}	3.17 $\times$ 10^{-4} °C^{-1}	3.0 $\times$ 10^{-4} °C^{-1}
	Specific heat, C_{Vw}	NA	1.0 kcal kg^{-1}°C^{-1}	1.0 kcal kg^{-1}°C^{-1}	1.0 kcal kg^{-1}°C^{-1}	1.0 kcal kg^{1}°C^{-1}
Rock	Mass density, ρ_s	NA	2.6 $\times$ 10^{3} kg m^{-3}	2.5 $\times$ 10^{3} kg m^{-3}	2.9 $\times$ 10^{3} kg m^{-3}	2.7 $\times$ 10^{3} kg m^{-3}
	Porosity, ϕ	2.0 $\times$ 10^{-1}	5.0 $\times$ 10^{-2}	2.0 $\times$ 10^{-1}	1.5 $\times$ 10^{-2}	1. $\times$ 10^{-2}
	Permeability, k	4.0 $\times$ 10^{-6} m/s	10^{-18} m^2	2.5 $\times$ 10^{-13} m^2	10^{-20} m^2	10^{-17} m^2
	Young's modulus, E	6 $\times$ 10^{3} Pa	51.3 GPa	50.0 GPa	35 GPa	60 GPa
	Poisson's ratio, v	4.0 $\times$ 10^{-1}	2.3 $\times$ 10^{-1}	2.5 $\times$ 10^{-1}	2.5 $\times$ 10^{-1}	2.5 $\times$ 10^{-1}
	Biot's coefficient, α	1.0	1.0	1.0	1.0	1.0
	Biot's coefficient, M	NA	5.0 GPa	5.0 GPa	1.3 GPa	40.0 GPa
	Thermal expansion coefficient, (linear) β	3.0 $\times$ 10^{-7} °C^{-1}	8.8 $\times$ 10^{-5} °C^{-1}	8.0 $\times$ 10^{-6} °C^{-1}	1.1 $\times$ 10^{-5} °C^{-1}	8.6 $\times$ 10^{-6} °C^{-1}
	Specific heat, C_{Vs}	$(\rho C)_M$ = 167 kJm^{-3} °C^{-1}	2.1 $\times$ 10^{-1} kcal kg^{-1} °C^{-1}	2.1 $\times$ 10^{-1} kcal kg^{-1} °C^{-1}	2.0 $\times$ 10^{-1} kcal kg^{-1} °C^{-1}	1.9 $\times$ 10^{-1} kcal kg^{-1} °C^{-1}
	Lumped thermal conductivity, K_M	8.36 $\times$ 10^{-1} KJ m^{-1} s^{-1} °C^{-1}	3.18 $\times$ 10^{-3} KJ m^{-1} s^{-1} °C^{-1}	7.65 $\times$ 10^{-4} KJ m^{-1} s^{-1} °C^{-1}	6.9 $\times$ 10^{-4} KJ m^{-1} s^{-1} °C^{-1}	8.6 $\times$ 10^{-4} KJ m^{-1} s^{-1} °C^{-1}
Fracture	Porosity, ϕ	NA	1.0	1.0	16.0	1.0
	Initial aperture, $2b$	NA	10^{-4} m	10^{-7} m	17 $\times$ 10^{-6} m	58 $\times$ 10^{-6} m
	Initial normal stiffness, k_n	NA	85 GPa^{-1}	100 GPa^{-1}	1620 GPa^{-1}	2000 GPa^{-1}
	Initial tangential stiffness, k_s	NA	0.085 GPa m^{-1}	10 GPa m^{-1}	3 GPa m^{-1}	1 GPa m^{-1}
	Biot's coefficient, α	NA	1.0	1.0	1.0	1.0
	Biot's coefficient, M	NA	5.0 GPa	1.0 GPa	2.0 GPa	2.0 GPa
	Friction angle, ψ	NA	30°	30°	30°	30°
	Cohesion, C	NA	0.0	0.0	0.0	0.0
	Boundary distance, Re	NA	NA	3.700 m	NA	10 m
	Initial normal stress, σ_n	NA	NA	67 MPa	NA	13.5 MPa

Stripa granite. The delayed responses of the extensometers in these experiments (Witherspoon, *et al.*, 1981a) may be explained by the closing of the fracture. Similarly, reduction leading to stoppage of the water inflows into the heater boreholes (Nelson, *et al.*, 1981) can also be explained by the same phenomenon.

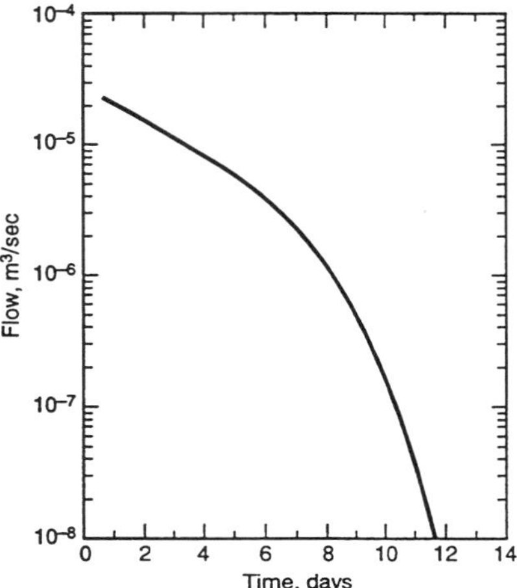

Fig. 3. Variation of fluid inflow to the heater borehole as a function of time.

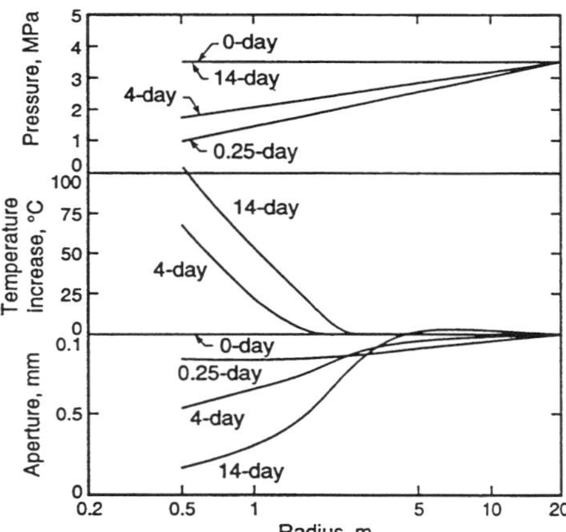

Fig. 4. Pressure and aperture profile in the fracture for various durations and temperature profiles along the heater midplane.

8.3 Example 3: Thermohydraulic fracturing

This work also provides a basis for understanding the role of thermal stresses in the thermohydroelastic phenomenon through inspection of the stress-strain relationships. As shown in the formulation of the THM phenomenon, presentation of the *temperature* term is similar to the *pressure* term, with Biot's coupling coefficient (Biot, 1941) replaced by $E\beta/(1-v)$, where E and v are elastic moduli and β is the linear thermal expansion coefficient. Solutions of uncoupled thermoelasticity, such as that of thermal stresses in an elastic thick-walled cylinder (Nowacki, 1962), provide a good insight.

The variation in the tangential stresses ($\delta_{\theta\theta}$) at the inner cylinder boundary, caused by a change in temperature ΔT, is given as

$$\Delta\sigma_{\theta\theta}\big|_{r=a} = -\frac{\beta E \Delta T}{1-v} \, , \tag{56}$$

where a is the inner cylinder radius and tension is assumed positive. A change in temperature of about 10°C can create stress variations from 1 to 10 MPa, depending on the magnitude of the elastic moduli used in the calculation. It is obvious that such stresses could exceed the tensile strength of rocks in certain cases. To investigate the role of thermal stresses in circumstances where transport of energy is enhanced by fluid flow, as well as in conjunction with the mechanical aspect of the flow of fluids, numerical techniques such as the code ROCMAS must be used.

The numerical simulation in this work is based on semiquantitative data on cold-water flooding experiments provided by an oil company. In these experiments, it was noticed that hydrofracing and/or reopening of existing fractures in the warm reservoir consistently took place at pressure gradients that were 1.5×10^{-3} MPa/m less than the expected values. For a reservoir at a depth of about 3000 m, the above reduction in gradient implies a shut-in pressure reduction of about 5 MPa. Equation (56) shows that this corresponds to a 10°C average cooling of the rocks near the well for a hard host rock. We constructed a hypothetical two-dimensional (x-y) model of the reservoir to study this problem. Table 1 provides the data used in this model.

Figure 5 shows a sketch of the geometry and the initial and boundary data. As can be seen, a fracture spans the geometry of the model. In the field experiments, the wells are pumped at constant rates until well pressure stabilizes; the rate is then increased by a constant amount, and the procedure continues for a period of a day or more, during which one or two hydrofracing episodes are observed. A realistic simulation of the experiment is not possible; instead, we seek a crude approximation. We do this by obtaining a thermohydroelastic response of the model through a series of steady-state hydromechanical calculations that use the results of a coupled transient thermal analysis. As explained earlier in the text, the approximation is justified because of the large difference between the time constants for fluid flow and heat flow. The fracture in the model is regarded to be closed initially (with aperture 10^{-7} m). Pressurization of the reservoir opens the fracture elastically against the sustained compressive stresses. This increase in the aperture allows further penetration of the pressure front until the fracture goes into a tension state and hydrofracing takes place. In the simulations, the occurrence of hydrofracing is marked by instability of the system in the solution. The presence of thermal stresses accelerates this phenomenon. Figure 6 shows the results of an isothermal (HM) calculation and a non-isothermal (THM) calculation of the model. As can be observed in the figure, the system becomes unstable at an injection pressure close to 10 MPa less than that of isothermal injection calculations.

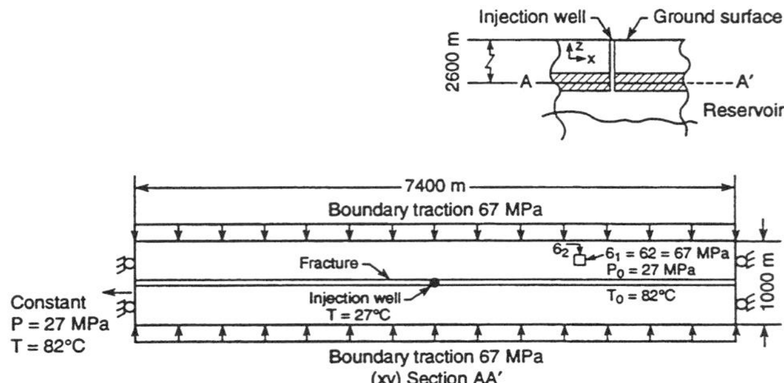

Fig. 5. Schematic geometry of the model and initial and boundary data.

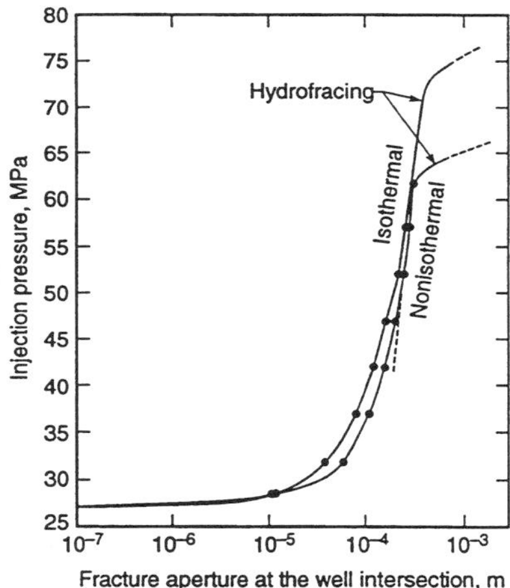

Fig. 6. Variation of fracture aperture in response to pressurization.

8.4 Example 4: The post-closure far-field effects in a HLNW repository

This attempt evolves from three main phases, namely, conceptualization, simulation strategy, and modeling, that we explain respectively in the following.

A. *Conceptualization of an HLNW Repository Setting*

A search of the literature concerning the gross characteristics of proposed high-level nuclear waste (HLNW) repositories revealed the general criteria that form the basis of the conceptualization used in this study. The repository is assumed to be located in granitic rocks at a depth of 600 m and has a length and width of 3 km. Our analysis will be performed on the two-dimensional vertical central section of the rock-repository system, which possibly emulates a worst-case scenario. The modeled section has a vertical dimension of 2.2 km and a horizontal dimension of 18 km. Figure 7 shows the cross-section with the vertical scale exaggerated by 5 times, which will be used here.

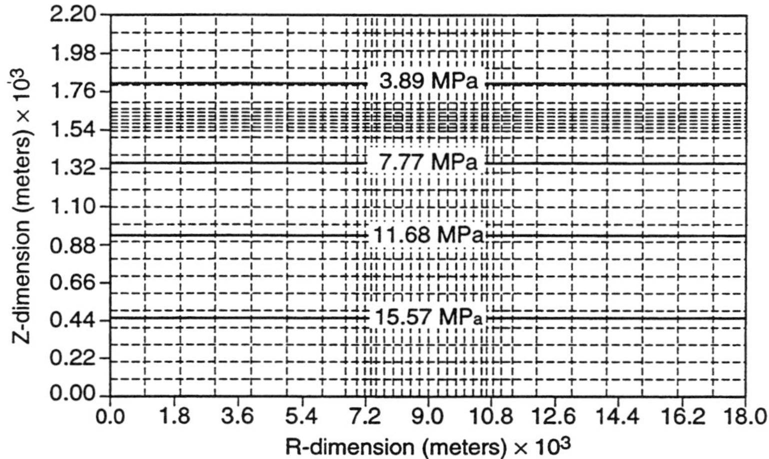

Fig. 7. Model sketch with vertical scale exaggerated five times.

We placed two major fractures in this vertical section: a horizontal fracture that intercepts the repository in its plane and a vertical fracture that cuts through the repository along its plane of symmetry. Both fractures extend to the model boundaries. A regional hydraulic gradient of 0.1 percent and a geothermal gradient of 30°C/km were assumed to prevail in the model. Material properties assumed are shown in Table 1.

The repository is represented by a flat plane coinciding with the horizontal fracture and possesses a decaying thermal power output. The power decay curve, as shown in Figure 8, is adapted from Wang and Tsang (1980) with a 10-year cooled-down power output of 14.4 W/m^2 gross thermal loading for the spent fuel. This selected power is chosen in some level of conformity with one of the waste-repository design considerations of the Atomic Energy of Canada (Acres, 1978). The modified Raleigh number for a laterally infinite system with characteristics similar to the repository model indicates possibilities of development of a natural thermal circulation pattern in the absence of a regional hydrologic gradient. In the laterally bounded system of the model, calculations with the initial pre-repository conditions indicated formation of two major circulation cells that bound two small central circulation patterns. Sensitivity studies of the model proved this thermal circulation phenomenon to be an artifact of the initial data setup of the model. Imposition of a small regional hydrologic gradient wiped out the thermal cells for two different spatial discretization schemes.

B. Simulation Strategy

Simulation times of interest for performance assessment of a HLNW repository range from orders of one to thousands of years. This fact immediately brings out the difficulty of obtaining a time-marched solution of the THM processes occurring in the host geologic system, because of the large discrepancy in the time constants of the heat transport and the

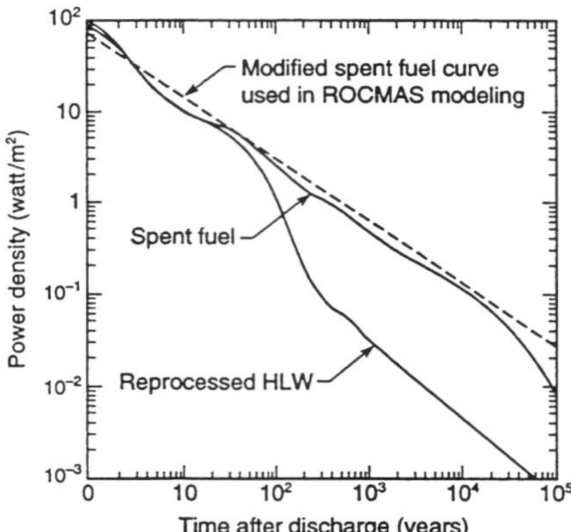

Fig. 8. Areal power density of the generated heat. Heavy lines are reproduced from Wang and Tsang (1980), and the dashed line reresents the criterion used in our modeling.

mass-flow phenomena. The flow processes take place rather quickly compared to the slowly changing temperature field. Following an approach used by Runchal (1980) to investigate the role of convection in the evolution of temperature field around a repository, we adapted the snapshot approach in this example. Within this strategy, we follow the evolution of the quasi-static thermomechanical processes by a time-marched simulation of the THM phenomena, which uses a steady-state solution of the coupled flow-deformation equations at each time increment. This approach not only provides a reasonably correct picture of the thermohydromechanical state of the problem but also significantly reduces the computation time.

C. Modeling Studies

In a first approximation to the repository behavior, we assume linear properties for the host medium. This assumption implies that we also ignore the two discontinuity features of the model. The results of this simpler model provide a basis for comparison with the results of the nonlinear model to be analyzed later. Also, the linearity assumption is a common feature of the early thermomechanical and hydrothermal studies (e.g., Acres, 1978; Runchal, 1980). In our first run of the model, with the linearity assumption, we decouple the hydraulic processes from the mechanical events. However, hydrothermal processes and thermomechanical processes remain intact. Thus convective and conductive processes, which control temperature distribution, also control differential thermal stresses in the rock. The simulation time is 100,000 years, although, because of space shortage, we only discuss the results at 4,000 years. Figures 9–11 show evolution of temperature, flow directions, and principal stresses for $t = 4,000$ years. Two important points emerge from this part of the analysis. The first is that among the processed times, the 4,000-year results correspond to the time period when the changes from the initial state seem to reach a maximum. The second finding is the development of a tension zone at a ground surface location on the top of the repository. These points confirm similar findings of other investigators (Acres, 1978; Runchal, 1980) in their assessment studies of HLNW repositories. This fact adds further confidence to the proper conceptualization in these results. We should add that this model, in spite of its chosen simplicity, is reproducing a more complete behavior compared to other scoping repository modeling reported in the literature. The advantage in this modeling so far, with the uncoupled flow and linear continuum, is in its consideration of convective-conductive thermomechanical analysis instead of the generally used conductive thermomechanical ones. Considering the development of circulation patterns around the repository, the convective transport may be an important mode of both energy and mass transport. This depends on the strength of the established velocity field. In this case, comparison of the temperature distribution results of runs with and without consideration of convection effects show no noticeable differences. Considering the low permeability of the rock in this conceptualization, this result does not imply we can ignore the role of convection. Parametric studies in each particular case should establish the significance of convective energy transport. This simulation depicts the far-field performance of a linear isotropic homogenous repository. However, with the basic conceptualization in place, it is possible to increase the complexity of the model, subject only to the capability of the computer.

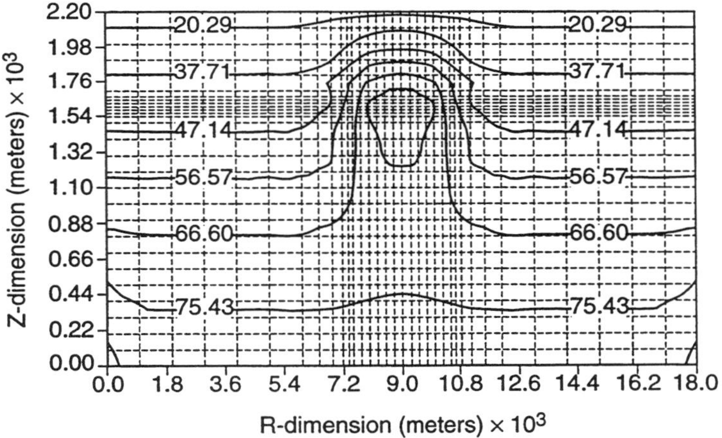

Fig. 9. Temperature contours at 4,000 years.

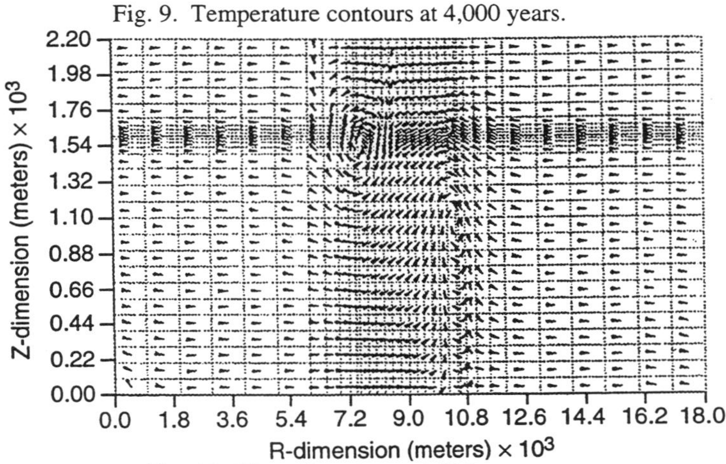

Fig. 10. Flow direction at 4,000 years.

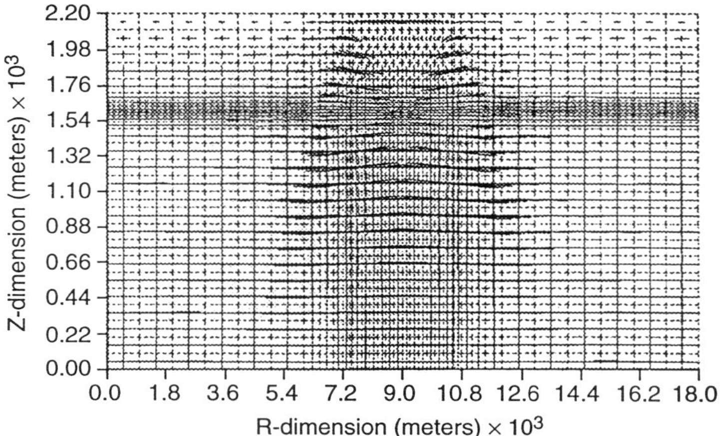

Fig. 11. Principal stresses at 4,000 years.

In a second attempt, the nonlinearity is introduced by the two extensive fractures that span the model. We had deactivated these features in calculations reported in the previous section. Comparison of most of the results of this simulation with that of preceding linear run reveals a noticeable difference. The vertical fracture influences the results as it accommodates part of the tensile stresses. However, the fractures are not effective in facilitating circulation in this case, as one might expect. The main reason is the fact that fractures, though by themselves high-permeability features, are choked by the neighboring very low-permeability rocks.

8.5 Example 5: Effects of placement of a HLNW canister and bentonite overpack

The problem* is three-dimensional (Figure 12), but here it is simplified to a two-dimensional plane strain configuration intersecting the deposition hole at about one "tunnel diameter" below the tunnel floor (Figure 13). The deposition hole is assumed to be located in a tunnel with a series of deposition holes which are spaced 6 m apart along the center line of the tunnel. The symmetry of the problem lends itself to the model geometry and boundary conditions shown in Figure 14.

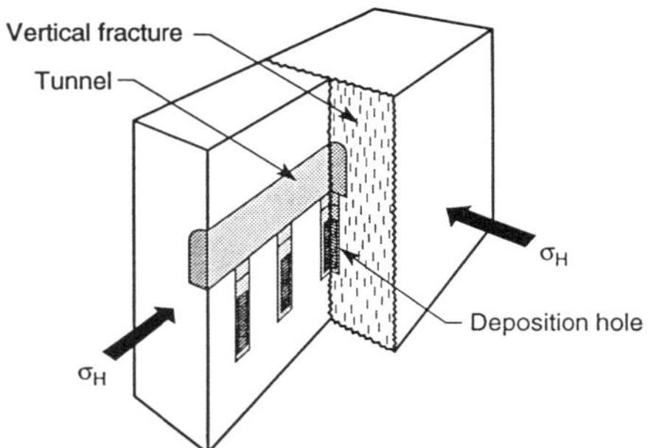

Fig. 12. Studied problem.

---

* For more details please see Rutqvist *et al.* (1992).

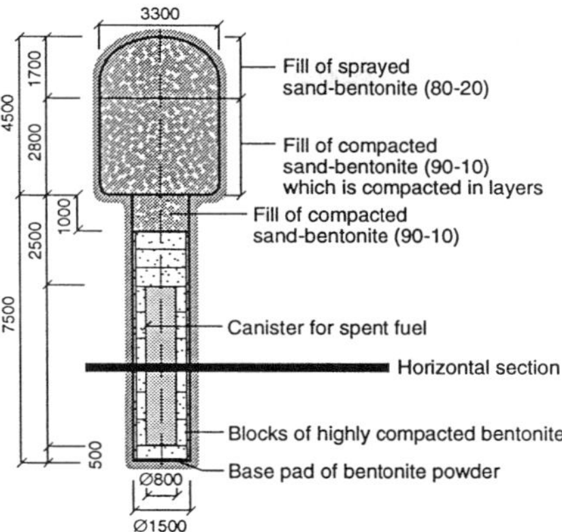

Fig. 13. Modeled horizontal section intersecting a deposition hole.

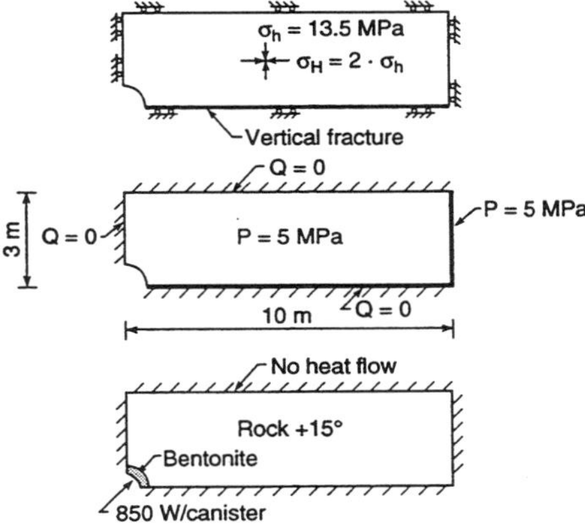

Fig. 14. Horizontal model section of the deposition hole: (a) mechanical boundary conditions, (b) fluid-flow boundary conditions, (c) heat-flow boundary conditions.

An initial hydrostatic fluid pressure of 5 MPa is assumed throughout the rock mass. Initial stresses are σ_h = 13.5 MPa (minor principal horizontal compressive stress) and $\sigma_H = 2\sigma_h$ (major principal horizontal compressive stress). σ_H corresponds to a stress level at a depth of 500 m below ground surface (see Stephansson, *et al.*, 1987), and σ_h equals the weight of the overburden rock. σ_H is assumed to be oriented perpendicular to the tunnel axis. The heat output from the nuclear waste is modeled as a constant heat flow of 850 watts per canister at the canister-bentonite interface. The initial temperature in the rock is assumed to be 15°C.

Material properties for the model, which correspond to crystalline rock with low permeability, are given in Table 1. Thermal conductivity of the bentonite is assumed to be 28% of that of the rock, and heat capacity is about 6% over the rock's heat capacity.

The modeling sequences are:

1) excavation of the deposition hole,
2) emplacement of the bentonite buffer in the deposition hole with no swelling pressure,
3) temperature increase of the rock mass due to heat output from the canister,
4) bentonite buffer swelling pressure applied in the deposition hole.

Excavation of the deposition hole: The deposition hole in the KBS-3 Swedish concept (Stephansson, *et al.*, 1980) is intended to be excavated downwards with full-face drilling equipment from the tunnel floor. This implies that stress and fluid pressure fields around the deposition hole are changing in a slow, continuous way to a final state when the entire hole is excavated. Therefore, no dramatic transient effects, such as instantaneous fracture closing, are expected during this sequence.

The wall of the open hole is a zero pressure boundary, which causes a hydraulic gradient in the surrounding rock and results in a groundwater flow into the hole. The effective normal stresses exceeds the initial value (8.5 MPa) along the entire fracture length due to the concentration of total stresses adjacent to the hole and the decrease in fluid pressure. The joint aperture at the wall face decreases from the initial 58 to 32 microns (Figure 15).

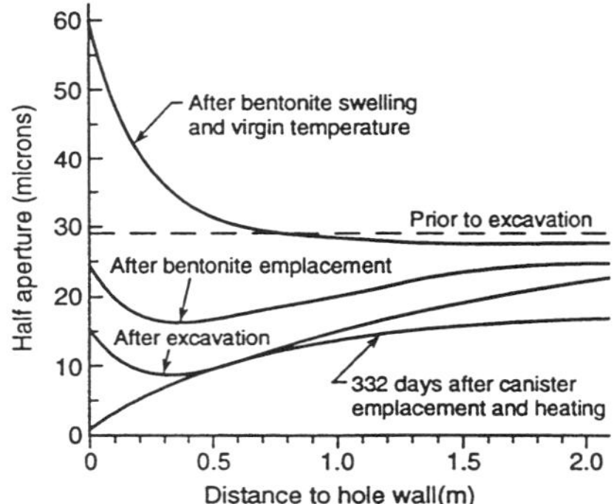

Fig. 15. Aperture profiles along the vertical fracture after each modeling sequence.

Emplacement of the canister and the bentonite buffer in the deposition hole: This sequence represents the situation right after the emplacement of the canister and the bentonite buffer, during which the process of saturation of the bentonite buffer takes place. The inflow from the rock into the compacted bentonite was calculated according to Pusch, *et al.*, (1985). However, in this simulation, the calculated rate of water uptake was so low that the hole-rock boundary can be considered as a zero-flow boundary. For this sequence, no noticeable swelling pressure is mobilized in the bentonite.

The zero-flow boundary condition at the hole wall results in re-establishment of the pre-excavation fluid pressure of 5 MPa in the rock and the fracture. The total principal stresses around the deposition hole become only slightly different compared to the excavation stage, while the effective stresses are changed due to the increased fluid pressure. These give conditions a new aperture profile in which the aperture at the hole wall face has increased to 48 microns, still less than the initial aperture of 58 microns, Figure 15.

Temperature increase in the rock mass: The bentonite starts to swell as it becomes gradually saturated with groundwater. At the same time, heat is flowing from the canister, causing a temperature increase in the surrounding rock mass. This gives rise to large thermal compressive stresses, and the aperture decreases less than one year after deposition of the nuclear waste canister. The swelling of the bentonite gives rise to a swelling pressure applied on the deposition hole wall in the radial direction. This contributes to relief of tangential compressive stresses adjacent to the deposition hole wall. However, the magnitude of the relief of compressive stresses due to the bentonite swelling is less than the stress increase caused by thermal expansion. Furthermore, it takes more than 15 years to attain complete saturation of the bentonite and a fully developed swelling pressure. Therefore, the fracture will be compressed until the temperature in the rock mass declines.

The temperature increase compresses the fracture further, especially adjacent to the deposition hole (Figure 15). At this stage (about 1 year after deposition), no considerable swelling pressure has been developed.

Swelling of the bentonite in the deposition hole: The Stripa buffer mass test (Pusch, *et al.,* 1985), showed that the initial inflow towards the hole through major discontinuities caused local filling of bentonite which sealed them off and directed the flow to finer fractures. Later, the finer fractures also became tight, causing redistribution of the flow to the intact rock matrix. This led to a steady, fairly uniform water uptake in the bentonite, and a relatively uniform swelling pressure developed at the rock/bentonite interface.

In this sequence, a fully developed swelling pressure of 10 MPa (effective stress) is applied to the hole wall; in this case, it is assumed that the rock mass has cooled down to virgin temperature, which takes about 100,000 years after deposition.

The effective stress normal to the fracture decreases to close to zero and the aperture increases to 118 microns, which is more than twice the initial pre-excavation aperture, as depicted in Figure 15.

In summary, the effective normal joint stresses changed in each of the four simulation sequences due to redistribution of total stresses and fluid pressure around the deposition hole. The fracture responded to the change in effective normal stress by aperture change, which took place in the zone lying between the wall of the hole and 5 m into the rock. Temperature increase of the rock mass caused further closure of the fracture. The largest aperture increase is obtained after fully developed swelling pressure and when virgin temperature again prevails. This aperture increase is located in a zone from the hole wall extending to about 0.5 m into the rock.

Concluding Remarks

The macroscopically based, coupled thermohydroelastic formulation and the proposed numerical scheme provide a robust method of analysis of the THM processes in fractured porous rocks under weak non-isothermal conditions. However, realistic applications of the methodology require alternative or proper constitutive modeling in cases where discrete modeling of fractures becomes intractable or when anisotropy and nonlinearity (inherent, induced) becomes the major concern. Questions of inelasticity of the continuum can be handled to a fair level. However, proper representation of fractured rocks or failing rocks in a three-prong process under changing saturation poses the greatest challenge. Also, strong non-isothermal conditions are prevalent enough to call for the development of an appropriate simulation tool for them. The formulation developed for the case described here is the first step towards such a goal.

Footnote

Biot's parameters, α and μ can be expressed in various ways in terms of measureable material parameters. Rice and Cleary (1976) express these parameters (for isotropic conditions) as

$$\alpha = \frac{3(v_u - v)}{B(1 - 2v)(1 + v_u)}$$

$$\mu = \frac{2GB^2(1 - 2v)(1 + v_u)^2}{9(v_u - v)(1 - 2v_u)}$$

where

B	=	Skempton's pore pressure coefficient (ratio of induced pore pressure over the variation of confining pressure under undrained conditions)
G	=	Shear modulus
v	=	drained Poisson ratios
v_u	=	undrained Poisson ratios

Acknowledgment

This work was jointly supported by a grant from the Swedish Power Inspectorate (SKI), Stockholm, Sweden and by the Director, Office of Energy Research, Office of Basic Energy Sciences, U.S. Department of Energy under Contract No. DE-AC03-76SF00098.

9. REFERENCES

Aboustit, B.L., S.H. Advani, J.K. Lee, and R.S. Sandhu, Finite element evaluation of thermo-elastic consolidation, *Proc. U.S. Symp. Rock Mech., 23rd,* 587–595, 1982.

Acres Consulting Company, Ltd., Radioactive Waste Repository Study—Technical memorandum 3-3, 4-1, AECL-6451. A disposal center for immobilized fuel, 1980.

Aktan, T., and S.M. Farouq Ali, Finite element analysis of temperature and thermal stresses induced by hot water injection, *Soc. Pet. Eng. J.,* 457–469, 1978.

Alonso, E.E., F. Batle, A. Gens, and A. Lloret, Consolidation analysis of partially saturated soils—application to earthdam construction, *Numerical Methods in Geomechanics,* G. Swoboda, ed., A.A. Balkema, Rotterdam, Netherlands, 1988.

Baca, R.G., Coupled geomechanical/hydrological modeling: An overview of BWIP studies, *Proceedings of the ONWI Workshop on Thermomechanical-Hydrochemical Modeling for a Hard Rock Waste Repository, Rep. LBL-11204,* Lawrence Berkeley Lab., Berkeley, CA, 1980.

Barton, N., and Bandis, S., Effects of block size on the shear behavior of jointed rock, Keynote Lecture, 23rd U.S. Symposium on Rock Mechanics, Berkeley, California.

Bear, J., and M.Y. Corapcioglu, A mathematical model for consolidation in a thermoelastic aquifer due to hot water injection or pumping, *Water Resources Research, 17(3),* 723–736, 1981.

Biot, M.A., and D.C. Willis, The elastic coefficients of the theory of consolidation, Paper No. 57-APM-44, presented at *The Applied Mechanic Division Summer Conf.,* Berkeley, CA, 1957.

Biot, M.A., General theory of three-dimensional consolidation, *J. Appl. Phys., 12,* 144–164, 1941.

Biot, M.A., Thermoelasticity and irreversible thermodynamics, *J. Appl. Phys., 27,* 240–253, 1956.

Bishop, A.W., The principle of effective stress, *Tek. Ukebl., 106(39),* 155–164, 1959.

Blight, G.E., A study of effective stress for volume change, *Symposium on Moisture Equilibrium and Moisture Changes in Soils Beneath Covered Areas,* Butterworths, Australia, 259–269, 1965.

Bodvarsson, G.S., K. Pruess, and M.J. Lippmann, Modeling of geothermal systems, *J. Petrol. Technol., 38,* 1007–1021, 1986.

Booker, J.R., and J. Small, Finite element analysis of primary and secondary consolidation, *Int. J. Solids Struct. (13),* 137–149, 1977.

Booker, J.R., Time-dependent strain following faulting of a porous medium, *J. Geophys. Res., 79,* 2037–2044, 1974.

Börgesson, L., and Hernelind, J., Coupled modelling of the thermal, mechanical, and hydraulic behaviour of water-unsaturated buffer material in a simulated deposition hole, DECOVALEX Report.

Cai, M., and H. Horii, A constitutive model and FEM analysis of jointed rock masses, *Int. J. Rock Mech., Min. Sci & Geomech. Abstr., 30, No. 4,* 351–359, 1993.

Cameron, J.T., A numerical model for coupled heat and moisture flow through porous media, Ph.D. thesis, Univ. of Calif., Berkeley, CA, 1986.

Carter, J.P., and J.R. Booker, Geomechanical applications of fully coupled, transient thermoelasticity, *Numerical Methods in Geomechanics,* G. Swoboda, ed., A.A. Balkema, Rotterdam, Netherlands, 1988.

Carter, J.P., Finite deformation theory and its application to elastoplastic soil, Ph.D. thesis, Univ. of Sydney, Sydney, 1977.

Carter, J.P., J.R. Booker, and J.C. Small, The analysis of finite elasto-plastic consolidation, *Int. J. for Numer. and Analy. Meth. in Geomech.,* 3, 107–129, 1979.

Chang, C.S., and J.M. Duncan, Consolidation analysis for partly saturated clay by using an elastic-plastic effective stress-strain model. *Int. J. for Numer. and Analy. Meth. in Geomech,* 7, 39–55, 1983.

Cheng, A.H-D., and J.A. Liggett, Boundary integral equation method for linear poro-elasticity with applications to fracture propagation, *Int. J. Numer. Meth. Eng.,* 20, 279–296, 1984.

Childs, E.C., The ultimate moisture profile during infiltration in a uniform soil, *Soil Sci.,* 97(3), 173–178, 1964.

Coleman, J.D. Stress/strain relations for partly saturated soil, *Geotechnique, 12,* 348–350, 1962.

Crochet, M.J., and P.M. Naghdi, On constitutive equations for flow of fluid through an elastic solid, *Int. J. Eng. Sci., 4,* 383–401, 1966.

Davis, S.N., and G.W. Moore, Semidiurnal movement along a bedrock joint in Wool Hollow Cave, California, *NSS Bull., 27(4),* 133–142, 1965.

Derski, W., and S.J. Kowalski, Equations of linear thermo consolidation, *Arch. Mech., 31,* 303–316, 1979.

Detournay, E., and A.H-D. Cheng, Poroelastic response of a borehole in a non-hydrostatic stress field, *Int. J. Rock Mech., Min. Sci. & Geomech. Abstr., 25, No. 3,* 171–182, 1988.

Detournay, E., and A.H-D. Cheng, Poroelastic solution of a plane strain point displacement discontinuity, *Proceedings of Am. Soc. Mech. Engr.* 1987.

Evans, D.M., The Denver area earthquake and Rocky Mountain Arsenal Well, *Mt. Geol., 3(1),* 23–36, 1966.

Falta, R.W., K. Pruess, I. Javandel, and P.A. Witherspoon, Numerical modeling of steam injection for the removal of nonaqueous phase liquids from the subsurface 1. Numerical formulation, *Water Resources Research, 28(2),* 433–449, 1992.

Fredlund, D. G., and H. Rahardjo, *Soil Mechanics for Unsaturated Soils,* John Wiley& Sons, 1993.

Fujita, T., Y. Moro, K. Hara, A. Kobayashi, and Y. Ohnishi, Full-scale test on coupled thermo-hydro-mechanical process in engineered barrier system, DECOVALEX report.

Gale, J.E., A numerical field and laboratory study of flow in rocks with deformable fractures, Ph.D. thesis, Univ. of Calif., Berkeley, 1975.

Garg, S.K., J.W. Pritchett, M.H. Rice, and T.D. Riney, U.S. Gulf Coast geopressure geothermal reservoir simulation, Final Report, Systems, Science and Software, Rep. SSS-R-77-3147, La Jolla, CA, 1977.

Gerrard, C.M., Joint compliances as a basis for rock mass properties and the design of supports, *Int. J. Rock Mech., Min. Sci. & Geomech. Abstr., 19,* 285–305, 1982.

Ghaboussi, J., and E.L. Wilson, Flow of compressible fluid in porous elastic media, *Int. J. Numer. Meth. Eng., 5,* 419–442, 1973.

Ghaboussi, J., Dynamic stress analysis of porous elastic solids saturated with compressible fluids, Ph.D. thesis, Univ. of Calif., Berkeley, 1971.

Goodman, R.E., R.L. Taylor, and T. Brekke, A model for the mechanics of jointed rock, *J. Soil Mech. Found. Div. Am. Soc. Civ. Eng., 94(SM3)*, 637–659, 1968.

Helm, D.C., Evaluation of stress-dependent aquitard parameters by simulating observed compaction from known stress history, Ph.D. thesis, Univ. of Calif., Berkeley, 1974.

Hocking, G., Thermomechanical modeling for hardrock: Status and needs, *Proceedings of the Workshop on Thermomechanical Modeling for a Hard Rock Waste Repository, Rep. UCAR-10043*, Lawrence Livermore Lab., Livermore, CA, 1979.

Horii, H., and H. Yoshida, constitutive modeling of rock masses containing fracturing joints and analysis of large-scale excavation, *Rock Mechanics*, 681–688, 1994.

Hwang, C.T., N.R. Morgenstern, and D.W. Murray, Application of the finite element method to consolidation, *Proceedings of the Symposium on Finite Element Methods in Geotechnical Engineering*, Vicksburg, MI, 1972.

International Formulation Committee, A formulation of the thermodynamic properties of ordinary water substance, IFC Secretariat, Düsseldorf, Germany, 1967.

Josa, A., A. Balmaceda, A. Gens, and E.E. Alonso, An elastoplastic model for partially saturated soils exhibiting a maximum of collapse, in *Computational Plasticity, Fundamentals and Applications*, D.R.J. Owen, E. Onate, and E. Hinton, eds., 815–826, Pineridge Press, Swansea, Wales, 1992.

Kaneko, K., and T. Shiba, Equivalent volume defect model for estimation of deformation behavior of jointed rock, *Mech. of Jointed and Faulted Rock,* 277–284, 1990.

Kawamoto, T., A modelling of jointed rock mass, *Numerical Methods in Geomechanics*, 107–120, G. Swoboda, ed., A.A. Balkema, Rotterdam, Netherlands, 1988.

Kyoya, T., Y. Ichikawa, and T.Kawamoto, A damage mechanics theory for discontinuous rock mass, *5th Int. Conf. on Numer. Methods in Geomech.*, Nagoya Univ., Nagoya, Japan, 469–480, 1985.

Lewis, R.W., and N. Karahanoglu, Simulation of subsidence in geothermal reservoirs, in R.W. Lewis, K. Morgan, and B.A. Schrefler, eds., *Numerical Methods in Thermal Problems, Vol. III*, 326–335, Pineridge Press, Swansea, 1981.

Lewis, R.W., C.E. Majorana, and B.A. Schrefler, A coupled finite element model for the consolidation of nonisothermal elastoplastic porous media, *Transport in Porous Media 1*, 155–178, 1986.

Lloret, A., and E.E. Alonso, Consolidation of unsaturated soils including swelling and collapse behavior, *Geotechnique, 30(4)*, 449–477, 1980.

Long, J.C.S., Investigation of equivalent porous medium permeability in networks of discontinuous fractures, Ph.D. thesis, Univ. of Calif., Berkeley, 1983.

Louis, C., and Y.N.T. Maini, Determination of *in-situ* hydraulic parameters in jointed rock, in *Proceedings Second Congress International Society for Rock Mechanics*, 1–32, Privredni Pregled, Belgrade, 1970.

Matyas, E.L., and H.S. Radhakrishna, Volume change characteristics of partially saturated soils, *Geotechnique, 18(4)*, 432–448, 1968.

McNamee, J., and R.E. Gibson, Displacement functions and linear transforms applied to diffusion through porous elastic media, *Quart. J. Mech. Appl. Math., 13*, 99–111, 1960.

Mercer, J.W., Finite-element approach to the modeling of hydrothermal systems, Ph.D. thesis, Univ. of Ill., Urbana-Champaign, 1973.

Milly, P.C.D., Moisture and heat transport in hysteretic, inhomogeneous porous media: a matric head-based formulation and a numerical model, *Water Resources Research, 18(3)*, 489–498, 1982.

Narasimhan, T.N., and P.A. Witherspoon, Numerical model for saturated-unsaturated flow in deformable porous media, 1, Theory, *Water Resources Research, 13(3)*, 657–664, 1977.

Nelson, P.H., R. Rachiele, and J.S. Remer, Water inflow boreholes during the Stripa heater experiments, *Rep. LBL-12574*, Lawrence Berkeley Lab., Berkeley, CA 1981.

Noorishad, J., C-F. Tsang, and P.A. Witherspoon, Coupled thermal-hydraulic-mechanical phenomena in saturated fractured porous rocks: numerical approach, *J. Geophys. Res., 89*, 10365–10373, 1984.

Noorishad, J., C.F. Tsang, P. Perrochet, and A. Musy, A perspective on the numerical solution of convection-dominated transport problems: a price to pay for the easy way out, *Water Resources Research, 28(2)*, 551–561, 1992.

Noorishad, J., Finite-element analysis of rock mass behavior under coupled action of body forces, flow forces, and external loads, Ph.D. thesis, Univ. of Calif., Berkeley, 1971.

Noorishad, J., M.S. Ayatollahi, and P.A. Witherspoon, A finite element method for coupled stress and fluid flow analysis of fractured rocks, *Int. J. Rock Mech., Min. Sci., & Geomech. Abstr., 19*, 185–193, 1982a.

Nowacki, W., Thermoelasticity, Pergamon, New York, 1962.

O'Sullivan, M.J., Geothermal reservoir simulation, *Energy Research, 9*, 313–332, 1985.

Oda, M., Fabric tensor for discontinuous geological materials, *Souls and Found., Vol. 22(4)*, 96–108, 1982.

Oda, M., Permeability tensor for discontinuous rock masses, *Geotechnique, 35(4)*, 483–495, 1985.

Pariseau, W.G., Equivalent properties of a jointed Biot material, *Int. J. Rock Mech., Min. Sci. & Goemech. Abstr., 30(7)*, 1151–1157, 1993.

Philip, J.R., and D.A. deVries, Moisture movement in porous materials under temperature gradients, Eos Trans., *AGU, 38(2)*, 222–232, 1957.

Polivka, R.M., and Wilson, E.L., Finite element analysis of nonlinear heat transfer, *UC SESM Report* 76-2, 1976.

Pritchett, J.W., S.K. Garg, D.H. Brownell, Jr., and H.B. Levine, Geohydrological environment effects of geothermal power production phase I, *Rep. SSS-R-75-2733*, Syst. Sci. and Software, La Jolla, CA, 1975.

Pruess, K., Development of the general purpose simulator MULKOM, Annual Report 1982, Earth Sciences Division, Lawrence Berkeley Lab., *Rep. LBL-15500*, 1983b.

Pruess, K., Modeling of geothermal reservoirs: fundamental processes, computer simulations, and field applications, *Geothermics, 19(1)*, 3–15, 1990a.

Pruess, K., TOUGH user's guide, Nuclear Regulatory Commission, report *NUREG/CR-4645* (also Lawrence Berkeley Lab., *Rep. LBL-20700*, Berkeley, CA, 1987).

Pusch, R. Final report of the buffer mass test, *Stripa Project, Technical Report, 85-14*, 1985.

Rice, J.R., and M.P. Cleary, Some basic stress-diffusion solutions for fluid saturated elastic porous media with compressible constituents, *Rev. Geophys. Space Phys. 14*, 227–241, 1976.

Rudnicki, J.W., Fluid mass sources and point forces in linear elastic diffusive solids, *Mechanics of Materials*, 1987.

Rudnicki, J.W., On "Fundamental solutions for a fluid-saturated porous solid," by M.P. Cleary, *Int. J. Solids Struct., 17*, 855–857, 1981.

Runchal, A., and T. Maini, The impact of a high-level nuclear waste repository on the regional groundwater flow, *Int. J. Rock Mech. Min. Sci. & Geomech. Abstr., 17*, 253–264, 1980.

Runesson, K., On nonlinear consolidation of soft clay, Ph.D. thesis, Chalmers Univ. of Techn., Dept. of Struct. Mech., *Pub. 78:1*, Göteborg, 1978.

Rutqvist, J., J. Noorishad, O. Stephansson, and C.F. Tsang, Modelling of hydro-thermo-mechanical response of rock mass around a nuclear waste deposition hole, Swedish Nuclear Power Inspectorale, *SKI-TR Report*, 1992.

Safai, N.M., and G.F. Pinder, Vertical and horizontal land deformation in a desaturating porous medium, *Adv. Water Resour., 2*, 19–25, 1979.

Sandhu, R.S., and E.L. Wilson, Finite-element analysis of seepage in elastic media, *J. Eng. Mech. Div. Am. Soc. Civ. Eng., 95(EM3)*, 641–652, 1969.

Schrefler, B.A., and L. Simoni, A unified approach to the analysis of saturated-unsaturated elastoplastic porous media, in *Numerical Methods in Geomechanics*, G. Swoboda, ed., 205–212, A.A. Balkema, Rotterdam, Netherlands, 1988.

Schrefler, B.A., and Z. Xiaoyong, A fully coupled model for water flow and airflow in deformable porous media, *Water Resources Research, 29*, 155–167, 1993.

Small, J.C., J.R. Booker, and E.H. Davis, Elasto-plastic consolidation of soils, *Int. J. Sol. Struct., 12*, 431–448, 1976.

Snow, D.T., Anisotropic permeability of fractured media, *Water Resources Research, 5(6)*, 1273–1289, 1969.

Snow, D.T., Fracture deformation and changes of permeability and storage upon changes of fluid pressure, *Q. Colo. Sch. Mines, 63(1)*, 201, 1968.

Stephansson, O., R. Blomquist, T. Groth, P. Jonasson, and T. Tarandi, Modelling of temperature fields and deformations for radioactive waste repositories in hard rock, *Proceedings of a symposium on waste repositories in hard rock*, 1980.

Stephensson, O., A. Myrvang, L.-O. Dahlström, O.K. Fjeld, K. Bergström, T.H. Hanssen, P. Särkkä, and A. Väätäinen, Fennoscandian rock stress data base - FRSDB, *Research Report TULEA* 1987:06, Division of Rock Mechanics, Luleá University of Technology, Sweden, 1987.

Sun, Z., C. Gerrard, and O. Stephansson, Rock joint compliance tests for compression and shear loads, *Int. J. Rock Mech. Min. Sci. & Geomech. Abstr., 22(4)*, 197–213, 1985.

Swoboda, G., M. Stumvoll, H. Beichuan, Damage tensor theory and its application to tunnelling, *Mech. of Jointed and Faulted Rock*, 1990.

Terzaghi, K., *Erdbaumechanik auf bodenphysikalischer Grundlage*, F. Deuticke, Leipzig, 1925.

Theis, C.V., The significance and nature of the cone of depression in groundwater bodies, *Econ. Geol., 33*, 889–902, 1938.

Van Poollen, H.K., Report of pumping tests, Rocky Mountain Arsenal disposal well, September-October 1968, U.S. Army Corps of Eng., Omaha, NB, 1969.

Wang, J.S.Y. and C.F. Tsang, Buoyancy flow in fractures intersecting a nuclear waste repository, in F.A. Kulacki and R.W. Lyczkowski, eds., *Heat transfer in nuclear waste disposal:* Winter Annual Meeting of ASME, Chicago, IL, HTD, v. 11, 105–112, *(LBL-11112)*, 1980.

Wang, J.S.Y., R. Sterbentz, and C.F. Tsang, The state of the art of numerical modeling of thermohydrologic flow in fractured rock masses, *Rep. LBL-10524,* Lawrence Berkeley Lab., Berkeley, CA, 1980.

Wijesinghe, M.A.A., A similarity solution for coupled deformation and fluid flow in discrete fractures, *Second Int. Radioactive Waste Management Conf.*, Winipeg, Canada 1986.

Wilson, C.R. and P.A. Witherspoon, An investigation of laminar flow in fractured rocks, *Geotech. Rep. 70-6*, Univ. of Calif. Berkeley, 1970.

Witherspoon, P.A., N.G.W. Cook, and J.E. Gale, Geologic storage of radioactive waste, field studies in Sweden, *Science, 211*, 894–900, 1981a.

Witherspoon, P.A., J. Long, Y. Tsang, J. Noorishad, New approaches to the problem of fluid flow in fractured rock masses, *Proc. U.S. Congr. Rock Mech., 22nd*, 1–20, 1981b.

Yurtzine, T., M. Panet, and A. Genut, Analysis of tunnels in strain-softening grounds, p. 635–644, *4th Int. Conf. on Numer. Methods in Geomech.*, Edmonton, Canada, 1982.

Zienkiewicz, O.C., *The Finite-Element Method in Structural and Continuum Mechanics,* McGraw-Hill, New York, 1976.

Zienkiewicz, O.C., and T. Shiomi, Dynamic behavior of saturated porous media; the generalized Biot formulation and its numerical solution, *Int. J. for Numer. and Analy. Meth. in Geomech., 8*, 71–96 (1984).

O. Stephansson, L. Jing and C.-F. Tsang (Editors)
Coupled Thermo-Hydro-Mechanical Processes of Fractured Media
Developments in Geotechnical Engineering, vol. 79
© 1996 Elsevier Science B.V. All rights reserved.

CONTINUUM REPRESENTATION OF COUPLED HYDROMECHANIC PROCESSES OF FRACTURED MEDIA: HOMOGENISATION AND PARAMETER IDENTIFICATION.

A. Stietel[a], A. Millard[a], E Treille[a], E. Vuillod[b], A. Thoraval[b], R. Ababou[c]

[a] CEA/DMT/SEMT, CEN Saclay,BP N°2
91191 Gif/Yvette Cedex FRANCE

[b] INERIS, Ecole des Mines, Parc de Saurupt
54042 Nancy Cedex FRANCE

[c] IMFT, Allée du Professeur Camille Soulas
31400 Toulouse FRANCE

Abstract

To provide equivalent properties for a rock mass which contains a large number of discontinuities, the discontinuous medium is treated as an anisotropic elastic porous medium. The equivalent properties are formulated on the assumptions that all discontinuities can be replaced by a set of parallel planar plates connected by two springs. The validity of the assumptions is studied, and the equivalent properties are compared with the actual properties calculated using a discrete approach.

1. INTRODUCTION

Geological discontinuities such as faults and fractures are of widespread occurrence in crystalline rock masses. For the purpose of numerical analysis, "joint elements" have been developed, but when the number of discontinuities is too large, it becomes impossible to take them all into account. Thus, homogenisation techniques must be applied to take into account all discontinuities.

The problem is then to determine the properties of the equivalent continuous medium, consisting of an ensemble of intact rock and discontinuities.

The description of discontinuities may be explicit (each discontinuity is known by its position and its geometric properties - length and aperture) or statistical (each family of discontinuities is known by its probability distribution, its probability of geometric properties and its probability of orientation). In both cases, identical assumptions have been made for the determination of the equivalent properties, with formulae expressed as discrete summations in the first case, and as integrals in the second case.

Since the authors have worked with bench mark test 3 - BMT3- formulae in terms of discrete summations are presented. This work is based on the work by Oda (1986).

In a first part, the hypotheses and formulation concerning each property (hydraulics, mechanics,...) are presented. In a second part, the validity of the assumptions made is evaluated by comparing the homogenised properties with the "actual" properties, derived from a calculation made with the discrete element code UDEC. Moreover, when the discontinuity behaviour is non-linear, the only way to obtain equivalent properties is to use "numerical experiments".

Furthermore both the homogenisation and discrete approaches can be used simultaneously: by determining equivalent continuous medium with a few joint-elements for the most significant discontinuities and by taking into account discontinuities in regions where the gradients are important.

2. DETERMINATION OF EQUIVALENT PROPERTIES

Here are the main assumptions concerning the intact rock and the discontinuities :
   -The intact rock is elastic and impermeable.
   -Only two dimensions are considered (thus the discontinuities are planar, and extend to
            infinity in the third dimension), under plane strains condition.
   -The mechanical behaviour of a discontinuity is given by :

$$\delta a_n = \frac{1}{k_n}\sigma_n$$
$$\delta a_t = \frac{1}{k_s}\sigma_t \tag{1}$$

where δa_n : normal aperture variation of discontinuity (m),
$\quad\quad\delta a_\tau$: tangential displacement of discontinuity (m),

$\quad\quad k_n$: normal stiffness of discontinuity (Pa/m),
$\quad\quad k_s$: shear stiffness of discontinuity (Pa/m),
$\quad\quad\sigma_n$: normal stress applied to discontinuity (Pa),
$\quad\quad\sigma_\tau$: shear stress applied to discontinuity (Pa).

The hydraulic behaviour of a discontinuity follows Poiseuille's law :

$$\bar{V} = a^2 \frac{\rho_{fl}\, g}{12\,\mu}\bar{J} \tag{2}$$

where V : mean velocity of flow,
$\quad\quad a$: aperture of discontinuity,
$\quad\quad\rho_{fl}$: density of fluid,
$\quad\quad\mu$: viscosity of fluid,
$\quad\quad J$: hydraulic gradient.

Furthermore, the mechanical aperture and the hydraulical aperture are assumed identical.

The behaviour of the porous equivalent continuous medium follows Biot's and Darcy's equations (constitutive laws) and conservation equations (momentum and mass).

The Biot's equations are:

$$\begin{cases} \sigma_{ij} = T_{ijkl}\,\varepsilon_{kl} - B_{ij}\,p \\ p = -G\left(B_{kl}\,\varepsilon_{kl} - \xi\right) \end{cases} \tag{3}$$

The Darcy's equation is:

$$\begin{cases} q_i = -\frac{g}{\mu}k_{ij}\left(\nabla p + \rho_{fl}\, g\frac{\partial z}{\partial x_i}\right)\end{cases} \tag{4}$$

The momentum equation is:

$$\frac{\partial \sigma_{ij}}{\partial x_i} + F_j = 0 \tag{5}$$

The mas conservation is:

$$\frac{\partial \xi}{\partial t} = - \nabla(q) \tag{6}$$

where σ_{ij} : stress tensor,
  ε_{kl} : strain tensor,
  T_{ijkl} : elasticity tensor,
  B_{ij} : Biot's tensor,
  p : pressure,
  G : Biot's coefficient,
  ξ : fluid production,
  q : fluid flow,
  g : gravitational acceleration,
  μ : fluid viscosity,
  k_{ij} : tensor of permeability,
  ρ_{fl} : density of fluid,
  z : depth below ground,
  F : body volumetric force.

In the equivalent continuous medium the properties can vary spatially.

2.1. Hydraulic equivalent properties

Snow (1969) was the first to propose a relation for the equivalent conductivity of fractured rock mass consisting of an impermeable matrix traversed by planar discontinuities. Since then, several authors have developed or applied similar formulas (discontinuity network, percolation threshold, etc.).

a) Porosity:

No particular assumptions are made for the determination of equivalent porosity. A discontinuity of length l_f and aperture a_f included in a domain of area A has a porosity ϕ_f, given by:

$$\phi_f = \frac{l_f\, a_f}{A} \tag{7}$$

For n discontinuities one has :

$$\phi = \frac{1}{A} \sum_{f=1}^{n} l_f a_f \tag{8}$$

In the case of BMT3 all the discontinuities are connected.

b) Permeability:

For a continuous medium the tensor of permeability relates the fluid flow velocity and the hydraulic gradient as follow:

$$v_i = \frac{\rho_{fl}\, g}{\mu} . k_{ij}\, \frac{\partial H}{\partial x_j}$$

or

$$v_i = \frac{\rho_{fl}\, g}{\mu} . k_{ij}\, J_j \tag{9}$$

where: v_i : i^{th} component of fluid velocity,

ρ_{fl} : density of fluid ,

μ : viscosity of fluid,

k_{ij} :   tensor of permeability of intact rock,

H : hydraulic head,

$J_j = \dfrac{\partial H}{\partial x_j}$: j^{th} component of hydraulic gradient (applied over the boundary of the

domain).

The number of discontinuities is so large that the hydraulic gradient may be considered uniformly distributed over the entire flow region. Thus the flows in each discontinuity are superimposed, giving by (see Figures 1 and 2) :

$$v_i^f = \frac{\rho_{fl} g}{12\mu} \cdot a_f^2 J_i^f \tag{10}$$

where v_i^f : i^{th} component of mean velocity in discontinuity f,

a_f : aperture of discontinuity f,

j_i^f : projection of the hydraulic gradient $\bar{j}$ over discontinuity f,

$$J_i^f = \left(\delta_{ij} - \cos\theta_i^f \cos\theta_j^f\right) J_j \tag{11}$$

The mean velocity over the domain is equal to the velocity in the equivalent continuous medium :

$$\bar{v}_i^f = \frac{1}{A}\int_A v_i^f \, dA$$

$$\bar{v}_i^f = \frac{1}{A} v_i^f \, l_f \, a_f \tag{12}$$

$$\bar{v}_i^f = \frac{\rho_{fl}\, g}{12\,\mu} \frac{a_f^3 \, l_f}{A} \left(\delta_{ij} - \cos\theta_i^f \cos\theta_j^f\right) J_j$$

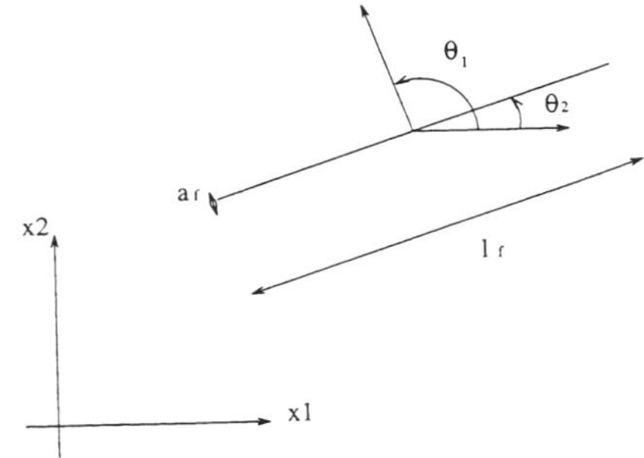

Figure 1 . Orientation of a discontinuity

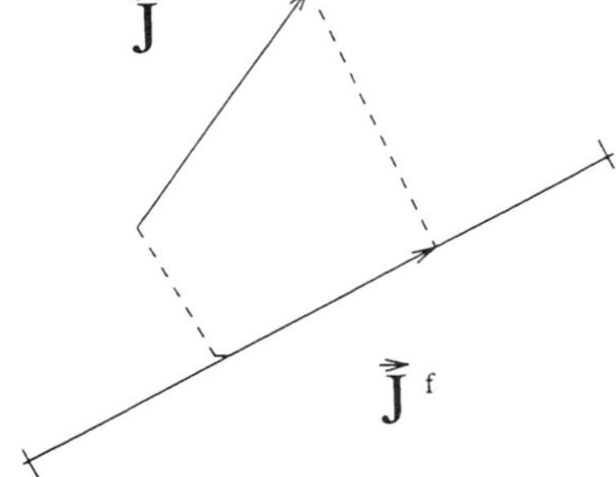

Figure 2 . Hydraulic gradient along a discontinuity

Thus the tensor of permeability is given by:

$$k_{ij}^f = \frac{\rho_{fl} g}{12\,\mu} \frac{a_f^3 \, l_f}{A} \left(\delta_{ij} - \cos\theta_i^f \, \cos\theta_j^f\right) \tag{13}$$

$$k_{ij} = \frac{1}{A} \frac{\rho_{fl} \, g}{12\,\mu} \sum a_f^3 \, l_f \left(\delta_{ij} - \cos\theta_i^f \, \cos\theta_j^f\right) \tag{14}$$

Using S_f for discontinuity spacing or density of discontinuity in the domain, ϕ_f for the porosity and σ_f for specific area this can be rewritten as:

$$\begin{aligned}
S_f &= \frac{l_f}{A} \\[4pt]
\sigma_f &= \frac{2(l_f + a_f)}{A} \cong \frac{2\,l_f}{A} \\[4pt]
k_{ij} &= \frac{\rho_{fl} \, g}{12\,\mu} \sum \frac{a_f^3}{S_f} \left(\delta_{ij} - \cos\theta_i^f \, \cos\theta_j^f\right) \\[4pt]
&= \frac{\rho_{fl} \, g}{3\,\mu} \sum \frac{\phi_f^3}{\sigma_f^2} \left(\delta_{ij} - \cos\theta_i^f \, \cos\theta_j^f\right)
\end{aligned} \tag{15}$$

Ababou (1991) has observed that this last expression of permeability is an anisotropic form of the Kozeny Carman relation ($k = \alpha\phi^\beta$)

If it is assumed that the hydraulic gradient is uniformly distributed, then all discontinuities have the same importance. Thus, the permeability is always overestimated (in particular in the "poorly connected" network) and the "actual equivalent tensor of permeability" depends on the direction of the flow.

The following examples illustrate the limitations of the previous assumptions:
For a straight discontinuity (Figure 3-a):

$$k_{eq} = \begin{pmatrix} k_{11} & 0 \\ 0 & 0 \end{pmatrix} \quad \text{where} \quad k_{11} = \frac{\rho_{fl} g}{12\mu} \cdot \frac{a^3 l}{A} \cdot (1 - \cos^2\theta) \tag{16}$$

This is the correct solution.

For the case of a continuous but non-planar discontinuity (Figure 3-b):

$$k_{eq} = \begin{pmatrix} k_{11} & k_{12} \\ k_{12} & k_{22} \end{pmatrix} \text{ where } \quad k_{11} = \frac{\rho_{fl} g}{12\mu} \cdot \frac{a^3}{A} \cdot \sum_f l_f (1 - \cos^2 \theta_1^f)$$

$$k_{22} = \frac{\rho_{fl} g}{12\mu} \cdot \frac{a^3}{A} \cdot \sum_f l_f \cdot (1 - \sin^2 \theta_1^f) \tag{17}$$

$$k_{12} = \frac{\rho_{fl} g}{12\mu} \cdot \frac{a^3}{A} \cdot \sum_f l_f \sin \theta_1^f \cos \theta_1^f$$

However, the exact equivalent permeability is :

$$k = \begin{pmatrix} k_{ex} & 0 \\ 0 & 0 \end{pmatrix} \quad \text{where} \quad k_{ex} = \frac{\rho_{fl} g}{12\mu} \cdot a^3 \sum_f l_f (1 - \cos^2 \theta_1^f) \tag{18}$$

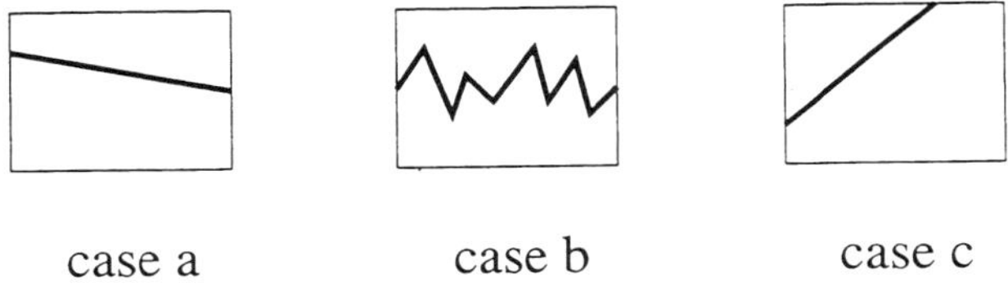

case a   case b   case c

Figure 3 . Examples of particular network

For the case of a single discontinuity with any orientation in space (Figure 3-c):

$$k_{eq} = \begin{pmatrix} k_{11} & k_{12} \\ k_{12} & k_{22} \end{pmatrix} \text{ where } \quad k_{11} = \frac{\rho_{fl} g}{12\mu} \cdot \frac{a^3}{A} \cdot l(1 - \cos^2 \theta_1)$$

$$k_{22} = \frac{\rho_{fl} g}{12\mu} \cdot \frac{a^3}{A} \cdot l \cdot (1 - \sin \theta_1) \tag{19}$$

$$k_{12} = \frac{\rho_{fl} g}{12\mu} \cdot \frac{a^3}{A} \cdot l \cdot \sin \theta_1 \cos \theta_1$$

However, the exact equivalent permeability is

$$k = \begin{pmatrix} 0 & k_{12} \\ k_{12} & 0 \end{pmatrix} \text{ with } \left(k_{12} = \frac{\rho_{fl}\, g}{12\, \mu} \frac{la^3}{A} \sin\theta_1 \cos\theta_1 \right) \quad \text{or} \quad k \equiv 0 \tag{20}$$

dependent on the direction of hydraulic gradient.

2.2. Mechanical and hydromechanical properties

The evaluation of the stiffness matrix (T_{ijkl}) and Biot's coefficients and modulus (B_{ij} and G) in equation (3) is now considered, based on the work of Singh(1973).

The basic principles of the elasticity of composite materials have been proposed by Hill (1963). The theory involves several assumptions concerning various phases of the material:

    (i)   all phases of the material are firmly bonded to each other,

    (ii) the material is macroscopically homogeneous.

Hill (1963) showed, with this hypothesis, that the average density of strain energy in any region can be calculated from the average of the stresses and strains. Thus the equivalent strain and stress are derived from the following energy equation :

$$\overline{W} = \frac{1}{2}\, {}^t\overline{\sigma}_{ij}\, \overline{\varepsilon}_{ji} = \frac{1}{2}\, {}^t\overline{\sigma}_{ij}\, D_{ijkl}\, \overline{\sigma}_{kk} \tag{21}$$

where W : energy,

    $\overline{\sigma}_{ij}$: tensor of equivalent stress (stress in the equivalent medium),

    $\overline{\varepsilon}_{ij}$: equivalent strain,

    D_{ijkl} : compliance tensor.

According to the principle of minimum complementary energy, the actual strain energy of an elastic body is less than the strain energy corresponding to any fictitious equilibrium state of stress. Using this principle one may obtain an upper bound estimate of the compliance of a rock mass with discontinuities by assuming the superposition of strain of rock, ε_{rock}, and strain of discontinuities, $\varepsilon_{fracture}$, i.e. $\varepsilon = \varepsilon_{rock} + \varepsilon_{fracture}$. This is known as the "Reuss" estimate (when discontinuities are not secant one must have $\sigma_{rock} = \sigma_{fracture}$ to assume the equilibrium of the body). A simple calculation shows that the actual stiffness of a heterogeneous material is less than or equal to the volumetric average stiffness of its phases. This lower bound estimate is known as the "Voigt" estimate.

The convention of negative compressive stress is used in this chapter. Over the discontinuities only the effective stress $\left(\sigma'_{ij} = \sigma_{ij} + p\,\delta_{ij}\right)$ has an effect because the parallel increase of both σ_n and p exerts no change on the elastic deformation of discontinuities.

The deformation of rock mass and discontinuities are superposed :

$$\varepsilon_{ij} = \varepsilon^r_{ij} + \varepsilon^f_{ij}$$

where $\quad\varepsilon_{ij}$ tensor of strain of equivalent continuous medium,

$\quad\quad\varepsilon^r_{ij}\quad$ tensor of strain of rock matrix,

$\quad\quad\varepsilon^f_{ij}\quad$ tensor of strain of discontinuities.

The stress is the same in the discontinuities and rock matrix (which is true when all discontinuities are isolated) and the effects of interaction between the discontinuities (in other words stress concentration factor) and the rotation of the blocks are neglected.

For the rock matrix which is supposed to be isotropic and elastic one has :

$$\varepsilon^r_{ij} = \left(\frac{1+v}{E}\,\delta_{ik}\,\delta_{jl} - \frac{v}{E}\,\delta_{ij}\,\delta_{kl}\right)\sigma_{kl} \tag{22}$$

$$M_{ijkl} = \frac{1+v}{E}\,\delta_{ik}\,\delta_{jl} - \frac{v}{E}\,\delta_{ij}\,\delta_{kl} \tag{23}$$

where $\quad v\quad\quad$: $\quad$ Poisson's ratio,

$\quad\quad\quad E\quad\quad$: $\quad$ Young's modulus,

$\quad\quad\quad M_{ijkl}$: $\quad$ tensor of stiffness.

For the discontinuities :

$$\varepsilon^f_{ij} = \frac{1}{A}\int_A \frac{1}{2}\left(\frac{\partial(\delta a_i)}{\partial x_j} + \frac{\partial(\delta a_j)}{\partial x_i}\right) dA \tag{24}$$

ε_{ij}^f : tensor of discontinuity strain ,

A : area of the domain,

δa_i : aperture variation of discontinuity in the ith direction.

δa_n and δa_τ are projected onto the reference axes and the average strain over the whole domain is calculated as:

$$\varepsilon_{ij}^f = \frac{1}{A} l_f \, \delta a \cos \theta_i^f \, \cos \theta_j^f$$

with σ' effective tensor of stress, one has:

$$\begin{aligned}
\delta a &= \delta a_n + \delta a_\tau \\
&= \frac{1}{k_n} \sigma_n' + \frac{1}{k_s} \sigma_\tau' \\
&= \frac{1}{k_n} \sigma_{kl}' \cos \theta_k^f \cos \theta_l^f + \frac{1}{k_s} \left(\delta_{kl} - \cos \theta_k^f \cos \theta_l^f \right) \sigma_{kl}'
\end{aligned}$$

(25)

$$\begin{aligned}
\varepsilon_{ij}^f &= \left(\frac{1}{k_n} - \frac{1}{k_s} \right) \frac{l_f}{A} \left(\cos \theta_i^f \cos \theta_j^f \cos \theta_k^f \cos \theta_l^f \right) \sigma_{kl}' \\
&+ \frac{1}{k_s} \frac{l_f}{A} \left(\cos \theta_i^f \cos \theta_j^f \delta_{kl} \right) \sigma_{kl}'
\end{aligned}$$

(26)

The compliance C_{ijkl} of the discontinuity is given by:

$$\begin{aligned}
C_{ijkl} &= \sum_{f=1}^{n} \left(\frac{1}{k_n} - \frac{1}{k_s} \right) \frac{l_f}{A} \cos \theta_i^f . \cos \theta_j^f . \cos \theta_k^f . \cos \theta_l^f \\
&+ \frac{1}{k_s} \frac{l_f}{A} \frac{1}{4} (\cos \theta_i^f \cos \theta_j^f \delta_{kl} \\
&\qquad + \cos \theta_j^f \cos \theta_l \delta_{ik} \\
&\qquad + \cos \theta_j^f \cos \theta_k \delta_{il} \\
&\qquad + \cos \theta_i^f \cos \theta_l^f \delta_{jk})
\end{aligned}$$

(27)

The total strain is the sum of strain from the matrix and the discontinuities and have the form:

$$\begin{aligned}
\varepsilon_{ij} &= \varepsilon_{ij}^r + \varepsilon_{ij}^f \\
&= M_{ijkl} \sigma_{kl} + C_{ijkl} \sigma_{kl}' \\
&= \left(M_{ijkl} + C_{ijkl} \right) \sigma_{kl} + \delta_{kl} C_{ijkl} \, p
\end{aligned}$$

(28)

Thus with :

$$D_{ijkl} = M_{ijkl} + C_{ijkl}$$
$$C_{ij} = C_{ijkl}\delta_{kl} = \frac{1}{k_n}\sum_f \frac{l_f}{A}\cos\theta_i^f \cos\theta_j^f \tag{29}$$

one has :

$$\sigma_{kl} = (D_{ijkl})^{-1}\varepsilon_{ij} - C_{ij}(D_{ijkl})^{-1}p \tag{30}$$

The total stiffness matrix T_{ijkl} and Biot's coefficient B_{ij}, can be expressed as:

$$T_{ijkl} = \left(D_{ijkl}\right)^{-1} = (M_{ijkl} + C_{ij}kl)^{-1}$$
$$B_{ij} = \left(D_{ijkl}\right)^{-1}.C_{ij} \tag{31}$$

The value of the coefficient of Biot has thus been determined. For determination of the modulus of Biot the fluid production term ξ is calculated, which is also the variation of volumetric water content : $\xi = \Delta\varphi$, if the fluid density is constant.. $\Delta\varphi$ is the variation of porosity due to the stress change (variation of discontinuities aperture), given by:

$$\Delta\varphi = \frac{1}{A}\sum_f l_f \,\delta a_f$$
$$\Delta\varphi = \frac{1}{A}\sum_f \frac{\Delta\sigma_n'}{k_n}l_f \tag{32}$$
$$\Delta\varphi = \frac{1}{A}\sum_f l_f \cos\theta_i^f \cos\theta_j^f \,\Delta\sigma_{ij}'$$

With the use of total stress one obtains:

$$\xi = \frac{1}{k_n}(\sum_f \frac{l_f}{A}\cos\theta_i^f \cos\theta_j^f \Delta\sigma_{ij} + \sum_f \delta_{ij}\frac{l_f}{A}\cos\theta_i^f \cos\theta_j^f p)$$
$$\xi = C_{ij}(\Delta\sigma_{ij} - \delta_{ij}p)$$
$$\xi = C_{ij}(T_{ijkl}\varepsilon_{kl} - B_{ij}p) \tag{33}$$

thus one obtains :

$$p = \frac{1}{C_{ij}(B_{ij} - \delta_{ij})}\left(\xi - B_{ij}\,\varepsilon_{kl}\right) \tag{34}$$

by identification the modulus of Biot is:

$$G = \frac{1}{C_{ij}(B_{ij} - \delta_{ij})} \tag{35}$$

2.3. Stress dependency of hydraulic properties

The aperture variation of discontinuities due to the stress on the porosity (second Biot equation) have been taken into account. However, the variation of permeability is more sensitive to the aperture variation (the permeability varies with the cube of the aperture).

$$a_f^n = a_f^{n-1} + \delta a_f \tag{36}$$

where a_f^n : aperture at time t^n
$\quad\quad a_f^{n-1}$: aperture at time t^{n-1}

The aperture increment is given as:

$$\delta a_f = \frac{1}{k_n}\,\sigma_n' = \frac{1}{k_n}\left(\delta\left(\sigma_{kl} + \delta_{kl}\,p\right)\right) \tag{37}$$

$$\begin{aligned}
k_{ij}^n &= k_{ij}^{n-1} + \alpha_{klij}\,\delta\left(\sigma_{kl} + \delta_{kl}\,p\right) \\
&\quad + \alpha_{klmnij}\cdot\delta\left(\sigma_{kl} + \delta_{kl}\,p\right)\cdot\delta\left(\sigma_{mn} + \delta_{mn}\,p\right) \\
&\quad + \alpha_{klmnopij}\cdot\delta\left(\sigma_{kl} + \delta_{kl}\,p\right)\cdot\delta\left(\sigma_{mn} + \sigma_{mn}\,p\right)\cdot\delta\left(\sigma_{op} + \sigma_{op}\,p\right)
\end{aligned} \tag{38}$$

where :

$$\alpha_{klij} = \frac{\rho_{fl}\, g}{12\, \mu} \frac{1}{k_n} \sum \frac{a_f^2}{s_f} \cos\theta_k^f \, \cos\theta_l^f \left(\delta_{ij} - \cos\theta_i^f \, \cos\theta_j^f \right)$$

$$\alpha_{klmnij} = \frac{\rho_{fl}\, g}{12\, \mu} \frac{1}{k_n} \sum \frac{a}{s_f} \cos\theta_k^f \, \cos\theta_l^f \, \cos\theta_m^f \, \cos\theta_n^f \left(\delta_{ij} - \cos\theta_i^f \, \cos\theta_j^f \right) \qquad (39)$$

$$\alpha_{klmnopij} = \frac{\rho_{fl}\, g}{12\, \mu} \frac{1}{k_n} \sum \frac{1}{s_f} \cos\theta_k^f \, \cos\theta_l^f \, \cos\theta_n^f \, \cos\theta_n^f \, \cos\theta_o^f \, \cos\theta_p^f \left(\delta_{ij} - \cos\theta_i^f \, \cos\theta_j^f \right)$$

If δa_f was small enough, one could linearize the equations. However, a simple calculation shows that δa_f is of the same order of magnitude as the aperture a_f.

Two problems appear (which also exist in the discrete codes) :

- The coefficients of the equations in Darcy's law, depend strongly on the solution. One must use a "small" time step (sufficiently small so that the coefficients vary weakly) or use an iterative scheme.
- The discontinuity cannot have a negative aperture but it is not possible to determine which discontinuity is closed.

With our experiments this stress dependence of permeability has a small effect on the stress and strain (which depend of the boundary conditions) but has a large effect on the flow of the fluid.

3. VALIDATION

In order to derive the equivalent properties of a mass of rock containing many discontinuities, some strong assumptions have been made. To asses the validity of this hypotheses, it is necessary to compare the homogenised properties with the actual properties of the medium. These can be evaluated through numerical simulations using the code UDEC. Such an approach has been used by E. Vuillod and was described in detail in her Ph.D. thesis.(Vuillod,1995)

This methodology is also appropriate when the properties of discontinuities are non-linear (for example the relation between σ_n, δa and σ_τ for the discontinuity).

It is based on the following principle: by giving a series of values to a variable, its associated variables can be determined directly, with all other quantities being kept constant. Thus, for mechanical properties, displacements are imposed and stresses can be determined, thereby yielding the value of the stiffness D_{ijkl}. For hydraulic properties, pressures are imposed and the flow rate can be determined, thereby yielding the permeability k_{ij}. For hydromechanical properties the pressure variation is imposed and the flow rate and stresses can be determined, yielding the Biot's coefficients and modulus.

The difficulty of the method lies in the choice of the location where values are measured. This could be the whole domain, a single line, or otherarrangements, depending on what use is made of the coefficients and also in the intuition of the user. In subsequent computations (numerical resolution of the problem), one can either choose a homogenisation length scale larger than the mesh size, or equal to it. In the former case, it seems logical to use averaged coefficients determined from averaged values, measured inside the domain of homogenisation. In the latter case, it seems more logical to use coefficients determined from values measured at the boundary of the domain of homogenisation. Comparing the results of the two methods gives an indication of the sensitivity of the coefficients. In the case where only the statistical properties of the discontinuities are known, it is interesting to evaluate the difference between averaged values (determined from the statistical properties) and a particular realisation.

The actual medium is not necessary ideal : the properties of symmetry are not always observed.

The fracture network is thus from BMT3. We have chosen test samples of 5 m x 5 m which comprises a large number of discontinuities. We have paid careful attention to bias such as closing of discontinuity or reaching the maximal aperture allowed by the code.

3.1. Hydraulics

The hydraulic gradient $\vec{J}$ and the fluid flow are linked by the relation :

$$\vec{Q} = - k_{ij} \, \vec{J} \tag{40}$$

To obtain a "constant" hydraulic gradient over the test sample, a linear pressure variation is imposed on its boundary (Figure 4), given by:.

$$P(\vec{X}) = \vec{J}.\vec{X} + C \tag{41}$$

with $X(x_1, x_2)$
$$P(x_1, x_2) = a_1 \, x_1 + a_2 \, x_2 + a_3 \tag{42}$$

We have :

$$Q_1 = K_{11} J_1 + K_{12} J_2$$
$$Q_2 = K_{21} J_1 + K_{22} J_2 \tag{43}$$

with : $\bar{Q} = (Q_1, Q_2)$ flow rate throughout the sample
$\quad\quad \bar{J} = (J_1, J_2)$ hydraulic gradient $\tag{44}$

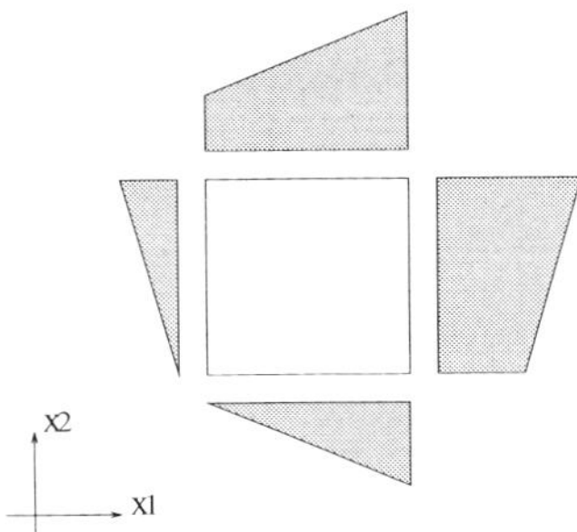

Figure 4 . Pressure boundary conditions over the sample

To determine the four unknowns (k_{ij}), one must write four equations. There are obtained by choosing two arbitrary directions of hydraulic gradient J and J'.

Q_1 is computed by averaging the flow rate through twenty five regularly spaced lines, each perpendicular to the direction 1.

Q_2 is computed by averaging the flow rate through twenty five regularly spaced lines, each perpendicular to the direction 2.

Six test samples have been chosen, and are shown in Figure 5. The homogenised and real tensor of permeability are given in Table 1, where the principal component and the direction of the greater component are also indicated.

Table 1

Comparison of real and equivalent tensor of permeability

Sample Number	Number of discontinuities	Actual permeability tensor (x 10^{10} m/s)	Equivalent permeability tensor (x 10^{10} m/s)	Principal components of actual permeability		Principal components of equivalent permeability	
				Values $x\ 10^{10}$ m/s	*Direction* (rad)	*Values* $x\ 10^{10}$ m/s	*Direction* (rad)
1	77	$\begin{pmatrix} 3{,}89 & 0{,}38 \\ 0{,}21 & 3{,}09 \end{pmatrix}$	$\begin{pmatrix} 5{,}29 & -1{,}77 \\ -1{,}77 & 5{,}01 \end{pmatrix}$	3,98 2,99	0,314	6,58 2,85	- 0,714
2	52	$\begin{pmatrix} 3{,}26 & 0{,}86 \\ 0{,}84 & 3{,}72 \end{pmatrix}$	$\begin{pmatrix} 8{,}48 & -2{,}19 \\ -2{,}19 & 4{,}55 \end{pmatrix}$	4,37 2,61	0,917	3,16 2,47	- 0,089
3	54	$\begin{pmatrix} 6{,}11 & -1{,}70 \\ -1{,}54 & 6{,}17 \end{pmatrix}$	$\begin{pmatrix} 9{,}34 & 6{,}77 \\ 6{,}77 & 15{,}1 \end{pmatrix}$	7,76 4,52	- 0,795	19,24 4,65	0,977
4	76	$\begin{pmatrix} 12{,}08 & 6{,}81 \\ 6{,}17 & 11{,}17 \end{pmatrix}$	$\begin{pmatrix} 10{,}9 & -5{,}31 \\ -5{,}91 & 15{,}5 \end{pmatrix}$	18,13 5,12	0,750	18,47 6,79	- 1,000
5	66	$\begin{pmatrix} 2{,}28 & 0{,}38 \\ 0{,}32 & 2{,}74 \end{pmatrix}$	$\begin{pmatrix} 10{,}7 & -6{,}30 \\ 6{,}30 & 10{,}2 \end{pmatrix}$	2,93 2,09	1,076	14,19 3,43	- 0,672
6	76	$\begin{pmatrix} 10{,}00 & 3{,}05 \\ 2{,}78 & 10{,}77 \end{pmatrix}$	$\begin{pmatrix} 8{,}90 & -3{,}92 \\ -3{,}42 & 7{,}15 \end{pmatrix}$	13,33 7,44	0,851	8,28 3,35	- 0,455

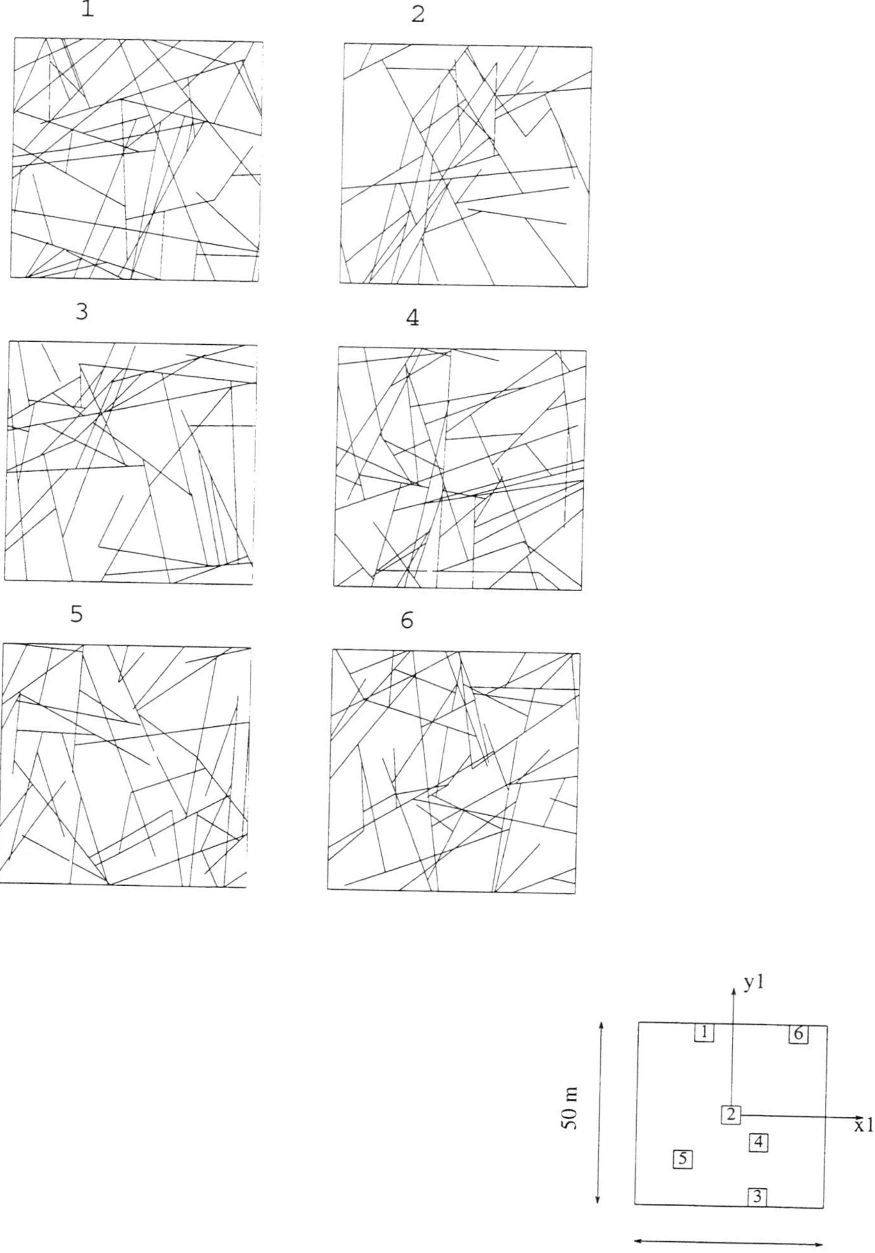

Figure 5 . Network of each test and location of the samples

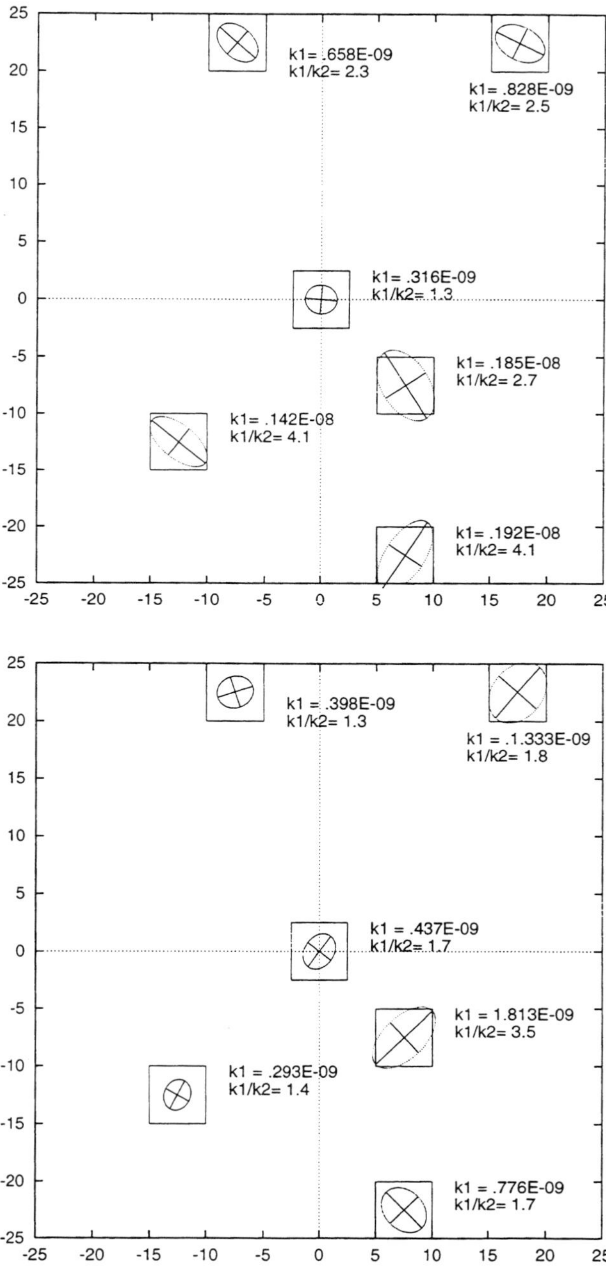

Figure 6 . Anisotropy of permeability ellipses (bottom actual and top equivalent permeability)

                                                              A. Stietel et al.

The anisotropy ellipses for each sample are shown in Figure 6. The anisotropy ellipses have a major axis of size $\sqrt{K_1}$ and a minor axis of size $\sqrt{K_2}$, K_1 and K_2 being the principal components of the permeability tensor.

Table 2

Effect of the sample length on permeability

Sample length	Number of discontinuities	Actual permeability tensor (x 10 m/s)	Equivalent permeability tensor (x 10 m/s)	Principal component of actual tensor		Principal component of equivalent tensor	
				Values $x\ 10\ m/s$	*Direction* *(ad)*	*Values* $x\ 10\ m/s$	*Direction* *(ad)*
5	52	$\begin{pmatrix} 3{,}26 & 0{,}86 \\ 0{,}84 & 3{,}72 \end{pmatrix}$	$\begin{pmatrix} 3{,}16 & -0{,}61 \\ -0{,}61 & 2{,}48 \end{pmatrix}$	4,64 2,93	0,744	3,16 2,47	- 0,089
6	79	$\begin{pmatrix} 3{,}31 & 0{,}03 \\ 0{,}16 & 2{,}69 \end{pmatrix}$	$\begin{pmatrix} 4{,}03 & 0{,}76 \\ +0{,}76 & 3{,}00 \end{pmatrix}$	3,32 2,38	0,097	4,43 2,50	0,489
7	115	$\begin{pmatrix} 1{,}97 & -0{,}11 \\ -0{,}15 & 1{,}81 \end{pmatrix}$	$\begin{pmatrix} 3{,}69 & 0{,}49 \\ +0{,}49 & 2{,}74 \end{pmatrix}$	2,04 1,74	- 0,509	3,90 2,53	0,397
8	152	$\begin{pmatrix} 5{,}51 & 1{,}20 \\ 1{,}19 & 5{,}31 \end{pmatrix}$	$\begin{pmatrix} 14{,}2 & -2{,}83 \\ -2{,}83 & 4{,}14 \end{pmatrix}$	6,61 4,21	0,744	14,96 3,40	- 0,256
9	203	$\begin{pmatrix} 4{,}07 & 1{,}27 \\ 1{,}55 & 4{,}24 \end{pmatrix}$	$\begin{pmatrix} 13{,}7 & -2{,}73 \\ -2{,}73 & 4{,}21 \end{pmatrix}$	5,57 2,74	0,816	14,44 3,48	- 0,261
10	239	$\begin{pmatrix} 3{,}27 & 0{,}41 \\ 0{,}41 & 3{,}38 \end{pmatrix}$	$\begin{pmatrix} 12{,}3 & -2{,}35 \\ -2{,}35 & 4{,}33 \end{pmatrix}$	3,74 2,91	0,852	12,90 3,68	- 0,267
12	343	$\begin{pmatrix} 4{,}57 & 0{,}45 \\ 1{,}25 & 5{,}36 \end{pmatrix}$	$\begin{pmatrix} 10{,}2 & -2{,}15 \\ -2{,}15 & 3{,}71 \end{pmatrix}$	6,12 3,81	0,959	10,83 3,06	- 0,293
15	576	$\begin{pmatrix} 5{,}96 & 0{,}75 \\ 1{,}01 & 6{,}69 \end{pmatrix}$	$\begin{pmatrix} 8{,}85 & -1{,}78 \\ 1{,}78 & 3{,}80 \end{pmatrix}$	7,28 5,37	0,981	9,42 3,23	- 0,307
16	644	$\begin{pmatrix} 2{,}69 & 0{,}28 \\ 0{,}34 & 2{,}73 \end{pmatrix}$	$\begin{pmatrix} 8{,}80 & -1{,}90 \\ -1{,}90 & 4{,}01 \end{pmatrix}$	3,02 2,40	0,818	9,46 3,35	- 0,336

The differences are smaller than an order of magnitude, but the anisotropy ratio and principal directions differ. Note that the real tensors are not quite symmetric.

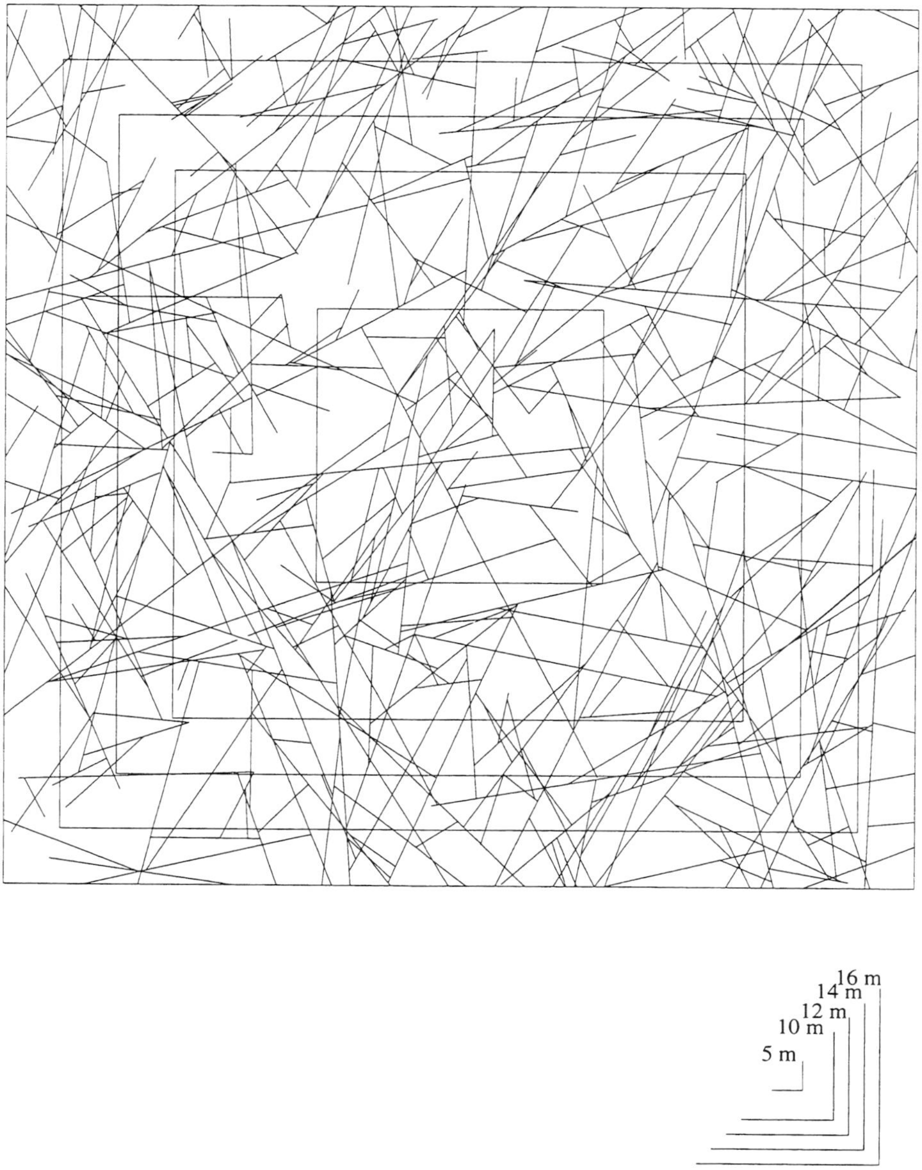

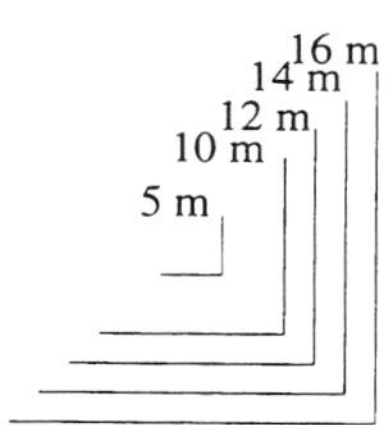

Figure 7 Fracture network to demonstrate the effect of the sample length.

The effect of the scale of the test sample has also been studied. A first sample located at the centre of the network has been considered, and the remaining samples have been obtained by increasing the length from 5 m to 16 m. The network of the different samples is shown in Figure 7 and the results in Table 2.

The variations seem to be erratic in both cases (real and equivalent). Only the principal direction seems to stabilise (but not at the same value).

Finally, the effect of the simplification of the network by neglecting the discontinuities of small apertures has been studied. The networks thus obtained are shown in Figure 8, and the comparison of the results is given in Table 3.

The variations of the equivalent permeability are more regular. At the beginning the variations are very small due to the fact that a discontinuity of small aperture has a far smaller permeability than a discontinuity of great aperture, i.e. for $\frac{a_1}{a_2} = 0,1$ on has $\frac{K_1}{K_2} = 10^{-3}$.

For the actual permeability, a discontinuity of small aperture can be connected with two other discontinuities.

For a_{min} greater than 4 μ the discontinuities of the network are poorly connected : the top and the bottom boundary are no longer connected.

For a_{min} greater than 5 μ the left and right side of the network are no longer connected.

The actual permeability depends on the direction of the hydraulic gradient and is very sensitive to the network.

In this case, the intensities (not the directions) are not too different.

If one applies a hydraulic gradient parallel to one axis and impervious the two faces parallel to this axis, one neglects, thus, the term K_{21}.

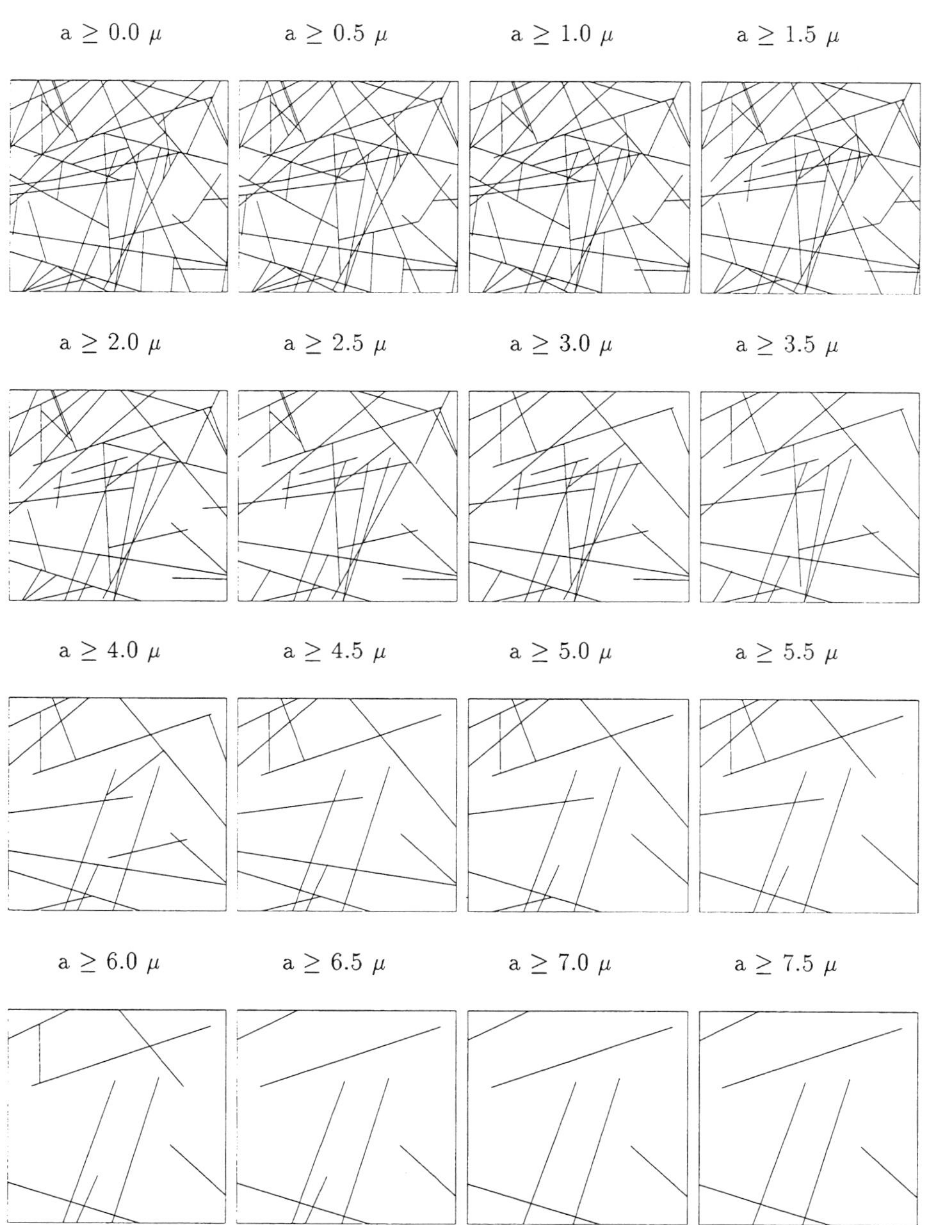

Figure 8 . Effect on simplifications of the discontinuity network

Table 3

Effect of the simplification of the discontinuity network

Minimal aperture (μ)	Number of discontinuities	Actual tensor of permeability ($\times 10^{10}$ m/s)		Equivalent tensor of permeability ($\times 10^{10}$ m/s)		Principal component of actual tensor		Principal component of equivalent tensor	
						Values $\times 10^{10}$ *m/s*	*Direction (rad)*	*Values* $\times 10^{10}$ *m/s*	*Direction (rad)*
0	77	$\begin{pmatrix} 3{,}29 & 0{,}38 \\ 0{,}20 & 3{,}09 \end{pmatrix}$		$\begin{pmatrix} 4{,}48 & -1{,}85 \\ -1{,}85 & 4{,}45 \end{pmatrix}$		3,98 3,00	0,314	6,58 2,85	− 0,714
0,5	77	$\begin{pmatrix} 3{,}29 & 0{,}38 \\ 0{,}20 & 3{,}09 \end{pmatrix}$		$\begin{pmatrix} 4{,}98 & -1{,}85 \\ -1{,}85 & 4{,}45 \end{pmatrix}$		3,98 3,00	0,314	6,58 2,85	− 0,714
1	73	$\begin{pmatrix} 3{,}95 & 0{,}39 \\ 0{,}16 & 3{,}04 \end{pmatrix}$		$\begin{pmatrix} 4{,}98 & -1{,}85 \\ -1{,}85 & 4{,}45 \end{pmatrix}$		4,02 2,96	0,268	6,58 2,85	− 0,714
1,5	66	$\begin{pmatrix} 4{,}60 & 0{,}26 \\ 0{,}11 & 3{,}62 \end{pmatrix}$		$\begin{pmatrix} 4{,}98 & -1{,}85 \\ -1{,}85 & 4{,}44 \end{pmatrix}$		4,63 3,59	0,176	6,52 2,84	− 0,714
2	56	$\begin{pmatrix} 4{,}47 & 0{,}28 \\ 0{,}03 & 3{,}38 \end{pmatrix}$		$\begin{pmatrix} 4{,}97 & -1{,}85 \\ -1{,}85 & 4{,}44 \end{pmatrix}$		4,49 3,36	0,134	6,57 2,83	− 0,715
2,5	49	$\begin{pmatrix} 4{,}78 & 0{,}63 \\ 0{,}23 & 3{,}66 \end{pmatrix}$		$\begin{pmatrix} 4{,}95 & -1{,}85 \\ -1{,}85 & 4{,}42 \end{pmatrix}$		4,93 3,51	0,327	6,55 2,82	− 0,714
3	42	$\begin{pmatrix} 4{,}66 & 0{,}43 \\ 0{,}71 & 4{,}41 \end{pmatrix}$		$\begin{pmatrix} 4{,}92 & -1{,}85 \\ -1{,}85 & 4{,}40 \end{pmatrix}$		5,39 3,67	0,712	6,53 2,79	− 0,716
3,5	36	$\begin{pmatrix} 4{,}23 & 0{,}79 \\ 0{,}56 & 3{,}97 \end{pmatrix}$		$\begin{pmatrix} 4{,}87 & -1{,}83 \\ -1{,}83 & 4{,}35 \end{pmatrix}$		4,77 3,40	0,682	6,45 2,76	− 0,713
4	27	$\begin{pmatrix} 4{,}37 & 0{,}32 \\ 0{,}21 & 4{,}16 \end{pmatrix}$		$\begin{pmatrix} 4{,}69 & -1{,}76 \\ -1{,}76 & 4{,}27 \end{pmatrix}$		4,54 3,98	0,593	6,26 2,71	− 0,725
4,5	24	$\begin{pmatrix} 3{,}55 & 0{,}21 \\ 0{,}02 & 3{,}32 \end{pmatrix}$		$\begin{pmatrix} 4{,}65 & -1{,}74 \\ -1{,}74 & 4{,}19 \end{pmatrix}$		3,59 3,27	0,321	6,18 2,66	− 0,720
5	21	$\begin{pmatrix} 2{,}26 & 0{,}05 \\ -0{,}06 & 2{,}64 \end{pmatrix}$		$\begin{pmatrix} 4{,}64 & -1{,}77 \\ -1{,}77 & 4{,}02 \end{pmatrix}$		2,64 2,61	1,28	6,13 2,54	− 0,699
5,5	17	$\begin{pmatrix} 1{,}70 & 0{,}63 \\ 0{,}58 & 1{,}61 \end{pmatrix}$		$\begin{pmatrix} 4{,}57 & -1{,}81 \\ -1{,}81 & 3{,}91 \end{pmatrix}$		1,83 0	1,22	6,08 2,40	− 0,696
6	13	$\begin{pmatrix} 1{,}55 & 0{,}56 \\ 0{,}43 & 1{,}29 \end{pmatrix}$		$\begin{pmatrix} 4{,}40 & -1{,}74 \\ -1{,}74 & 3{,}62 \end{pmatrix}$		1,92 0,91	0,656	5,79 2,22	− 0,675
6,5	11	$\begin{pmatrix} 0{,}92 & 0{,}37 \\ 0{,}34 & 0{,}83 \end{pmatrix}$		$\begin{pmatrix} 4{,}18 & -1{,}83 \\ 1{,}83 & 3{,}54 \end{pmatrix}$		1,23 0,52	0,721	5,72 2,00	− 0,700
7	10	$\begin{pmatrix} 0{,}92 & 0{,}37 \\ 0{,}34 & 0{,}83 \end{pmatrix}$		$\begin{pmatrix} 4{,}08 & -1{,}79 \\ -1{,}79 & 3{,}53 \end{pmatrix}$		1,23 0,52	0,721	5,61 1,99	− 0,708
7,5	10	$\begin{pmatrix} 0{,}92 & 0{,}37 \\ 0{,}34 & 0{,}83 \end{pmatrix}$		$\begin{pmatrix} 4{,}08 & -1{,}79 \\ -1{,}79 & 3{,}53 \end{pmatrix}$		1,23 0,52	0,721	5,61 1,99	− 0,708

3.2. Mechanical properties

In the absence of hydraulic pressure the strain and stress are linked by the relation: :

$$\varepsilon_{ij} = D_{ijkl}\, \sigma_{kl}$$

Or conversely,

$$\sigma_{kl} = T_{ijkl}\, \varepsilon_{ij} \tag{45}$$

The stiffness tensor is of 4th rank and it is difficult to compare two tensors.

For mechanical calculations, planes strain assumption is used and thus one cannot determine all the coefficients, since one has:

$$
\begin{aligned}
\sigma_{11} &= T_{11}\,\varepsilon_{11} + T_{12}\,\varepsilon_{22} + T_{16}\,\varepsilon_{12} \\
\sigma_{22} &= T_{21}\,\varepsilon_{11} + T_{22}\,\varepsilon_{22} + T_{26}\,\varepsilon_{12} \\
\sigma_{33} &= T_{31}\,\varepsilon_{11} + T_{32}\,\varepsilon_{22} + T_{36}\,\varepsilon_{12} \\
\sigma_{12} &= T_{61}\,\varepsilon_{11} + T_{61}\,\varepsilon_{22} + T_{66}\,\varepsilon_{12}
\end{aligned}
\tag{46}
$$

There are twelve unknowns which can be determined by considering three different loading conditions.

The loads (boundary conditions) are described in Figure 9. For each load a column of T_{ij} is determined and the σ_{ij} are averaged over the whole test sample.

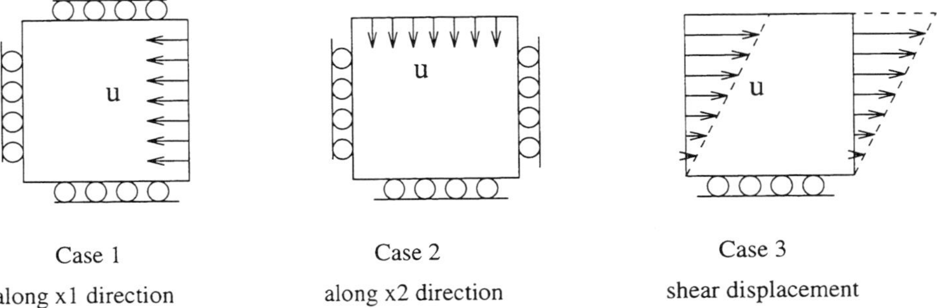

Figure 9 . Displacements prescribed to the specimen

Three test samples have been chosen and are shown in Figure 10 The equivalent and real tensors obtained in each case are gathered in Table 4.

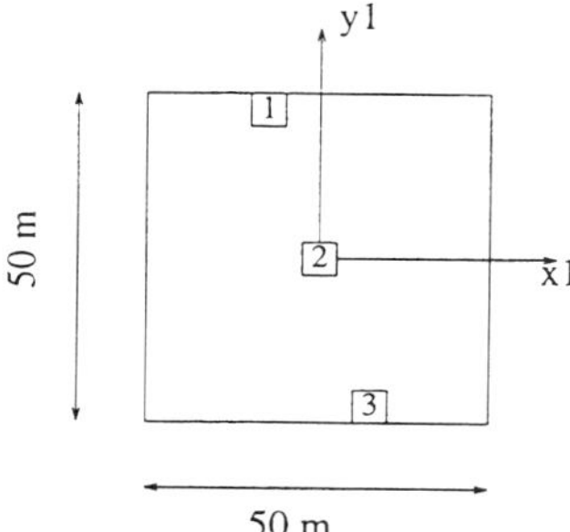

Figure 10 . Location of the samples in the BMT3 network

The real tensors T_{ij} are not symmetric, in particular the term T_{6x}. With other loads, the values would perhaps be modified.

The terms of the actual tensors are generally greater than those of equivalent tensor, but the differences are smaller than 10 %. This is comparable to the dispersion of results with mechanical experiments.

3.3. Hydromechanical properties

The Biot's modulus and coefficients appear in the two equations of Biot: (cf. equation(3)).

If the pressure is modified, and the displacements are kept constant, one has :

$$\Delta p = - G \Delta\xi$$
$$\Delta\sigma_{ij} = - B_{ij} \Delta p \tag{47}$$

where Δ stands for variations.

Between two equilibrium states, without displacement variations, the total fluid flow gives the Biot modulus and the stresses give the Biot coefficients.

The loads are shown in Figure 11.

Table 4

Tensors T_{ij} (MPa) for case 1-3

	Equivalent stiffness tensor						Actual stiffness tensor					
N° 1	17200	10400	6330	0	0	−306	20165	9513	x	x	x	2141
	10400	17200	6340	0	0	−53	9915	19447	x	x	x	1967
	6330	6340	62900	0	0	−83	6918	6661	x	x	x	945
	0	0	0	3460	−263	0	0	0	x	x	x	0
	0	0	0	−263	3400	0	0	0	x	x	x	0
	−306	−53	−83	0	0	29100	109	698	x	x	x	1816
N° 2	21400	12500	7810	0	0	−229	24176	11725	x	x	x	2653
	12500	19900	7460	0	0	−610	11719	27774	x	x	x	3223
	7810	7460	63500	0	0	−190	8256	9085	x	x	x	1351
	0	0	0	3890	−650	0	0	0	x	x	x	0
	0	0	0	−650	5020	0	0	0	x	x	x	0
	−229	−610	−190	0	0	39000	158	−86	x	x	x	2342
N° 3	21700	11200	7570	0	0	−272	23093	6787	x	x	x	2504
	11200	20200	7320	0	0	−513	9568	19102	x	x	x	1883
	7570	7320	63400	0	0	−55	7512	5955	x	x	x	1009
	0	0	0	3770	−255	0	0	0	x	x	x	0
	0	0	0	−255	4260	0	0	0	x	x	x	0
	−272	−513	−55	0	0	32100	568	−432	x	x	x	2594

x : value not determined

Using an untrained formulation, (i.e. $\xi = 0$, the boundaries are impervious),

$$\sigma_{ij} = \left(T_{ijkl} + G\, B_{ij}\, B_{kl} \right) \varepsilon_{kl} - \underbrace{B_{ij}\, G\, \xi}_{=0} \tag{48}$$

one has:

$$T^* = T_{ijkl} + G\, B_{ij} B_{kj} \tag{49}$$

just as for T By difference G B_{ij} B_{kl} is obtain. This method is very difficult to apply because the errors (on T and T*) are cumulated.

The analytical and numerical results are shown in Table 5.

Table 5
Biot module G(Pa) and Biot coefficients B_{ij}

	Equivalent **BIOT** coefficients and modulus	Actual **BIOT** coefficients and modulus
N° 1	$B_{ij} = \begin{bmatrix} 0.695 & 0.040 & 0 \\ 0.040 & 0.641 & 0 \\ 0 & 0 & 0.313 \end{bmatrix}$ $G = 66.3\ 10^9$	$B_{ij} = \begin{bmatrix} 0.850 & 0.017 & 0 \\ 0.017 & 0.750 & 0 \\ 0 & 0 & 0.400 \end{bmatrix}$ $G = 52.2\ 10^9$
N° 2	$B_{ij} = \begin{bmatrix} 0.624 & 0.009 & 0 \\ 0.009 & 0.641 & 0 \\ 0 & 0 & 0.271 \end{bmatrix}$ $G = 76.6\ 10^9$	$B_{ij} = \begin{bmatrix} 1.050 & 0.005 & 0 \\ 0.005 & 0.900 & 0 \\ 0 & 0 & 0.450 \end{bmatrix}$ $G = 72.25\ 10^9$
N° 3	$B_{ij} = \begin{bmatrix} 0.636 & 0.030 & 0 \\ 0.030 & 0.652 & 0 \\ 0 & 0 & 0.289 \end{bmatrix}$ $G = 71.8\ 10^9$	$B_{ij} = \begin{bmatrix} 1.050 & 0.100 & 0 \\ 0100 & 1.100 & 0 \\ 0 & 0 & 0.5001 \end{bmatrix}$ $G = 9.85\ 10^9$

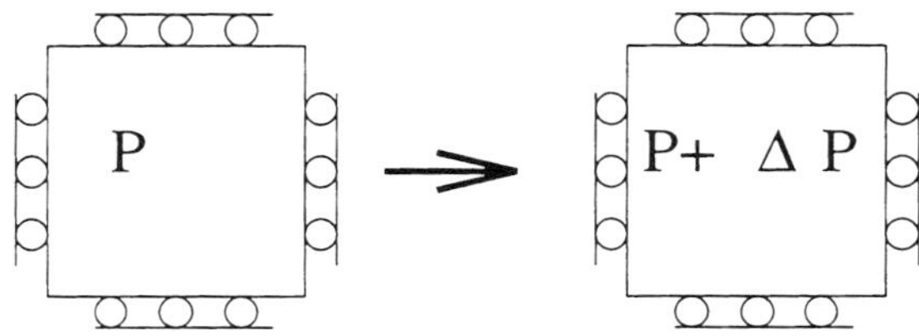

Figure 11 . Boundary conditions prescribed to the sample

Biot's coefficients are always less or equal to one. The values of the calculated actual Biot's coefficients give an idea of the error made in the determination of the actual values. This is due to the small variations of stresses (0.01 MPa). For the Biot modulus the difference between actual and equivalent values depends on the sample.

3.4. Stress dependency of permeability

The stress dependency of permeability has an important effect on the fluid flow.

In the analytical relation (cf. equation(39)) there appear tensors of 4th, 6th and 8th rank. It is not possible to determine each term of the tensors.

The only way to assess the importance of the stress dependency of the permeability is to compare the results with a complete load (displacement and pressure). Such comparisons have not yet been made.

To determine, with non-linear laws of discontinuities, the behaviour of permeability tensor, one must determine a tensor α which depends on the intensity of the stress variation. This tensor (variable with $\Delta\sigma'$) takes into account the variation of permeability with the cube of the aperture :

$$K_{ij}^n = K_{ij}^{n-1} + \alpha_{ijkl}\left(\Delta\sigma'\right).\left(\sigma_{ij}'^n - \sigma_{ij}'^{n-1}\right) \tag{50}$$

4. CONCLUSIONS

Starting from the work of Oda, formulae for the equivalent physical properties (permeability tensor, stiffness tensor, coefficients and modulus of Biot) of a medium consisting of intact rock and discontinuities have been derived. The validity of these formulae has been assessed by performing a series of numerical simulations with the UDEC code.

The permeability tensor of the equivalent media differ from of the real tensor by less than one order of magnitude. However, difference in the degree of anisotropy and the principle directions are more substantial. For the stiffness tensor, the difference is less than 10%. Conclusions concerning the coefficients and modulus of Biot are less clear.

Results show that it is not necessary to have many discontinuities to obtain accurate equivalent properties. The choice of the homogenisation length scale (size of the volume for which equivalent properties are determined) depends on the scale of the physical phenomenon, and not on that of a representative elementary volume (REV). Further studies are needed. Also, there remain an important problem which has yet to be to resolved, namely the variation of the permeability tensor due to the stress variations.

5. REFERENCES

1 DECOVALEX secretariat, Specification of a benchmark test of Decovalex phase 2, Decovalex Doc 92 113 (1992).

2 M. Oda, An equivalent continuum model for coupled stress and fluid flow analysis in jointed rock masses. Water Resources Research, 22, 1986, 1845-1856.

3 P. Cundall, UDEC a computer model for simulating progressive, large scale movements in blocky rock systems. Proc. Int. Symp. Rock Fractures Nancy (1991).

4 A. Biot, Theory of elasticity and consolidation for a porous anisotropic solid. J. Appl. Phys., 26, 1955, 182-185.

5 DJ. Snow, Anisotropic permeability of fractured media.Water Resources Research, 5,1969, 1273-1289.

6 R. Ababou, Approaches to large scale unsaturated flow in heterogeneous stratified and fractured geologic media. Report NUREG/CR-5743, US. Nuclear Regulatory Commission, Washington DC.(1991).

7 B.Singh, Continuum characterisation of jointed rock masses. Part I constitutive equations.Int. J. Rock Mech. Min. and Geomech. Abstr, 10, 1973,311-335.

8 R.Hill, Elastic properties of reinforces solids. Some theoretical principles.J. Mech. Phys. Solids, 11, 1963 , 357-372.

9 R.Hill, A self-consistent mechanics of composite materials. J. Mech. Phys. Solids, 13, 1965, 213-222.

10 E.Vuillod, Modélisation thermo-hydro-mécanique de massifs rocheux discontinuities. Thèse INPL Nancy (1995).

O. Stephansson, L. Jing and C.-F. Tsang (Editors)
Coupled Thermo-Hydro-Mechanical Processes of Fractured Media
Developments in Geotechnical Engineering, vol. 79
© 1996 Elsevier Science B.V. All rights reserved.

165

FEM analysis of coupled THM processes in fractured media with explicit representation of joints

S-M. Tijani   G.Vouille

Ecole Nationale Supérieure des Mines de Paris
Centre de Géotechnique et d'Exploitation du Sous-sol
35, Rue Saint Honoré
77305 Fontainebleau CEDEX (France)

Abstract

The well-known governing equations for thermo-hydro-mechanical processes in continua and joints are presented with emphasis on the difference between fundamental laws (balance equations) and empirical relationships (constitutive laws). The assumptions commonly used to simplify the balance equations are described underlining the neglected terms.

The classical variational formulation of THM processes requires some additional terms to take into account the rock joints. The hydraulic and mechanical added terms are detailed. The use of Finite Element Method to solve THM problems with explicit representation of joints is examined.

Coupled THM processes are usually analysed using "fully coupled" numerical techniques which often need restrictive assumptions on the governing equations. To avoid such a restriction an iterative algorithm is proposed. This algorithm does not depend neither on the theoretical equations to be solved nor on the numerical method chosen.

1. INTRODUCTION

The increasing importance of modeling in the analysis of coupled Thermo-Hydro-Mechanical processes in geomechanics has induced a lot of work and publications dealing with various methods. They differ not only on the numerical techniques used, but also on the theoretical formulations of the physical problem. When the industrial applications are taken into account, the main criterion to choose a numerical procedure must be the computational rigour in solving the mathematical formulation with as less restriction as possible. More precisely, one must not be obliged to simplify the theoretical formulation only for reasons of efficiency of the numerical methods. As far as the mathematical formulation is concerned, it is well-known that the governing equations can be put into two families :

A) Balance Laws: there is a general consensus on these laws in their original forms, but they are often simplified with some application-specific hypothesises.

- mass balance: this concerns the balance of fluid mass flow throughout the skeleton of the porous continuum and inside the rock joints. In geomechanics the classical simplification is to assume that the fluid mass density has small variations in space so that the continuity equation is equivalent to the fluid volume balance law.

- momentum balance: generally the inertia forces are neglected (static approach) and, in the equations governing the fluid motion, one assumes that the mechanical actions of the skeleton are reduced to body forces depending upon the intrinsic permeability of the porous medium (or the rock joint aperture) and upon the fluid viscosity (Darcy's law).

- energy balance: the common way to reduce the first law of thermodynamics is to neglect all kinds of energy except the thermal one. In this case the thermal problem can be solved independently but we must keep in mind that this simplification is not valid when there is either high fluid velocity or finite deformation of the porous medium or non negligible energy dissipation.

B) Constitutive Laws: all these state laws are restricted only by the Clausius-Duheim inequality (second law of thermodynamics). The weakness of such a restriction is the reason why, sometimes, several laws are proposed to describe the same physical phenomenon. The only valid way to choose the appropriate law is to compare its predictions to actual results in situations where all other physical laws and properties are known.

In any case (with or without simplifications), establishing governing equations must be the task of geomechanics experts who might be allowed to impose any desired assumptions on all the governing equations as well as on the values of the attached parameters. The numerical methods are only tools used to help these experts. This idea will be the major guide for this short presentation of finite element method analysis of coupled thermo-hydro-mechanical processes in fractured media with explicit representation of joints.

2. GOVERNING EQUATIONS OF THM PROCESSES FOR CONTINUA

2.1 Problem unknowns

The unknowns are time dependent fields defined on a domain Ω (current configuration) occupied at time t by the skeleton of the saturated porous medium. The principal unknowns are the temperature T (the same for solid and fluid phases), the pore pressure P and the deformation vector $\vec{u}$. The related secondary unknowns (Figure 1) are the heat flux vector $\vec{\Psi}$, the fluid volume flux vector $\vec{q}$ and the total Cauchy's stress tensor $\tilde{\sigma}$. Other internal time dependent fields may have to be determined when required by the constitutive law of the skeleton: non-elastic strain tensor $\tilde{\epsilon}^p$, hardening variables ξ ... In the initial configuration ($t = 0$) the fields T, P, $\tilde{\sigma}$ and the internal variables are necessarily assumed to be known.

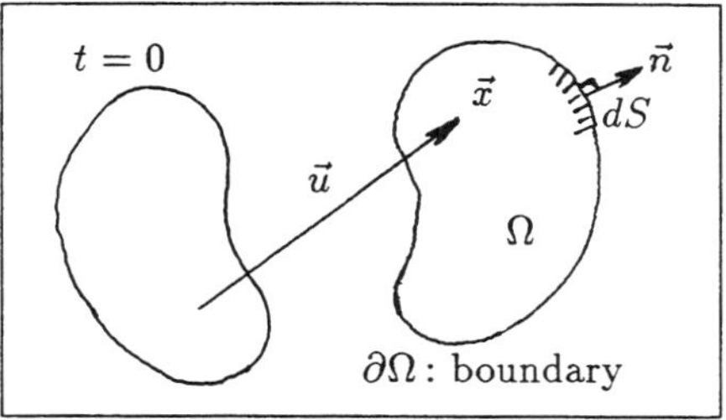

$\vec{\Psi}.\vec{n}dS$: thermal energy rate which leaves Ω

$\vec{q}.\vec{n}dS$: fluid volume rate which leaves Ω

$\tilde{\sigma}\vec{n}dS$: total force vector acting on Ω

Figure 1. : Secondary unknown fields

2.2 Balance laws

2.2.1 Mass balance

Let V be a small volume of the skeleton. At each time t, the fluid and solid masses in V are respectively $\phi\rho_f V$ (because the porous medium is saturated) and $(1 - \phi)\rho_s V$ where ρ_f and ρ_s are the mass densities of fluid and solid and ϕ is the connected porosity. Since the porous medium deformation is analysed in a convected co-ordinate system attached to the deformed skeleton (current configuration) the solid mass is constant and the fluid mass rate is $-div(\rho_f\vec{q})V$ where differential operator $div = \vec{\nabla}$ uses the position vector $\vec{x}$ at time t. On the other hand, the volume rate $\dot{V}$ is related to the volumetric strain rate of the skeleton $\dot{\epsilon} = div(\dot{\vec{u}})$ since $\dot{V}/V = \dot{\epsilon}$. The fluid and solid volumetric strain rates are $\dot{\epsilon}_f = -\dot{\rho}_f/\rho_f$ and $\dot{\epsilon}_s = -\dot{\rho}_s/\rho_s$, respectively. Then the fluid and solid mass continuity laws can be written :

$$div(\rho_f\vec{q})/\rho_f + \phi\dot{\epsilon} - \phi\dot{\epsilon}_f + \dot{\phi} = 0 \tag{1}$$

$$(1 - \phi)\dot{\epsilon} - (1 - \phi)\dot{\epsilon}_s - \dot{\phi} = 0 \tag{2}$$

Eliminating the porosity rate by adding the two equations, we obtain :

$$div(\rho_f\vec{q})/\rho_f + \dot{\epsilon} - \phi\dot{\epsilon}_f - (1 - \phi)\dot{\epsilon}_s = 0 \tag{3}$$

Since $div(\rho_f\vec{q})/\rho_f = div(\vec{q}) + \vec{q}.\vec{grad}(\rho_f)/\rho_f$, if we assume that either the fluid volume flux $\vec{q}$ or the gradient $\vec{grad}(\rho_f)$ of the fluid mass density is negligible, the fluid mass balance law may be simplified into :

$$div(\vec{q}) + \dot{\epsilon} - \phi\dot{\epsilon}_f - (1 - \phi)\dot{\epsilon}_s = 0 \tag{4}$$

2.2.2 Momentum balance

The equilibrium of any part of the saturated porous medium is satisfied if and only if the total Cauchy's stress tensor $\tilde{\sigma}$ is symmetric and :

$$\vec{div}\tilde{\sigma} + \rho\vec{g} \;=\; \text{inertia forces} \tag{5}$$

$$\rho \;=\; \phi\rho_f + (1-\phi)\rho_s \tag{6}$$

where ρ is the satured porous medium density and $\vec{g}$ is the gravity forces mass density [1].

The Cauchy's stress tensor in the fluid is $-P\tilde{1} + \tilde{\zeta}$ where P is the pore pressure, $\tilde{1}$ is the unit tensor and $\tilde{\zeta}$ is the "dynamic" stress tensor which is due to the relative motion of the viscous fluid [2]. The equilibrium of any part of the fluid phase is then governed by the following equation (8) where the volumetric force $\vec{div}\tilde{\zeta}$ is the action of the skeleton due to the fluid relative motion. This volumetric force is often assumed to be a function of $\vec{q}$ using the dynamic fluid viscosity η_f and the intrinsic permeability tensor $\tilde{\kappa}$ (equation 9).

$$\vec{div}\tilde{\zeta} - \vec{grad}(P) + \rho_f\vec{g} \;=\; \text{inertia forces} \tag{7}$$

$$\vec{div}\tilde{\zeta} \;=\; -\eta_f\tilde{\kappa}^{-1}\vec{q} \tag{8}$$

In a static approach (when inertia forces are neglected) the assumption concerning the viscous forces implies the simplified momentum balance law :

$$\vec{q} \;=\; -(1/\eta_f)\tilde{\kappa}(\vec{grad}(P) - \rho_f\vec{g}) \tag{9}$$

When $\vec{grad}(\rho_f)$ is neglected we obtain the classical Darcy's law $\vec{q} = -\tilde{k}\vec{grad}(H)$ where $H = (P/\rho_f/g - \vec{g}.\vec{x}/g)$ is the hydraulic head and $\tilde{k} = (\rho_f g/\eta_f)\tilde{\kappa}$ is the permeability tensor which depends on the properties of both the skeleton and the fluid.

2.2.3 Energy balance

Let $C = \phi\rho_f c_f + (1-\phi)\rho_s c_s$ where c_f and c_s are the fluid and solid specific heat capacities. Then the energy balance law can be written [2]:

$$div(\vec{\Psi}) + C\dot{T} \;=\; \ldots \tag{10}$$

The right hand side of equation (10) represents several terms generally neglected such as :

- dissipative energies due to the viscosity of the fluid and the irreversible energy dissipation in the skeleton.

- the energies associated with thermal expansion (coupled thermomechanical phenomena)

– the transported energy due to the fluid flow.

2.3 Constitutive laws

2.3.1 Fluid state law

The thermodynamic state of the fluid is described by the couple (P, T) so that ρ_f, η_f and c_f are functions of P and T [2]. For instance, if we introduce the compressibility modulus β_f of the fluid and its linear thermal expansion coefficient α_f (both are functions of P and T), we obtain:

$$\dot{\epsilon}_f = -\beta_f \dot{P} + 3\alpha_f \dot{T} \tag{11}$$

2.3.2 Solid state law

It is more difficult to describe the deformation of the solid. Following Terzaghi and Biot's works some authors ([3],[4]) proposed sophisticated thermodynamic interpretations, others used simple micro-models to explain the macroscopic constitutive laws. Nevertheless, in any case, it is only a matter of assumptions and the unique way to establish the validity of the suggested governing equations is to verify that they account for the actual observations and measurements.

In all these assumptions, the main difficulty is due to the fact that if we consider an elementary volume V of the satured porous medium during a drained ($\dot{P} = 0$) and isothermal ($\dot{T} = 0$) deformation, we have no theoretical possibility to know how to relate the variation of V to the variations of its two parts: the pore volume $V_\phi = \phi V$ and the solid volume $V_s = (1 - \phi)V$. The Terzaghi's theory assumes that the solid (grains) is rigid. So $dV_s = 0$ and $dV_\phi = dV$. In Biot's approach the last equation is generalized in such a way as the ratio $b = dV_\phi/dV$ can range from 0 to 1. The latter value of this Biot's coefficient corresponds to the particular case of Terzaghi's theory. Since $dV_s = (1 - b)dV$, $\dot{\epsilon}_s = \dot{V}_s/V_s$ and $\dot{\epsilon} = \dot{V}/V$, we obtain $\dot{\epsilon}_s = [(1 - b)/(1 - \phi)]\dot{\epsilon}$. But this last equation, valid only in isotropic drained and isothermal conditions, has to be generalized. Let $\tilde{\dot{\epsilon}}$ be the symmetric part of the gradient of the velocity vector $\vec{u}$ which is called the strain rate tensor. Its trace is $\dot{\epsilon} = \tilde{1}.\tilde{\dot{\epsilon}}$ because $tr(\tilde{\dot{\epsilon}}) = div(\vec{\dot{u}})$ and $\tilde{1}$ is the unit Kronecker's tensor. To avoid the restriction corresponding to the isotropy the coefficient b is replaced by a symmetric tensor $\tilde{B}$: $\dot{\epsilon}_s = [(\tilde{1} - \tilde{B})/(1 - \phi)].\tilde{\dot{\epsilon}}$. The Biot's tensor $\tilde{B}$ which is reduced to $b\tilde{1}$ in the isotropic case depends only on the state of the skeleton and is generally constant. To generalize the relationship between $\dot{\epsilon}_s$ and $\tilde{\dot{\epsilon}}$ for any deformation the common way is to assume that this equation concerns only the mechanical parts of the solid and the skeleton deformations which are respectively $\dot{\epsilon}_s - 3\alpha_s \dot{T} + \beta_s \dot{P}$ and $\tilde{\dot{\epsilon}} - \dot{T}\tilde{\alpha}$ where α_s is the linear thermal expansion coefficient of the solid, β_s is its compressibility modulus and $\tilde{\alpha}$ is the symmetric tensor of the linear thermal expansion coefficients of the skeleton. The generalized relationship is then:

$$(1 - \phi)\dot{\epsilon}_s - \dot{\epsilon} = [3(1 - \phi)\alpha_s - (\tilde{1} - \tilde{B}).\tilde{\alpha}]\dot{T} - (1 - \phi)\beta_s \dot{P} - \tilde{B}.\tilde{\dot{\epsilon}} \tag{12}$$

Substituting equations (11) and (12) into equation (4) we obtain the expression:

$$div(\vec{q}) = 3\alpha_m \dot{T} - \dot{P}/M - \tilde{B}.\dot{\tilde{\epsilon}} \tag{13}$$

where $\alpha_m = \phi\alpha_f + (1-\phi)\alpha_s - (1/3)(\tilde{1} - \tilde{B}).\tilde{\alpha}$ is the linear differential expansion coefficient and $M = [\phi\beta_f + (1-\phi)\beta_s]^{-1}$ is the Biot's modulus.

2.3.3 Skeleton constitutive law

From equation (13), the mechanical energy rate per unit skeleton volume which is $\tilde{\sigma}.\dot{\tilde{\epsilon}} - P div(\vec{q})$ can be written $\tilde{\sigma}_{eff}.\dot{\tilde{\epsilon}} - P(3\alpha_m \dot{T} - \dot{P}/M)$ where $\tilde{\sigma}_{eff} = \tilde{\sigma} + P\tilde{B}$ is the effective symmetric Cauchy's stress tensor ([5],[6],[7]). The behaviour of the skeleton is then governed by constitutive laws using only T and $\tilde{\sigma}_{eff}$. For instance, the strain rate tensor can be the sum of three parts:

- an elastic part related to the Jauman's rate of the effective stress using Hooke's elasticity tensor $\tilde{\tilde{H}}$ which may depend on the temperature T.

- a non-elastic part $\dot{\tilde{\epsilon}}^p$ governed by plastic and viscoplastic laws using the temperature T, the effective stress tensor $\tilde{\sigma}_{eff}$ and all internal variables $\tilde{\epsilon}^p$ and ξ.

- a thermal part $\dot{T}\tilde{\alpha}$ where thermal expansion coefficients tensor $\tilde{\alpha}$ may depend on the temperature T.

3. GOVERNING EQUATIONS OF THM PROCESSES FOR JOINTS

3.1 The geometry of a joint

3.1.1 Discontinuities

In three-dimensional problems the joint is a surface Σ where the deformation vector $\vec{u}$ has two values (discontinuity) one for each of the two parts of the skeleton Ω separated by the surface. Let $\vec{n}$ be the unit vector normal to the surface Σ at a point $\vec{x}$. The variables attached to such a point will be superscripted by a $+$ when the material point $\vec{x}$ belongs to the region Ω^+ pointed to by the vector $\vec{n}$. Otherwise, the superscript $-$ will be used. For instance the displacement discontinuity vector $\sharp\vec{u}\sharp = \vec{u}^+ - \vec{u}^-$ represents the relatif movement of Ω^+ referred to the part Ω^-. The normal component $u_n = \sharp\vec{u}\sharp.\vec{n}$ is the variation of the mechanical aperture of the joint which is e_m^0 at time 0 and $e_m = e_m^0 + u_n$ at time t. The hydraulic aperture is often assumed to have the same variation but with a different initial value: $e_h = e_h^0 + u_n$. The last equation can be generalized using a coefficient f which ranges from 0 to 1: $e_h = e_h^0 + fu_n$. The initial apertures e_m^0 and e_h^0 may of course vary in space (on Σ) but they are known (given data). Across the surface Σ, the temperature T, the pore pressure P, the scalar thermal flux $\vec{\Psi}.\vec{n}$ and the total stress vector $\tilde{\sigma}\vec{n}$ are continuous (We whall see further what are the reasons for the continuity of

$\vec{\Psi}.\vec{n}$ and $\tilde{\sigma}\vec{n}$). The normal component of the last vector is $\sigma_n = \tilde{\sigma}\vec{n}.\vec{n}$ which is the total normal stress of the joint. The total tangential stress is the vector $\vec{\sigma}_t = \tilde{\sigma}\vec{n} - \sigma_n\vec{n}$ the norm of which is the shear stress.

3.1.2 Tangential gradients

Let f be a function of $\vec{x}$ in Ω and let $\vec{grad}_\Sigma(f)$ denotes the tangential part of the vector $\vec{grad}(f)$. The new differential operator $\vec{grad}_\Sigma(.) = \vec{grad}(.) - [\vec{n}.\vec{grad}(.)]\vec{n}$ has the same mathematical properties as $\vec{grad}(.)$. Let $\vec{Q}$ be a vector function of $\vec{x}$ on Σ which is everywhere tangent to Σ (i.e. $\vec{Q}.\vec{n} = 0$ at all points on Σ). Let us construct a volume Ω_Σ surrounding the surface Σ with a small uniform thickness $2a$:

$$\Omega_\Sigma \; = \; \{\vec{X} = \vec{x} + \zeta\, a\vec{n} \; ; \; \vec{x} \in \Sigma \; ; \; \zeta \in [-1, +1] \; ; \; \vec{n} \text{ normal to } \Sigma \text{ at point } \vec{x}\} \quad (14)$$

We can then extend the Σ-field $\vec{Q}$ to all the volume Ω_Σ: $\vec{Q}^*(\vec{X}) = \vec{Q}(x)$. The scalar field $div(\vec{Q}^*)$ is well defined at every point $\vec{X}$ in Ω_Σ and we can put as a definition: $div_\Sigma(\vec{Q}) = div(\vec{Q}^*)$ for all points on Σ. It can easily be proved that the new operator has similar properties as the classical one. For instance: $div_\Sigma(f\vec{Q}) = f div_\Sigma(\vec{Q}) + \vec{grad}_\Sigma(f).\vec{Q}$

3.2 Joint unknowns

The unknowns attached to a joint are time dependent fields defined on Σ in the current configuration of the skeleton [8]. All joint governing laws will be established as equations valid at each point of the surface Σ. Since the temperature T, the pore pressure P, the scalar thermal flux $\vec{\Psi}.\vec{n}$ and the total stress vector $\tilde{\sigma}\vec{n}$ are continuous, the only unknowns related to a joint are the discontinuity $\llbracket\vec{u}\rrbracket$ of the skeleton deformation vector $\vec{u}$ and the fluid volume flow vector $\vec{Q}$ inside the joint, the exact definition of which is that the average absolute fluid velocity is $\vec{Q}/e_h + (1/2)(\dot{\vec{u}}^+ + \dot{\vec{u}}^-)$. That is to say that the average (in the thickness of the joint) of the relative fluid velocity is $\vec{Q}/e_h$, reffered to the mean position of the skeleton which has two parts moving independently.

3.3 Balance laws

3.3.1 Mass balance

Since there is no solid phase inside the joint, the only continuity equation is related to fluid mass balance law where we must take into account the fluid flow in the surrounding porous continuum. The derived equation is similar to equation (3) and it can be simplified in the same way:

$$div_\Sigma(\rho_f\vec{Q})/\rho_f + \dot{e}_h - e_h\dot{\epsilon}_f + \llbracket\vec{q}\rrbracket.\vec{n} \; = \; 0 \quad (15)$$

After simplification neglecting $\vec{grad}_\Sigma(\rho_f).\vec{Q}/\rho_f$ we obtain:

$$div_\Sigma(\vec{Q}) + \dot{e}_h - e_h\dot{\epsilon}_f + \llbracket\vec{q}\rrbracket.\vec{n} \; = \; 0 \quad (16)$$

3.3.2 Momentum balance

When all the fluid body forces inside a joint are neglected (gravity as well as inertia forces), the only global momentum balance law we obtain is the continuity of the total stress vector $\tilde{\sigma}\vec{n}$. But we must complete this law assuming a governing equation similar to the Darcy's law. The usual proposed equation is: $\vec{Q} = -\tilde{k}_J \vec{grad}_\Sigma(H)$ where the joint permeability tensor $\tilde{k}_J$ depends upon the fluid properties as well as upon the hydraulic joint aperture. Usually the joint permeability tensor is assumed to be isotropic ($\tilde{k}_J = k_J \tilde{1}$) and k_J obeys the well-known cubic law [9]: $k_J = (\rho_f g/12/\eta_f) * e_h^3$.

3.3.3 Energy balance

If all kinds of energy inside the joint are neglected the energy balance law is reduced to the continuity of $\vec{\Psi}.\vec{n}$.

3.4 Constitutive laws

Since the fluid state law is the same as for the porous continuum (equation 11) and since there is no solid phase inside the joint, we only need to define the joint constitutive law governing its mechanical behaviour. Moreover the only question we have to answer concerns the joint effective stress vector which is generally assumed to be $\vec{\sigma}_{eff} = \tilde{\sigma}\vec{n} + b_J P\vec{n}$ where b_J is the joint Biot's coefficient. The general joint constitutive laws are based on a sharing of the displacement discontinuity $\sharp\vec{u}\sharp$ into two parts ([10], [11], [12], [13], [14], [15]). The first part depends linearly (elasticity) on the effective stress variation (normal and tangential stiffnesses are used). The second part is defined in such a way as to account for the actual normal law (hyperbolic equation for the closure and no tension material for the opening) and for the irreversible sliding behaviour with possible dilatancy phenomenon.

4. TIME INTEGRATION

4.1 Finite deformation

There are two kinds of difficulties due to the finite deformation of the geometry of a porous medium. The first kind is related to the large displacement which induces the fact that the current unknown position $\vec{x}$ cannot be replaced in balance equations (equation 5 for instance) by the given initial position. The second kind of difficulties concerns the constitutive laws where the stress rate must be taken in Jauman's sense and where all the material tensors (Hooke's tensor $\tilde{H}$, Biot's tensor $\tilde{B}$ ...) must be updated upon the rotation of the co-ordinate system attached to the deformed skeleton ([1], [16]).

The finite deformations can be approached numerically using a step by step method called "Updated Lagrangian Method". In such a technique, the actual deformation is divided into infinitesimal deformations at the start of which all the problem variables are known. When a small step is done, the geometry is updated ($\vec{x} = \vec{x} + \vec{\delta u}$), the obtained stress $\tilde{\sigma}_{eff}$ is corrected using the last rotation tensor in order to become the initial stress $\tilde{\sigma}_{eff}^I$ for the next step and all material tensors are updated in the rotated co-ordinate system. During each step, the classical constitutive law can be used: $\tilde{\sigma}_{eff} =$

$\tilde{\sigma}^I_{eff} + \tilde{\tilde{H}}(\tilde{\delta\epsilon} - \tilde{\delta\epsilon}^p - \delta T\tilde{\alpha})$ where $\tilde{\delta\epsilon}$, the Green's strain tensor, is the symmetric part of the gradient of the small deformation $\vec{\delta u}$ knowing that all space differentiations use the current known position $\vec{x}$ at the beginning of the step.

When this simple algorithm is used together with the Finite Element Method, only the node co-ordinates are modified (the connectivity matrix is unchanged). But at some level of deformation the mesh needs, some times, to be updated (adaptive griding) not only to avoid distorted elements but also to take into account the large change on the geometry of the joint elements.

4.2 Skeleton constitutive law

For each small deformation, the rheological laws for the skeleton and the joints are integrated by a step by step method in the case of viscoplasticity and by an iterative process for elastoplastic materials [17]. In both cases, within each step (or iteration), a simple linear elastic problem has to be solved (Initial Stress Method).

4.3 Transient phenomena

The thermal energy balance law can be integrated in time by an implicit Euler method. Let T^I be a known temperature at time t (start of a step) and let T be the unknown temperature at time $t + h$ where h is the time step. The equation (10) becomes:

$$div(\vec{\Psi}) + (C/h)T \;\; = (C/h)T^I \tag{17}$$

The transient problem is then transformed into a steady-state problem. This technique is applied also to the hydraulic problem: in equation (13) $\dot{P}$ is replaced by $(P - P^I)/h$.

5. SPACE INTEGRATION

5.1 Separated variational formulations

After all the time integrations we obtain three classical problems (linear elasticity, steady-state heat conduction and steady-state fluid diffusion) where the principal unknowns are fields ($V = T$ or H or $\vec{\delta u}$) defined on a known domain Ω and have to satisfy some differential equations (local formulation). These equations can be integrated using any numerical method such as Finite Difference Method, or Finite Volume Method, or Boundary Element Method ...

The use of the Finite Element Method to solve such a problem is easier when the local formulation is transformed into a variational one. The unknown field V must then satisfy some boundary conditions (prescribed values of V on $\partial\Omega$) and an equality $\varpi(V, V^*) = 0$ for any virtuel field V^*. The theorem of virual works in continuum mechanics is the most common example of this kind of formulation. The sign of the function ϖ is chosen in such a way as, in the linearized formulation, $\varpi(V, V^*)$ is the sum of a linear function of V^* and a bilinear function of V and V^* which is symmetric and positive. This convention is needed below (added terms).

Since this procedure is well known in the case of continua, we shall restrict this section to the terms which must be added to the functions ϖ in order to take into account the joints [18]. The theoretical background on which the determination of all added terms is based uses the fact that each function ϖ is the sum of integrals on all continuous parts (Ω^+ and Ω^-) of the whole domain Ω. Using the mathematical properties of the operator div, some of these integrals are transformed into integrals on the boundaries $\partial\Omega^+$ and $\partial\Omega^-$ which have a common part Σ (joint surface).

5.1.1 Hydraulic problem

There is no added term in the thermal function $\varpi(T, T^*)$ where the virtual temperature field T^* is continuous in the whole domain Ω.

The virtual hydraulic head field H^* is continuous too, but the hydraulic function $\varpi(H, H^*)$ must be modified using equation (16). The term to add is then:

$$\int_\Sigma -\vec{Q}.\vec{grad}_\Sigma(H^*)dS + \int_\Sigma (\dot{e}_h - e_h\dot{\epsilon}_f)H^*dS + \int_{\partial\Sigma} \vec{Q}.\vec{m}H^*dl$$

The unit vector $\vec{m}$ is tangent to the surface Σ and normal to its boundary $\partial\Sigma$ which is a line. The term $\vec{Q}.\vec{m}dl$ is the fluid volume rate which leaves the joint Σ from its boundary $\partial\Sigma$ per unit length. This term is of course known. It is a part of the given data (boundary conditions). Let us now use Darcy's law of the joint and equation (11) after time integration. The added term becomes:

$$\int_\Sigma [\tilde{k}_J \vec{grad}_\Sigma(H).\vec{grad}_\Sigma(H^*) + (e_h\beta_f\rho_f g/h)HH^*]dS +$$

$$\int_\Sigma [\dot{e}_h - 3e_h\alpha_f\dot{T} - (e_h\beta_f\rho_f g/h)H^I]H^*dS + \int_{\partial\Sigma} \vec{Q}.\vec{m}H^*dl$$

At the start of each time step the hydraulic head H^I is known. So, when the mechanical (variable e_h : hydraulic aperture) and the thermal (variable T : temperature) problems are solved, the factor of H^* in the second integral is then known.

5.1.2 Mechanical problem

Since the displacement vector is discontinuous throughout the joint surface Σ, the virtual field $\delta\vec{u}^*$ has a discontinuity $\sharp\vec{u}^*\sharp$ which operates on the added term to the mechanical function ϖ. This added term is the work of stresses acting on the joint:

$$\int_\Sigma \tilde{\sigma}\vec{n}.\sharp\vec{u}^*\sharp dS$$

5.2 Coupled variational formulation

The exact coupled variational formulation consists of adding together the three functions ϖ to obtain a function Π associated to the coupled thermo-hydro-mechanical process ([3],[16],[19],[20],[21]). The result is that the unknown complex field $W = (T, H, \delta\vec{u})$ must satisfy some boundary conditions and the equality $\Pi(W, W^*) = 0$ for any virtual field

$W^* = (T^*, H^*, \delta\vec{u}^*)$. When a numerical technique is based on the coupled variational formulation, the central part of the algorithm consists in a linear algebraic system as usual. But this system is not symmetric due to the fact, for instance, that the temperature T is present in the hydraulic function ϖ while the hydraulic head H does not operate in the thermal function ϖ. Some authors oblige the system to be symmetric by using a weak form of the coupled variational formulation; this is done exactly in the same way as for solving elastoplastic problem by the Tangent Stiffnes Matrix Method in the case of non-associated material (i.e. for which the yield and the potential functions differ) [22]. To correct the effect of this forced symmetrization an iterative process is then used. The use of the coupled variational formulation has another undesirable effect: the increase of the size of the linear system due to the growth of the number of the degrees of freedom (DOF). We shall describe in the last section another iterative method that we have developed on the occasion of the DECOVALEX project and which avoids all these difficulties.

5.3 Finite element method and joints

In some numerical codes the joints are modeled by conventional volumetric elements with given thicknesses. The material associated to such a kind of element has, of course, particular constitutive laws to represent the actual behaviour of the physical joint. But, the most common approach is to use a surface joint element [18] in which each node, for a mechanical problem, has 6 DOF (deformation vectors: $\vec{u}^-$ and $\vec{u}^+$). In some codes such a 6 DOF node is replaced by a couple of 3 DOF nodes (n^-, n^+) where n^- is connected to the associated node in Ω^- and n^+ is connected to the associated node in Ω^+.

In all cases, the realization of the mesh is the most difficult operation when using Finite Element Method to model fractured continua. It is well known that the three-dimensional automatic meshing needs sophisticated computer routines, even when there is no joint surface, the presence of which increases considerably this sophistication. We give a brief description of the principles of some usual helpful tools used in griding codes:

- The ideal solution would be an automatic meshing code for continua which is able to represent the geometry of all given surfaces (not only the boundary $\partial\Omega$ but also the internal surfaces Σ corresponding to the joints).

- When this ideal meshing code does not exist, the total mesh generating operation may be done in two steps. First all surfaces (boundary and joints) are meshed. The second step needs a tool able to mesh continuous parts of the whole domain Ω with prescribed surface mesh of its external boundary. The last problem is often impossible to solve and needs an iterative process between the two steps.

- If the user has no ideal tool he can use a poor technique which works in a step by step process. First, a mesh is realized for the whole continum Ω. Then, a surface joint Σ is introduced to cut some of the volumetric elements of the mesh. Each of these elements is locally refined to take into acount the surface Σ without destroing the consistency of the surrounding three-dimensional mesh. The new mesh is modified in the same way by introducing a new surface joint Σ if any. This simple technique requires some precaution because a surface can cut an element into two parts, one of which has too small a volume. In such a case some nodes need to be moved but their new positions must be in accordance with prescribed geometrical data.

6. ITERATIVE METHOD FOR THM PROCESSES

6.1 Reformulation of the problem

Let $\underline{T}$ be a complex unknown including all time dependent thermal unknown fields. To be more explicit let us say that $\underline{T}$ is the time dependent temperature field which can be replaced, when Finite Element Method is used, by the discret set of all nodal temperatures at given times chosen close together. In the same way the hydraulic $\underline{H}$ and mechanical $\underline{M}$ unknowns are defined. Suppose now that $\underline{H}$ and $\underline{M}$ are known, solving the thermal problem (energy balance, even without any simplification and regardless the chosen method) is to determine $\underline{T}$ as a function of $\underline{H}$ and $\underline{M}$ using given data. Let Ψ_T be such a function which is fully known for each given actual problem even when it cannot be defined explicitely. The solution of a thermal problem (derived either from its local formulation or from its own variational formulation regardless of the coupled phenomena) can then be defined by the following equality:

$$\underline{T} = \Psi_T(\underline{H}, \underline{M}) \tag{18}$$

Of course, when the simplified energy balance law is used, the thermal function Ψ_T does not depend on the hydraulic variable $\underline{H}$ and it depends on the mechanical variable $\underline{M}$ only in the case of finite deformation. But all these particular cases are fully included in the general equation (18). Similar general equalities can be defined concerning the hydraulic (function Ψ_H) and mechanical (function Ψ_M) problems, respectively:

$$\underline{H} = \Psi_H(\underline{M}, \underline{T}) \tag{19}$$

$$\underline{M} = \Psi_M(\underline{T}, \underline{H}) \tag{20}$$

The functions Ψ_T, Ψ_H and Ψ_M can be constructed using any analytical approach (closed form) or numerical method. The proposed iterative algorithm described in the following section has only two restrictions:

- Since the methods used for solving each of the three problems (T, H and M) can differ, the representations of the three complex unknowns must be homogeneous.

- The chosen method for each problem (equations 18, 19 and 20) has to be able to produce the left hand side of the equation for any given "values" of the arguments of the associated function.

Thanks to the weakness of these restrictions the functions Ψ_T, Ψ_H and Ψ_M do not need to be explicitly constructed.

6.2 Proposed iterative algorithm

6.2.1 Algorithm

1. **Initialization**: starting "values" of $\underline{H}$ and $\underline{M}$ are chosen. Generally, the hydraulic head is chosen constant in time and equal to the given initial field and the skeleton is assumed to remain undeformed (no variation of the initial given geometry). These starting "values" are put in current "values" stack.

2. **Thermal problem**: solve the thermal problem using all given data and the current known "values" of $\underline{H}$ and $\underline{M}$. Determine the difference between the calculated "values" of $\underline{T}$ and the previous current "values". This difference D_T is a scalar using any chosen norm. The calculated "values" of $\underline{T}$ are put in current "values" stack.

3. **Hydraulic problem**: solve the hydraulic problem using all given data and the current known "values" of $\underline{M}$ and $\underline{T}$. Determine the difference between the calculated "values" of $\underline{H}$ and the previous current "values". This difference D_H is a scalar using any chosen norm. The calculated "values" of $\underline{H}$ are put in current "values" stack.

4. **Mechanical problem**: solve the mechanical problem using all given data and the current known "values" of $\underline{T}$ and $\underline{H}$. Determine the difference between the calculated "values" of $\underline{H}$ and the previous current "values". This difference D_M is a scalar using any chosen norm. The calculated "values" of $\underline{M}$ are put in current "values" stack.

5. **Convergence**: as long as the iteration number is less then a given maximum value and as long as D_T, or D_H or D_M is yet higher than some given levels of accuracy, continue the iterative process restarting from the thermal problem (step 2 above).

6.2.2 Advantages of the method

All the advantages of the method are due to its simplicity.

- **Easy to install**: the method can use three separate codes (T, H and M) and an external process to perform the loop (T -> H -> M -> T ...). For instance the codes T, H and M may be binary FEM executable and the external process is a simple UNIX script shell or a command file under DOS as Operating System ...

- **General purpose**: the algorithm does not need any simplification neither in the formulation of each of the three problems nor in the governing equations for coupled THM processes.

- **Validity**: if the THM problem is consistent (i.e. it has a solution) and if the iterative process converges then the obtained solution is a good solution of the THM problem. The mathematical proof of this assertion is based only on the "continuity" of the three functions Ψ_T, Ψ_H and Ψ_M in the sense of a chosen norm in the space of the complex variable $(\underline{T}\,,\,\underline{H}\,,\,\underline{M})$.

- **Convergence**: the technique belongs to a wide familly of classical iterative methods used to solve the well known problem [*find a, a=f(a), f is a given function*].

The iterative process which consists in constructing a series (a_n) using the recurrent relationship: $a_n = f(a_{n-1})$ is widely used to solve algebraic systems. It can be proved easily that if the function f has a LIPSCHITZ's constant less than 1, then the problem is consistent and the iterative process is convergent.

6.2.3 Weakness of the method

For the time being this simple algorithm has been validated only experimentally (heuristic approach). A few cases more or less sophisticated, for which the exact solutions were known ([23],[24]), have been treated with this numerical technique which appears to be efficient and quite fast (even in the case of non-linear materials the external iterative loop needed a maximum of 10 iterations to converge towards the exact solution with a high level of precision).

We believe that one of the main reasons of the efficiency of the proposed numerical technique lies in the fact that at each step of the process and for each time considered, we can easily derive a good approximation of the rates of all the variables which operate in the governing equations.

However, a lot of research work has yet to be done in order to construct a rigourous theoretical background to proof the absolute convergence of the proposed iterative method. The study of the mathematical properties of the THM problems may show that the LIPSCHITZ's condition is always satisfied. Perhaps also such a work will allow either to establish some necessary conditions to the convergence of the technique or to propose some modifications of the process to ensure this convergence.

7. ACKNOWLEDGMENTS

The authors are grateful to the Institut de Protection et de Sureté Nucléaire of the Commissariat à l'Energie Atomique who has supported their participation to the DECO-VALEX Project and has enabled them to carry out the research work which was needed to develop and validate the iterative method of analysis of coupled THM processes.

8. REFERENCES

1. C. Truesdell, Introduction à la Mécanique Rationnelle des Milieux Continus, Masson et Cie, Paris (1974).

2. F. Fer, Thermodynamique Macroscopique, Gordon and Breach, Paris (1971).

3. O. Coussy, Mécanique des Milieux Poreux, Editions Technip, Paris (1991).

4. P. Charlez (ed.), Mechanics of Porous Media, A.A. Balkema, Rotterdam, Brookfield (1995).

5. L.W. Morland, Simple Constitutive Theory for Fluid-saturated Porous Solid, J. Geophys. Res., 77 (1972) 890.

6. S.K. Gard and A. Nur, Effective Stress Law for Fluid-saturated Porous Rocks, J. Geophys. Res., 78(26) (1973).

7. R.M. Bowen, Compressible Porous Media Models by use of Theory of Mixtures, Int. J. Eng. Sci., 20(6) (1982) 697.

8. N. Barton, S. Bandis and K. Bakhtar, Strength, Deformation and Conductivity Coupling of Rock Joints, Int. J. Rock Mech. & Min. Sci., 22(3) (1985) 121.

9. P. Witherspoon, J. Wang, K. Iwai and J. Gale, Validity of Cubic Law for Fluid in a Deformable Rock Fracture, Wat. Resour. Res., 16(6) (1980) 1016.

10. S. Bandis, A. Lumsden and N. Barton, Fundamentals of Rock Joint Deformation, Int. J. Rock Mech. & Min. Sci., 20 (1983) 249.

11. H. Benjelloun, Etude Expérimentale et Modélisation du Comportement Hydroméca-nique des Joints Rocheux, Thèse de Doctorat, Université Joseph Fourier, Grenoble (1991).

12. A. Bougnoux, Modélisation Thermo-Hydro-Mécanique des Massifs Fracturés a Moyenne ou Grande Echelle, Thèse de Doctorat, Ecole des Mines de Paris (1995).

13. R. Goodman, Methods of Geological Engineering in Discontinuous Rocks - Ch. 5: Mechanical Properties of Discontinuities, West Pub. Company, San Francisco (1976).

14. W. Leichnitz, Mechanical Properties of Rock Joints, Int. J. Rock Mech. & Min. Sci., 22(3) (1985) 313.

15. M. Plesha, Constitutive Models for Rock Discontinuities with Dilatancy and Surface Degradation, Int. J. Num. & An. Meth. in Geomech., 11 (1987) 345.

16. S.H. Advani, T.S. Lee, J.K. Lee and C.S. Kim, Hygrothermomechanical Evaluation of Porous Media Under Finite Deformation, Int. J. Num. Meth. Eng., 36(1) (1993) 147.

17. S-M. Tijani, Résolution numérique des problèmes d'élastoviscoplasticité - Application aux cavités de stockage de gaz en couches salines profondes, Thèse de Docteur-Ingénieur. P. & M. Curie. Paris VI (1978).

18. G. Beer, an Isoparametric Joint/Interface Element for Finite Element Analysis, Int. J. Num. Meth. Eng., 21(4) (1985) 585.

19. P.M. Cleary, Fundamentals Solutions for a Fluid-saturated Porous Solid, Int. J. Solids Structures, 13 (1977) 785.

20. O.C. Zienkiewicz and T. Shiomi, Dynamic Behaviour of Saturated Porous Media - The Generalized BIOT Formulation and its Numerical Solution, Int. J. Num. & An. Meth. in Geomech., 8 (1984) 71.

21. R.S. Sandhur and S.J. Hong, Dynamics of Fluid-saturated Soils - Variational Formulation, Int. J. Num. & An. Meth. in Geomech., 11 (1987) 241.

22. H.R. Thomas and S.D. King, A Non-linear, Two-dimensional, Potential-based Analysis of Coupled Heat and Mass Transfer in Porous Medium, Int. J. Num. Meth. Eng., 37(21) (1994) 3707.

23. B. Amadei and T. Illangasekare, Analytical Solutions for Steady and Transient Flow in Non-homogeneous Rock Joints, Int. J. Rock Mech. & Min. Sci.,29(6) (1992) 561.

24. J. Booker and C. Savvidou, Consolidation Around a Point Heat Source, Int. J. Num. & An. Meth. in Geomech., 9 (1985) 173.

O. Stephansson, L. Jing and C.-F. Tsang (Editors)
Coupled Thermo-Hydro-Mechanical Processes of Fractured Media
Developments in Geotechnical Engineering, vol. 79

Distinct element models for the coupled T-H-M processes: Theory and implementation

Mikko P. Ahola [a], Alain Thoraval [b], and Asadul H. Chowdhury [a]

[a]Center for Nuclear Waste Regulatory Analyses, Southwest Research Institute, 6220 Culebra Road, San Antonio, TX, 78238, USA

[b]Laboratoire de Mecanique des Terrains, Institut National de L'Environment Industriel et des Risques, Parc de Saurupt (INERIS), 54042, Nancy, France

Abstract

Numerical assessment of the design and long-term performance of deep underground disposal facilities for high-level radioactive waste in jointed rock has led to the increased use of discontinuum modeling approaches to evaluate the coupled T-H-M effects within the immediate vicinity of the waste emplacement drifts. This is particularly true with regard to accurate determinations of the long-term mechanical deformations of the fractured rock mass around the emplacement drifts/boreholes due to thermal and other loads, as well as accurate determination of fluid fluxes into the drifts which could take place along selected preferential fracture pathways. This chapter presents the basic theoretical background and numerical formulation of the distinct element method for each of the individual as well as two-component (e.g., T-M, M-H, and T-H) processes, and approaches for T-H-M modeling of fractured media for either transient or steady state conditions. Particular emphasis is placed on extension of the distinct element method to simulate the thermal-hydrologic coupling in fluid filled fractures. Results are presented to show that thermal convection in flowing fractures can have a very important effect on the fluid and adjacent rock temperatures depending on the fracture hydraulic aperture, fluid velocity, and fluid viscosity.

1. INTRODUCTION

In the case of a fractured rock medium where the discontinuities play a critical role in determining the deformation and rigid body motion [1-3], the distinct element method (DEM) [4-10] is a rational method for modeling the medium. In this method, properties of both the joints and the intact rocks are explicitly modeled. This is in contrast to continuum methods such as finite element and finite difference which, in most cases, homogenize the properties of joints and intact rock into a pseudocontinuum. The DEM has two distinguishing features compared to continuum methods: (i) the behavior of the geologic system is described by both a continuum material description of the intact rock and a discontinuum material representation for discontinuities (i.e., joints, faults, etc.), and (ii) the deformation mechanisms include large displacement (i.e., joint slip and separation) and block rotation. The geometry of the blocks is generally constrained by the spacing and orientation of the discontinuities in the rock mass, thereby allowing blocks to interact with (or disconnect from) neighboring blocks. The DEM includes not only continuum theory representation for the blocks but also force-displacement laws which specify forces between blocks and motion law which specifies motion of each block due to unbalanced forces acting on the block.

Blocks may be treated as rigid or deformable in the DEM. For many applications (e.g., analysis of deep underground excavations), the deformation of individual blocks cannot be reasonably ignored [11], i.e., blocks cannot be assumed to be rigid. In this case, arbitrary deformation of blocks is permitted through internal discretization of blocks into finite difference zones, in addition to the rigid-body modes associated with each block of the jointed rigid block deformation.

The DEM was initially developed primarily for mechanical analysis of jointed or blocky systems involving large rigid body block motions and/or large relative displacements along the block interfaces or joints [11]. These types of deformations commonly arise in hard rock mining as well as other related industries [2,3]. With the current international effort to permanently dispose of radioactive wastes in underground repositories, it has become necessary to expand DEM capabilities to take into account the coupled thermal-hydrological-mechanical (THM) processes in jointed rock in order to assure that the repository performance requirements are met. In many cases, the rock matrix itself has a very low permeability such that the majority of the flow occurs within the fractures. Also, depending on the flow rates and aperture distributions within fractured rock mass, convection as a result of fluid flowing in the fractures can have an impact on the temperature profile and corresponding strains within the rock. Thus, the ability to explicitly model fluid flow through fractures in a distinct element model is of importance especially within the near-field waste emplacement region.

In codes utilizing the DEM (e.g., UDEC), the equations governing the mechanical, thermal, and hydrologic response are not fully coupled. Rather, the coupling is achieved within the solution process in which explicit or implicit time marching is done on one process while the other is held fixed, and vice versa. For mechanical processes, the governing equations are the equations of motion, while for hydrological processes, Darcy's parallel plate law is applied to flow within the fractures and the matrix is assumed to be impermeable. Thermal processes are governed by the heat diffusion equation in which the flux is related to temperature through Fourier's law. Fractures are assumed to have no effect on heat transfer.

2. GOVERNING EQUATIONS

2.1. Mechanical Behavior

There are numerous references which describe the theoretical background and numerical formulation for mechanical behavior used in UDEC. These are briefly discussed in this section.

2.1.1. Equations of Motion

The motion of an individual block is determined by the magnitude and direction of resultant out-of-balance moment and forces acting on it as a result of changes in loading states on a system of blocks. In this section, the equations of motion, which describe translation and rotation of the block about its centroid, are described [6,12]. Consider the motion of a single mass with viscous damping acted on by a varying force $F^{(t)}$. The equation of motion, including viscous damping, can be written as:

$$\ddot{u} = \frac{F^{(t)}}{m} - \alpha \dot{u} + g \tag{1}$$

where
$$\begin{aligned}
\dot{u} &= \text{velocity (m/s)} \\
t &= \text{time (s)} \\
F^{(t)} &= \text{force acting on mass (N)} \\
m &= \text{mass (kg)} \\
g &= \text{gravitational acceleration (m/sec}^2) \\
\alpha &= \text{damping coefficient (1/sec)}
\end{aligned}$$

The central difference scheme for the left-hand side of Eq. (1) at time t can be written as:

$$\ddot{u} = \frac{\dot{u}^{(t+\Delta t/2)} - \dot{u}^{(t-\Delta t/2)}}{\Delta t} \tag{2}$$

A difference equation equivalent to (1) can be written as:

$$\frac{\dot{u}^{(t+\Delta t/2)} - \dot{u}^{(t-\Delta t/2)}}{\Delta t} = \frac{F^{(t)}}{m} - \alpha \left[\frac{\dot{u}^{(t+\Delta t/2)} + \dot{u}^{(t-\Delta t/2)}}{2} \right] + g \tag{3}$$

Note that the damping force in the equation is centered at time t.

Rearranging Eq. (3) yields:

$$\dot{u}^{(t+\Delta t/2)} = \left[\dot{u}^{(t-\Delta t/2)}\left(1 - \frac{\alpha \Delta t}{2}\right) + \left(\frac{F^{(t)}}{m} + g\right)\Delta t \right] \Big/ (1 + \alpha \Delta t/2) \tag{4}$$

With velocities stored at the half-timestep point, it is possible to express displacement as:

$$u^{(t+\Delta t)} = u^{(t)} + \dot{u}^{(t+\Delta t/2)}\Delta t \tag{5}$$

Because the force depends on displacement, the force-displacement calculation is done at one time instant. The acceleration is also given by the force at this time instant (i.e., t+ Δt) and the mass. Figure 1 illustrates the central difference scheme with the order of calculation indicated by the arrows.

For blocks which are acted upon by several forces as well as gravity, the velocity equations become:

$$\dot{u}_i^{(t+\Delta t/2)} = \left[\dot{u}_i^{(t-\Delta t/2)} \left(1 - \frac{\alpha \Delta t}{2} \right) + \left(\frac{\Sigma F_i^{(t)}}{m} + g_i \right) \Delta t \right] \bigg/ \left(1 + \alpha \Delta t/2 \right) \tag{6}$$

where $\dot{u}_i$ = velocity components of centroid in ith direction

$\qquad \Sigma F_i^{(t)}$ = summation of forces acting on centroid in ith direction

Similiary, the equation of motion for rotation of a block acted upon by several forces is given by:

$$\dot{\theta}^{(t+\Delta t/2)} = \left[\dot{\theta}^{(t-\Delta t/2)} \left(1 - \frac{\alpha \Delta T}{2} \right) + \frac{\Sigma M_i^{(t)}}{I} \Delta t \right] \bigg/ \left(1 + \frac{\alpha \Delta t}{2} \right) \tag{7}$$

where ΣM = total moment acting on block centroid
            I = moment of inertia of block, and
            $\dot{\theta}$ = angular velocity of block about centroid

The velocity obtained from Eqs. (6) and (7) are used to determine the new block location according to

$$u_i^{(t+\Delta t)} = u_i^{(t)} + \dot{u}^{(t+\Delta t/2)} \Delta t \tag{8}$$

$$\theta^{(t+\Delta t)} = \theta^{(t)} + \dot{\theta}^{(t+\Delta t/2)} \Delta t \tag{9}$$

where θ =   rotation of block about centroid, and
         u_i =   coordinates of block centroid

Thus, each iteration produces new block positions which generate new contact forces. Resultant forces and moments are used to calculate linear and angular accelerations of each block. Block velocities and displacements are determined by integration over incremental timesteps. The procedure is repeated until a satisfactory state of equilibrium or mode of failure results.

Two forms of viscous damping are available in the distinct element formulation: mass proportional damping and stiffness proportional damping [6,7]. Mass proportional damping has an effect similar to that of immersing the block assembly in a viscous fluid,

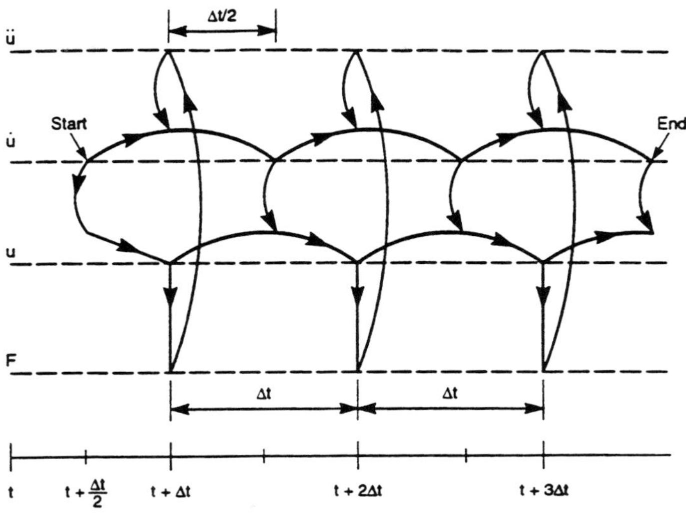

Figure 1. Interlaced nature of the calculation cycle used in the distinct element formulation [6].

i.e.,absolute motion relative to the frame of reference is damped. Stiffness proportional damping is physically equivalent to damping across contacts and serves to damp block relative motion. Damping across the contacts operates both in the shear and normal directions; the shear damping is "switched off" during sliding. For an elastic continuous system (one in which there is no slip or breaking and making of new contacts), the damping scheme described above is termed Rayleigh damping. For a discontinuous system that dissipates energy in slip, the theory does not apply, but damping still occurs and can be understood in terms of the physical effects of each type of damping. Either type of damping can be used separately or together. Mass proportional damping is effective in reducing low-frequency motion where the whole block assembly "moves" from side to side. Stiffness proportional damping is more effective against the high-frequency noises of individual blocks "vibrating" against their neighbors.

2.1.2. Interface Constitutive Relations

The deformability of the discontinuities or interfaces between blocks and the frictional characteristics are represented in the DEM by spring-slider systems with prescribed force-displacement relations which allow evaluation of shear and normal forces between blocks [6]. In the model, spring-slider systems are located at contact points between blocks. The amount of penetration or overlap between two adjacent blocks can be defined directly from block geometry and block centroid translation and rotation. The force-displacement relation at one contact is thus uncoupled from that at another on the same block.

After each timestep, incremental stresses acting across a contact are calculated by dividing the incremental force by the associated contact length. These incremental stresses are added to the existing stresses, and the constitutive criteria are checked. In general, the joint constitutive relations must provide the stress increments as a function of the displacement increments, current stresses, and possibly other state parameters

$$\Delta \sigma_n, \Delta \sigma_s = f(\Delta u_n, \Delta u_s, \sigma_n, \sigma_s, \ldots) \tag{10}$$

UDEC uses several joint behavior relations to describe both the normal and shear mechanical behavior at the interface. These are Mohr-Coulomb, Barton-Bandis, and Continously-Yielding joint models. The Mohr-Coulomb model is a linear deformation model in which the joint assumes perfect plastic deformation if the shear strength of the joint is exceeded. In its basic form, the Mohr-Coulomb model does not consider joint wear and dilation behavior. However, the dilation behavior may be added to the joint behavior [6]. The Barton-Bandis model was proposed to take into consideration the effect of various joint material properties [e.g., joint roughness coefficient (JRC) and joint wall compressive strength (JCS)] as well as applied normal loading on joint deformation and strength [13,14]. Attrition of the surface roughness or reduction of the JRC is represented in a piece-wise linear manner. Once the JRC becomes zero, the joint shear essentially resumes the Coulomb model type of behavior. The dilation of the joint in the Barton-Bandis model is also a function of the JRC value, and also decreases along with the JRC during shearing. Another joint model, the Continuously-Yielding model [6,15] is also intended to simulate the progressive damage of the joint under shear and display irreversible nonlinear behavior from the onset of shear loading. As shear damage accumulates, the joint friction angle in the Continuously-Yielding model is continually reduced. Both the normal and shear joint stiffness are specified as functions of the joint normal stress. The formulation of the joint dilation angle in the Continuously-Yielding model is given by [6].

2.1.3. Deformability of Fully Deformable Blocks

Fully deformable blocks are internally discretized into finite difference triangles [6,7]. The vertices of these triangles are gridpoints, and the equations of motion for each gridpoint are formulated as follows:

$$\ddot{u}_i = \frac{\int_s \sigma_{ij} n_j \, ds + F_i}{m} + g_i \tag{11}$$

where s is the surface enclosing the mass m lumped at the gridpoint,

n_j is the unit normal to s

F_i is the resultant of all external forces applied to the gridpoint (from block contacts or otherwise), and

g_i is the gravitational acceleration

During each timestep, strains and rotations are related to nodal displacements in the usual fashion:

$$\epsilon_{ij} = \frac{1}{2}(u_{i,j} + u_{j,i})$$
$$\theta_{ij} = \frac{1}{2}(u_{i,j} - u_{j,i})$$

(12)

The constitutive relations for deformable blocks are used in an incremental form, so that implementation on non-linear problems can be accomplished easily. The actual form of the equations is:

$$\Delta \tau^e_{ij} = \lambda \, \Delta \epsilon_v \delta_{ij} + 2\mu \, \Delta \epsilon_{ij}$$

(13)

where λ, and μ are the Lame's constants,

$\Delta \tau^e_{ij}$ = are the elastic increments of the stress tensor

$\Delta \epsilon^e_{ij}$ = are the incremental strains

$\Delta \epsilon_v$ = $(\Delta \epsilon_{11} + \Delta \epsilon_{22})$ is the increment of volumetric strain in two dimensions, and

δ_{ij} = Kronecker delta function

2.1.4. Calculation Sequence

The calculations performed in the DEM (e.g., UDEC) alternate between application of a force-displacement law at the contacts and Newton's second law of motion at the blocks. The force-displacement law is used to find contact forces from displacements. Newton's second law gives the motion of the blocks resulting from the forces acting on them. If the blocks are deformable, motion is calculated at the gridpoints of the triangular finite-difference (constant-strain) elements within the blocks. Then, the application of the block material constitutive relations gives new stresses within the elements. Figure 2 shows schematically the calculation cycle for the DEM.

This numerical formulation conserves momentum and energy by satisfying Newton's laws of motion exactly. Although some error may be introduced in the computer programs by the numerical integration process, this error may be made arbitrarily small by the use of suitable timesteps.

2.2. Hydrologic Behavior

In many cases hydrologic flow through rock masses has been observed both in the laboratory and field to be fracture dominated [16]. As the fractures or discontinuities in a rock mass will be several orders of magnitude more permeable than the rock matrix itself, the flow of fluid in a saturated rock mass can be expected to be concentrated along the discontinuities. Thus, in distinct element formulations it is reasonable to consider only fluid flow within the fractures, as is the case with UDEC. Flow of fluid in an unsaturated rock mass may not be dominated by fractures, depending on the infiltration, since the matrix suction (potential) would cause the fluid to avoid large pores/fractures.

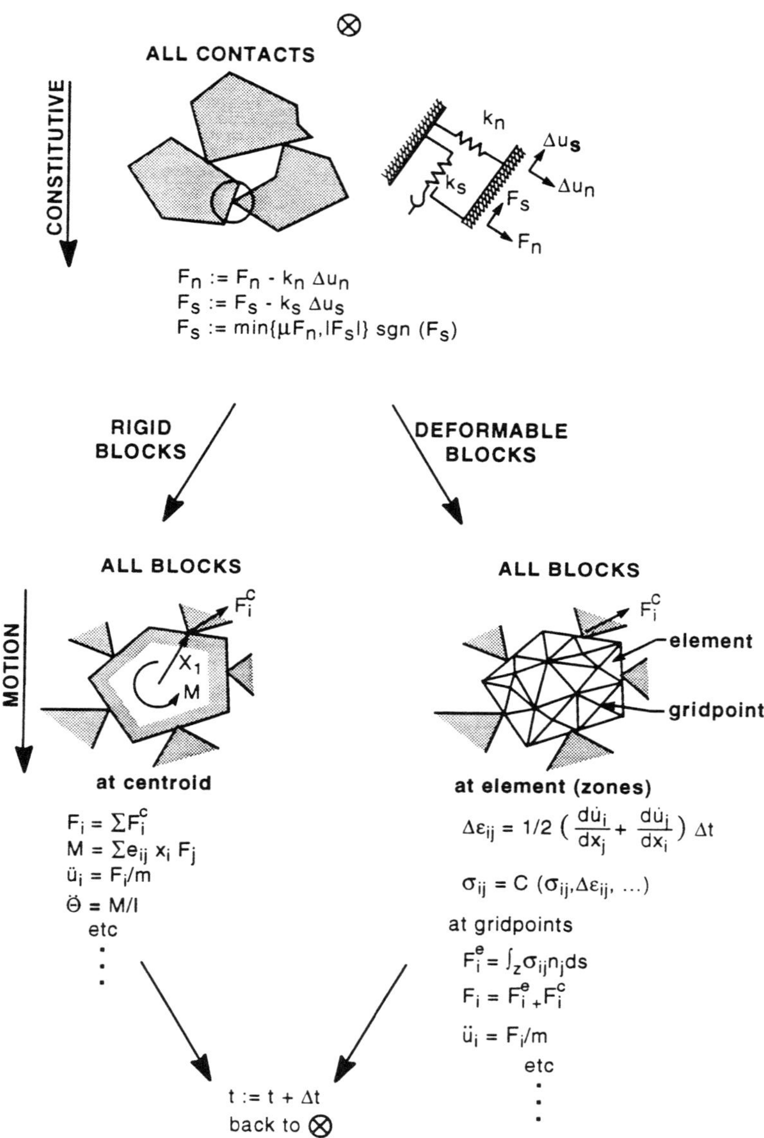

Figure 2. Calculation cycle for the distinct element method [17].

Flow in planar rock fractures is idealized as laminar viscous flow between parallel plates. In this model, the flow rate per unit width, q, is given by

$$q = C a^3 \left[\frac{\Delta p}{l} \right] \qquad (14)$$

$$C = \frac{1}{12\mu} \qquad (15)$$

where C is the fluid flow joint property which is assumed to remain constant

a   =   joint hydraulic aperture
μ =   dynamic viscosity of the fluid
Δp =   change in pressure across a contact between adjacent domains, and
l   =   length assigned to the contact

The joint permeability can be defined as $K = \frac{ga^2}{12\nu}$ where g is the acceleration due to gravity, and ν is the kinematic viscosity of the fluid.

The rate of fluid flow thus is assumed to be dependent upon the cubic power of the aperture. In actual rock fractures, the fracture walls are far from smooth and Eq. (14) does not truly represent the real case. The effect of roughness may cause a reduction in flow from that predicted using Eq. (14), however, this can be accounted for by applying an empirical correction factor to Eq. (14) to account for fracture roughness [18]. Witherspoon [19] tested both open and closed joints and concluded that the cubic law is still valid for the latter, provided that the actual mechanical aperture is used.

2.3. Thermal Behavior

Heat transfer can take place through either conduction, convection, or radiation. Convection can take place within the rock mass via groundwater flow and redistribution due to gravity, heating, or other mechanisms. This is discussed in more depth in Section 3.3. Convection may also take place from a surface of, for example, a tunnel or waste canister as a result of air circulation. Thermal radiation heat transfer can be the dominant heat transfer mechanism from solid surface to solid surface within a spent fuel assembly [20] and perhaps also important from the waste package to the surrounding borehole rock as well as across tunnel openings (e.g., from hot floor to cool roof) depending on whether the borehole/tunnel is backfilled or not. Within the rock medium, depending on amount of fluid movement within the fractures, conductive heat transfer most often dominates. It has been observed in studies that the existence of fractures can have some effect on conductive heat transfer through the rock by lowering the thermal conductivity [21]. However, such thermal properties (e.g., thermal conductivity and specific heat), can have a strong dependence on the degree of saturation and porosity within the rock medium [22].

The heat transfer in UDEC is based on conductive transfer within a continuous medium (i.e., neglecting the presence of fractures) with the provision for temperature, flux, convective or radiative boundaries. The standard equations for transient heat conduction can be found in many texts (e.g. [23]), and are reviewed here. The basic equation of conductive heat transfer is Fourier's law, which can be written in one dimension as

$$Q_x = -k_x \frac{\partial T}{\partial x} \tag{16}$$

where Q_x =   flux in the x-direction (W/m^2), and

    k_x =   thermal conductivity in the x-direction (W/m- °C)

A similar equation can be written for Q_y. Also, for any mass, the change in temperature can be written as

$$\frac{\partial T}{\partial t} = \frac{Q_{net}}{C_p M} \tag{17}$$

where Q_{net} = net heat flow into mass (M)

    C_p =   specific heat (J/kg- °C), and

    M =   mass (kg)

These two equations form the basis of the governing heat flow logic in UDEC. Eq. (17) can be written as

$$\frac{\partial T}{\partial t} = \frac{1}{C_p \rho} \left[\frac{\partial Q_x}{\partial x} + \frac{\partial Q_y}{\partial y} \right] \tag{18}$$

where ρ is the mass density.

Combining this with Eq. (16) gives,

$$\frac{\partial T}{\partial t} = \frac{1}{C_p \rho} \frac{\partial}{\partial x} \left[k_x \frac{\partial T}{\partial x} \right] + \frac{\partial}{\partial y} \left[k_y \frac{\partial T}{\partial y} \right] = \frac{1}{\rho c_p} \left[k_x \frac{\partial^2 T}{\partial x^2} + k_y \frac{\partial^2 T}{\partial y^2} \right] \tag{19}$$

if k_x and k_y are constant. This is the standard two-dimensional heat diffusion equation.

The method suggested by St. John [24] can be applied to determine the radius of influence of a single heat source or waste container on rock temperatures as a function of time in order to determine the size of the area required in a model for heat transfer analysis. The equation for temperature change at a distance, R_o, from a decaying point source of initial strength, Q_o, is given by Christianson [25]

$$\Delta T = \frac{Q_0}{\pi^{3/2}} \exp(-At) \frac{\sqrt{\pi}}{4\kappa} \exp(-R_o^2/4\kappa t) \, \text{Re} \left[w \left(\sqrt{At} + \frac{iR_o}{\sqrt{4\kappa t}} \right) \right] \qquad (20)$$

where
$$
\begin{aligned}
i &= \text{imaginary number } \sqrt{-1} \\
A &= \text{thermal constant} \\
\kappa &= \text{thermal diffusivity} \\
t &= \text{time (s)} \\
w(z) &= \text{complex error function in which z is the complex argument} \\
\text{Re}(\ ) &= \text{real part of argument}
\end{aligned}
$$

It is seen that the temperature change decays from the point source approximately proportional to

$$\exp\left(-R_o^2/4\kappa t\right) \qquad (21)$$

St. John [24] suggested that $R_o^2/4\kappa t = 4$ is sufficient to ensure a small temperature change. This expression requires that

$$R_o \geq 4\sqrt{\kappa t} \qquad (22)$$

where t is time in years.

3. COUPLED PROCESSES SIMULATED BY DEM

This section concentrates on the interactions among the thermal, hydrological and mechanical (THM) processes in fractured rock masses, simulated by DEM.

3.1. One-Way Thermo-Mechanical Behavior in Fractured Media

Heat generated by emplacement of radioactive wastes in underground excavations will, given time, spread throughout the surrounding rock mass and, consequently, cause the rock mass to expand. It is expected that, in most cases, heat transfer through a rock mass will be dominated by the conduction process [26]. Restriction of the expansion by the surrounding rock will result in thermally induced mechanical stresses. This thermally induced stress field, in addition to the *in situ* stresses and the stresses induced by excavation and repeated seismic effects, can induce normal and shear displacements of the rock joints. This increases the potential for rock mass failure resulting from excessive joint shear displacement. It may also induce microcracks in the rock which could reduce the stiffness of the rock and may lead to the formation or extension of a fractured network through coalescence and propagation of individual microcracks. The mechanical processes are thought to influence thermal processes primarily by changing fracture apertures, hence changing the effective thermal conductivity. This coupling is not modeled in UDEC.

Temperature changes given by Eq. (19) cause stress changes for fully deformable blocks according to the equation

$$\Delta \sigma_{ij} = -\delta_{ij} K\beta \; \Delta T \tag{23}$$

where $\Delta\sigma_{ij}$ = change in ij stress component
δ_{ij} = Kronecker delta function
K = bulk modulus (N/m^2)
β = volumetric thermal expansion coefficient (1/ °C), and
ΔT = temperature change

Note that $\beta = 3\alpha$ where α is the linear thermal expansion coefficient.

Equation (23) assumes a constant temperature in each triangular zone which is interpolated from the surrounding gridpoints. The incremental change in stress is added to the zone stress state prior to application of the constitutive law. The procedure for running a coupled thermomechanical simulation is shown in Figure 3. The fundamental requirement in performing the simulation is that temperature increases between successive thermal timesteps cause only "small" out-of-balance forces in blocks. Out-of-balance forces are small if they do not adversely affect the solution. For nonlinear problems, some experimentation may be necessary to obtain a sense of what small means in the particular problem being solved. This is performed by trying different allowable temperature increases when running the problem.

3.2 Two-Way Hydro-Mechanical Analysis for Fluid Flow in Fractured Media Simulated by DEM

Mechanical processes can affect the flow of fluids in the rock mass by changing the joint aperture and the bulk porosity of the rock matrix. Changes in aperture, in turn, would change the permeability of the joints. The change in joint aperture may be due to both normal and shear displacements of the joints. Shear displacement causes dilation which increases the joint aperture. The aperture of a joint also increases with the decrease of normal stress acting on it.

Fluid pressure increments within UDEC are calculated from the joint volume variation and the new inflow into the domain as follows [6]:

$$P = P_o + K_w Q \frac{\Delta T}{V} - K_w \frac{\Delta V}{V_m} \tag{24}$$

where P_o = domain pressure in the proceeding timestep
Q = sum of flowrates into the domain
K_w = bulk modulus of the fluid, and
ΔV = $V - V_o$, where V and V_o are new and old domain volumes, respectively
V_m = $(V + V_o)/2$

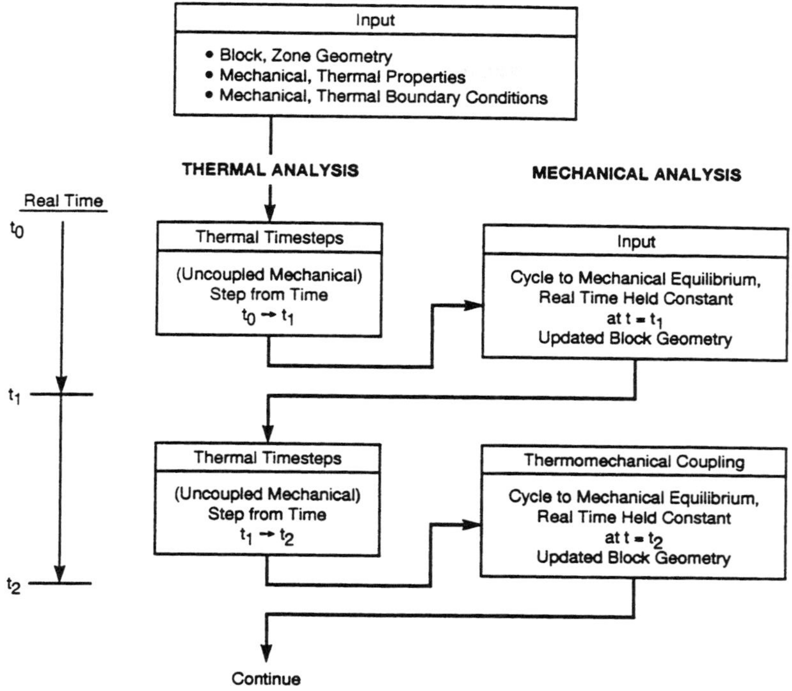

Figure 3. Method of running a coupled thermomechanical simulation with UDEC [27].

The fluid timestep, which is calculated by

$$\Delta t_f = \min \left[\frac{V}{K_w \sum_i K_i} \right] \qquad (25)$$

where V is the domain volume and the summation of permeability factors K_i is extended to all contacts surrounding the domain, is inversely proportional to the bulk modulus and joint conductivity. Small fluid timesteps are thus required in situations where large contact apertures (i.e., high permeability) and small domain areas exist. For typical joint apertures, fluid timesteps on the order of milliseconds are obtained. Therefore, using this current explicit algorithm, transient analyses can only be efficiently carried out for short-durations. In addition, fluid filling joints with small apertures increases the apparent joint stiffness, which may in addition require a reduction in the mechanical timestep.

A new scheme was proposed by INERIS and incorporated into UDEC to better allow long duration transient analyses to be conducted [6,28]. This development utilizes the facts that significant changes in fluid volume in the rock fractures do little to change rock

                                                                 M. P. Ahola et al.

stresses, and that the bulk modulus of the fluid is relatively unimportant. The flow rate is first calculated from pressure difference in the usual way, as expressed previously by Eq. 14. At each "domain " (intersection of several joints or the middle part of a joint), the flow contributed by each joint is added algebraically and multiplied by the fluid timestep to obtain the net fluid volume entering the domain [6]:

$$\Delta V_f = \sum q \Delta t_f \tag{26}$$

Instead of trying to translate this volume immediately into rock displacements, it is assumed that the excess fluid is stored in a "balloon " attached to the domain. The flow-time is then held constant, and the contents of each balloon are allowed to leak into its associated domain. This leakage stops when the increase in domain volume equals the volume stored in the balloon. The process involves the usual dynamic relaxation of the equations of motion of the gridpoints, but with an additional pressure boundary condition supplied by the leakage of fluid into the domain. The following leakage scheme was found to be satisfactory in terms of providing similar results with much greater numerical efficiency:

$$p' = p^o + F_p (\Delta V_{stored} - \Delta V_{domain}) \tag{27}$$

where p' and p^o are the domain pressures at the new and old (mechanical) timesteps, respectively, ΔV_{stored} is the volume originally stored in the balloon, ΔV_{domain} is the volume increase of the domain, and F_p is a constant factor.

The algorithm proceeds by performing a sequence of fluid cycles, the timestep being defined by the user. For each time step i (see Figure 4):

— UDEC computes the flow rate $Q_{(j)/(j+1)}$ between the domain (j) and the domain (j + 1) as a function of the (unbalanced) pressure $P_{(j)}$ in each domain.

— UDEC computes the initial volume $\Delta Vb_{(j)}$ attached to each domain (j), and a series of mechanical relaxation steps are performed in order to achieve continuity of flow at each domain. Given the assumption of fluid incompressibility, the net flow into a domain during a fluid step must equal the increment of domain volume. The unbalanced fluid volume, being the difference between the two, is gradually reduced during the relaxation procedure. For this purpose, the domain pressure is increased or reduced proportionally to the unbalanced volume for each domain. The proportionality factor F_p is controlled by an adaptive scheme and, therefore, varies during the iterations to provide better convergence. The pressures are assumed to be balanced after n mechanical relaxation steps if all the ratios between the balloon volume (j) and domain volume (j) are lower than VOLTOL (fixed by the users). Figure 5 depicts what is done at each relaxation step.

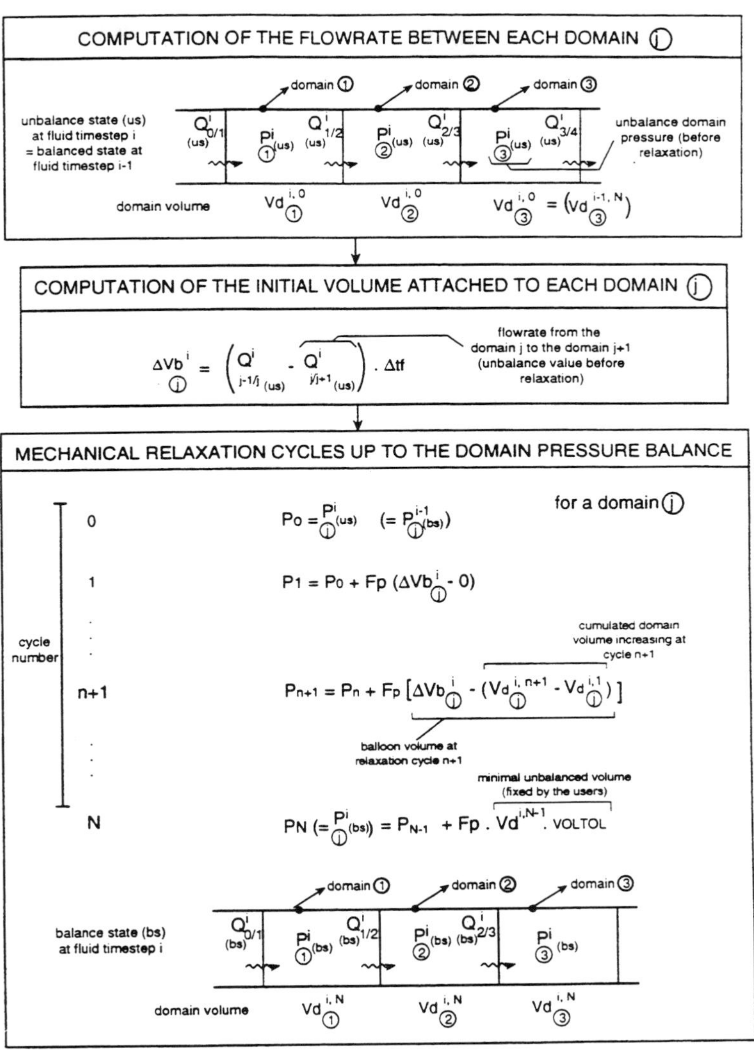

Figure 4. Transient hydro-mechanical scheme computation algorithm for each fluid cycle.

3.3. Thermal Convection Due to Fluid Flow in Fractures

Convective heat transfer related to the circulation of fluid in the rock fractures can be substantial if the fluid pressure gradient is sufficiently high. In this case, the temperature field (and hence the induced stress field) would be modified. Including this phenomenon in a modeling process led INERIS to propose a modification of the UDEC code [28].

3.3.1 Theoretical Model of Thermal Convection

The theoretical approach permits an analytical expression of the changes in the thermal flux (conductive and convective) entering an elementary volume of fluid of unit thickness in the out of plane direction and of surface area (dx*dy), circulating between two blocks (Figure 6).

— *for a fluid/fluid convective transfer, the variation in flux is expressed as follows:*

$$dQ_{conv,x} = dQ_1 = Q_1 - Q'_1 = -\int_o^a \left[\rho^f \, C_p^f \left(v\frac{\partial T^f}{\partial x} + T^f\frac{\partial v}{\partial x} \right) dx \right] dy \qquad (28)$$

In this expression, ρ^f is the density of the fluid, C_p^f is the specific heat of the fluid, v is the velocity of the fluid along x direction which is supposed to be the axis of the fracture, and T^f is the temperature of the fluid.

— *for a fluid/fluid conductive transfer, the variation in flux is expressed as follows:*

$$dQ_{cond,x} = dQ_2 = Q_2 - Q'_2 = \int_o^a \left[\lambda^f \, \frac{\partial}{\partial x} \left(\frac{\partial T^f}{\partial x} \right) dx \right] dy \qquad (29)$$

where λ^f is the thermal conductivity of the fluid which is assumed to be constant [W/m ·K].

— *for a rock/fluid convective transfer, the general expression is:*

$$dQ = h \, (T^r - T^f) \, dx \qquad (30)$$

where h is the rock/fluid convective heat transfer coefficient [W/m^2·K], and T^r is the temperature of the rock. Therefore we can define that on the surface $AD : dQ_3 = h(T^{r2} - T^f) dx$, while on the surface $BC : dQ'_3 = h(T^{r1} - T^f) dx$.

The change in the temperature of the hydraulic domain ABCD of mass m is obtained from the algebraic sum of all the heat transfer fluxes entering and leaving this domain.

The following equality can be written which reflects the heat balance in this domain:

$$mC_p^f \frac{dT^f}{dt} = dQ_1 + dQ_2 + dQ_3 + dQ'_3 \qquad (31)$$

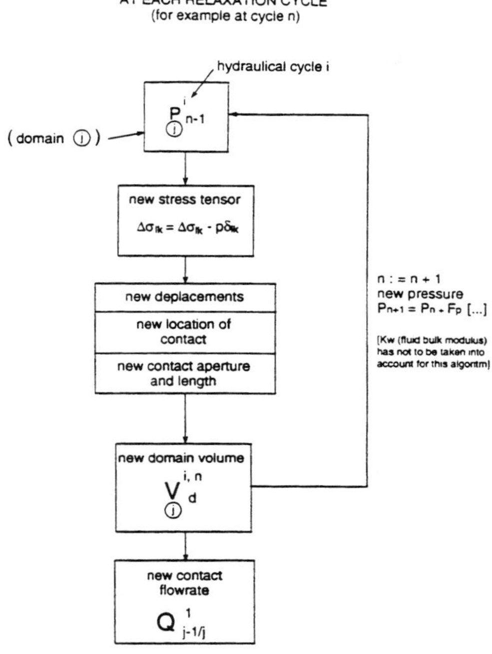

Figure 5. Transient hydro-mechanical scheme computation algorithm for each relaxation cycle.

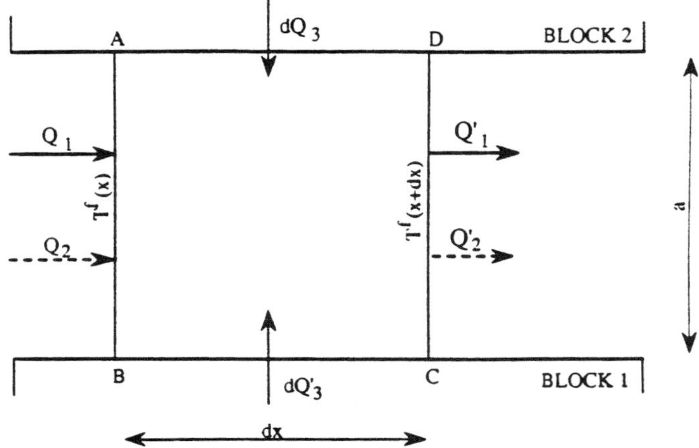

Figure 6. Heat balance of fluid in the domain ABCD of an opening between two blocks.

3.3.2. Procedures for Introducing Thermal Convection into UDEC

3.3.2.1. Data Structure Modifications

UDEC requires data of different types. A distinction is made between those related to the blocks: the nodes and corners (nodes at the boundary of the blocks) and those related to the joints: the "mechanical " contacts (corner/corner c-c: between 2 corners, corner/edge c-e: between a corner and an edge), the "hydraulical " contact, and the domains (located between two or more hydraulical contacts). Figure 7 shows the representation of notation used in UDEC. The distinction between "hydraulical " and "mechanical " contact has been introduced in UDEC Version 2.0.

In order to discretize Eq. (28) through (31), the case of a domain defined between two contacts shall be taken. To simplify the problem, the domains involving more than three contacts (areas of intersections between the joints) have (temporarily) been set aside. Consider a domain, i, of length d_i, defined by the two "hydraulical " contacts $K_{1,i}$ and $K_{2,i}$ where the hydraulic apertures of the joint are $a_{1,i}$ and $a_{2,i}$ respectively. The contact $K_{1,i}$ is located between hydraulical corners $S_{1,i}$ and $S'_{1,i}$ while $K_{2,i}$ is located between $S_{2,i}$ and $S'_{2,i}$ (Figure 8). The distance between the "hydraulic " corner and the "mechanical " corner (d_{1i}, d_{2i}, d_{3i}, d_{4i}) are computed as a function of "hydraulical " and "mechanical " contact locations. For each domain, i, a heat transfer coefficient h_i, a temperature T_i^f, and a velocity v_i equal to the mean of two velocities calculated in the two sections of the two "hydraulical " contacts defining the domain shall be assigned.

3.3.2.2. Assumptions

For introducing thermal convection into UDEC, two assumptions have been made:
— Since it can be assumed that the longitudinal temperature gradient is low compared with the lateral gradient for fractures whose thicknesses are small compared with their lengths, the conductive heat transfer term between the two surfaces AB and CD can be neglected (Figure 6). The heat balance expressed in Eq. (31) becomes:

$$mC_p^f \frac{dT^f}{dt} = dQ_1 + dQ_3 + dQ'_3 \tag{32}$$

— The convective heat transfer is neglected for domains defined at joint intersections. This has been assumed because of the very small areas of those domains.

3.3.2.3. Discretization of the Heat Transfer Terms

Discretizing all the terms in the heat balance gives the following expressions:

$$\Delta Q_{l,i} = -\rho^f C_p^f a_i d_i \left[v_i \left(\frac{T_i^f - T_{i-1}^f}{d_i} \right) + T_i^f \left(\frac{v_{k_{2,i}} - v_{k_{1,i}}}{d_i} \right) \right] \tag{33}$$

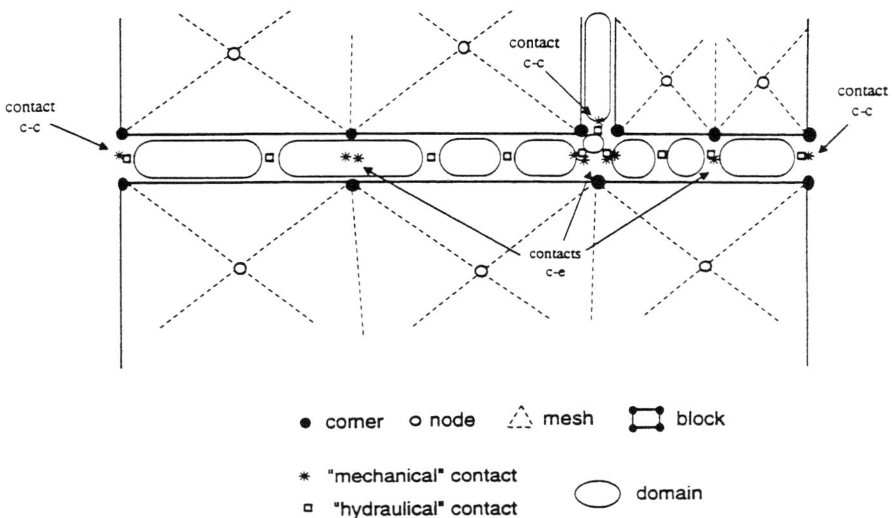

Figure 7. Representation of data in UDEC.

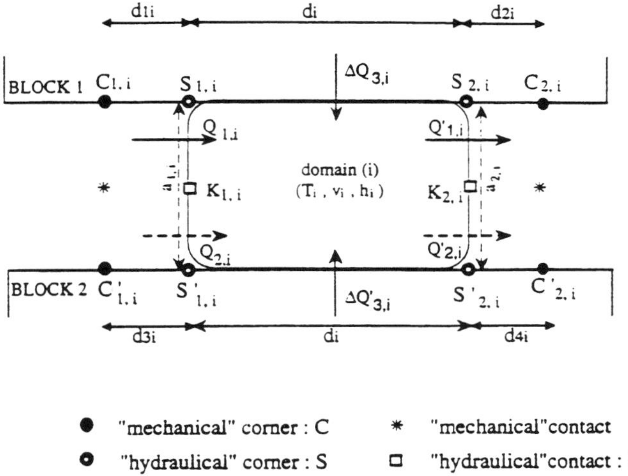

Figure 8. Discretization of the heat transfer at the interfaces of a domain in UDEC.

$$\Delta Q_{3,i} = h_i d_i \left(\frac{T_{S_{1,i}} + T_{S_{2,i}}}{2} - T_i^f \right) \tag{34}$$

$$\Delta Q'_{3,i} = h_i d_i \left(\frac{T_{S'_{1,i}} + T_{S'_{2,i}}}{2} - T_i^f \right) \tag{35}$$

$$m_i C_P^f \frac{\Delta T_i^f}{\Delta t} = \rho^f C_P^f \vartheta_i \frac{\Delta T_i^f}{\Delta t} = \rho^f C_p^f a_i d_i \frac{\Delta T_i^f}{\Delta t} \tag{36}$$

where m_i, and ϑ_i are the mass and volume of domain, i. In Eq. (33), the difference $(T^f_{2,i} - T^f_{1,i})$ has been replaced by $(T^f_i - T^f_{i-1})$ as Patankar [29] proposed in order to enhance the numerical stability.

3.3.2.4. Calculation Algorithm

The problem is to determine, at each time step, the change in temperature due to heat transfer from the domain, i, and from all the surrounding corners. The change in temperature in the domain, i, during each thermal cycle is calculated as follows:

$$m_i C_p^f \frac{\Delta T_i'}{\Delta t} = \Delta Q_{1,i} + \Delta Q_{3,i} + \Delta Q'_{3,i} \tag{37}$$

When replacing each term by its value, we obtain the expression for the change in the temperature in a domain, i, during a time step Δt:

$$\Delta T_i^f = \left[\frac{2h_i}{\rho_f a_i C_P^f} \left(\frac{T_{S_{1,i}} + T_{S_{2,i}} + T_{S'_{1,i}} + T_{S'_{2,i}}}{4} - T_i^f \right) \right.$$

$$\left. - \left(v_i \left(\frac{T_i^f - T_{i-1}^f}{d_i} \right) + T_i^f \left(\frac{v_{k_{2,i}} - v_{k_{1,i}}}{d_i} \right) \right) \right] \Delta t \tag{38}$$

The temperature change in a "hydraulical" corner related to a domain is given, here for example for "hydraulical" corner $S_{1,i}$, by the following expression (already introduced into the UDEC code to model forced convection as a boundary condition):

$$\Delta T_{S_{1,i}} = h_i \left(\frac{d_i}{2}\right) \left(T_i^f - T_{S_{1,i}}\right) thm_{S_{1,i}} \, \Delta t \tag{39}$$

where $thm_{S_{1,i}}$ is the thermal capacity of the "hydraulical" corner $S_{1,i}$ given as

$$thm_{S_{1,i}} = \left(\frac{thm_{C'_{1,i}} - thm_{C_{1,i}}}{d_{1i} + d_i + d_{2i}}\right) \cdot d_{1i} + thm_{C_{1,i}} \tag{40}$$

where

$$thm_{C_{1,i}} = \frac{1}{\rho^r C_p^r A_{C_{1,i}}} \tag{41}$$

and ρ^r is the density of the rock, C_p^r is the specific heat of the rock, and $A_{c_{1,i}}$ is the third sum (there are 3 nodes per element) of the surface areas of all the mesh elements to which the corner $C_{1,i}$ belongs. The temperature change of a "mechanical" corner is given by linear interpolation between the two surrounding "hydraulical" corners (belonging to the same block).

The changes of the temperature of a domain or corner are based on the same time step Δt. The first applications of the model showed that the calculations diverged for time steps previously accepted by UDEC in purely conductive calculations. In cases of numerical stability (shorter time steps), the temperature of the domains was found to change rapidly, while the temperature in the rock matrix changed more slowly.

These two problems led INERIS to propose the calculation algorithm displayed in Figures 9 and 10. In this algorithm, the conductive time step has been subdivided into a number of convective time steps. In a time step Δt_{cond}

— for n $'$ convective cycles or for each Δt_{conv}, the change in temperature (related to convective heat transfer) in the domains and the surrounding corners is updated;

— the temperature in the rock matrix is updated (conductive heat transfer).

The ratio of time steps can be defined as:

$$R = (\Delta t_{conv} / \Delta t_{cond}) \times 100 \tag{42}$$

3.3.3. Sensitivity and Verification of the Thermal Convection Model

3.3.3.1. Convection Sensitivity to Hydraulic Aperture, Flow Velocity, and Fluid Viscosity

A simple model consisting of two discrete blocks of dimensions 1 m long by 0.5 m high, separated by a single horizontal fracture was used to test the sensitivity of thermal

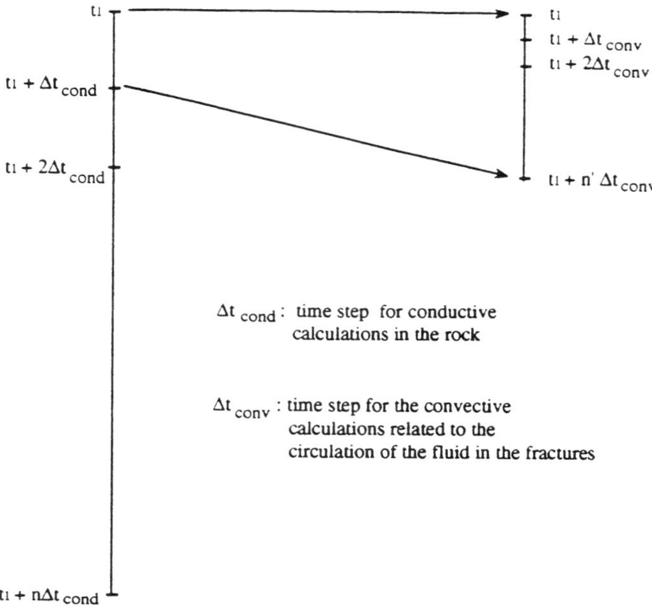

Figure 9. Conductive and convection timestep.

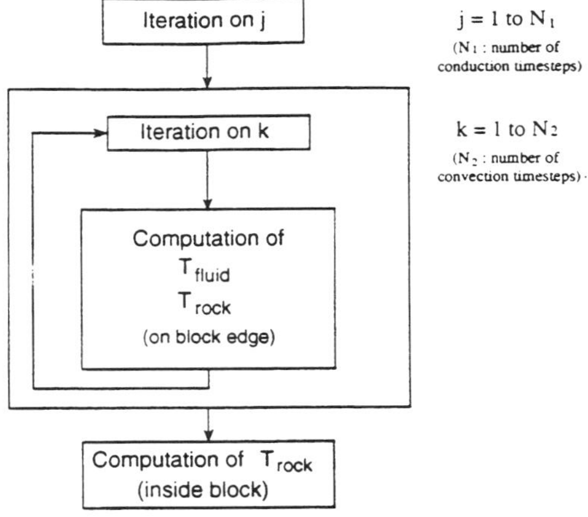

Figure 10. Calculation algorithm.

convection as a result of varying the joint hydraulic aperture, flow velocity, and fluid viscosity. A full description of the test model including block discretization, geometrical characteristics, and the thermal and hydraulic parameters is given by Abdallah [28].

The importance of the hydraulic aperture, a, was investigated for four cases (a = 0.5; 1; 1.5; and 2 mm). Figure 11a shows the temperature profile in the rock along the fracture in the different cases, and Figure 11b shows that of the fluid circulating along the fracture.

It can be seen that the result of increasing the hydraulic aperture (and hence of doubling the amount of fluid flowing in the fracture) is:
— greater cooling of the rock, especially near the fracture: the maximum change in temperature resulting from convection (i.e., maximum difference between initial temperature state before convection and final temperature state after convection) reaches 72 °C for an aperture of 2 mm while it did not exceed 10 °C for an aperture of 0.5 mm. It can also be seen that the zone of cooling dominates that of the zone of heating.
— a smaller increase in the fluid temperature: this does not exceed 17 °C for an aperture of 2 mm while it reached 83 °C for the aperture of 0.5 mm. It will be noted that the differences are less (the reverse is true when a = 0.5mm) close to the zone of heating.
— a smaller thermal gradient within the fluid and the rock near the fracture.

It can be seen that increasing the fluid flow velocity (Figure 11c and 11d) as well as a decreasing of the fluid viscosity (Figure 11e and 11f) have the same effect. Thus, it appears that the amount of convection increases as the hydraulic aperture and fluid velocity in the fractures increase and as the dynamic viscosity of the fluid decreases.

3.3.3.2. Comparison with a Finite Difference Solution

The heat equation in the model was solved by a finite difference code using an implicit central finite difference scheme (FDS) algorithm. The full development of the FDS algorithm is given by Abdallah [28].

Comparison of the UDEC results (for the base case) with those obtained using the FDS (See Figure 12) show that:
— The new development of the UDEC code gives values that are quantitatively in agreement with FDS both in the rock and in the fluid.
— The maximum temperature difference between the two codes does not exceed 10 °C and this exists in the cooling zone. The temperature differences between the two codes in the rock and in the fluid begins to decrease appreciably as the fluid flows through the heating zone. It is possible this difference is partly due to the mesh used in the two codes (16 domains in UDEC and 23 in the FDS solution). This may be caused by the limited lateral dimension (normal to the fracture) of the rock, between the top and the bottom, where adiabatic boundary conditions are applied.

3.3.4. Estimation of the Importance of the Coupling Between Thermics and Hydromechanics

In the previous section the consistency of the model for thermo-hydrological computation was tested. The thermo-hydro-mechanical computation is now realized, allowing the hydraulic aperture to vary with the block strain.

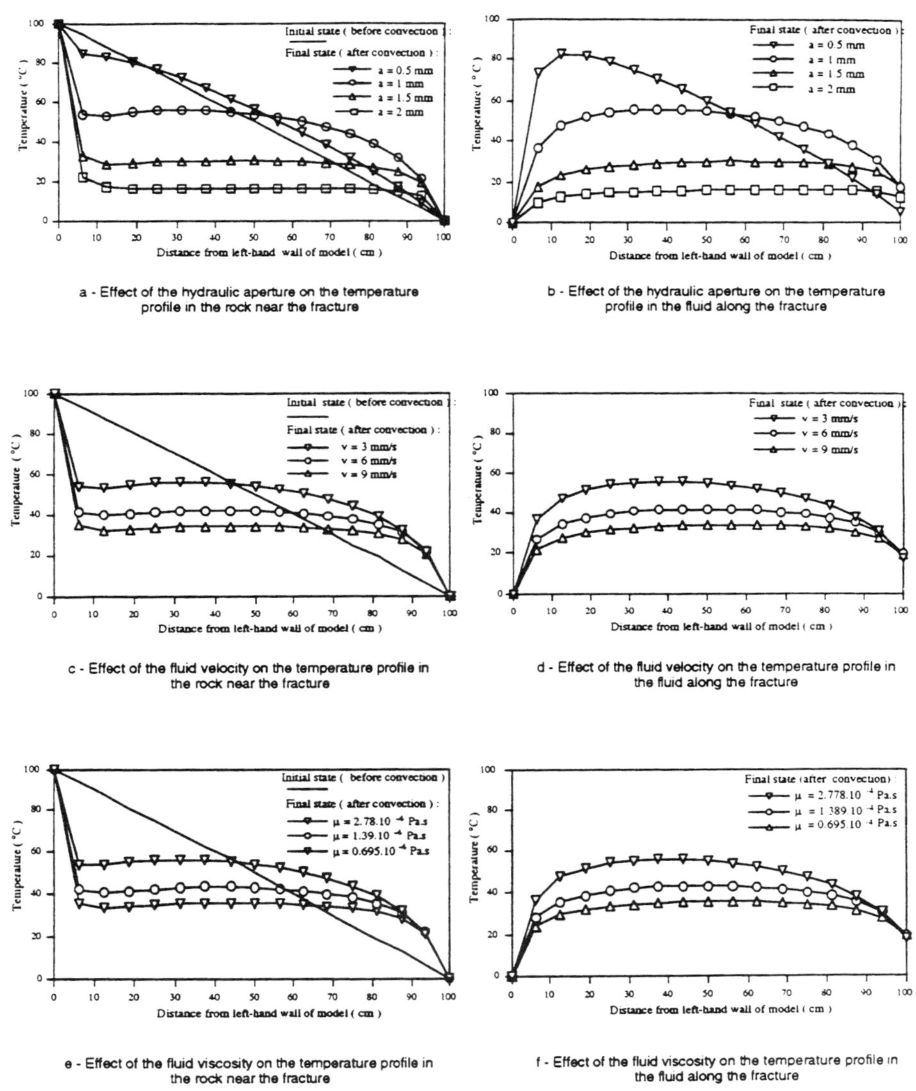

Figure 11. Effect of hydraulic aperture, fluid velocity, and fluid viscosity on temperature profiles.

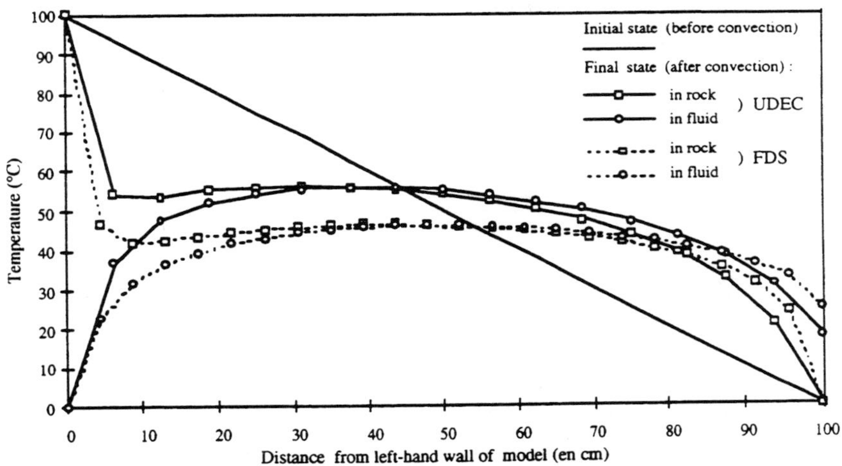

Figure 12. Comparison of results from UDEC with FDS solution.

The computation algorithm is depicted in Figure 13a. After realizing at first successively a HM computation followed by a T conductive computation, HM computations (using the steady-state algorithm) and T sequences (using the conductive/convective algorithm depicted in Figure 10) are alternated.7

The realism of THM coupling simulations depends on the alternation number between HM and T calculations. The effect of thermal sequence duration Dt on the calculation (Dt is related to the thermal timesteps as follows: $Dt = N_1 \Delta t_{cond} = N_1 N_2 \Delta t_{conv}$) was analyzed. Changing the sequence duration Dt it can be seen in Figure 13b that:

— long thermal sequences (up to 1,000 hours) are acceptable for steady-state analysis,

— short thermal sequences (lower than 4 hours) are necessary for transient analysis.

3.4. Thermo-Hydro-Mechanical Analysis in Fractured Media

Figure 14 shows the current extent of THM coupling in UDEC. The thermomechanical and THM couplings are unidirectional. There is no coupling in terms of energy changes due to the mechanical deformation, and the convective heat transfer between the fluid flowing through the joints and the rock matrix is not yet available in the commercial version of UDEC. For the case in which the steady state algorithm for the hydromechanical (HM) portion is used, a THM analysis proceeds by conducting a transient thermal computation between time t_i and t_{i+1} and an HM computation at time t_{i+1} as shown in Figure 15. This approach speeds up the computational time, and works

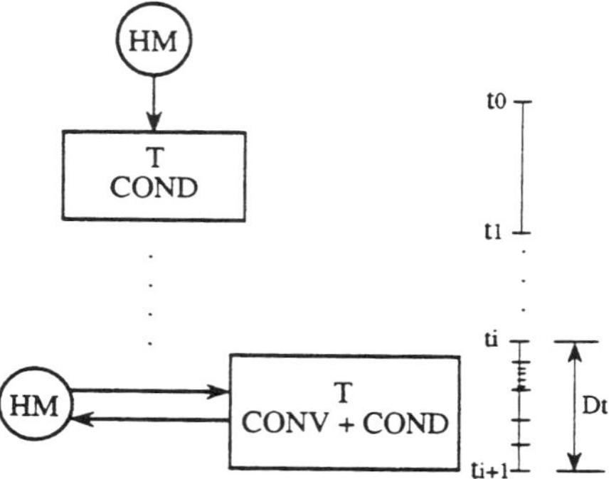

b - Results of the computations

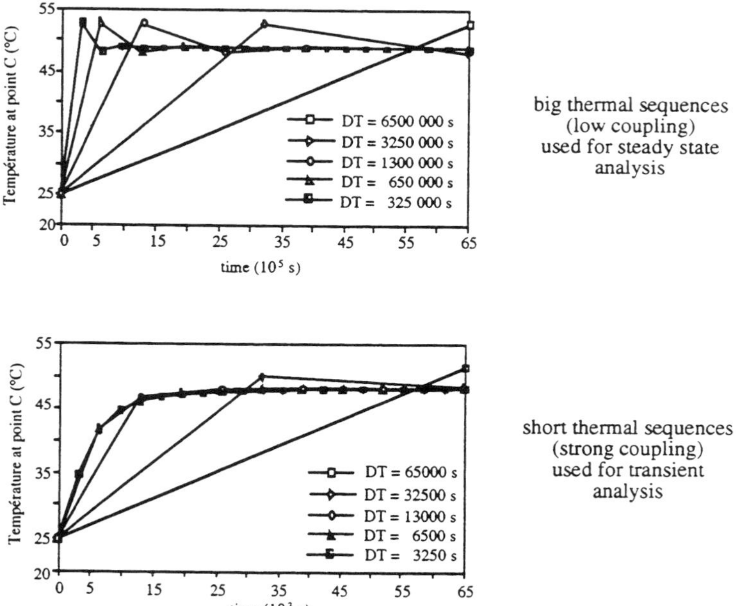

Figure 13. Estimation of the importance of the coupling between thermics and hydromechanics.

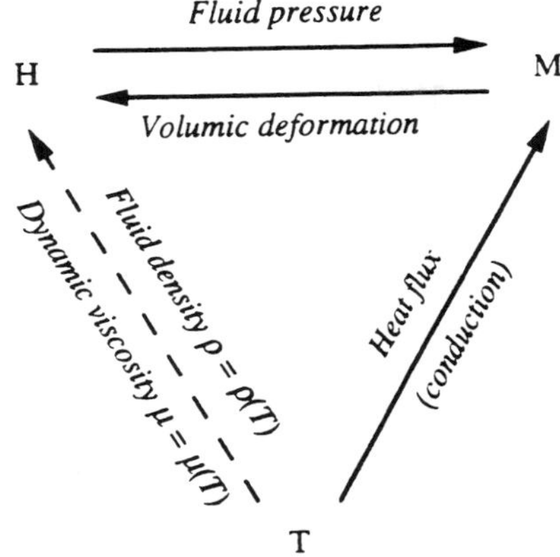

Figure 14. Existing couplings in UDEC.

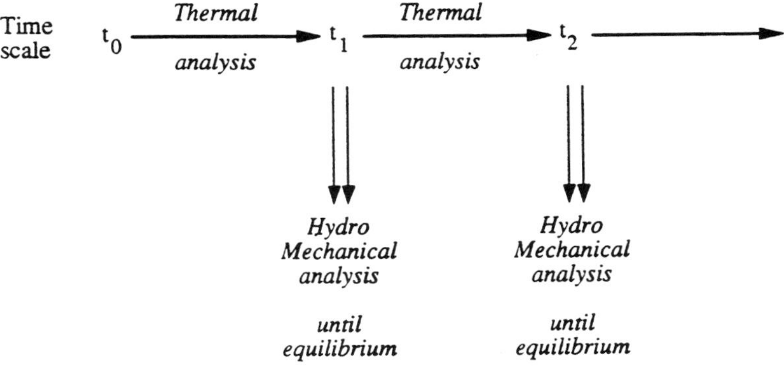

Figure 15. THM processes in UDEC with steady state HM analysis.

well for problems in which the mechanical and thermal processes equilibrate faster than the hydrological process. For the case in which the transient algorithm for the HM coupling is used, a THM analysis proceeds by successively doing transient thermal computation and transient HM computation as shown in Figure 16.

4. DISCUSSION

In many situations, the geologic region to be modeled contains too many fractures to be incorporated into the distinct element model. Due to the explicit nature of the solution scheme, simulation of a large number of deformable blocks, especially when thermal and hydrologic effects are considered, becomes practically and computationally infeasible. In most cases, a rock mass consists of only a few joint sets in fairly well defined orientations. Provided that these joint sets are not so narrowly spaced compared to the domain being analyzed, most if not all can be included to capture the mechanical and hydrological responses within the rock mass. In the case where the fractures are narrowly spaced or more or less randomly distributed as discussed in a later chapter regarding BMT3, this type of approach is not possible. As a result, simplifications to the fracture network are necessary while at the same time trying to retain the important mechanical, hydrological, and thermal characteristics of the system. One such approach is to subdivide the problem into different domains according to scale. For example, the very near-field model domain around a tunnel could closely approximate the actual fracture distribution, whereas the far-field model domain could contain a reduced

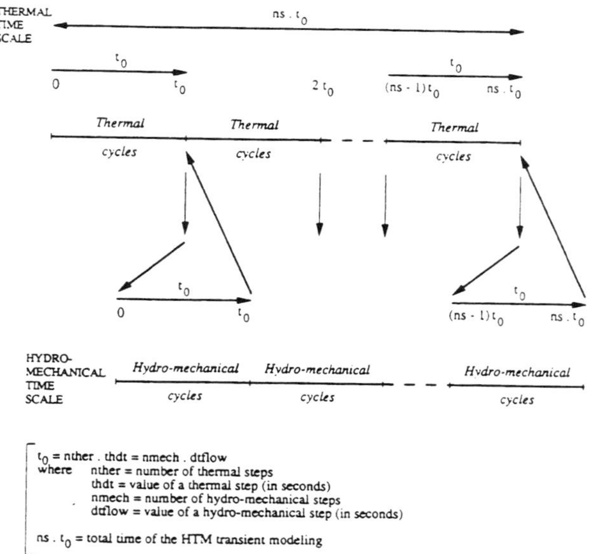

Figure 16. TMH process in UDEC with the transient HM analysis.

equivalent fracture distribution so as to minimize the computational effort [30,31]. Other approaches include simplifying the fracture network by eliminating fractures below a certain length or aperture. Numerous such approaches were utilized in the DEM modeling of BMT3, as discussed by Jing [31]. It was determined from that analysis that there was very good to reasonable agreement among the various discrete element as well as continuum models in the predicted temperatures, displacements and stresses around a heated tunnel in a highly fractured rock mass. The major difference in the results occurred in the water flux calculations. One to two orders of magnitude differences were found to exist between the various discontinuum models, as well as between the discontinuum and continuum models. The discontinuum approach was found to be very sensitive to simplification approaches used on complex fracture geometries in assessing the hydraulic response of the fractured rock mass. More research effort is needed in refining these approaches, since this is a crucial problem concerning the validity of numerical models for radioactive waste repositories and safety assessment.

5. REFERENCES

1　D.D. Kana, B.H.G. Brady, B.W. Vanzant, and P.K. Nair, Critical Assessment of Seismic and Geomechanics Literature Related to a High-Level Nuclear Waste Underground Repository, NUREG/CR-5440, San Antonio, TX: Center for Nuclear Waste Regulatory Analyses (1991).

2　S.M. Hsiung, A.H. Chowdhury, W. Blake, M.P. Ahola, and A. Ghosh, Field Site Investigation: Effect of Mine Seismicity on Jointed Rock Mass, CNWRA 92-012. San Antonio, TX: Center for Nuclear Waste Regulatory Analyses (1992).

3　S.M. Hsiung, W. Blake, A.H. Chowdhury, and T.J. Williams, Effects of Mining-Induced Seismic Events on a Deep Underground Mine, Pure and Applied Geophysics, 139 (1992) 741–762.

4　P.A. Cundall, and R. Hart, Development of Generalized 2-D and 3-D Distinct Element Programs for Modeling Jointed Rock, U.S. Army Engineers Waterways Experiment Station, Final Report, Vicksburg, MS, Misc. paper SL-85-1 (1985).

5　Applied Mechanics, Inc, Use and Modification of the Universal Distinct Element Code (UDEC) for Basalt Block Test Analysis, SD-BWI-TD-020, Rev. 0, for Rockwell Hanford Operations, Richland, WA (1985).

6　Itasca Consulting Group, Inc, UDEC Universal Distinct Element Code Version 2.0 Volume I: User's Manual, Minneapolis, MN (1993).

7　P.A. Cundall, J. Marti, P. Beresford, N. Last, and M. Asgian, Computer Modeling of Jointed Rock Masses, U.S. Army Engineers Waterways Experiment Station, Technical Report N-78-4, Vicksburg, MS (1978).

8　Itasca Consulting Group, Inc, Sensitivity of the Stability of a Waste Emplacement Drift to Variation in Assumed Rock Joint Parameters in Welded Tuff, NUREG/CR-5336, Washington, DC (1989).

9　R.D. Hart, P.A. Cundall, and J. Lemos, Formulation of a Three-Dimensional Distinct Element Model-Part II: Mechanical Calculations for Motion and Interaction of a System Composed of Many Polyhedral Blocks, International Journal of Rock Mechanics and Mining Sciences & Geomechanics Abstracts, 25 (1988) 117-126.

10 J.R. Williams and G.G. Mustoe, Modal Methods for the Analysis of Discrete Systems." Computers and Geotechnics, 4, (1987) 1–19.

11 P.A. Cundall, A Computer Model for Simulating Progressive, Large Scale Movements in Blocky Rock Systems, Paper II-8, Proceedings of the International Symposium on Rock Fracture. Organized by International Society for Rock Mechanics, Nancy, France (1971).

12 L.J. Lorig, A Hybrid Computational Model for Excavation and Support Design in Jointed Media, Department of Civil and Mineral Engineering, Minneapolis, MN: University of Minnesota (1984).

13 N.R. Barton, S.C. Bandis, and K. Bakhtar, eds, Strength, Deformation and Conductivity Coupling of Rock Joints, International Journal of Rock Mechanics and Mining Sciences & Geomechanics Abstracts, 22(3)d (1985) 121-140.

14 N.R. Barton, and S.C. Bandis, Effects of Block Size on the Shear Behavior of Jointed Rock, Proceedings of the 23rd U.S. Symposium on Rock Mechanics, Berkeley, CA (1982).

15 P.A. Cundall, and J.V. Lemos, "Numerical Simulation of Fault Instabilities With the Continuously-Yielding Joint Model, Proceedings of the Second International Symposium on Rockbursts and Seismicity in Mines, Minneapolis, MN: University of Minnesota (1988).

16 J.J. Nitao, T.A. Buscheck, and D.A. Chestnut, The Implications of Episodic Non Equilibrium Fracture-Matrix Flow on Site Suitability and Total System Performance, Proceedings of the International High-Level Radioactive Waste Management Conference, ANSI, La Grange Park, IL (1992) 279-296.

17 R.D. Hart, An Introduction to Distinct Element Modelling for Rock Engineering Proceedings of the 7th International Congress on Rock Mechanics, Aachen, Germany 3, Rotterdam, Netherlands: A.A. Balkema (1991) 1,881-1,891.

18 C. Louis, A Study of Groundwater Flow in Jointed Rock and its Influence on the Stability of Rock Masses, Imperial College, Rock Mechanics Research Report No. 10 (1969).

19 P.A. Witherspoon, J.S.Y. Wang, K. Iwai, and J.E. Gale, Validity of Cubic Law for Fluid Flow in a Deformable Rock Fracture, Water Resources Research, 16(6), (1980) 1106-1024.

20 R.D. Manteufel, Heat Transfer in an Enclosed Rod Array, Department of Mechanical Engineering, Cambridge, MS: MIT (1991).

21 T.C. Sandford, E.R. Decker, and K.H. Maxwell, The Effect of Discontinuities, Stress Level, and Discontinuity Roughness on the Thermal Conductivity of a Maine Granite, Proceedings of the 25th U.S. Symposium on Rock Mechanics, Evanston, IL (1984) 304-311.

22 H.A. Wollenberg, J.S.Y. Yang, and G. Korbin, An Appraisal of Nuclear Waste Isolation in the Vadose Zone in Arid and Semi-Arid Regions, NUREG/CR 3158, Washington, DC: Nuclear Regulatory Commission (1983).

23 B.V. Karlekar and R.M. Desmond, Heat Transfer, 2nd edition, West Publishing Co., St. Paul, MN (1982).

24 C.M. St. John, Thermal Analysis of Spent Fuel Disposal in Vertical Displacement Boreholes in a Welded Tuff Repository, SAND84-7207, Albuquerque, NM: Sandia National Laboratories (1985).

25 M. Christianson, TEMP3D: A Computer Program for Determining Temperatures Around Single of Arrays of Constant or Decaying Heat Sources User's Guide and Manual, Minneapolis, MN, Univeristy of Minnesota, Department of Civil & Mineral Engineering (1979).

26 R.D. Manteufel, M. P. Ahola, D.R. Turner, and A.H. Chowdhury, A Literature Review of Coupled Thermal-Hydrologic-Mechanical-Chemical Processes Pertinent to the Proposed High-Level Nuclear Waste Repository at Yucca Mountain, NUREG/CR-6021, Washington, D.C: Nuclear Regulatory Commission (1993).

27 M. Board, UDEC (Universal Distinct Element Code) Version ICGI.5 Vols. 1–3, NUREG/CR-5429, Washington, D.C., Nuclear Regulatory Commission (1989).

28 G. Abdallah, A. Thoraval, A. Sfeir, and J.P. Piguet, Thermal Convection of Flow in Fractured Media, International Journal of Rock Mechanics and Mining Sciences & Geomechanics Abstracts, in publication (1995).

29 S.V. Patankar, Numerical Heat Transfer and Fluid Flow, Washington, DC: Hemisphere (1980).

30 M.P. Ahola, L.J. Lorig, A.H. Chowdhury, and S.M. Hsiung, Thermo-Hydro-Mechanical Coupled Modeling: Near-Field Repository Model, BMT3, DECOVALEX-Phase II, CNWRA 93-002. San Antonio, TX: Center for Nuclear Waste Regulatory Analyses (1993).

31 L. Jing, J. Rutqvist, O. Stephansson, C.-F. Tsang, and F. Kautsky, DECOVALEX-Mathematical Models of Coupled T-H-M Processes for Nuclear Waste Repositories, Report of Phase II, SKI Report 94:16, Stockholm, Sweden: Swedish Nuclear Power Inspectorate (1994).

O. Stephansson, L. Jing and C.-F. Tsang (Editors)
Coupled Thermo-Hydro-Mechanical Processes of Fractured Media
Developments in Geotechnical Engineering, vol. 79
© 1996 Elsevier Science B.V. All rights reserved.

213

Modelling approaches for discrete fracture network flow analysis

A.W.Herbert

School of Earth Sciences, University of Birmingham, Edgbaston, Birmingham, B15 2TT, United Kingdom

Abstract

In many geological formations being considered for the disposal of radioactive waste, the primary flow system is through a connected network of discrete fractures. Such systems are very heterogeneous and the fracture network geometry can lead to dispersion of any contaminants being transported through the formation. The orientation of the fractures with respect to the rock stress field may result in significant coupling between groundwater flow in the fractures and mechanics of the fractures. This chapter considers the approaches available for modelling fracture network flow. First the motivation for using discrete modelling approaches is discussed in the context of radioactive waste disposal. The approaches to representing such discrete flow systems are reviewed and we consider how to obtain the parameters required to characterise the fracture network. The fracture networks constructed in this way are very complex flow systems and sophisticated algorithms are required to solve the flow problem, particularly in three dimensions. The approaches taken by several leading numerical codes are briefly discussed, focusing on the NAPSAC code that was used in DECOVALEX. Next, the approach to modelling more complex physics within these models is considered. In particular, the way in which mechanics can be incorporated in flow models such as NAPSAC is compared with the approach taken by fully coupled models such as UDEC. The limitations of each of these approaches are assessed.

1.   MOTIVATION FOR USE OF DISCRETE REPRESENTATIONS

In many geological formations being considered for the disposal of radioactive waste, the primary flow system is through a connected network of discrete fractures. This provides a very heterogeneous system and the fracture network geometry can also lead to dispersion of any contaminants being transported through the formation. The orientation of the fractures with respect to the rock stress field may result in significant coupling between groundwater flow in the fractures and the mechanics of the fractures.

In order to provide an assessment of the geological barrier to the release of radionuclides from a radioactive waste disposal facility situated in such a fractured rock, it is necessary to show an understanding of flow in fracture networks [1]. In particular, one should show sufficient understanding of the flow system to give confidence that predictions of the large scale properties of the flow system can be made from the results of field scale investigations, and that predictions can be made of the flow and transport over the very long timescales associated with radioactive waste disposal. In order to build confidence in such an understanding, it is important to show a very detailed understanding of field experiments which are generally on scales at which the influence of the fracture network geometry is significant [2]. The geometry and connectivity of the fracture system and the possibility of hydraulically important pathways through the network can play an important role in determining the scale dependence of the effective properties of the system. Indeed one of the early motivations for the development of the direct fracture network approach was to develop an understanding of the scale-dependence of the effective dispersion parameters for radionuclide transport through fractured rock which had been inferred from field data [for example, 3].

This chapter concerns the use of direct representations of the fracture network. In such approaches, the geometry of the fracture network is accounted for explicitly. Figure 1 illustrates a typical realisation of a small three dimensional fracture network model in a cuboid region that illustrates the complexity of the flow geometry.

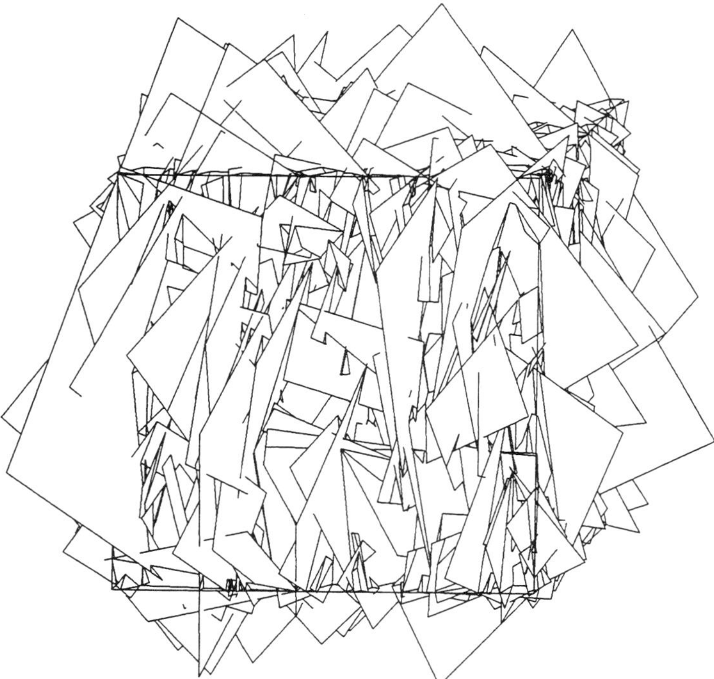

Figure 1 Typical realisation of a three-dimensional fracture network model.

These representations are needed for describing or predicting aspects of the performance of the fractured system where the geometry of the fracture network plays a significant role. Some examples of such circumstances are:

- representations of any flow experiments where the fracture connectivity is important, which in practice means almost all interpretations of field experiments where a detailed understanding is needed;

- representation of the anisotropy and detailed distribution of the stress field around an opening where rock deformation might depend on the relative location of individual fractures or their connectivity;

- prediction of the effective flow properties of the fracture network system and of the scale dependence of effective properties;

- prediction of the effect of the fracture network geometry on the effective dispersion for solute transport;

- prediction of the effect of the fracture network geometry on the effective hydraulic diffusivity of the pressure field in response to a pressure change and the inferred radius of influence of pressure tests;

- the impact of coupled processes involving the combination of a heterogeneous pore pressure distribution together with the discrete stress field.

From the above list, it can be seen that such an understanding of the role of the fracture geometry can be important in almost all aspects of an investigation of a fractured rock system. The two main reasons that such discrete models are not more commonly used are the complexity of the models and the fact that stochastic models inevitably require uncertainty to be addressed formally.

The complexity means that a large quantity of data is required to characterise fracture systems adequately. Whilst there are still issues to be resolved in the experimental characterisation of fracture network flow geometry, a number of research projects for the radioactive waste industry have demonstrated the feasibility of collecting suitable basic input data [4-7]. Understanding fracture channelling and the extent of the flow wetted surface of the fracture are still research tasks, but simple assumptions can be made and the other data interpreted consistently so that the resulting fracture network geometry reproduces key features of the physical network. In many cases however there will be a balance between the benefits of a more detailed representation of the system, and the increased cost of collecting data for which there may be significant uncertainty. The other aspect of using such a complex approach is the difficulty of representing the system. An earlier chapter in this book considers the approach to modelling the fracture network using an effective media approach. Such an approach captures many aspects of the system, but cannot fully account for the influence of fracture connectivity. In general numerical models must be used and, again, until recently the feasibility of simulating such a complex numerical model had not been proven. This chapter will describe how such a complex numerical problem is being addressed by computer codes used in the DECOVALEX project and elsewhere [see also further references given in 8].

The second reason why the direct fracture network approach is not more widely used is the need to treat predictions in a probabilistic framework and consider the uncertainty due to the details of the fracture geometry directly. Fracture network models are necessarily stochastic since it is not possible to determine the location and extent of each flow conducting or mechanical break in the rock. Instead a stochastic approach is used in which the statistics of the fracture system are determined and realisations of the fracture network geometry that exhibit the same statistics as the physical system are generated and used for simulation. This means that a discrete fracture network approach does not predict the result of a given experiment. Instead, it predicts a probability distribution of equally likely results given the stochastic description of the fracture geometry and properties. This realisation dependent uncertainty corresponds to a lack of knowledge of the precise fracture geometry. In many respects this is an advantage of the approach over deterministic models since the uncertainty is real and unavoidable. Conventional models which make single valued predictions are simply hiding this feature of our understanding.

2. STOCHASTIC MODELLING AND DEFINITION OF THE FRACTURE NETWORK GEOMETRY

2.1. Stochastic versus deterministic modelling

Stochastic modelling is based on a statistical description of the system to be represented. The fracture network system is not described deterministically, with the location and orientation of all the fractures incorporated into the model. This is clearly not possible for almost all networks of interest, since the details of the fracture network in the rock away from exposures or boreholes cannot be known. Instead, the statistical properties of the fracture network system are measured and fracture networks are generated that exhibit the same statistics. This means that our models are not exact representations of the real physical fracture network, and one would not expect any individual model to give an accurate prediction of the detailed flow in the real network. However, if one simulates many different realisations of the fracture network flow system, each having the same statistical properties as the real network, then the range of model results should bound the behaviour of the real network (if a good statistical description of the fracture network has been used). For this to be the case, it is important that sufficiently many realisations of the fracture network have been generated and simulated. If only a few realisations are used then the distribution of possible behaviour will not be accurately predicted and in particular the likelihood of more extreme behaviour will not be known. Ideally, several hundred realisations may be necessary to determine this probability distribution of equally likely results and to predict, say 95% confidence limits. In practice, it is not always possible to simulate sufficiently many realisations and often more qualitative bounds are estimated from a smaller sample of model results.

In contrast, deterministic modelling involves specifying the problem explicitly. All important features are incorporated in the model and their properties specified. Where a continuum model is used this may be appropriate, and a best estimate made of the effective properties that should apply to the region. The uncertainty due to the random location of many small features is no longer considered since only larger scale average results are being predicted. Deterministic fracture network models are more difficult to justify since in general it is not possible to know the fracture flow geometry within the rock mass, and the deterministic approach simply avoids consideration of a real uncertainty. Thus when the details of a fracture flow system form part of the prediction, the stochastic approach has advantages over deterministic approaches. The uncertainty in the prediction of the stochastic model is a real uncertainty and reflects the uncertainty in the consequence of the details of the fracture system that have not been determined.

The stochastic approach does lead to problems when one attempts to 'validate' a model of a particular fracture network flow by comparison with field experiments [9-10]. The fracture network model predicts a distribution of possible results, and there are well established hypothesis tests to determine whether a sample of results is a sample from the same distribution as the sample of predictions. However, there is generally only a single physical experiment and it is difficult to compare alternative predictions of the distribution of results from two rival fracture network models when there is only one physical result to distinguish between them. The task is to decide whether the physical result might correspond to a single sample from the distribution of predicted results for each alternative conceptual model of the fracture network. It is quite likely that this will be true, albeit with a relatively small likelihood for a wide range of alternative descriptions. The distribution of predictions from different conceptual models may well overlap, further complicating the task of deciding whether a given model is a valid description of the fracture network flow system. The best approach to validating stochastic models is still being developed but it is likely to rely on a measure of judgement based on a range of quantitative and qualitative measures of the predicted behaviour.

2.2. Two-dimensional network models

Early network flow modelling [3,11] and much current modelling of coupled physical processes in fracture networks [8,12] are based on a two-dimensional approximation. This is a common approach that can often be justified in continuum representations of groundwater flows. The difference between the value of two-dimensional approximations of continuum flows and two-dimensional approximations of discrete network flows is that the key reason for choosing a network approach should be a need to incorporate the effect of the flow geometry. Much of the justification for the use of a network approach for flow prediction is compromised by using such a poor approximation to the geometry as is given by a two-dimensional representation. Nevertheless, this is a necessary approximation for fully coupled models given the current computational limitations. With careful choice of the network geometry parameters, many key aspects of the network flow characteristics can be taken into account in a qualitative way.

This section first considers the key differences between two- and three-dimensional networks, and then briefly discusses an approach to developing appropriate two-dimensional approximations.

One of the main characteristics of a fracture network that controls the behaviour of the flow is the connectivity of the network. The most basic measure of connectivity is whether the fracture network has or has no connection across the region. This depends on the fracture density and the change from unconnected to connected networks is predicted by the percolation threshold. The percolation threshold gives the density at which the size of connected clusters of fractures suddenly increases from a relatively small typical cluster size to the existence of a percolating cluster that spans the region. The percolation threshold is quite a sharp transition: a small increase in fracture densities will change the network from one for which no realisations have connections across the model region to one for which all realisations are well connected [11]. This percolation threshold depends upon the statistical properties of the network, but for random networks there is a much more significant dependence on the dimension of the network geometry. Three-dimensional networks become well connected at much lower fracture densities than two-dimensional networks. In addition to the simple percolation problem of connected versus unconnected, the issue of connectivity affects the basic flow characteristics of the fracture network. If one considers only the most transmissive fractures and asks what proportion of the most transmissive fractures are needed to obtain a high transmissivity connection across the network, this too is a percolation problem and is very different in two-dimensional as compared to three-dimensional networks. Thus, the least transmissive link in the most transmissive pathway through the network will be quite different in a two-dimensional system to in a three-dimensional system. This issue of connectivity is fundamental to the influence of the network geometry on the flow solution.

Given that two-dimensional approximations are needed for coupled problems, it is important to develop techniques to derive a two-dimensional representation of the network geometry that matches the connectivity characteristics of the real system as closely as possible. This is discussed briefly in the chapter of this book describing DECOVALEX Benchmark Test 3 [13], where a two-dimensional benchmark problem was defined that is related to inflow experiments in the OECD Stripa project. The starting point was to consider a two-dimensional section through one realisation of the three-dimensional interpreted network geometry. Whilst the three-dimensional network was well connected, the simulated trace map was much less well connected. In order to define a network geometry that would reproduce a more realistic flow behaviour, various modifications were made to enhance the connectivity to match our understanding of real fracture network characteristics. These modifications include a simple extension of the fracture trace lengths, together with rules to ensure that very close fractures connect and unphysical block geometries are avoided. There is no accepted methodology for deriving two-dimensional equivalent networks with comparable connectivity to real networks, but the example given in the chapter on Benchmark Test 3 illustrates one possible approach and describes the modifications to the fracture trace geometry [see also 14].

2.3. Three-dimensional network characterisation

This section describes the main methods for inferring fracture network geometries from field measurements of the fracture network properties [15,16]. This is the starting point for three-dimensional simulations, and also for the derivation of appropriate two-dimensional equivalent networks.

The key parameters used to characterise a fracture network are:

- the distribution of fracture orientations and identification of independent fracture sets;
- the statistical process for generating the fracture locations in space;
- the fracture density;
- the distribution of fracture lengths;
- and the distribution of fracture transmissivities.

In addition to the parameters listed above, one might also identify correlations between these parameters. The parameters are addressed in turn below.

When characterising the fracture orientation distribution, it is generally found that the fractures can be divided into a number of distinct fracture sets. These sets of fractures comprise fractures that can be characterised by common distributions of parameters, and which have a common origin and history. These fracture sets are often defined in terms of their orientation distributions which tend to be clustered around preferred orientations on a lower hemisphere projection of the poles to the fracture planes. This definition of the characteristic orientation is best achieved by using conventional statistical methods to identify distinct clusters. The fractures can then be separated into their distinct sets and further parameters inferred for each set independently.

The distribution of fractures has commonly been assumed to be uniform in space with just a single fracture density being used to specify how many fractures to generate. An equivalent approach to using a fracture number density is to generate fractures up to a specified area density of fracture surfaces. The fractures are then generated by sampling the distributions of the other parameters and using a Poisson process to generate values for the coordinates of the fracture centres. Care must be taken to avoid edge effects, and this is usually accomplished by generating the fracture network in a larger region than that to be simulated. The fracture density may be obtained from the spacing of fractures along a scan line on a mapped exposure, or from a fracture log along a borehole or core. Each distinct set of fractures has its own characteristic distributions of properties, and the density of each of these fracture sets is usually determined independently. For a given fracture set, the number density, ρ, is given in terms of the mean spacing of intersections along a straight line, $\bar{s}$, by:

$$\bar{s} = \left(\rho \bar{X}\right)^{-1}, \tag{1}$$

where $\overline{X}$ is the mean projected area of the fractures onto a plane perpendicular to the measurement line.

The fracture set length distribution is one of the more difficult parameters to infer since we have only one- or two-dimensional data from which to infer a length distribution which will only be fully determined by a three-dimensional description. A number of assumptions need to be made at this stage. First, it is difficult to characterise the shape of the transmissive area of the fracture plane. It is generally assumed that this surface has a simple geometry. In the case of the NAPSAC code [17] it is assumed to be rectangular, in the case of other codes it may be assumed circular or polygonal. There are no reliable data available to suggest whether any of these models might be better than the others. Once the fracture shape has been fixed, then one can use analytical results giving the relationship between the distribution of fracture lengths to the distribution of fracture trace lengths as measured on a large two-dimensional trace plane intersecting the network. For example, for square fractures of side length distribution, L, the moments of the length distribution, L_i, are related to the moments of the corresponding distribution, t, of fracture traces measured on a large trace mapping plane by

$$t_1 = \frac{\pi}{4} \frac{L_2}{L_1},$$

$$t_2 = \left[\ln\left(1+\sqrt{2}\right) - \frac{\sqrt{2}-1}{3} \right] \frac{L_3}{L_1}, \tag{2}$$

where L_i are the 'i'th moments of the length distribution and t_i are the 'i'th moments of the trace length distribution. Similar formulae can be obtained for higher moments and for circular fractures. In fact the common approach is to make an assumption as to the mathematical form of the distribution of fracture lengths and then either use these simple formulae between the means and second moments of the distribution, or to simply calibrate against statistics from a specific trace map. In fact the trace length to fracture length relationship is quite insensitive to the precise shape assumed for the fractures and there is relatively little difference between the results for circular or square fractures. A more significant assumption is the choice of the mathematical form of the fracture length distribution. This is generally taken to be a log-normal distribution which will often result in a good fit between the main part of the simulated trace length distributions and the main part of the measured trace length distribution. However, the goodness-of-fit of the tails of the two distributions is often less good. A poor match in the tail of the distribution may result in the existence of extreme, unphysical fractures with very long traces. These are quite unimportant to many of the statistics used to infer parameters but may have a much more important role in the network flow. Finally, assumptions regarding the likelihood of fractures terminating against each other or not will influence fracture network connectivity.

Finally, the hydraulic properties of the fractures need to be defined. The usual assumption is that some form of the parallel plate law for plane fracture flow applies, but rather than measure a distribution of apertures directly, a more reliable approach is to infer a distribution of fracture transmissivities. This too, generally relies on an assumption as to the form of the probability distribution of fracture transmissivities. Generally the log-normal distribution is used. With this distribution and a specified fracture spacing, then the mean and standard deviation of fracture transmissivities can be related to the mean and standard deviation of short interval packer tests in boreholes so long as it is assumed that the transmissivities add. Strictly, fracture connectivity away from the borehole will affect the packer test results, but for short tests, the radius of influence of the test will be small and the measurements can be taken to correspond to the summation of local transmissivities. The fitting process involves typically using maximum likelihood estimators and in general will require numerical evaluation of the best estimates. Again, the results of the fracture property interpretation should be checked by simulation of the measurement process and it may be appropriate to infer the parameters of the distribution by calibrating directly against the experimental data [see 16 for more detail].

An alternative approach to generating the fracture network that is often used is to generate fractures using an initial approximation and test the resulting network by simulating the experimental measurement procedures. Then the network is modified to improve the correspondence between, for example, the numerically simulated log and the physical log. This calibration procedure is particularly appropriate when more complex correlations between the different parameters are being simulated, and the assumptions used in deriving the formulae given above are not valid. Such simulated measurements should in any case be used to check the validity of the interpretation of the network parameters [see 18 for example].

Once the fracture network has been generated, the next step is to calculate all the fracture intersections. This allows an interpretation of the fracture network connectivity. In principle this is a straightforward task. The intersections are generally evaluated by solving the equation for the intersection of the two fracture planes using elementary geometry. For large networks it is worthwhile optimising the search for intersections by sorting the planes into subregions and only testing planes in the same subregion for intersections. In this way the asymptotic cost of the calculation of the intersections will be proportional to the number of planes rather than the square of the number of planes. The other difficulty in the calculation is a common difficulty in stochastic fracture network calculations. When generating large networks, the algorithms need to be very robust since what might be a very unlikely 'pathological' case of the arrangement of the planes whereby the algorithm used might fail, will actually be quite likely to occur in a large network. Examples of such 'pathological' cases include planes that only just intersect, very nearly parallel planes, and so forth. Whilst the treatment of these extreme cases should not make a big difference to the overall results of the calculations, the way in which different codes address them may result in slightly different models for different codes. This in turn makes verification of the flow models more difficult. It is important that the fracture generation and connectivity calculations are verified before addressing verification of the flow calculations [19].

3.   FLOW SIMULATION FOR LARGE NETWORKS

Once the fracture network geometry and the connectivity of the network have been evaluated, the next task is to evaluate the flow through the fracture network for sufficiently many realisations to determine the distribution of possible groundwater flows.

A simple law such as Snow's parallel plate law is used to relate a fracture aperture to a transmissivity, and then the volumetric flux, $\mathbf{q}$, is given by:

$$\mathbf{q} = -\frac{e^3}{12\mu}\left(\nabla P - \rho\mathbf{g}\right), \tag{3}$$

where q is the volume flux of water, e is the fracture aperture, P is the pressure, ρ is the fluid density, μ is the kinematic viscosity and g is the acceleration due to gravity. The equation for flow is the mass conservation equation:

$$\frac{S}{\rho g}\frac{\partial P}{\partial t} = -\nabla.\mathbf{q}, \tag{4}$$

where S is the fracture storage parameter. S is usually related to the fracture transmissivity either by an empirical correlation from measured pressure diffusivities, or by a compressibility term derived from fracture normal stiffness and compressibility of the water in the fracture opening.

Stochastic, discrete fracture network models were first used to simulate two-dimensional flow problems, and here the numerical problem is quite straightforward. For steady-state constant density flows, with flow restricted to the fracture network, the problem is reduced to a mass balance equation at intersections. Each intersection will have at most four fracture segments connected to it and the problem reduces to a sparse banded matrix inversion with a narrow bandwidth of five. For transient calculations then the storage properties of the fracture need to be taken into account and near sources and sinks, the fractures need to be broken into small segments to discretise the evolution of the pressure field over scales smaller than the intersection spacing. In practice, for many problems, the resolution of the intersection spacing in a well-connected network may well be adequate. In such cases then again the problem reduces to a sparse linear matrix equation with a known narrow bandwidth. The main issue for solving the two-dimensional transient equation is the possibly large ratio of timescales between the most open and the tightest fractures in the network (actually the fractures with the largest ratios of pressure diffusivities). This can make the choice of timestep difficult, and lead to artificial overshoots of pressure in untransmissive sections of the network at early times if too small a timestep is taken for the pressure to be adequately resolved by the chosen discretisation.

A more demanding task is the solution of equations (3) and (4) for large three-dimensional networks. The complex flow geometry makes this a very computationally intensive task, but it is now routinely performed for issues relating to radioactive waste disposal in low permeability fractured formations. Typical fracture network models may have to consider flow through tens of thousands of fractures, each of which must be adequately discretised. This requires very efficient and robust algorithms.

In principle, the solution of the flow equation is like any other groundwater flow problem and conventional finite-element techniques could be used to solve the problem numerically. This is the approach taken by some codes. However, the numerical problem will have a very complex geometry leading to difficulties in developing a good finite-element mesh. It will very likely involve large transmissivity changes at fracture intersections. Further, the scale of interest will generally involve very many fractures and this leads to a large numerical problem. For this reason, several computer codes have incorporated special algorithms to solve the flow equation for fracture networks. In the following paragraphs, three different approaches are discussed, and the relative advantages and disadvantages identified.

The most straightforward approach is to treat the network as any other flow geometry and simply use finite elements to discretise the system [20]. This can result in very large finite-element meshes. The size of model that can be simulated may be smaller than if more sophisticated algorithms were used, however, the scale of problem that can be simulated will increase in the future as computers become more powerful. The advantage of this simple approach is that with less specialised solution algorithms, it is more straightforward to generalise the equations to address more complex flow physics. The alternative approaches can be difficult or impossible to generalise. The method does however require sophisticated automatic grid generators to create the finite element mesh on each fracture plane. This is likely to be a triangulation scheme, but care needs to be taken to ensure that good meshes are created for the worst cases of fracture intersection geometry and property contrasts. It is a similar problem to that discussed above for the accurate calculation of all the intersection lines in difficult network configurations. The second difficulty, and most serious drawback, is to achieve and demonstrate mesh convergence. The cost of these runs can increase dramatically if fine discretisation is used on each fracture plane. Whilst the basic behaviour of the fracture network can be determined, it may be difficult to demonstrate that large runs are fully grid converged.

The second method that has been used to solve the flow problem is to use analytic or semi-analytic results for the flow between intersections on each plane [21]. The analytical response of a parallel-sided circular fracture to a flux or pressure at an intersection can be calculated, and for a linear problem, can be superimposed with other such fluxes from all the intersections on the fracture. The problem is then reduced to a solution for the mass balance for the intersections. The method essentially deals with the resolution of flow on the individual fractures analytically, and avoids the need for fine meshes covering each fracture. This makes the calculation of the flow solution much smaller than if the basic approach to discretising the network had been used.

The third approach is similar but a numerical model of the response of fracture plane can be used. This is the approach taken in the NAPSAC code. It has the advantages of the analytic response function scheme in terms of efficiency in very large networks, but allows much more flexibility [17]. For example, models of aperture variation or changes in properties in response to effective stress can be accommodated by modifying the local properties of the finite element mesh on each fracture in turn. By choosing a simple discretisation scheme, the grid on each fracture can be made very fine and grid convergence demonstrated. It has been found that acceptable results can be obtained for models of tens of thousands of fractures with individual fracture meshes each comprising a few thousand finite elements. This scale of problem may be solved on workstation computers quite easily.

In the NAPSAC approach, the flow response of each fracture to a unit pressure at each of a number of nodes is calculated and these fluxes are integrated over each intersection line. The response functions are analogous to basis functions on finite elements, with the complex fracture plane taking the place of the finite element. The response functions have to be calculated with a simple flow calculation for each node on each intersection of the network. For this to be efficient, a very simple regular mesh is used on each fracture. This sacrifices some resolution in the precise location of intersections, but this can be made arbitrarily small by increasing the resolution of the fracture response mesh, and in practice the model can be shown to be grid converged.

The details of this algorithm, as incorporated in the NAPSAC code, are presented in an appendix to this book.

For transient problems, the interpretation of the response functions as being analogous to finite-element basis functions can be continued, with the evaluation of the integral of the product of basis functions over the fracture plane. This leads to a solution that is locally a steady-state solution on any given fracture plane. This may be compared to the use of locally quadratic basis functions on regular finite elements. Whilst the use of the response functions means that the results are better for the complex fracture flows than simple basis functions, the transient flow is not well resolved for scales of a single fracture or smaller. This will be acceptable in regions of the network away from sources or sinks, where the pressure is slowly changing, but will not enable the model to predict the pressure field near such a source or sink. This will be important near a tunnel or borehole. A finer resolution is required here and so the NAPSAC code uses the local scale fracture mesh in such regions. Consistent flux and pressure continuity conditions are applied where this region of fine resolution is joined to regions of the main network which use the default resolution. This allows very well discretised solutions for a small overhead on the cost of the simulation.

As with the generation of the fracture geometry, it is important to verify the algorithms used. This can be done against exact solutions for very simple regular networks, but must generally be achieved by cross-comparison of independent codes. However, there are some difficulties in comparing the results from independent codes. For example, the codes must address identical networks for the flow results to be compared in detail. Thus both codes must use an identical realisation of the network geometry (so as to avoid difficult issues associated with cross comparison of statistics from many realisations, when any differences would be very difficult to evaluate), have found identical intersections, and, for example, consider square fractures. Such a comparison was undertaken as part of the international OECD/NEA Stripa project [19]. In DECOVALEX, the range of different models used to address the benchmarks makes such a direct comparison of numerical accuracy difficult as different approximations are inherent in the various codes. Instead, a more qualitative evaluation of the approaches adopted for realistic problems was made.

4. COUPLED MODELLING OF HYDRO-MECHANICAL PROCESSES

4.1. Fully coupled models

Developing an understanding of the coupled mechanical and hydraulic behaviour of fractured rock requires discrete models to complement more conventional continuum modelling, as was discussed in the context of flow modelling above. The coupled problem is, however, much harder since it involves predicting the behaviour of the rock mass as well as the pressure of the fluid flowing within the fractures. The matrix between the fractures can no longer be ignored.

The main consequence of the increased complexity of the coupled hydro-mechanical problem is that the current state-of-the-art for fully-coupled, large, discrete models uses two-dimensional representations. The models are also restricted by current computational limits to smaller network systems than the simple uncoupled flow models. Finally, when modelling these coupled problems, the equations are in general non-linear and the system involves changes occurring on a range of timesteps. To properly resolve the solution, the space and time discretisation is controlled by the shortest length scales and timescales of the problem. The limitation to small, two-dimensional problems means that the fracture network connectivity is only approximately represented as discussed above.

The physics to be solved for the coupled hydro-mechanical problem are as follows. In the rock matrix, the equations for conservation of momentum are solved for the movement and deformation of the rock matrix. Conventional fluid flow equations such as (3) and (4) are solved for the movement of water in the fractures. Finally, constitutive laws such as Barton-Bandis [22] describe mechanical behaviour of the fractures, and the normal displacement of the matrix surrounding the fracture defines the fracture aperture. The deformation of the matrix is governed by constitutive laws such as linear elasticity.

There are two components to the coupling of the fluid and rock matrix behaviour. First, the coupling from hydraulic processes to the mechanical behaviour is provided by the pore pressure acting on the boundary of the matrix blocks, in addition to forces at block contacts. This changes the stress-field and the corresponding deformation of the rock mass. The modified stress, taking the effect of the pore water pressure into account is the effective stress.

The second component to the coupling is the corresponding impact of the rock mass deformation on the fluid. Usually the timescale for stress changes and deformation is sufficiently long that the rock mass is in pseudo-equilibrium with the water in the fractures and the main effect of the stress field is to modify the fracture aperture as the pore water pressure changes modify the effective stress. The deformation of the fracture aperture by normal and shear displacements of the fracture sides due to the rock matrix strain changes the effective flow aperture and hence, using equation (3), the fracture transmissivity field. This in turn affects the fracture hydraulic behaviour. The full set of equations used by discrete hydro-mechanical codes are given in the earlier chapters of this book and in the code appendices.

More extreme couplings do exist, where deformation causes changes to the pore pressure and actually drives the water through the fracture network. This might occur for example during seismic pumping due to earthquakes, or due to deep burial in a sedimentary basin. These processes were not investigated in the DECOVALEX project.

The use of discrete models to predict stress fields and solve the mechanical problem by itself is well established. The representation of the coupled problem that incorporates the effect of the pore water pressures on the stress field is also quite well established and the empirical laws describing this coupling work well [23]. The stress field calculated in the discrete models and measured in fractured rocks is often not significantly influenced by the scale of variability of pore pressures due to the discrete fracture flow field away from sources and sinks. In this case, the results are quite similar to continuum approximations.

The impact of the coupling in the opposite direction, that is the effect of the stress field on the fracture apertures and hence on the flow, is less well understood. Results from the DECOVALEX study of Test Case 1 illustrate the difficulty that still exists in describing accurately how the aperture of a fracture responds to changes in the mechanical stress applied [24]. In this testcase, the constitutive laws gave a reasonable description of how an experiment that measured the flow on a single fracture progressed under a range of applied normal stresses. However, there was poor agreement between constitutive laws to describe the behaviour of the flow following shear displacement and the corresponding experimental results. For the issues relating to the long term safety case for radioactive waste disposal, the influence of the stress field on the flow is potentially much more important than the impact of the fracture geometry on the stress field. Experimental evidence for the important influence of the stress field on flow was provided by the Stripa Project D-hole and Validation Drift experiments [25]. These measured the difference between flow to an excavated tunnel compared to the flow to an array of boreholes designed to simulate the same pressure drawdown. These two experiments involved essentially identical pressure boundary conditions to the flow, but the mechanical stress changes are much greater around a tunnel than around the borehole array. The reduction in flow to the tunnel compared to flow to the boreholes was a factor of ten overall and more in the relatively good rock away from a small fracture zone. There is no consensus as to the role of the stress differences in this flow reduction, but stress and two-phase flow effects are both believed to contribute to the differences in the flow fields.

Coupled models have so far not been successful in predicting the flow fields where there are significant mechanical stress changes, and the DECOVALEX studies represent the current state-of-the-art in this field. One of the main problems is that the models are two-dimensional, whereas the problem is three-dimensional. Two-dimensional models show the development of very high aperture sections of fractures immediately adjacent to the corners of matrix blocks undergoing displacement. These would correspond to highly transmissive 'pipe-like' conduits perpendicular to the plane of the model were the results to be simply translated into three-dimensions. This is believed to be at least partially an artefact of the two-dimensional representation of the rock block geometry, which will be physically better supported and less able to move in three dimensions. The problems are, in this respect, very similar to those faced 10 years ago for discrete flow modelling, when two-dimensional modelling was used to illustrate the importance of network connectivity and structure, but before three-dimensional models were able to accurately predict the behaviour of specific fracture network flows. The coupled codes are currently being developed to extend their capabilities to three-dimensional problems, for example 3DEC can now address small network stress problems with thousands of rock blocks and is being developed to address the full three-dimensional coupled problem.

4.2. Approximate methods

In order to develop an understanding of the role of coupled processes in three-dimensional networks, fully coupled codes have been developed and the capability of these codes is being extended from two dimensions to three dimensions. Currently the three-dimensional connectivity and the three-dimensional support of the rock mass is neglected. An alternative way to approach the fully coupled problem is to look for simple approximate representations of the physical processes that can be used in models that represent more realistic fracture geometries and fracture densities. Since the stress field is less sensitive to the flow processes than the corresponding coupled effect of the stress on the fracture flows, it is reasonable to consider the changes to the flow due to a specified rock stress field. The rock stresses will be approximate where the block geometry plays a significant role, but the qualitative differences between the results of such one-way coupled models in three-dimensional networks and the fully coupled two-dimensional models are useful. If the stress-field in the rock mass is specified, then the effective stresses in the fractures can be computed and changes to the fracture properties following, for example, the excavation of a tunnel can be evaluated. Care must be taken in interpreting such approximations since they will only be appropriate for small changes to the fracture properties where the stress field is unaffected by the flow and the effective stress distribution in the fractures. Nevertheless this is a feasible approach that allows the direction of fracture property changes and qualitative results to be obtained from realistic fracture geometries. Using the algorithms in NAPSAC, the change in effective stress and in the fracture apertures over a fracture plane near a borehole of tunnel can be resolved in great detail. Where the stress field is accurate to within a factor of two (continuum approximations appear to be this accurate or better almost everywhere) and where the changes in fracture properties are not highly sensitive to small errors in the local stress, the model will be good.

Given the uncertainties in the dependence of the fracture flow properties on the stresses and strains on the fractures, this approach is probably as good as fully coupled two-dimensional modelling for many applications, is more robust, and can be applied to predict the impact of changes in effective stress within a network where the regional stress is well known.

5.   SUMMARY

This chapter has discussed the discrete approaches available to represent groundwater flow and coupled flow and mechanics in fractured rock. The discrete approach offers a better understanding of the influence of the fracture network geometry and the inherent uncertainty in our understanding of fractured sites. This is important for a number of issues related to flow alone, and here, the fracture network method is a proven tool in developing a detailed understanding. Where the rock stresses remain nearly constant, then the flow properties can be empirically characterised and the flow field predicted.

The coupled hydro-mechanical problem is more difficult and remains a research task. The currently available coupled models are helping to develop a qualitative understanding, but are not yet proven. They are research models rather than predictive tools. This means that there will be greater uncertainty where is it necessary to predict coupled changes, for example where the drawdown in flow experiments is sufficient to lead to significant changes in effective stress, or where the flow to excavated tunnels and caverns is to be predicted. The DECOVALEX project has made progress in understanding these issues and has identified further work that is required.

ACKNOWLEDGEMENTS

This chapter is based on work undertaken by the author when employed by AEA Technology. Funding for that work was provided through contracts from UK Nirex Limited to AEA Technology.

REFERENCES

1   A.Hooper, D.E.Billington and A.W.Herbert, Modelling framework for groundwater flow at Sellafield, Proceedings of NEA SEDE/PAAG workshop on conceptual model uncertainty, OECD, Paris, 1993.
2   A.W.Herbert and G.W.Lanyon, The application of a fracture network modelling approach to field experiments, Proceedings of Fourth North Sea Chalk Symposium, Deauville, 1992, AEA Technology Report.
3   L.Smith and F.W.Schwartz, An analysis of the influence of fracture geometry on mass transport in fractured media, Water Resour. Res., 20(9), 1241, 1984.
4   J.E.Bolt, P.J.Bourke, N.L.Jefferies, R.D.Kingdon, D.M.Pascoe and V.M.B.Watkins, The application of fracture network modelling to the prediction of groundwater flow through highly fractured rock, UK Nirex Report NSS/R281, 1995.
5   J.E.Geier, C-L.Axelsson, L.Hässler, and A.Benabderrahmane, Discrete fracture modelling of the Finnsjön rock mass: Phase 2, SKB Technical Report 92-07, 1992.

6   A.W.Herbert, G.W.Lanyon, J.E.Gale and R.MacLeod, Discrete fracture network modelling for Phase 3 of the Stripa Project using NAPSAC), Proceedings of the 4th international symposium on the NEA/OECD Stripa Project, OECD, 1992.

7   O.Olsson and J.E.Gale, Site assessment and characterisation for high-level nuclear waste disposal: results from the Stripa Project Sweden, Q.J.Eng.Geol, 28, S17, 1995.

8   L.Jing, J.Rutqvist, O.Stephansson, C.-F.Tsang and F.Kautsky, DECOVALEX - Mathematical models of coupled T-H-M processes for nuclear waste repositories: report of phase 1, SKI Technical Report 93:31, 1993.

9   D.Hodgkinson, A comparison of measurements and predictions for the Stripa tracer experiments, Stripa Project Technical Report 91-10, SKB, 1991.

10 D.Hodgkinson and N.Cooper, A comparison of predictions and measurements for the Stripa Simulated Drift Inflow experiment, Stripa Project Technical Report 92-20, SKB, 1992.

11 P.C.Robinson, Connectivity, flow and transport in network models of fractured media, D.Phil thesis, Oxford University, 1984.

12 M.Board, UDEC (Universal Distinct Element Code) Version ICG1.5, Vols. 1-3, NUREG/CR-5429, NRC, Washington, 1989

13 DECOVALEX Secretariat, Bench-Mark Test 3, Near-field repository model, DECOVALEX Document Doc 92/112, Royal Institute of Technology, Stockholm, 1992.

14 P.M.Wilcock, The results of applying the NAPSAC fracture network code to model BMT3: the near-field test case, AEA Technology Report AEA D&W 0640, 1993.

15 W.S.Dershowitz, Rock joint systems, PhD thesis, MIT, 1984.

16 A.W.Herbert and B.A.Splawski, Prediction of inflow to the D-holes at the Stripa mine, Stripa Project Technical Report 90-14, SKB, 1990.

17 A.W.Herbert, NAPSAC (Release 3.0) Summary Document, AEA Technology Report AEA D&R 0273, 1993.

18 W.Dershowitz, P.Wallman and S.Kindred, Discrete fracture modelling for the Stripa site characterisation and validation drift inflow predictions, Stripa Project Technical Report 91-16, SKB, 1991.

19 F.W.Schwartz and G.Lee, Cross verification testing of fracture flow and mass transport codes, Stripa Project Technical Report 91-29, SKB, 1991.

20 W.Dershowitz, G.Lee, J.Geier, S.Hitchcock and P.LaPoint, FracMan User Documentation, Golder Associates Report, Seattle, 1993.

21 J.C.S.Long, P.Gilmour and P.A.Witherspoon, A model for steady fluid flow in random three-dimensional networks of disc shaped fractures, Water Resour. Res. 21(8), 1105, 1985.

22 N.Barton, M.Bandis and K.Bakhtar, Strength, deformation and conductivity coupling of rock joints, Int. J. Rock Mech. Min. Sci. & Geomech. Abstr., 22(3), 1985.

23 K.Monsen, A.Makurat and N.Barton, Disturbed zone modelling of the SCV Validation Drift, Stripa Project Technical Report 91-05, SKB, 1991.

24 A.Makurat, M.Ahola, K.Khair, J.Noorishad, L.Rosengren and J.Rutqvist, DECOVALEX Test Case 1, Int. J. Rock Mech. Min. Sci. & Geomech. Abstr., Special Issue Thermo-Hydro-Mechanical coupling in rock mechanics, 339, 1995.

25 O.Ollson (ed.), Site characterisation and validation - final report, Stripa Project Technical Report 92-22, SKB, 1992.

O. Stephansson, L. Jing and C.-F. Tsang (Editors)
Coupled Thermo-Hydro-Mechanical Processes of Fractured Media
Developments in Geotechnical Engineering, vol. 79
© 1996 Elsevier Science B.V. All rights reserved.

231

Influence of fictitious outer boundaries on the solution of external field problems

G. Rehbinder

Dept of Civil and Environmental Engineering, The Royal Institute of Technology,
S-100 44 Stockholm, Sweden

Abstract
2D and 3D harmonic potentials in an external domain are not compatible, and possibilities of approximating results for 3D problems with 2D analysis are limited. The fictitious outer boundaries for 2D problems, which are always used in numerical calculations, must be located at a distance that far exceeds the dimensions of the problem. The location of the outer fictitious boundaries and the boundary conditions at them can hardly be set in such a way that a solution thus obtained agrees acceptably with the correct solution for the infinite domain.

1. INTRODUCTION

The theoretical calculation of motion of ground water, the conduction of heat, the stresses and displacement of rock and soil can involve difficulties. The difficulties have various origins.

1. The constitutive equations are uncertain and consequently so are the constitutive parameters.

2. The constitutive equations are reliable, but the constitutive parameters are not sufficiently well known.

3. The geometrical configuration is complicated.

4. Mutual coupling occurs between the ground water flow, the heat transport and the mechanical deformations.

The validity of the continuum approximation is sometimes questionable at any length scale, which implies that neither the conservation nor the constitutive equations are applicable.

If the continuum approximation is valid and the constitutive equation are reliable, all variables like stress, deformation, pore pressure, temperature etc. satisfy a set of differential equations. Together with proper boundary conditions, this set of equations has a solution.

The common way to obtain such a solution, is to resort to sophisticated computer codes. Codes and computers of today offer an excellent possibility to solve *mathematical* problems that were too difficult to solve thirty years ago. Unfortunately the codes treat only established and well posed mathematical problems and *not* physical or technical problems. It is very common that codes are identified with concepts called "modelling", "validation", "calibration" etc. of physical problems. There is a clear risk that this misunderstanding might obscure rather than illuminate the important question what a theoretical model is. These issues have been brought forward by Konikow & Bredehoeft (1992). Their viewpoints have been challenged by de Marsily & al. (1992).

A kind of problem that has not been discussed explicitly by Konikow & Bredehoeft deserves attention. The problem is the choice of boundaries and boundary conditions. It is important to note that the boundaries as such and not only the conditions on them, are important. If the domain is external, i. e. if the domain is infinite or semi-infinite and contains an inner boundary, a numerical calculation often includes conditions at a fictitious outer boundary, which is supposed to represent the state at infinite distance from the inner boundary. In solid mechanics the displacements and stresses strongly depend on the location of outer boundaries and the choice of boundary conditions. An illustrative example is given by Rehbinder (1995).

The aim of this presentation is to show that for external problems the choice of a fictitious outer boundary is decisive for the *entire* solution. The aim is also to show that the possibility of matching two and three dimensional external solutions is severely limited.

If the boundary value problem is internal, i.e. if the domain is limited, two and three dimensional solutions can show great similarity, and two dimensional solutions can be excellent approximations of slender three dimensional problems in the whole domain except for the "end" regions where three dimensional effects are important.

An example of an external problem is the one of a rock cavern used to store crude oil. The cavern is kept at atmospheric pressure and ground water seeps continuously into the cavern. The water is discharged from the cavern at the same rate as it flows into it. At the phreatic surface the pore pressure is atmospheric and far from the cavern the pore pressure is undisturbed by the cavern. Nothing is known about any possible impervious or highly conductive zone in the vicinity of the cavern.

A theoretical analysis of the above problem implies difficulties of two kinds.

1. A two dimensional approximation of a real three dimensional problem does not always represent the system adequately.

2. Formulation of the correct boundary conditions for the field equations can be difficult.

Another complication is that the shape and motion of a part of the boundary, like a phreatic surface, are unknown and are a part of the problem. This kind of complication is avoided through a linearization process, implying that the phreatic surface is undisturbed.

The rest of the presentation is devoted to the simplest possible application dealing with steady flow of ground water in isotropic and homogeneous but unbounded ground. Despite the simplicity of the problem, the analysis shows the decisive role that is played by the boundaries themselves and the corresponding boundary conditions.

In the present case the pore pressure p and consequently the potential $\phi = z + p/\rho g$ are harmonic functions, i.e. they satisfy Laplace's equation (ρ is the density)

$$\nabla^2 \phi = 0 \tag{1}$$

The character of the solution of (1) in a given domain D depends on two factors. The first one is whether the domain is a two or a three dimensional space. The second one is the combination of the prescribed boundary values of ϕ at the boundary of D. The choice of potential at the boundary is *arbitrary* and not governed by the theory leading to equation (1).

The possibility of solving (1) analytically depends on the shape of the domain. For a number of simple geometries, which can be described with appropriate orthogonal co-ordinate systems, separation solutions exist. These solutions are available in standard textbooks on applied mathematical physics, notably by Morse & Feshbach (1953).

2. HARMONIC FUNCTIONS IN A HALF SPACE WITH A CAVITY

The problem is to find a harmonic function ϕ, i.e. a function that satisfies (1), in a domain D which is a half space with a cylindrical or spherical cavity. Hence the boundary of the domain consists of three parts. Firstly the free surface, which is denoted C_s, secondly the surface of the cavity, which is denoted C_0 and finally "infinity", which is denoted C_∞. The situation is shown in figure 1.

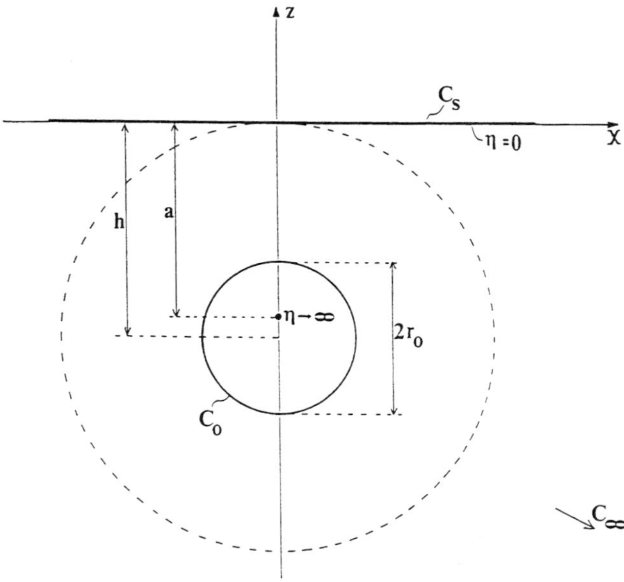

Figure 1. Definition of geometrical parameters in a semi-infinite space with a cavity.

The boundary values at the boundary $C = \{C_s \cup C_0 \cup C_\infty\}$ is of the Dirichlet type, i.e. the potential is given at the entire boundary and no part of the boundary involves the normal derivative. For the sake of mathematical simplicity, the flux is reversed, i.e. the cavity is pressurized instead of evacuated. Then

$$\phi(C_0) = \phi_0$$

$$\phi(C_s) = 0 \tag{2}$$

$$\phi(C_\infty) = 0$$

If the cavity is a cylinder, the problem is two dimensional, whereas if the cavity is a sphere, the problem is three dimensional. The system (1) - (2) can be solved in closed form in both of these cases. They are treated separately.

1. Two dimensional flow (cylindrical cavity).

   In Cartesian coordinates the boundary value problem is

$$\frac{\partial^2 \phi}{\partial x^2} + \frac{\partial^2 \phi}{\partial z^2} = 0$$

$$\phi(z = 0) = 0 \tag{3}$$

$$\phi(x^2 + (z + a)^2 = r_0^2) = \phi_0$$

$$\phi(x^2 + z^2 \to \infty) = 0$$

If the problem is described with bipolar coordinates (η, ψ), it is greatly simplified. The relation between Cartesian and bipolar coordinates are

$$x = \frac{a \sin \psi}{\cosh \eta - \cos \psi}$$

$$\tag{4}$$

$$z = \frac{a \sinh \eta}{\cosh \eta - \cos \psi}$$

where $-\infty < \eta < \infty$ and $-\pi \leq \psi < \pi$. The merit of the bipolar representation is that the separation solution of Laplace's equation is one dimensional. Equations (4) are introduced in (3). Then (3) becomes

$$\frac{\partial^2 \phi}{\partial \eta^2} = 0$$

$$\phi(\eta = \eta_0) = \phi_0 \tag{5}$$

$$\phi(\eta = 0) = 0$$

The last condition is equivalent to $\phi(-\eta_0) = -\phi_0$. The value η_0 is related to the dimensions in figure 1,

$$\eta_0 = \operatorname{arcsinh} \frac{a}{r_0} \tag{6}$$

Since Laplace's equation has degenerated to an ordinary equation, the solution is

$$\frac{\phi}{\phi_0} = \frac{\eta}{\eta_0} \tag{7}$$

The flux per axial length L is

$$\frac{Q}{L} = K \int_C \nabla \phi \, dA = K \int_{-\pi}^{\pi} \frac{\partial \phi}{\partial \eta} d\psi = \frac{2\pi \phi_0 K}{\eta_0} \tag{8}$$

K is the conductivity. If $h/r_0 > 1.5$, $h \approx a$, which simplifies the solution. The flux Q is normalized with the volume of the cavern $V = \pi r_0^2 L$. Then

$$\frac{Q r_0^2}{V \phi_0 K} = \frac{2}{\eta_0} = \frac{2}{\operatorname{arcsinh} \frac{h}{r_0}} \tag{9}$$

An often used approximation to the above, in itself quite a simple solution, is the solution in an annular domain (corresponding to the broken line in Figure 1). Then the solution yields the following potential

$$\frac{\phi}{\phi_0} = 1 - \frac{\ln(r/r_0)}{\ln(h/r_0)} \tag{10}$$

and the corresponding flux

$$\frac{Q r_0^2}{V \phi_0 K} = \frac{2}{\ln(h/r_0)} \tag{11}$$

2. Three dimensional flow (spherical cavity).

In Cartesian coordinates the boundary value problem is

$$\frac{\partial^2 \phi}{\partial x^2} + \frac{\partial^2 \phi}{\partial y^2} + \frac{\partial^2 \phi}{\partial z^2} = 0$$

$$\phi(z = 0) = 0 \tag{12}$$

$$\phi(x^2 + y^2 + (z + a)^2 = r_0^2) = \phi_0$$

$$\phi(x^2 + y^2 + z^2 \rightarrow \infty) = 0$$

This problem exhibits similaries with the previous one, but here bispherical coordinates (η, θ, ψ) must be used. They are related to the Cartesian ones by

$$x = \frac{a \sin \theta \cos \psi}{\cosh \eta - \cos \theta}$$

$$y = \frac{a \sin \theta \sin \psi}{\cosh \eta - \cos \theta} \tag{13}$$

$$z = \frac{a \sinh \eta}{\cosh \eta - \cos \theta}$$

where $-\infty < \eta < \infty$, $0 \leq \theta < \pi$ and $0 \leq \psi < 2\pi$. Due to the symmetry, $\partial/\partial\psi = 0$. The separation of Laplace's equation is more complicated here than in the previous case. The equation is

$$\frac{\partial}{\partial \eta} \left(\frac{1}{\cosh \eta - \cos \theta} \frac{\partial \phi}{\partial \eta} \right) + \frac{\partial}{\partial \theta} \left(\frac{\sin \theta}{\cosh \eta - \cos \theta} \frac{\partial \phi}{\partial \theta} \right) = 0 \tag{14}$$

The Dirichlet conditions are

$$\begin{aligned} \phi(\eta_0) &= \phi_0 \\ \phi(0) &= 0 \end{aligned} \tag{15}$$

The nontrivial solution is expressed in terms of Legendre polynomials $P_n(x)$.

$$\frac{\phi}{\phi_0} = \sqrt{2(\cosh \eta - \cos \theta)} \sum_{n=0}^{\infty} \frac{\sinh(n + 1/2)\eta}{\sinh(n + 1/2)\eta_0} e^{-(n+\frac{1}{2})\eta_0} P_n(\cos \theta) \tag{16}$$

The flux Q is most easily evaluated through the free surface $\eta = 0$.

$$Q = 2\pi\sqrt{2}a\phi_0 K \sum_{n=0}^{\infty} \left(n + \frac{1}{2}\right) \frac{e^{-(n+1/2)\eta_0}}{\sinh(n+1/2)\eta_0} \int_0^\pi \frac{\sin\theta}{\sqrt{1-\cos\theta}} P_n(\cos\theta)d\theta \quad (17)$$

The integral above is transformed by the substitution $t = \cos\theta$. Then

$$\int_0^\pi \frac{\sin\theta}{\sqrt{1-\cos\theta}} P_n(\cos\theta)d\theta = \int_{-1}^1 \frac{P_n(t)}{\sqrt{1-t}}dt = \frac{2\sqrt{2}}{2n+1} \quad (18)$$

As in the two dimensional case

$$\eta_0 = \operatorname{arcsinh} \frac{h}{r_0} \quad (19)$$

and $a = h$ if $h/r_0 > 1.5$.

In this case the volume of the cavity is $V = 4\pi r_0^3/3$, and the normalized flux is

$$\frac{Qr_0^2}{V\phi_0 K} = 3\frac{h}{r_0}\sum_{n=0}^{\infty} \frac{2}{e^{(1+2n)\eta_0} - 1} \quad (20)$$

If $h \to \infty$, the solution degenerates into the spherical one

$$\frac{Qr_0^2}{V\phi_0 K} = 3 \quad (21)$$

3. COMPARISON BETWEEN THE TWO AND THREE DIMENSIONAL SOLUTIONS

The two cases are compared in figures 2 & 3. Figure 2 shows how the flux is affected by the presence of the free surface. The discrepancy between the cases is sizable; they seem to have only one feature in common, that the flux decreases monotonically with the depth. The most remarkable deviation occurs if the depth is infinite; the two dimensional flux is zero whereas the three dimensional flux is finite. If a two dimensional model is expected to simulate a real three dimensional case, the simple formula (10) with $h/r_0 \leq e^{3/2} \approx 1.95$, gives the best value of the flux.

Figure 3 shows the potentials right above and right below the cavity for a given depth of the cavern. Even for the potential a considerable discrepancy occurs; the two dimensional potential is larger than the three dimensional one. As for the flux above, it turns out that the simple solution (10) for $h/r_0 \leq e \approx 2.72$ gives the best approximation of the three dimensional potential in the vicinity of the cavern.

The choice of the normalization volume V implies that the above comparison of the flux is meaningfull if the cavity is not too slender ($L/r_0 < 5$). If the cavity is very slender ($L/r_0 \gg 10$), on the other hand, the agreements between both the two and three dimensional potentials in the vicinity of the cavity and the fluxes are good. The potentials far from the cavity deviate however.

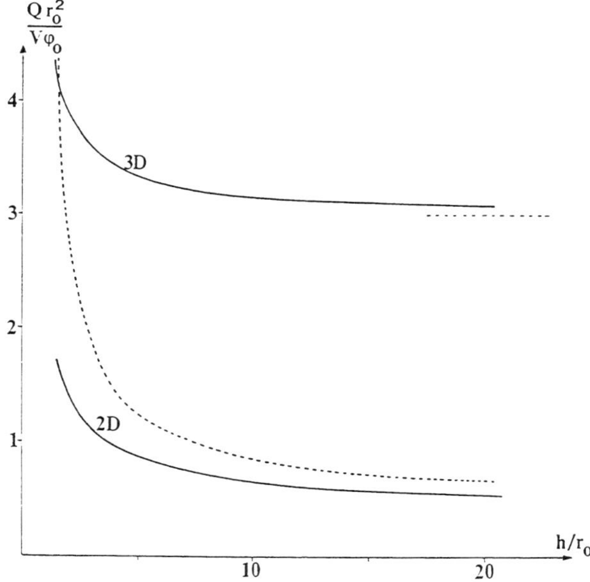

Figure 2. The flux from the cavity as a function of the depth of the cavity. 2D and 3D stand for the two and three dimensional solutions respectively. The dotted line represents the two dimensional solution for an annular domain that is indicated with the dotted line in figure 1.

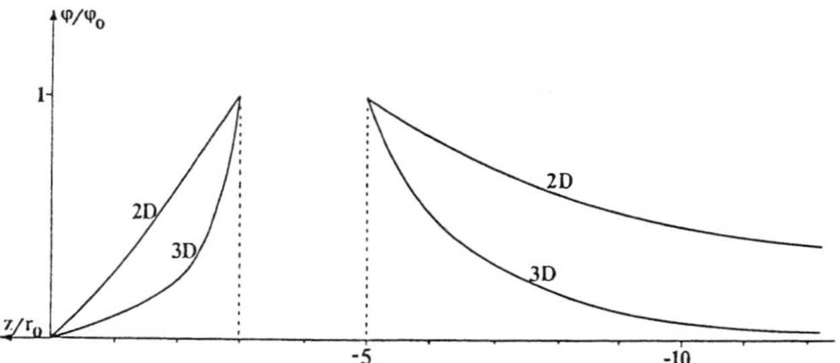

Figure 3. The potential $\phi(x = 0, z)$ above and the below the cavern for $h/r_0 = 4$.

4. NUMERICALLY COMPUTED TWO DIMENSIONAL POTENTIALS IN AN FINITE DOMAIN WITH A CYLINDRICAL CAVITY

If the cavern is neither circular, nor rectangular, a simple separation solution of the Laplace equation does not exist, and one has to resort to numerical methods. A disadvantage with some numerical methods, viz. the Finite Element Method and the Finite Difference Method is that they require an outer boundary with appropriate boundary conditions. The fictitious outer boundary, which is supposed to represent "infinity", is normally chosen as a rectangle, symmetrically enclosing the cavity. The inner boundary is, as in the previous section, a circular cylinder. The arrangement is shown in figure 4.

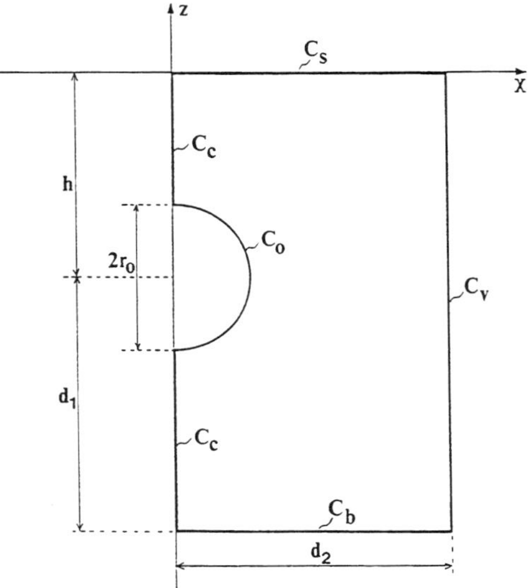

Figure 4. Definition of the fictitious rectangular domain which is supposed to simulate a semi infinite domain.

The question is now: How far from the cavity should the outer fictitious boundary be located ? What are the appropriate boundary conditions at the artificial outer boundary ? The questions cannot be answered unless the following questions have been answered: Is the purpose of the calculation to approximate a typically three dimensional problem with a two dimensional problem ? If so, is the purpose to calculate Q or ϕ ? Is the purpose to approximate the solution in a half space ? If the answer to the first question is yes, the development of the previous section applies. If the answer of the last question is yes, the following development applies.

Equation (1) is solved numerically in the domain $D' : (x \le d_2 \; - h - d_1 \le z \le 0) \cap (x^2 + (z + h)^2 \le r_0^2)$. The boundary of D' is $C' : C_s \cup C_v \cup C_b \cup C_c \cup C_0$. The domain and the boundaries are shown in figure 4. For symmetry reasons D' includes only $x \ge 0$. The values of ϕ at the boundary C_0 of the cavity and at the free surface C_s are the same as previously. i.e.

$$\begin{aligned} \phi(x^2 + (z + h)^2 = r_0^2) &= \phi_0 \\ \phi(z = 0) &= 0 \end{aligned} \tag{22}$$

At C_c

$$(\partial \phi / \partial x)_{x=0} = 0 \tag{23}$$

for reasons of symmetry.

At the fictitious outer boundaries C_v and C_b are two options, either a given potential or a vanishing flux. Therefore four possible combinations of boundary value problems exist.

1. No flux restrictions.

$$\begin{aligned} \phi(z = -h - d_1) &= 0 \\ \phi(x = d_2) &= 0 \end{aligned} \tag{24}$$

2. No flux across the bottom.

$$\begin{aligned} (\partial \phi / \partial z)_{z=-h-d_1} &= 0 \\ \phi(x = d_2) &= 0 \end{aligned} \tag{25}$$

3. Flux only through the free surface.

$$\begin{aligned} (\partial \phi / \partial z)_{z=-h-d_1} &= 0 \\ (\partial \phi / \partial x)_{x=d_2} &= 0 \end{aligned} \tag{26}$$

4. Flux through the free surface and the bottom.

$$\begin{aligned} \phi(z = -h - d_1) &= 0 \\ (\partial \phi / \partial x)_{x=d_2} &= 0 \end{aligned} \tag{27}$$

The last alternative does not seem to appear in the literature and will not be considered henceforth.

Figure 5 & 6 show a comparison between the potentials at the vertical line through the center of the cavity. In Figure 5 the domain is small ($d_1 = d_2 = h = 4r_0$) and in Figure 6 the domain is large ($d_1 = d_2 = 4h = 4r_0$).

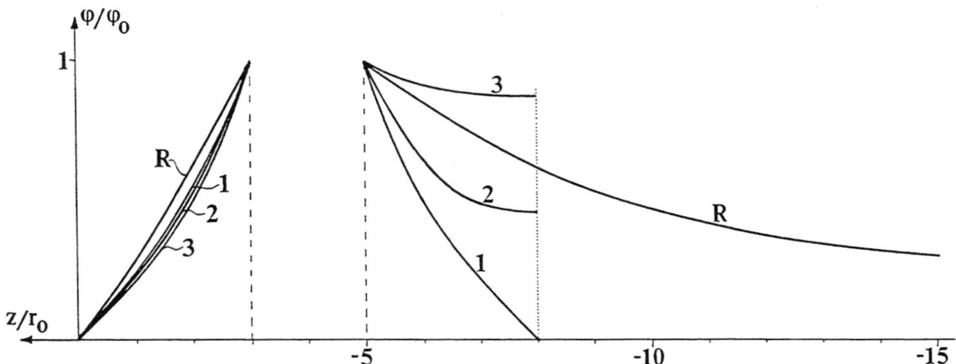

Figure 5. Comparison between the exact and the numerically computed two dimensional potentials above and below the cavity. $d_1 = d_2 = h = 4r_0$.
    R: The exact reference solution.
    1: The solution of the first alternative of boundary conditions.
    2: The solution of the second alternative of boundary conditions.
    3: The solution of the third alternative of boundary conditions.

It is not at all clear which one of the boundary value alternatives that gives the best approximation to the exact solution in the infinite domain.

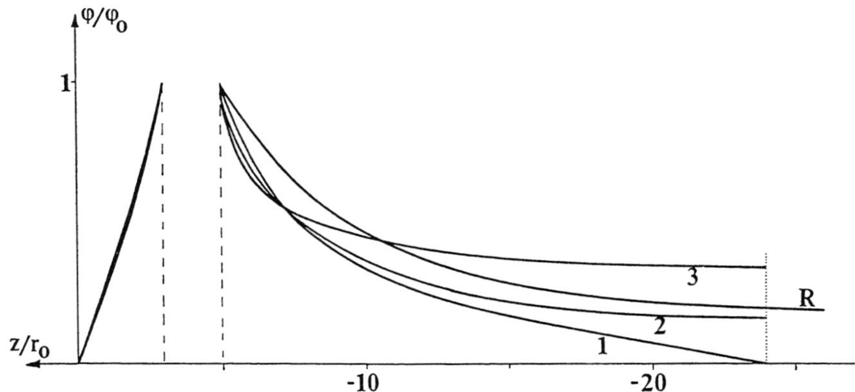

Figure 6. Comparison between the exact and the numerically computed two dimensional potentials above and below the cavity. $d_1 = d_2 = 4h = 4r_0$.
    R: The exact reference solution.
    1: The solution of the first alternative of boundary conditions.
    2: The solution of the second alternative of boundary conditions.
    3: The solution of the third alternative of boundary conditions.

The comparison becomes even more confusing if the corresponding fluxes in Table 1 are taken in consideration. Table 1 shows a comparison between the exact closed solution (at the top line) and the numerical solutions corresponding to the three combinations of fictitious boundaries and their boundary values.

The question if a numerical solution in a fictitious domain accurately approximates an exact one in a semi-infinite domain cannot be answered uniquely as mentioned previously. Table 2 shows that the different numerical solutions approximate the exact one more or less well depending on the purpose.

Table 1.

Comparison between th.. rmalized flux $Qr_0^2/V\phi_0$ for the exact two dimensional solution and the numerical solut.. ..s corresponding to the three alternative choices of fictitious outer boundary conditions, equations (24) - (26).

Case	Flux
Exact solution	0.95
1. Large domain	0.99
1. Small domain	0.79
2. Large domain	0.98
2. Small domain	0.89
3. Large domain	0.95
3. Small domain	0.48

Table 2.

Agreement between the exact solution and the numerical solutions of far field potential, the near field potential and flux, corresponding to the three alternative choices of fictitious outer boundary conditions, equations (24) - (26).

Case	Far field	Near field	Flux
1. Large domain	far too small	acceptable	acceptable
1. Small domain	far too small	too small	too small
2. Large domain	too small	good	acceptable
2. Small domain	too small	too small	acceptable
3. Large domain	far too large	far too small	acceptable
3. Small domain	too large	acceptable	far too small

It is obvious that if one numerical solution is a good approximation in one respect, it is less good in another. It is also clear that one alternative, which is good in one respect, may be less good if the extent of the domain is increased or decreased. All solutions seem however to give good approximations to the exact solution above the cavity, i.e. for $x = 0$, $0 \leq z \leq -h + r_0$.

One interesting question, which has not been studied, remains to be answered. What does the comparison look like, if the lateral extent of the domain is variable, i.e. if $d_1 \neq d_2$? Variation of the parameter d_1/d_2 is likely to change the results in Table 1.

5. CONCLUSIONS

The conclusions that can be drawn from the present investigation, which is based on simple applied mathematics, are the following:

1. Two and three dimensional external potentials are incompatible in the general case. If the cavity is very slender, i.e. if $r_0 \ll L$, the two dimensional potential approximates the three dimensional potential well in the vicinity of the cavity, but not at some distance from it.

2. If the analysis is two dimensional a priori, the exact semi-space solution can be approximated by a solution in a limited space whose fictitious outer boundaries are located at a distance that is much greater than the diameter and depth of the cavity. The exact solution for an infinite domain is completely defined by the parameter r_0/h. The solutions in a fictitious finite domain can be fitted to the exact solution in a real infinite domain in three ways. The first one is to adjust the *extent* of the domain, viz. d_1/r_0. The second one is to adjust the *shape* of the domain, viz. d_1/d_2. The third one is to choose the *type of boundary conditions*, viz. ϕ or $\partial\phi/\partial\hat{n}$. Thus, lots of combinations are available and possibly one, or maybe several, of them would yield "acceptable" or even "good" in all columns in Table 2. Such a combination is then *ad hoc*, a result that seems unsatisfactory from a scientific point of view.

These conclusions hold even if the geometry is complicated and if the permeability is not constant, i.e. even for a field variable that is not harmonic. Moreover, the conclusions hold for other field variables. They are of course valid for heat flux problems since temperature and hydraulic potential satisfy identical equations. If the variables are deformations of a solid, the equations are not identical but very similar. As mentioned previously, some illustrative examples have been given by Rehbinder (1995).

6. REFERENCES

1. Konikow, L.F. & Bredehoeft, J.D.: Ground-water models cannot be validated. Advances in Water Resources. **15** (1992) 75 - 83.

2. de Marsily, G.: Comment on ' Ground-water models cannot be validated. by L.F. Konikow & J.D. Bredehoeft. Advances in Water Resources. **15** (1992) 367 - 369.

3. Morse, P.M. & Feshbach, H.: Methods of theoretical Physics. McGrawHill N.Y. 1953

4. Rehbinder, G.: Analytical solutions of stationary coupled thermo–hydromechanical problems. Int. J. Rock Mec. Min. Sci. (1995)

O. Stephansson, L. Jing and C.-F. Tsang (Editors)
Coupled Thermo-Hydro-Mechanical Processes of Fractured Media
Developments in Geotechnical Engineering, vol. 79
© 1996 Elsevier Science B.V. All rights reserved.

245

Generic study of coupled THM processes of nuclear waste repositories as far-field initial boundary value problems (BMT1)

A. Millard[a], A. Stietel[a], A. Bougnoux[b], E. Vuillod[c]

[a] CEA/DMT/SEMT, CEN Saclay
    91191 GIF/YVETTE Cedex, FRANCE

[b] ENSMP-CGES, 35, rue Saint-Honoré
    77305 FONTAINEBLEAU Cedex, FRANCE

[c] INERIS, Ecole des Mines, Parc de Saurupt
    54042 NANCY Cedex, FRANCE

Abstract

The investigation of the thermal, hydraulic and mechanical effects of an underground waste repository on its far-field environment is a major objective. For this purpose, a two dimensional benchmark exercise, called BMT1, based on a generic study of a repository in a hypothetical regularly fractured rock, has been organised. Two possible approaches have been used, one based on a discrete representation of each fracture, and the other on an equivalent continuum. The methodologies used by the various research teams who participated in the exercise are presented in detail, and their results are compared. The influence of the fractures spacing as well as the couplings between phenomena are further investigated.

1. PRESENTATION OF THE BENCHMARK TEST BMT1

1.1. Introduction

The overall objective of the DECOVALEX project is to increase the understanding of various thermo-hydro-mechanical processes of importance for waste storage and its influence on transport of radionuclides from a repository to the biosphere and how these processes can be described by mathematical models.

The most important transport mechanism of radionuclides from a repository to the biosphere is to be by advection with the groundwater present in the rock. The host rock, where the repository is located, will be subjected to in-situ stresses, excavation-induced stresses and thermomechanical stresses resulting from radiogenic heat. These stresses can cause changes in hydraulic properties of the jointed rock masses and of the fracture zones. This could in turn lead to changes in the groundwater flow pattern of the repository area.

The Benchmark Test BMT1 simulates the thermo-hydro-mechanical effects, in the geosphere, surrounding an underground repository of high level nuclear waste.

Its primary aim was to compare and evaluate two different approaches in modelling the coupled THM response of a large fractured rock mass:
- a discrete approach, where all the discontinuities are explicitly represented, with their individual THM properties;
- a continuum approach where the discontinuities and the rock mass are represented by an "equivalent continuum", with THM properties derived from specific homogenization techniques.

For this purpose, a generic study of a repository in a hypothetical regularly fractured rock, has been defined. Four research teams have participated to the exercise: CEA/DMT (France), ENSMP (France), INERIS (France) and KPH (Japan).

1.2. Presentation of the benchmark test hypothesis

A large scale (km) rock mass is considered, with a waste repository located at 500 meters in depth. The model is bidimensional (vertical section) and measures 3000 x 1000 meters. It contains two sets of fractures, intersecting at right angles, with three possible spacings: 25, 50 and 100 m. The calculation of the thermo-hydro-mechanical response of the rock mass, to a heat source originated by the wastes, must be done over a 500 years period.

The original specification of the benchmark test BMT1 is given in Jing et al. [1]. It is detailed hereafter for sake of completeness.
- Model geometry:

The general dimensions, as well as the inclination of the fractures are shown on figure 1. A small rectangle shows the position of the wastes within the rock-mass.

The set of fractures was generated, starting from a given point located on the left of the repository.

Some specific monitoring points and profiles are identified on figure 2 in order to display and compare the calculated results.
- Physical properties:

The same THM properties are assumed for the repository zone and the surrounding rock mass.

The rock matrix is supposed to be impermeable and elastic, having isotropic properties, characterized by the following data:

Table 1

Rock matrix properties	Symbol	Specified value
Young's modulus	E_r	60 000 MPa
Poisson's ratio	v_r	0.23
Lineic thermal expansion coefficient	α_r	$9.10^{-6}/°K$
Density	ρ_r	2 670 kg/m^3
Specific heat	C_r	900 J/kg/°K
Thermal conductivity	λ_r	3 W/m/°K

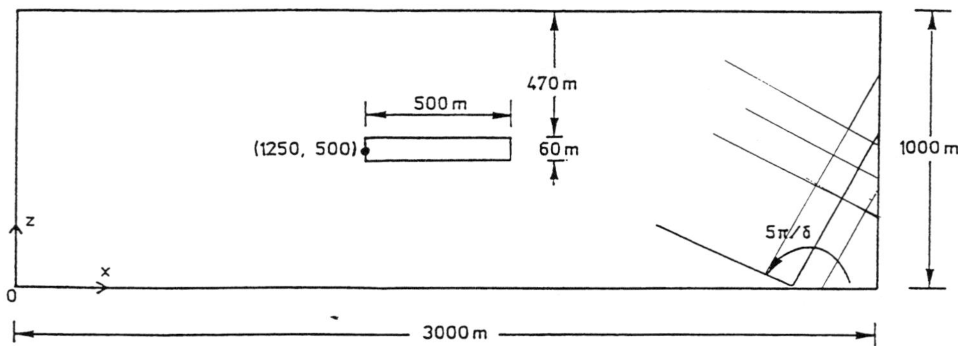

Figure 1. Model geometry and fracture sets of the far-field THM model

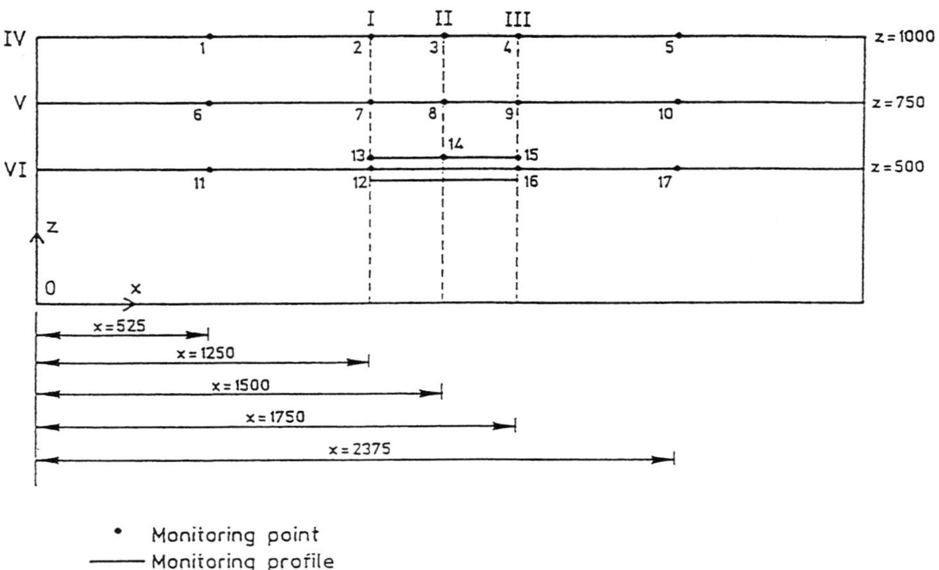

Figure 2. Locations of monitoring points and profiles

The water is flowing only in the fractures, which are each defined by two parallel planar surfaces. The separation of these planar surfaces corresponds to an effective hydraulic aperture α. Darcy's law is supposed to be valid for the flow. The water density ρw and dynamic viscosity μ are assumed to be temperature dependent, such as:

$$\rho_w = \rho_{w_0} \left[1 - \beta_w \left(T - T_0 \right) \right] \tag{1}$$

$$\frac{1}{\mu} = \frac{1}{\mu_0} \left[1 + \theta \left(T - T_0 \right) \right] \tag{2}$$

In these equations, ρ_{w_0}, μ_0 and T_0 are reference values and θ, β_w are two constants given below.

Table 2

Water properties	Symbol	Specified value
Reference density	ρ_{w_0}	1 000 kg/m^3
Reference dynamic viscosity	μ_0	10^{-3} Ns/m^2
Reference temperature	T_0	293 °K
Specific heat	C_w	4 200 J/kg/°K
Coefficient	β_w	6.10^{-4}/°K
Coefficient	θ	3.2 10^{-2}/°K

The mechanical behaviour of the fractures is supposed to be described by an elastoplastic model, obeying a Mohr-Coulomb failure criterion with zero tensile strength, and no dilatancy. Heat is supposed to be perfectly exchanged across the fractures.

Table 3

Fractures properties	Symbol	Specified value
Normal stiffness	k_n	100 GPa/m
Shear stiffness	k_s	10 GPa/m
Tensile strength	σ_f^T	0 MPa
Cohesion	C_f	0.1 MPa
Friction angle	φ_f	30°
Dilatancy angle	i_f	0°

It must be noted that the initial aperture a_0 of the fractures is not given, and must be determined, according to the approach followed, from the specified initial permeability of the rock mass: $K_0 = 10^{-9}$ m/s.
- Boundary conditions:
. For thermal conditions, a zero flux is imposed on the two vertical boundaries. A constant temperature (T = 283 °K) is prescribed on the top surface, and a geothermal flux on the bottom surface corresponding to a vertical thermal gradient of 0.03 °K/m.
. For hydraulic conditions, a variable head distribution, based on topographic considerations, is imposed at the top surface:

$$H(x) = H_o + h \cos\left(\frac{\pi(x - x_o)}{L}\right) \tag{3}$$

where $H_o = 100$ m, $h = 25$ m, $x_o = 1\ 250$ m and $L = 3000$ m.

A zero water flux is assumed elsewhere on the boundary.

. For mechanical conditions, zero normal displacements are prescribed on the lateral and bottom surfaces, while a vertical stress is imposed at the top surface, simulating an overburden load:

$$\sigma^s_{zz}(x) = -H(x)\,\rho_r\,g \qquad \text{(g is the gravity acceleration)} \tag{4}$$

Moreover, a plane strain hypothesis is assumed. Boundary conditions are summarized on figure 3.
- Initial conditions and loading:

The initial temperature and head distributions shall be determined from the specified boundary conditions, by means of a steady state calculation. The initial in-situ stresses are prescribed as:

$$\sigma^o_{xx} = \sigma^o_{zz} = -\left[1000 - z + H(x)\right]\rho_r\,g \quad \text{for } 0 \leq x \leq 3000 \text{ and } 0 \leq z \leq 1000. \tag{5}$$

Concerning the loading, a volumic heat source originated by the wastes, has the following form:

$$Q_s(t) = Q_o\,e^{-\beta t} \quad \text{with } Q_o = 0.5 \text{ W/m}^3 \text{ and } \beta = 0.02/\text{year}. \tag{6}$$

- Results required:

The results to be produced are the histories of temperature, displacement and stress components, fractures openings, water head and flow rates, over a 500 years period. They will be compared at monitoring points and along selected profiles.

2. COMPARISON OF DISCRETE AND CONTINUUM APPROACHES

2.1. General considerations

The most direct way to model the fractured rock mass is to represent each of the fractures separately. This is called the discrete or discontinuum approach. The rock mass consists of blocks separated by discontinuities. Two methods have been used with this approach:
- the Distinct Element Method (DEM) in which the blocks can be rigid or deformable, and the discontinuities are simulated by joint elements. For the mechanical aspects, the equations of motion are considered, and integrated using an explicit time marching scheme. This method has been used by the INERIS research team;
- the Finite Element Method (FEM) in which the blocks are discretized by means of standard continuum finite elements, and the discontinuities by means of joint elements similar to the DEM. For the mechanical aspects, a quasi-static version of the equilibrium equations is used. This method has been used by the ENSMP research team.

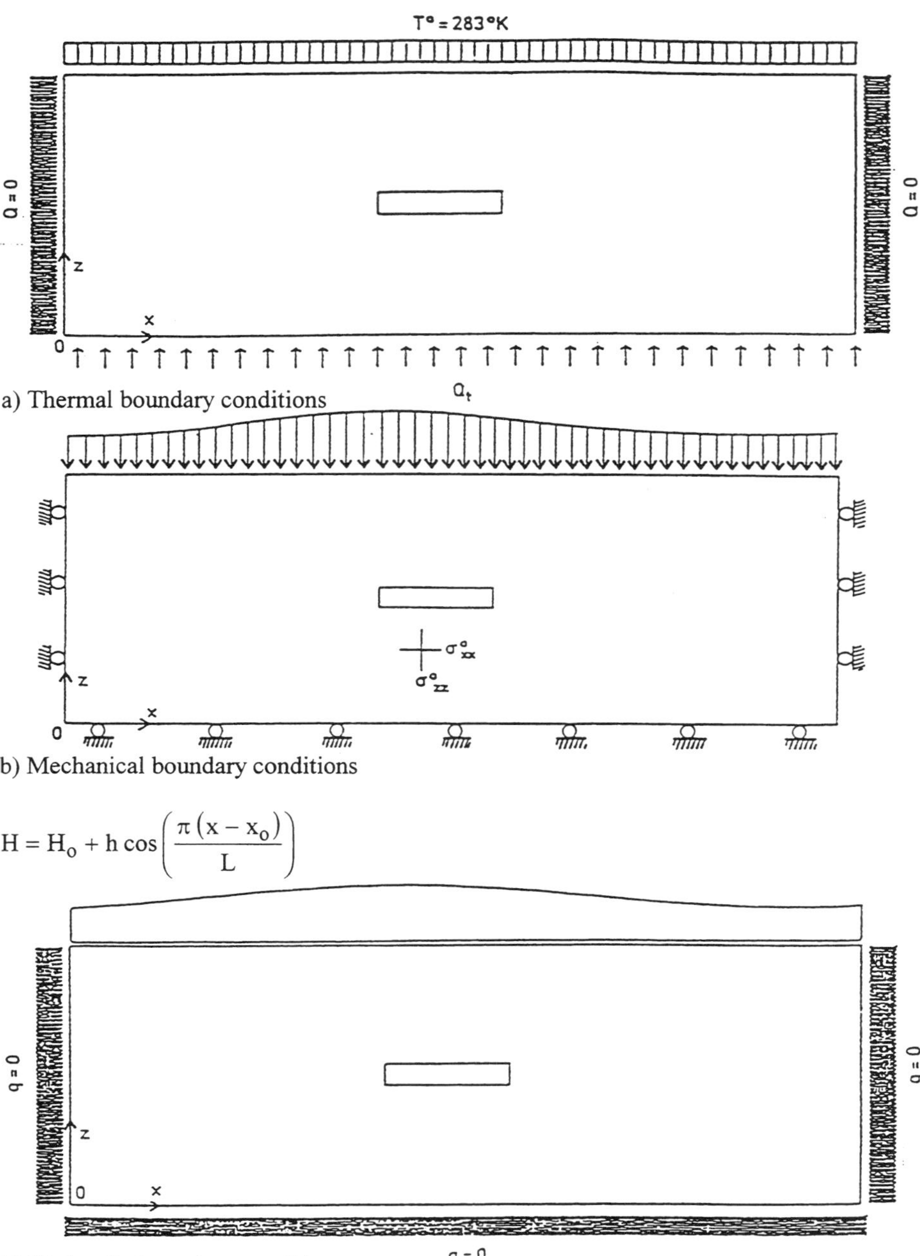

Figure 3. Boundary conditions

In both methods, geometrical and material non-linearities can be easily incorporated.

An alternative approach is to replace the blocks and discontinuities by an equivalent continuum. It is not always possible to do so, because a Representative Elementary Volume (REV) of continumm does not necessarily exist. When it does, its size may depend on the phenomenon under investigation (mechanics, hydraulics). In the benchmark⋅exercise, the existence of a unique Representative Elementary Volume has been assumed. There exists several methods to derive the equivalent continuum properties. Here, the same hypothesis were used by the CEA/DMT and KPH research teams concerning mechanics (homogeneous state of stresses in the REV) and hydraulics (homogeneous hydraulic gradient in the REV).

These approaches are detailed and compared in the following sections.

2.2. Discrete approaches

INERIS research team

The INERIS team has used the UDEC code (version 1.80), based on the distinct element method, pioneered by Cundall [2]. The rock mass is composed of an assembly of deformable blocks separated by discontinuities. The blocks are discretized by means of triangular zones where the strains are assumed to be constant (finite difference elements). The discontinuities are defined by contacts between blocks.

<u>Heat equation</u>

The heat transfer equation is:

$$\rho_r \; C_r \; \frac{\partial T}{\partial t} = \mathrm{div} \left(\lambda_r \; \mathrm{grad} \; T \right) + Q_s \tag{7}$$

<u>Mechanical equations for the blocks and the discontinuities</u>

Following a finite difference approach, the mechanical equations are expressed at nodes, within the blocks and at the boundaries of the blocks:

The equations of motion for each node i of the boundary of a block are:

$$m_{(i)} \; \ddot{U}_{(i)} + \alpha \; \dot{U}_{(i)} = F_{(i)}^{ext} + F_{(i)}^{int} + F_{(i)}^{c} \tag{8}$$

, The internal forces $F_{(i)}^{int}$ are related to the stresses in the block, which are derived from the displacements through the classical equations:

$$\Delta\varepsilon = \frac{1}{2} \left(\mathrm{grad} \; U + \left(\mathrm{grad} \; U \right)^{t} \right) \tag{9}$$

$$\Delta\sigma = H_r \left(\Delta\varepsilon - \alpha_r \; \Delta T \; \delta \right) \tag{10}$$

where H_r depends only on the Young's modulus E_r and Poisson's ratio ν_r of the rock matrix.

The contact forces are determined from the contributions of the fracture mechanical behaviour and the water pressure.

Concerning the mechanical behaviour, the increments of effective stresses are calculated from the normal and tangential displacements increments following the fracture behaviour laws:
-   the elastic behaviour is described by the normal and shear stiffnesses k_n and k_s:

$$\Delta \sigma'_{nn} = k_n \cdot \Delta U_n$$
$$\Delta \sigma'_{sn} = k_s \cdot \Delta U_s \tag{11}$$

-   the inelastic behaviour is described by means of a Mohr-Coulomb criterion, without dilatancy, together with zero traction strength of the fracture:

$$\sigma'_{sn} \leq C_f - \sigma'_{nn} \cdot tg\ \varphi_f \tag{12}$$
$$\sigma'_{nn} \leq \sigma_f^T = 0 \tag{13}$$

<u>Hydraulic equations for the discontinuities</u>
Concerning the hydraulic equations, the following assumptions have been made:
-   The fractures conductivity K_f is derived from Poiseuille's formula:

$$q = - \frac{1}{12\ \mu}\ a^3\ J \tag{14}$$

and is given by:

$$K_f = \frac{a^3\ \rho_w\ g}{12\ \mu} \tag{15}$$

where ρ_w and μ depend on temperature, and a depends on the mechanical deformations.
-   The hydraulic gradient J is approximated as:

$$J = \frac{\Delta p}{1} \tag{16}$$

where l represents the length of the contact zone of a particular node under consideration.
-   The hydraulic aperture is updated according to:

$$a = a_0 + \Delta U_n \tag{17}$$
$$a_{max} \geq a \geq a_{res} \tag{18}$$

Below the minimum value a_{res}, the mechanical closure of the discontinuity has no effect on the hydraulic aperture. For the calculation, the following values have been adopted:

$$a_{res} = \frac{a_0}{10} \tag{19}$$
$$a_{max} = 10\ a_0 \tag{20}$$

- The pressure variation is obtained from:

$$\Delta p = K_w \left(Q \frac{\Delta t}{V} - \frac{\Delta V}{V_m} \right) \tag{21}$$

where Q represents the sum of the water flows coming into a given domain and V denotes the present volume of this domain. If V_o represents the initial volume, V_m is the average: $V_m = \dfrac{V_o + V}{2}$ and ΔV the variation: $\Delta V = V - V_o$.

Treatment of couplings and numerical algorithms
UDEC simulates the coupled thermal-hydraulic-mechanical phenomena in an uncomplete manner. In the analysis, the following couplings were omitted:
- variations of solid properties with temperature,
- variations of thermal properties with mechanical strains,
- heat produced by mechanical intrinsic dissipation,
- heat convection by the fluid.

As a consequence, the thermal analysis is not influenced by the hydromechanical behaviour and can be performed independently. The heat transfer equation (7) is approximated by means of finite differences in space and time. The resulting set of algebraic equations is solved using an implicit Crank-Nicholson scheme [10], allowing for large time steps.

The one-way thermo-mechanical coupling (T → M) simply consists of accounting for the thermal stresses in the mechanical equations since the mechanical properties do not depend on temperature, according to the benchmark specification.

The one-way thermo-hydraulic coupling (T → H) is accounted for by changing the water density and viscosity with temperature, according to the benchmark specification.

The hydromechanical problem is fully coupled and is solved as follows: the equations are solved separately for the blocks and for the discontinuities, and the interactions are treated in an iterative way until equilibrium is reached. The process starts with a procedure for the blocks.

An explicit time integration scheme of the equations of motion is used for the blocks. Therefore, for quasi-static applications like the BMT1 exercise, some damping must be introduced in order to dissipate the kinetic energy as quickly as possible in the computation. The size of the time step is limited by a stability condition and is, in general, very small.

ENSMP research team
The ENSMP research team has developed three two-dimensional finite element codes: CHEF, HYDREF and VIPLEF devoted respectively to thermal, hydraulic and mechanical analysis [3]. The medium is modelled by means of classical 6-node isoparametric triangular finite elements for the rock matrix and 6-node joint finite elements.

Heat equation
The heat transfer equation is identical to equation (7).

Mechanical equations for the rock matrix and the joints
Starting from the discretization of the displacement field U, the classical principle of virtual work is used to derive the discretized equilibrium equations:

$$\int_\Omega \Sigma.\delta\varepsilon \; d\Omega = \int_\Omega \rho f.\delta U \; d\Omega + \int_{\partial\Omega} \phi.\delta U \; d\partial\Omega \tag{22}$$

These equations can be written either for the rock matrix, in which case Σ stands for the stresses σ in the rock, or for the joints, in which case Σ stands for the effective stresses σ' in the joints. $\delta\varepsilon$ is obtained from equation (9). The rock matrix behaviour follows equation (10). For the joints, the properties are similar to those used by the INERIS research team, except the following differences:
- the normal stiffness k_n of the joint depends on the compressive strain, according to figure 4,
- no residual hydraulic aperture has been considered,
- no distinction is made between the thickness e of the joint and its hydraulical aperture a.

Hydraulic equations for the joints
        Poiseuille's formula (14) is also used to obtain the joint conductivity K_f, and the flow in the joint is assumed to obey Darcy's law:

$$V_s = - K_f (grad \; h)_s \tag{23}$$

The hydraulic aperture a has been taken equal to the joint thickness, which is updated according to equation (17), ρ_w and μ depend on temperature T, and the water is assumed to be incompressible. The hydraulic head h has been discretized in the joints, leading to the minimization of the following functionnal:

$$\underset{h}{Min} \; \iint_\Omega (grad \; h)^t.K_f.(grad \; h) \; d\Omega - \int_{\partial\Omega} \bar{q}.h \; d\partial\Omega \tag{24}$$

Treatment of the couplings and numerical algorithms
        In the analysis, the same couplings as for INERIS research team have been disregarded, and the thermal analysis was performed independently of the hydromechanical aspects.

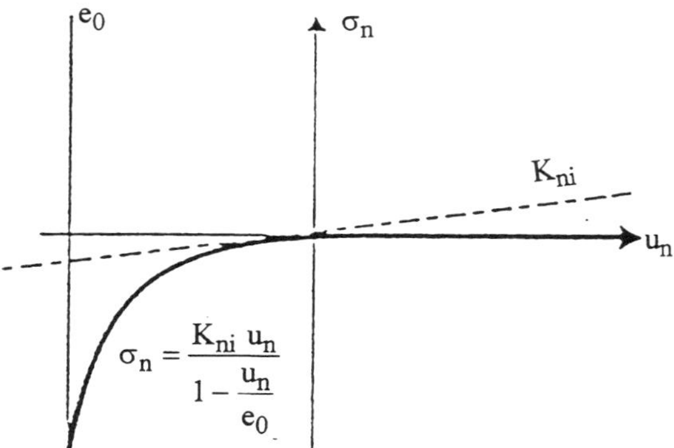

Figure 4. Normal behaviour of joints in VIPLEF

The heat equation is solved using an Euler implicit scheme [9] for time integration. The treatment of one way couplings $T \to M$ and $T \to H$ is identical to the one used by the INERIS research team.

The hydromechanical problem is solved using an iterative separate approach: at each iteration, the whole mechanical history is calculated and then used to recalculate the complete hydraulic history. The iterations stop when stabilized histories are obtained.

The mechanical equations are solved using a standard initial stress iterative algorithm [10].

2.3. Continuum approaches

KPH research team

The KPH research team has developed a fully coupled thermal, hydraulic and mechanical code, called THAMES [9], based on a representation of the fractured rock mass by an equivalent porous continuum. The medium is decomposed into 8-node isoparametric finite elements, in which the displacements, temperature and hydraulic head are discretized. The major difficulty consists in determining the properties of the equivalent continuum. For this purpose, the model proposed by Oda [4] is used.

<u>Heat equation</u>
The following heat transfer equation is solved for the continuum:

$$\rho c \frac{\partial T}{\partial t} = \text{div}\left(\tilde{\lambda}\ \text{grad}\ T\right) - n\ \rho_w\ C_w\ V.\text{grad}\ T + (1-n).\alpha T.\delta.H.\frac{\partial \varepsilon}{\partial t} + Q_s \tag{25}$$

where $\rho c = n\ \rho_w\ C_w + (1-n)\ \rho_r\ C_r$ and $\tilde{\lambda} = \lambda_r$ $\hspace{2cm}$ (26)

(λ_w was not given in the specification and has been taken equal to λ_r).

In equation (25), the second term in the right hand side is the convection term and the last term is the coupling term of the deformation process to the heat transport process, which is usually neglected.

<u>Mechanical equations</u>
The discretized equilibrium equations (22) are used, with an equivalent density $\tilde{\rho}$ for the continuum:

$$\tilde{\rho} = n\ \rho_w + (1-n)\ \rho_r \hspace{2cm} (n \text{ is the porosity}) \tag{27}$$

The material behaviour law is obtained from Oda's method [4] which assumes that the stress state is uniform in the volume under consideration. The elastic strains are decomposed into a contribution from the rock matrix and the fractures, respectively:

$$\varepsilon = D^{-1}\ \sigma + C\ \sigma' \tag{28}$$

D stands for the elasticity matrix of the rock. The fractures compliance matrix C can be written as:

$$C_{ijkl} = \left(\frac{1}{k_n} - \frac{1}{k_s}\right) F_{ijkl} + \frac{1}{k_s} G_{ijkl} \tag{29}$$

where F_{ijkl} and G_{ijkl} are geometric tensors, which depend only on the number of fractures M in the volume, and of their orientations and lengths:

$$F_{ijkl} = \frac{1}{V} \sum_{m=1}^{M} l_{(m)}^2 \; n_{i_{(m)}} \; n_{j_{(m)}} \; n_{k_{(m)}} \; n_{l_{(m)}} \tag{30}$$

$$G_{ijkl} = \frac{1}{4}\left(\delta_{ik} \; F_{jl} + \delta_{jk} \; F_{il} + \delta_{il} \; F_{jk} + \delta_{jl} \; F_{ik}\right) \tag{31}$$

$$F_{ij} = F_{ijkk} \tag{32}$$

where V is the volume of the region and $\bar{n}_{(m)}$ is the vector normal to the fracture (m).

By expressing the effective stresses σ' in terms of the total stresses σ and the hydraulic head h, and accounting for the thermal strains, the following expression can be obtained:

$$\Delta\sigma = \tilde{H}\left[\Delta\varepsilon - \alpha_r \; \Delta T \; \delta - \tilde{C} \; \rho_w \; h\right] \tag{33}$$

where $\tilde{C}$ is a second order coupling tensor and $\tilde{H}$ the fourth order elastic tensor of the equivalent continuum. The non-linear behaviour of the fractures is handled simply by a reduction of the stiffnesses when the failure criteria are reached:

$$k_n' = \frac{k_n}{100} \quad \text{when} \quad \sigma_{nn}' > \sigma_f^t \tag{34}$$

$$k_s' = \frac{k_s}{100} \quad \text{when} \quad \left|\sigma_{ns}'\right| > C_f - \sigma_{nn}' \; \text{tg} \; \varphi_f \tag{35}$$

<u>Hydraulic equations</u>

Darcy's law is supposed to be valid for the equivalent porous medium. Accounting for fluid mass conservation, the final head governing equation is:

$$\text{div}\left[\rho_w \; \frac{\tilde{K}}{\mu} \; \text{grad} \; (p + \rho_w \; g \; z)\right] - \rho_w \; \frac{\partial n}{\partial t} + \rho_{w_0} \; n \; \beta \; \frac{\partial T}{\partial t} = 0 \tag{36}$$

The density ρ_w and viscosity μ depend on temperature according to the benchmark specification, and the water is assumed to be incompressible. The permeability tensor $\tilde{K}$ of the equivalent continuum is also obtained by Oda's method [4], assuming that the hydraulic gradient is uniform in the region under consideration. By assuming that Poiseuille's formula is applicable for each fracture, and adding the contributions of all fractures, the following equivalent conductivity tensor can be obtained:

$$\tilde{K}_{ij} = \frac{\rho_w \, g}{12 \, \mu} \sum_{m=1}^{M} \frac{a_{(m)}^3}{l_{(m)}} \left[\delta_{ij} - n_{i_{(m)}} \, n_{j_{(m)}} \right] \qquad (37)$$

which depends on the number of fractures, M, their orientations, lengths $l_{(m)}$ and hydraulic apertures $a_{(m)}$.

The initial conductivity K_{f_0} of a fracture is obtained from Poiseuille's formula (14). The change of the fracture aperture is related to the strain ε_{nn} along the normal to the fracture, and the conductivity can thus be updated according to the following equation:

$$K_f = K_{f_0} \left(1 + \varepsilon_{nn}\right)^3 \qquad (38)$$

Treatment of the couplings and numerical algorithms

All couplings, except the heat produced by mechanical intrinsic dissipation, have been included in the analysis by the KPH research team. The whole set of thermal, mechanical and hydraulic equations are discretized using a Galerkin technique [10] and the resulting set of non-linear equations is solved using a Picard iterative method [10]. Because of the couplings, the properties of the equivalent continuum $\left(n, \tilde{H}, \tilde{C}, \tilde{K}\right)$ vary with time and are recalculated for each iteration of every time step.

CEA/DMT research team

The CEA/DMT research team has applied a continuum approach, which consists in representing the fractured rock mass by an equivalent homogenized mechanical and hydraulic continuum. The method used is based on one hand on the hypothesis of Reuss [5] for the elasticity tensor and on the other hand on works initiated by Snow [6] for the hydraulic conductivity tensor.

This method is not fundamentally different from Oda's method [4], for the particular case under consideration. All field equations are solved by means of 4-node quadrangular finite elements using the CASTEM 2000 [7] computer program for thermal and mechanical analysis and TRIO-EF [8] for hydraulic analysis.

Heat equation

The heat transfer equation used is an approximation of equation (25) and has the form:

$$\rho c \, \frac{\partial T}{\partial t} = \mathrm{div}\left(\tilde{\lambda} \, \mathrm{grad} \, T\right) + Q_s \qquad (39)$$

where ρc, given by equation (26), has been approximated by the rock matrix specific heat $\rho_r \, C_r$. The low value of the porosity n allows for this simplification.

Mechanical equations

The effect of the water pressure on the mechanical behaviour is accounted for by assuming that the total stresses can be decomposed as:

$$\sigma = \sigma' - p_0 \, \delta \qquad (40)$$

where p_0 represents the initial water pressure in the rock mass, thus neglecting the effects of pressure variations Δp as already mentioned.

Thus, the principle of virtual work (22) is written directly in terms of the effective stresses σ', using an effective density: $\rho' = \rho - \rho_{w_0}$

The material behaviour is described by an elasto-plastic model:
- for the elastic part, the following equation is valid:

$$\Delta \sigma' = \tilde{H}\left[\Delta \varepsilon - \alpha_r \, \Delta T \, \delta\right] \tag{41}$$

- for the plastic part, along each direction of the two sets of fractures, a Mohr-Coulomb criterion with a non associated flow rule is assumed:

$$\text{Criterion} \qquad : \quad f\!\left(\sigma'_{ij}\right) = \left|\sigma'_{sn}\right| + \sigma'_{nn} \, \text{tg} \, \varphi_f - C_f \le 0 \tag{42}$$

$$\text{Flow potential} \quad : \quad g\!\left(\sigma'_{ij}\right) = \left|\sigma'_{sn}\right| \tag{43}$$

<u>Hydraulic equations</u>

Darcy's law is supposed to be valid for the equivalent porous medium. The equation governing the head can be seen as a simplified version of equation (36):

$$\text{div}\left\{\tilde{K} \, \text{grad}\left[h + \left(\frac{\rho_w}{\rho_{w_0}} - 1\right)z\right]\right\} - \frac{\partial n}{\partial t} = 0 \tag{44}$$

Along one direction of discontinuity, the conductivity is taken as:

$$\tilde{K}_{ss} = \frac{\rho_{w_0} \, g}{12 \, \mu \, d}\left(\alpha_0 + \Delta\alpha\right)^3 \tag{45}$$

which is analogous to the expression presented in equation (38).

<u>Treatment of couplings and numerical algorithms</u>

The CEA/DMT research team has taken into account only the couplings which were judged to be relevant. Therefore, in view of the problem to be solved, the following couplings were disregarded:
- variations of solid properties with temperature,
- variations of thermal properties with mechanical strains,
- heat produced by mechanical intrinsic dissipation,
- heat convection by the fluid,
- mechanical effect of fluid pressure variations.

The validity of these assumptions has been checked a posteriori, on the basis of additional calculations (see section 4). Consequently, the thermal, mechanical and hydraulic analysis are performed sequentially, the couplings being thus of one way type. The time integration of the heat transfer equation is performed using a classical θ-method [10].

The system of non linear algebraic equations derived from the principle of virtual work, is solved by means of a modified Newton-Raphson algorithm [10].

The hydraulic calculation is performed for each time step of the mechanical calculation. The mechanical results are used to update the porosity:

$$\Delta n = \mathrm{Tr}\left(< \Delta \varepsilon^p > \right) \tag{46}$$

and the conductivity tensor $\widetilde{K}$. The Darcy velocity field is then obtained from the head field.

3. COMPARISON OF THM RESULTS FOR 100 M FRACTURE SPACING

Because of the cost of the computations based on the discrete approach, the comparisons of the results has been possible only in the case of a fracture spacing of 100 m.

The research teams have used different meshes, based on their engineering judgement. The following table indicates the number of elements used for the finite elements method and blocks zones used for the discrete elements method, together with the element or block types.

Table 4
Characteristics of computational models

Team	Approach	Code & Method	Element types	Number of elements or blocks	Number of nodes
INERIS	Discrete	UDEC (DEM)	Triangular zones	Blocks: 369 Zones: 69957	38987
ENSMP	Discrete	CHEF VIPLEF HYDREF (FEM)	6 noded	2071	16326
KPH	Continuum	THAMES (FEM)	8 noded (M) 4 noded (TH)	720	2277
CEA/DMT	Continuum	CASTEM 2000 TRIO-EF (FEM)	4 noded	2188	2233

3.1. Thermal results

The maximum temperature is reached in the repository after about 100 years of storage and is approximately 385°K.

Figure 5 shows the isotherms at 150 years. Clearly, the temperature distributions predicted by the various teams are very similar. This agreement is further confirmed by the comparison of the temperature evolution at points 8 and 12, as shown in figure 6. The discrepancies observed can be explained as follows:
- The temperatures are underestimated by the ENSMP and KPH teams, who use coarser meshes in the vicinity of the heat source but higher order interpolation functions.
- The KPH team follows a fully coupled THM approach, including convective effects.

Probably, this effect does not explain the lower temperature observed. Indeed, some additional calculations have been performed by the CEA/DMT team in order to investigate the influence of the convective effects. A fully coupled thermo-hydraulic calculation gives a maximum temperature at point 14 superior to the one predicted by a simple thermal calculation, by about 15°C. This indicates a tendancy opposite to the one obtained by the KPH team.
- Some truncation errors, due to the accuracy of the computer (single or double precision) and to the time units used (years or seconds) may lead to minor differences, in particular for the input of the heat source.

3.2. Mechanical results

The overall deformation pattern of the rock mass at 150 years is shown in figure 7. The non-symmetry induced by the anisotropic properties of the rock mass is qualitatively identical for the various research teams. Vertical displacements history at points 9 and 13 is presented in figure 8.

Although zero horizontal displacements are prescribed on the lateral boundaries of the domain, the non homogeneous behaviour of the medium leads to large displacements, e.g. at point 4 where the predicted value is between 80 mm and 105 mm.

All research teams, except KPH, find displacements histories which are qualitatively in agreement.

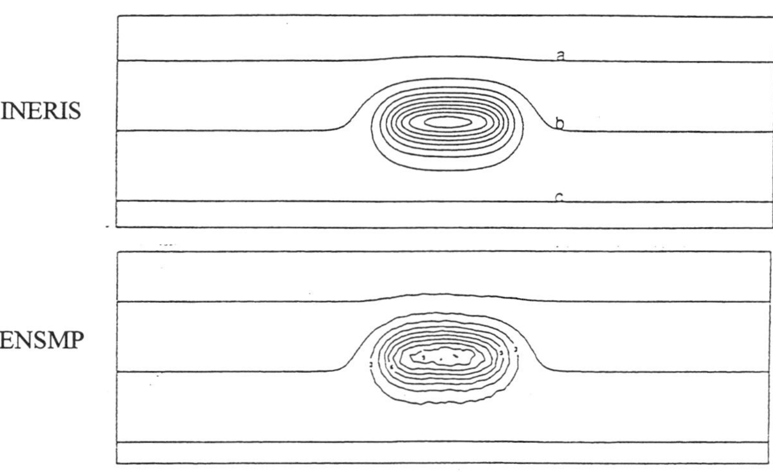

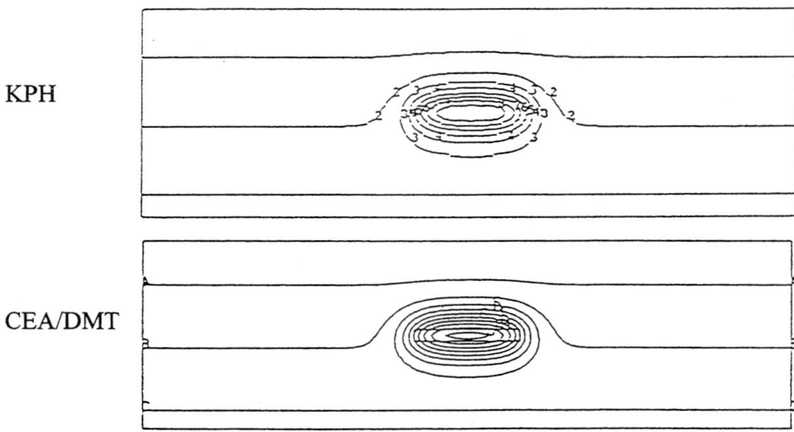

Figure 5. Isotherms at 150 years (one isocurve every 10°K the upper isocurve corresponds to 290°K)

Point number 8

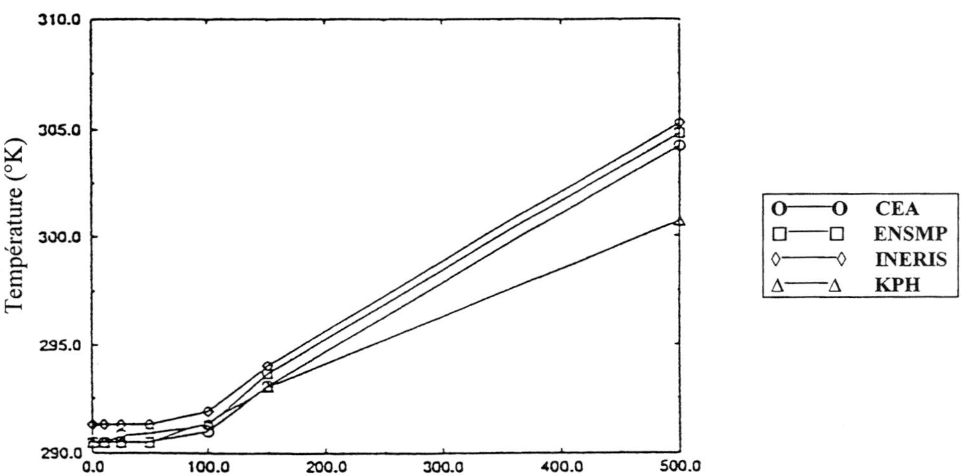

Time (years)

Point number 12

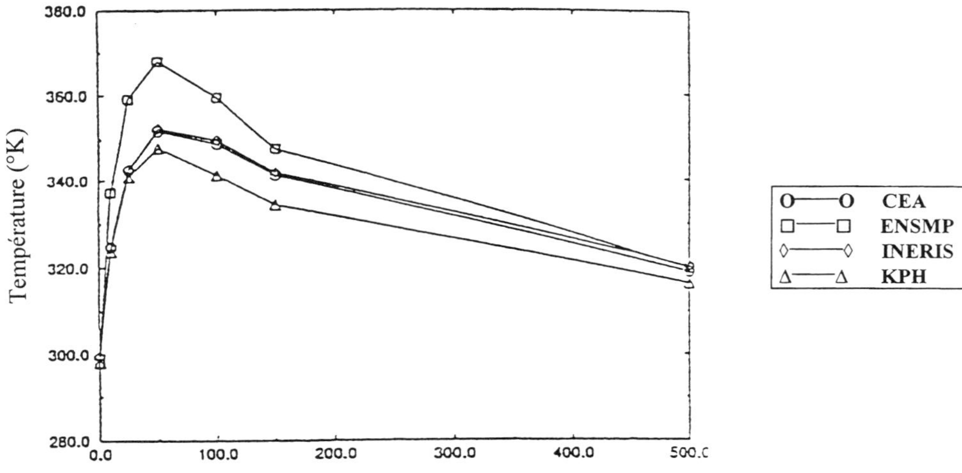

Time (years)

Figure 6. Temperature history at points 8 and 12 (note that the temperature scales are different)

A very good agreement can be observed for the vertical displacements obtained by INERIS and CEA/DMT teams. The ENSMP team values are always lower, due to the lower temperatures predicted and the fact that the deformation of the rock mass is induced by the thermal strains.

When analysing the stress histories, presented in figures 9 and 10, the maximum stress is obtained in the horizontal direction, governed by the lateral boundary conditions, while in the vertical direction, the rock mass is free to expand. Indeed, at a certain distance from the heat source, like at point 6, the vertical stress remains nearly constant, as predicted by all teams. The stress histories determined by the KPH team are more or less constant, while the other teams find large variations with time.

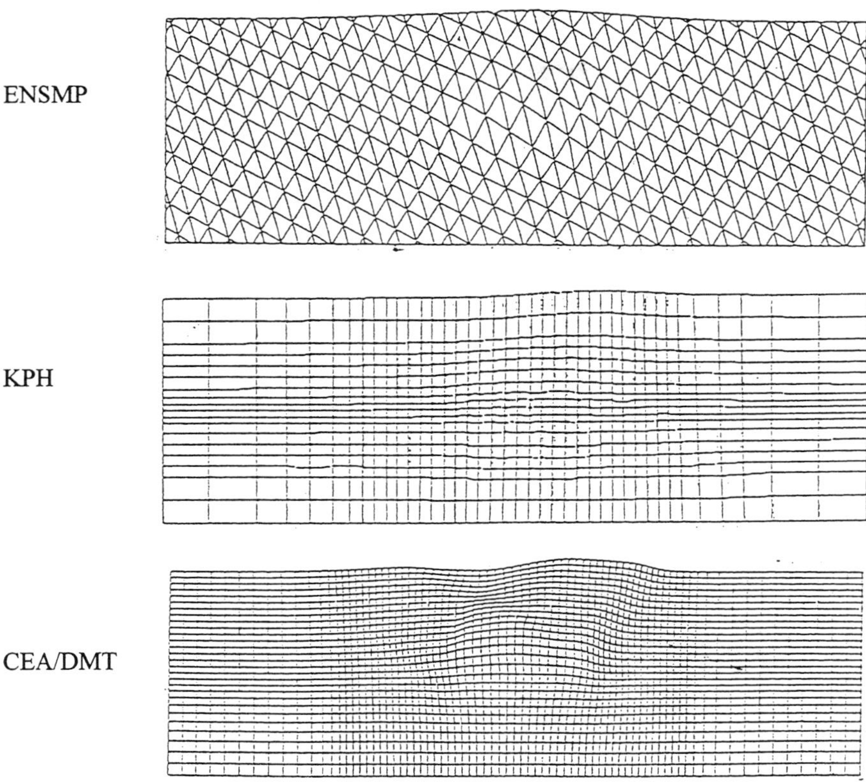

Figure 7. Overall deformation of the rock mass at 150 years (amplification factor 500)

In general, the stress variations obtained by the discrete approaches (INERIS and ENSMP teams) are larger than those obtained by the continuum approach of CEA/DMT. Two main reasons can possibly explain this:
- in a discrete approach, the blocks may undergo relative movements and rotations which can result in localized contacts which causes stress concentrations. This cannot happen in a continuum approach because of the kinematic continuity,
- the homogenization technique has a smoothing effect on the local stress fields and the calculated stresses are averaging the stresses in the rock mass and in the joints. Indeed, in the discrete approaches, only the stresses in the joints are limited by the Mohr-Coulomb criterion, while the stresses in the blocks may be higher. In the continuum approaches, the stresses are limited by the failure criterion everywhere. This difference is emphasized by the large distance between fractures.

Some additional observations about the stresses can be made:
- firstly, the stresses are not interpolated within the finite or discrete elements, as it is the case for displacements, and are not so accurately calculated. Moreover, it is always difficult to calculate a stress at a given point, since, for example in finite elements, stresses are generally computed at Gauss points and must then be extrapolated to the given point. In the continuum approaches, these points coincide with nodal points, while it is not necessarily the case in the discrete approaches,
- secondly, there are some discrepancies about the initial values of the stresses which can originate from the different methods used by the various teams: for some teams, initial stresses are simply given as input data, while for others, they are resulting from a first computational step,
- finally, the differences between the stress histories as predicted by INERIS and ENSMP teams are more important in the zone located above the repository. Indeed, this is the region where the joints are responding plastically.

Concerning the non-linear behaviour of joints, after 50 years the number of joints working in the plastic regime decreases, which corresponds for example to open joints which are closing. This situation is not reproduced by the continuum approaches, for which the positive plastic strains $\Delta\varepsilon^p_{nn}$, in the directions perpendicular to the fractures, are not recovered during unloading.

Point number 13

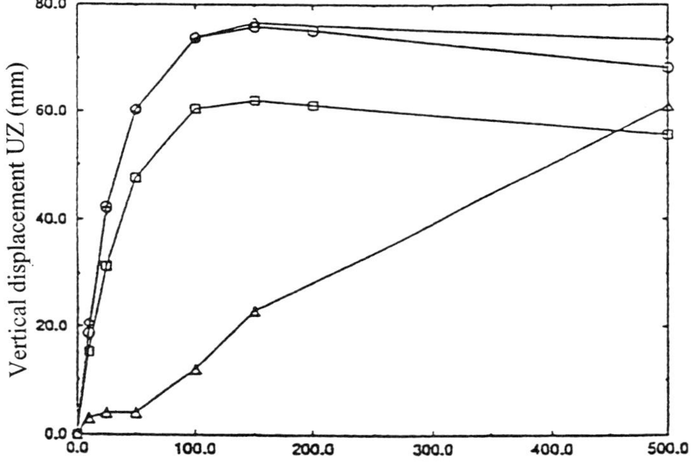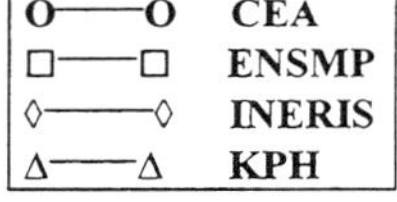

Time (years)

Point number 9

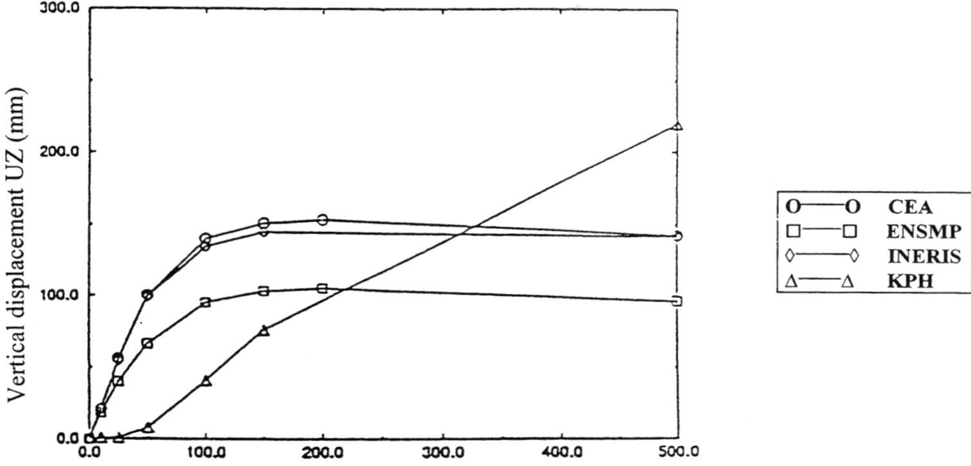

Time (years)

Figure 8. Vertical displacement history at points 9 and 13 (note that displacement scales are different)

The differences of behaviour in the zone located above the repository can also be seen on the joints openings along profile IV (figure 11).

The difference between ENSMP and INERIS models is more pronounced along profile IV which corresponds to the top surface. This is due again to the large movements of the blocks. Moreover, point 4 is located in the middle of a very small block surrounded by joints and is therefore not very much connected to the rest of the rock mass.

Point number 14

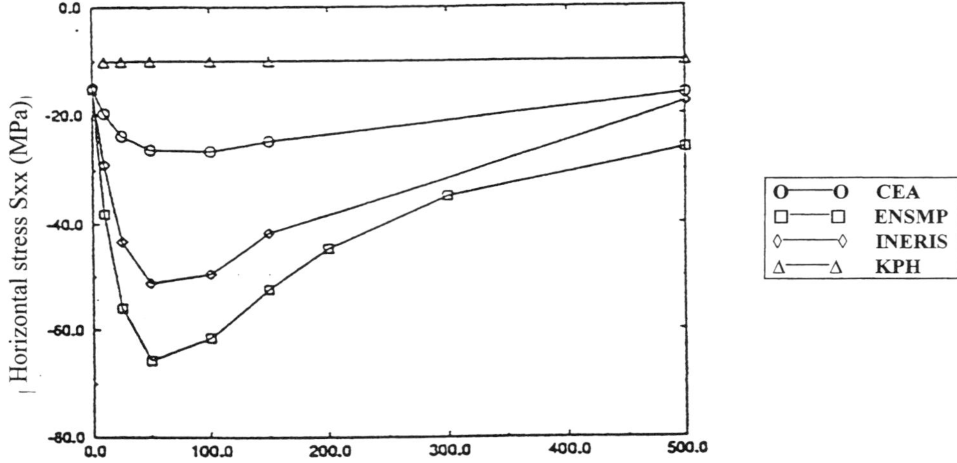

Time (years)

Point number 9

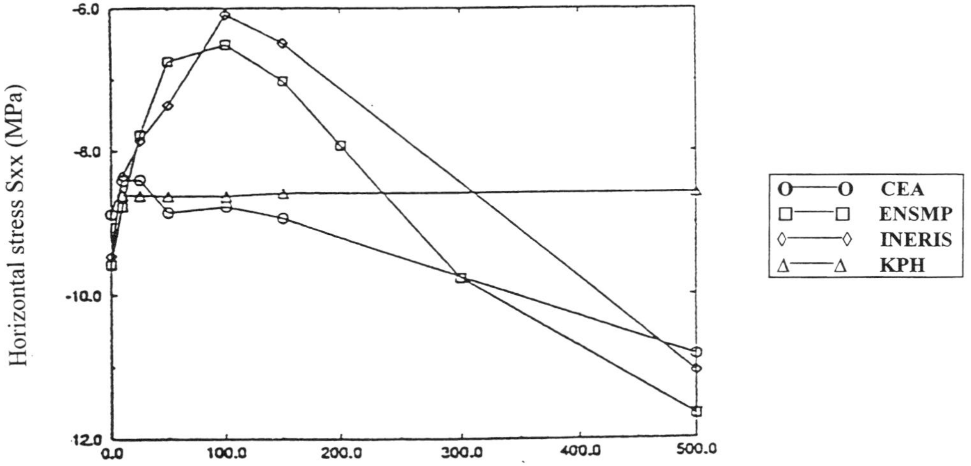

Time (years)

Figure 9. Horizontal stress history at points 9 and 14 (note that the stress scales are different)

The openings determined from the CEA/DMT continuum approach are one order of magnitude larger and are more localized than the discrete ones. Indeed, they follow the localization band which is clearly visible on figure 7, and which is not predicted so much by the other teams.

Point number 14

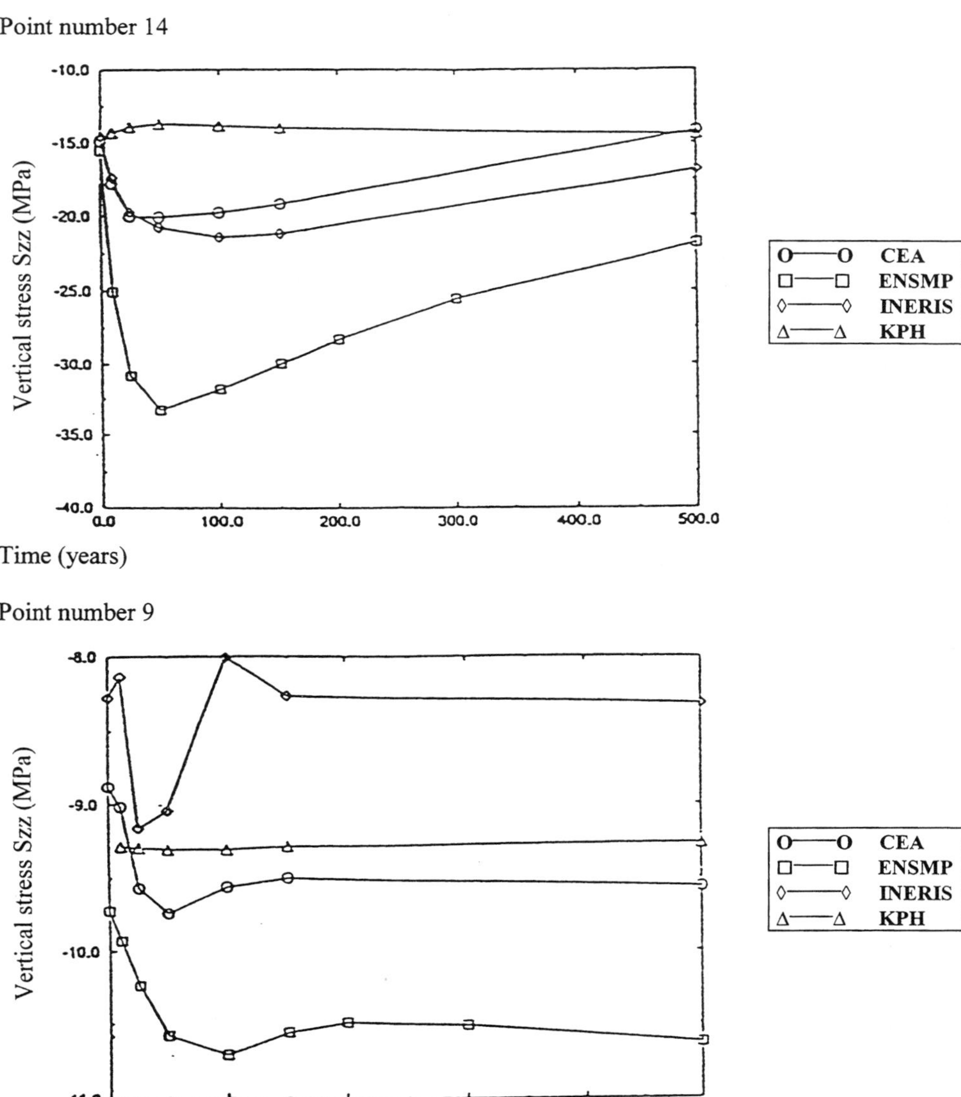

Time (years)

Point number 9

Time (years)

Figure 10. Vertical stress history at points 9 and 14 (note that the stress scales are different)

Profile IV - family 1 - (joint 60)
ENSMP - INERIS

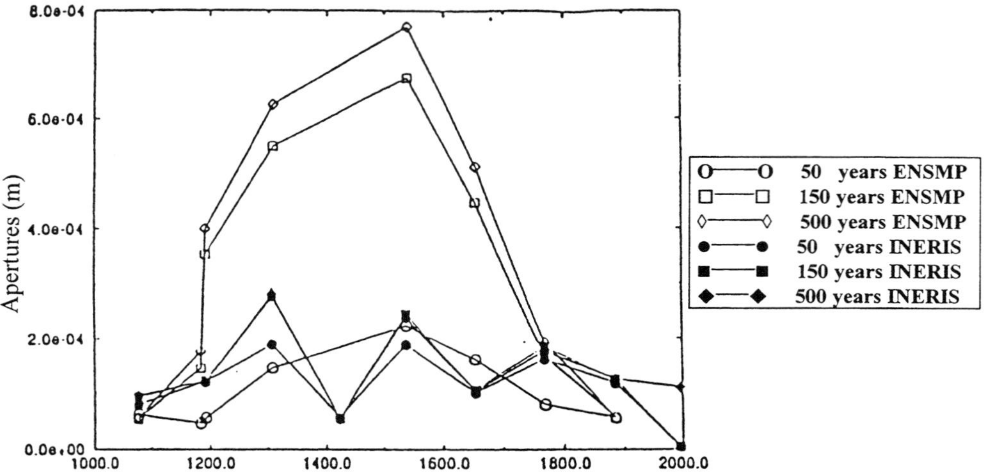

Abscissa (m)

Profile IV - family 2 - (joint 150)
ENSMP - INERIS

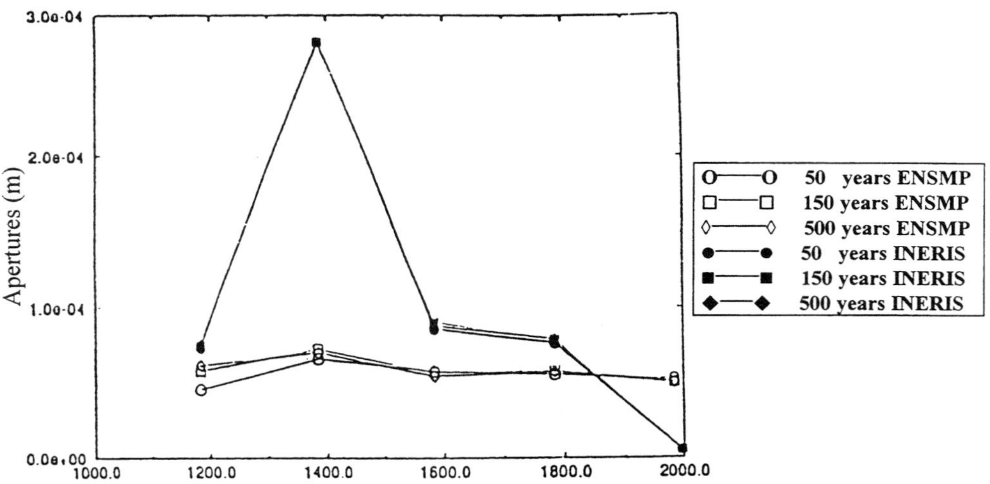

Abscissa (m)

Figure 11. Joints openings distributions, from x = 1000 m to x = 2000 m, along profile IV

3.3. **Hydraulic results**

Hydraulic results differ significantly between the research teams, both qualitatively and quantitatively.

Considering first the hydraulic head predictions, the spatial distributions at a given time are different, in the region surrounding the heat source, as can be seen on figure 12. Moreover, the intensity of the head along profiles V and VI is different for ENSMP, KPH and CEA/DMT teams and do not evolve in the same way with time.

Comparison of the flow rates across some given segments, per unit length of segment is shown in figures 13 and 14. There is a fair agreement in magnitude, between INERIS, ENSMP and KPH teams, but the shape and sign are different. The signs predicted by the INERIS team are in agreement with the water table profile, i.e. a negative flow rate through segment (2-3) and a positive flow rate through segments (3-4) and (3-8). The lower values of the flow rate through segment (8-14) can be attributed to the closure of the joints in the vicinity of the heat source.

The CEA/DMT team overpredicts the flow rates, and in particular across segments (2-3) and (8-14). This is of course directly related to the larger openings which occur in the localization band, the effect of which being amplified by the cubic law.

Note that for discrete approaches the flow rates are derived directly by summing the contributions of the various fractures intersecting the segment under consideration, independently of the fractures directions. For the continuum approaches, the flow rate is normal to the segment.

Some tentative explanations are proposed to understand the major differences observed on the heads and flow rates:
- firstly, in the ENSMP approach, no residual hydraulic aperture has been considered and therefore the joints can be closed and impermeable. Such a situation is likely to occur in particular where the blocks undergo large movements and rotations. This effect cannot be reproduced by the continuum approaches. Moreover the increased porosity and permeability induced by the joints openings are considered as irreversible by the continuum approaches. These two factors are cumulative and can partly explain the observed differences,
- secondly, the strong localization of joints in the plastic regime, predicted by the CEA/DMT team, induces an important modification of the hydraulic regime in the zone above the repository, and causes high flow rates,
- finally, the homogenization techniques used in the continuum approaches take into account the number and size of the fractures but not their connectivity, while this aspect is properly handled by research teams using discrete approaches.

ENSMP 

KPH

CEA/DMT

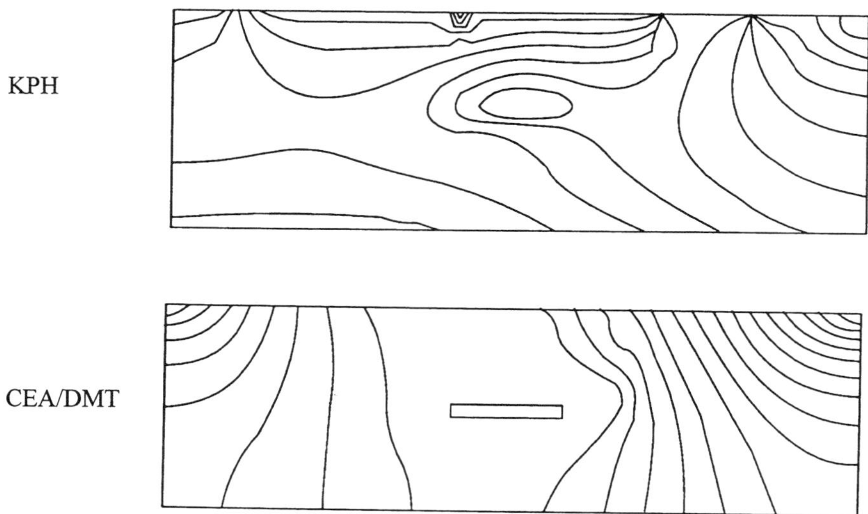

Figure 12. Iso-heads at t = 50 years (one isocurve every 2 m)

FLOW RATES - INERIS

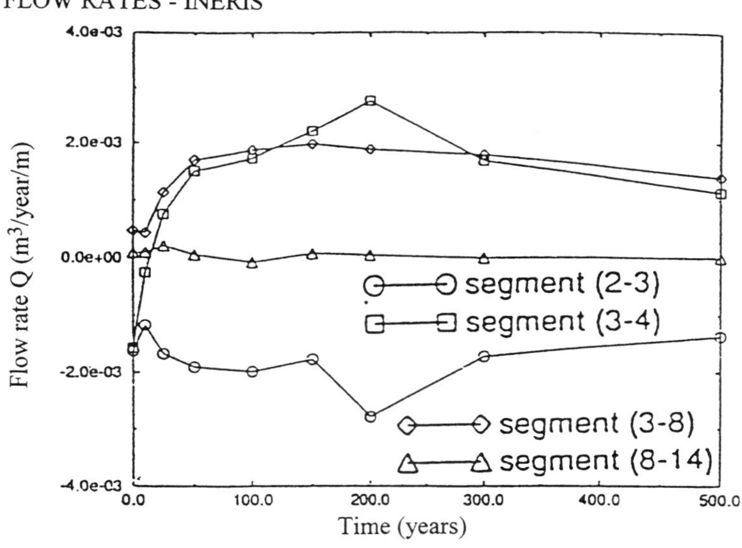

ENSMP

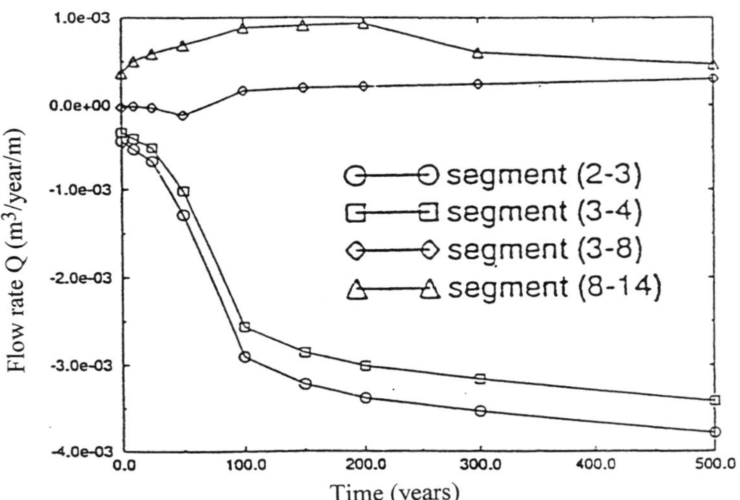

Figure 13. Flow rates across segments (2-3), (3-4), (3-8), (8-14). Discrete approaches (note that of flow rates scales are different)

FLOW RATES - KPH

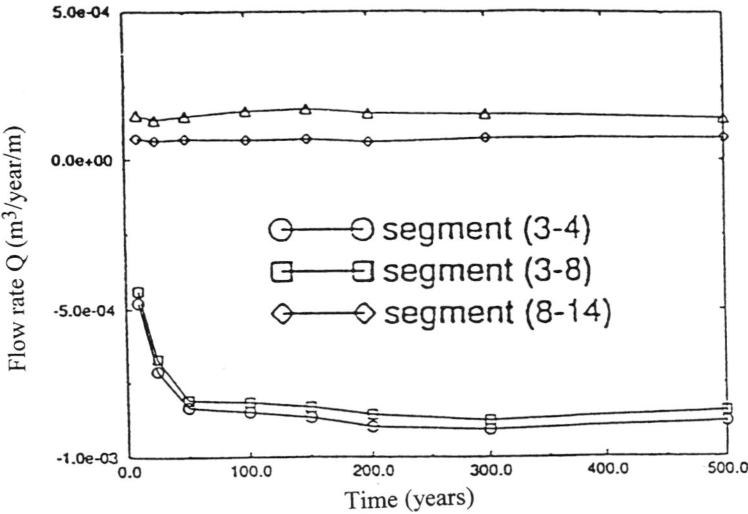

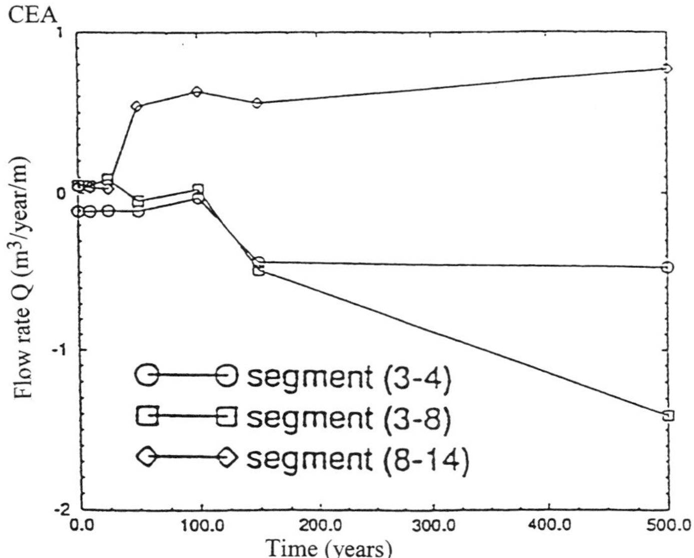

Figure 14. Flow rates across segments (2-3), (3-4), (3-8), (8-14). Continuum approaches (note that of flow rates scales are different)

4. INFLUENCE OF SOME COUPLINGS

In this section, the importance of two coupling effects is investigated: the effect of the water on the thermo-mechanical response of the rock mass, and the effect of convection on the temperature distribution.

4.1. Effect of hydraulic coupling

The ENSMP and INERIS research teams have performed additional calculations, in the case of a 100 m fracture spacing, in which they disregarded the hydraulic effects.

The following observations can be made from the comparison of displacement and stress components:
- In both cases (THM and TM) the displacement and stress distributions are similar.
- Neglecting the hydraulic coupling leads to overestimated stresses (for ENSMP as well as INERIS) and underestimated displacements (for ENSMP).

Thus, accounting for the water in the fractures leads to an increased flexibility of the rock mass. It may be attributed to the fact that the water diffusion due to the hydraulic gradient tends to diminish the pressure and consequently the total stresses. This tendency is confirmed by a similar comparison done by the INERIS team, for a 50 m fracture spacing, up to 50 years.

The ENSMP and INERIS results show that, in the vicinity of the repository, the variations of water pressure may be important and may invalidate the decomposition assumed by the CEA/DMT team: $\sigma = \sigma' - p_0 \delta$.

4.2. Effect of thermal convection

The CEA/DMT research team has investigated the influence of the thermal convection effect, by performing additional thermal analysis, for the three fracture spacings, accounting for convective terms [11]. The velocity fields have been taken from the hydraulic calculations. The results are displayed on figure 15 for t = 200 years.

It can be seen that the effect of convection increases with the fracture spacing and that the convective terms might be significant for a 100 m spacing, since they lead to a higher temperature (about 30 °K) just above the repository.

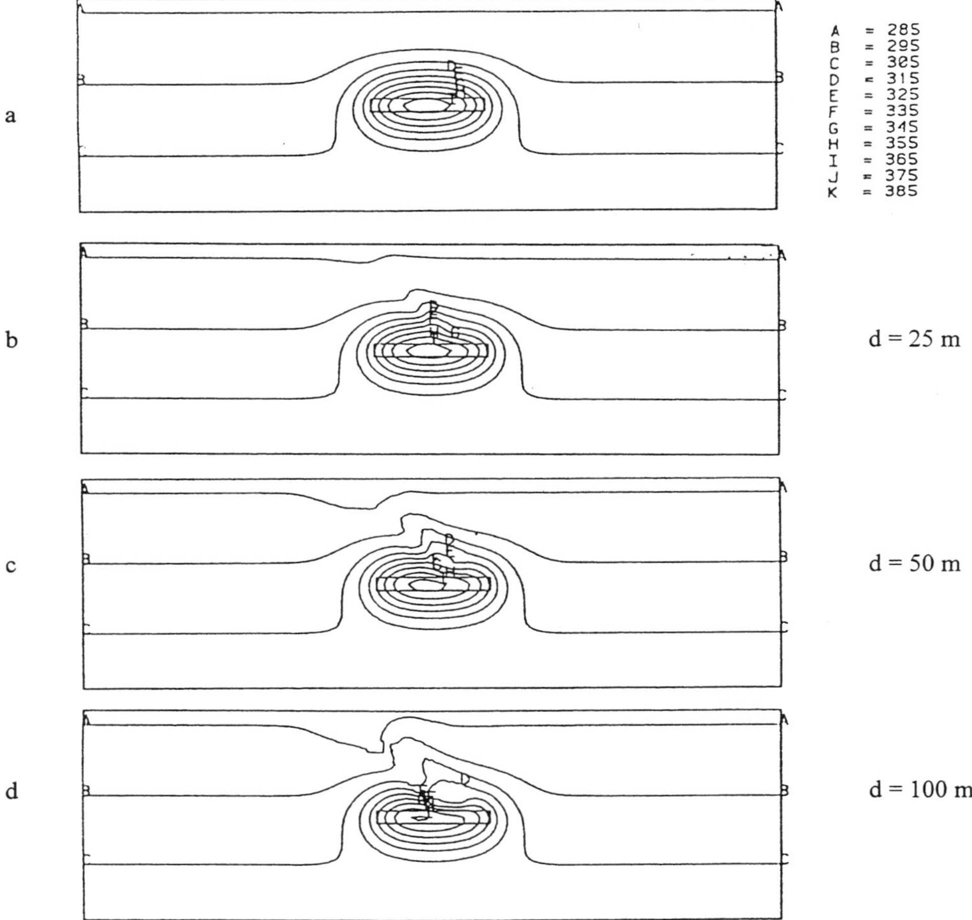

Figure 15. Effect of convective terms on temperature fields to t = 200 years - a - without convection. b, c, d - with convection (d = 25, 50, 100 m)

5. NUMERICAL ASPECTS

The two approaches used for the BMT1 (continuous/discrete) do not only lead to different systems of equations, but also to different ways of dealing with numerical processes. Two major points will be exposed here: coupling logics used and problems related to the mesh and the discretization.

5.1. Coupling logics

There are two ways of treating a THM problem: solving a complex THM system of equations, or cutting it into "simple" classical problems easier to solve (one for the thermal analysis, one for mechanics and one for hydraulics). In the last case, as each phenomenon influences the others, there must be a coupling logic between them. Only the KPH team has chosen the first method, i.e. the simultaneous computation of temperatures, strains and hydraulic flux (mathematical coupling). It is also the only fully coupled model. The three other teams chose to separate the problems and to realise a *numerical coupling*. They made the assumption that there is no influence of the mechanics and the hydraulics on the temperature field. The CEA/DMT has also neglected the influence of water pressure on the strains, and has checked this hypothesis *a posteriori*. The coupling logics are presented on figure 16.

ENSMP

INERIS

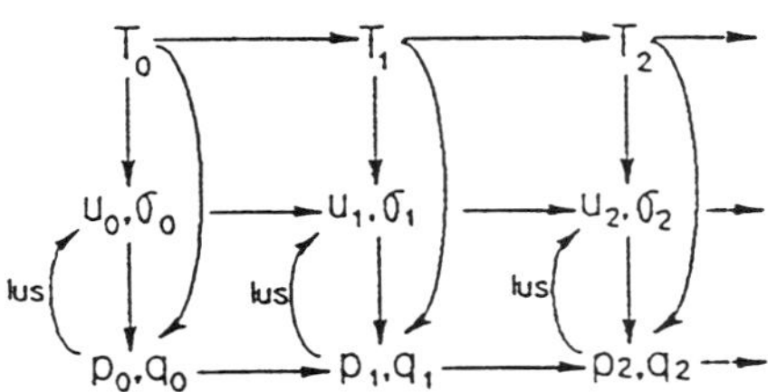

KPH

CEA/DMT

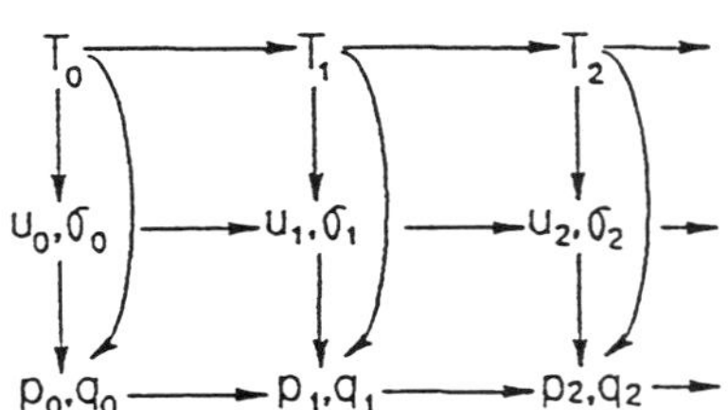

lus: loop until stabilization

Figure 16. Coupling logic of each team

For the two teams who have chosen to take into account a back influence of the pressures on strains (ENSMP and INERIS), the process has to be iterated between hydraulics and mechanics, until an equilibrium is reached. Both teams begin with the mechanical analysis and use the strains to compute the new fractures conductivity. In order to judge for a good convergence of this process, they both look at the evolution of a given variable (vertical strain, maximum velocity, mean stress ...) at a given point (near the repository zone): when it stabilizes, convergence is achieved and it is possible to increment to the next time step.

The two discrete approaches don't take the fractures into account in the same way. For the INERIS team, the equations of motion are first solved for the blocks and then the influence of contacts and water pressures is added; this process is iterated until equilibrium. For the ENSMP team, as fractures are special finite elements, they can be treated simultaneously with blocks.

Except for the INERIS team, all teams worked with the small displacements hypothesis and for all teams, mechanical results are used to modify permeability and porosity (continuous models) or fractures conductivity (discrete approaches).

5.2. Problems related to the mesh

A very important aspect of numerical models is the mesh. Actually, the larger the phenomena variations are, the more refined the mesh must be. The fundamental numerical difference between the two approaches is the discretization: in discrete models, each joint has to be explicitly represented, while in continuous approaches their behaviour is integrated to the one of the rock matrix itself.

The two teams who have chosen a continuous approach (CEA/DMT and KPH) have worked with FEM codes. They have chosen quadrangular elements. These elements are parallel to the model boundaries, which allows an easy automatic meshing. Their mesh are refined near the repository zone. As they have chosen a continuous approach, each element supports information on the fractured medium (i.e. both the rock matrix and the joints).

For the discontinuous approaches, creating the mesh is more complex. Actually, joints are individually represented either as special 6-node elements for the FEM codes (ENSMP) or as contacts between blocks for the DEM code (INERIS), while the rock matrix (cut into blocks) is represented by 6 or 8-node finite elements (ENSMP) or by triangular finite difference elements of given length (INERIS). The number of elements is then obviously higher than for the continuous approaches, which leads to increased computing times and files sizes. Furthermore, the fact that the two sets of fractures are inclined has required new developments from the ENSMP team, in order to mesh automatically such sets, adjusting elements size when they are too small at the intersections with boundaries. Figure 17 shows the principal differences between the four types of meshing.

ENSMP   INERIS

KPH   CEA/DMT

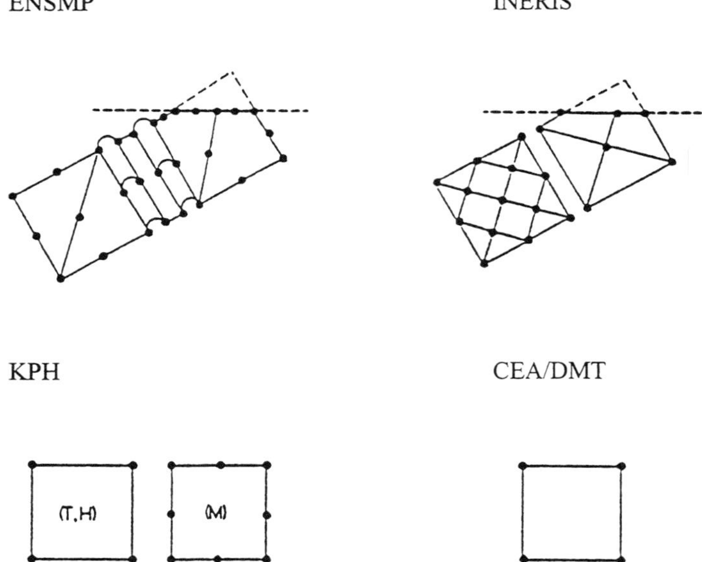

Figure 17. Different types of meshing

The table 5 indicates the principal characteristics of the meshes, for the different spacings, and table 6 the computer ressources required for the corresponding computations.

Table 5
Characteristics of the meshes (b = blocks, z = triangular zones)

Team	d = 100 m		d = 50 m		d = 25 m	
	Number of elements	Number of nodes	Number of elements	Number of nodes	Number of elements	Number of nodes
ENSMP	2071	16326	6539	59444	22705	223858
INERIS	b: 370 z: 69950	38986	b: 1338 z: 66305	41316	b: 5075 z: 71234	55577
KPH	720	2277	720	2277	720	2277
CEA/DMT	2188	2233	2188	2233	2188	2233

Table 6
Computer ressources

Team	Computer	Number of time steps	d = 100 m		d = 50 m		d = 25 m	
			Number of iterations	Total time	Number of iterations	Total time	Number of iterations	Total time
ENSMP	IBM 570	56	3	2 h 57	3	20 h 43	none	none
INERIS	HP 9000/750	6000 (T) or 6 (HM)		27 h 30		80 h		166 h 40
KPH	SUN SPARC 10	64	none	48 h 40	none	48 h 40	none	48 h 40
CEA/DMT	IBM RS6000	157 (T, M) 107 (H)		4 h 20		4 h 20		4 h 20

The accuracy of the results is particularly critical for the temperature calculation. The lack of refinement of the ENSMP mesh for d = 100 m is one explanation for the differences with the other teams at points 12 to 16 because elements in the center of the model can have a part inside the repository zone (with heating) and another one in the rest of the domain (figure 18). As the algorithm for the temperature calculation considers that only the nodes that are in the repository zone are heating, and as elements are too big (for d = 100 m), the results are not accurate. This is confirmed by the results for d = 50 m, that are in better agreement with those of the other teams.

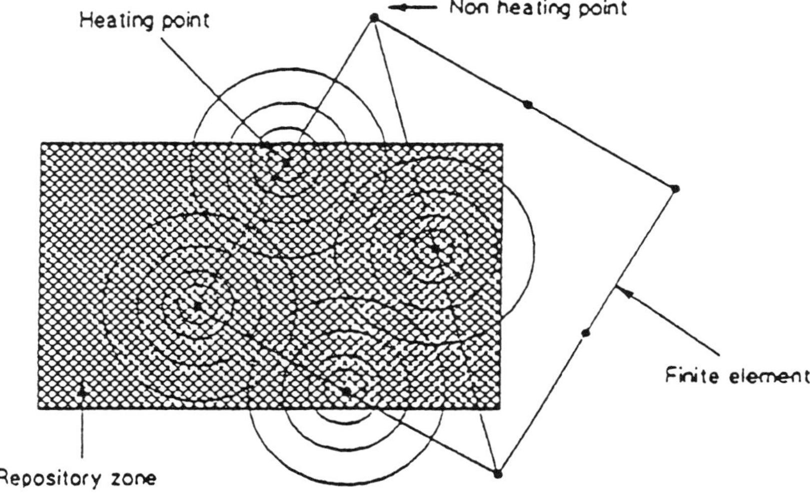

Figure 18. Intersection of the elements with the repository zone (ENSMP team)

6. CONCLUSIONS

The benchmark test BMT1 of the DECOVALEX project has been a fruitful exercise in the understanding of coupled thermo-hydro-mechanical processes in a fractured rock mass.

The research teams have gained a better knowledge of their computing tools and their domain of validity.

The two possible approaches (discrete or continuum) have been compared, and the influence of some couplings has been evaluated. However a clear understanding of the reasons for the differences observed is difficult because the research teams have done different hypothesis, disregarded different discretization methods (FEM or DEM) and different numerical algorithms. Thus, one of the outcomes of this exercise is a better understanding of what the specifications of a large scale benchmark should be in order to assess the influence of the various couplings.

The following general conclusions can be drawn:
- provided the meshes are fine enough temperature fields are well predicted by all research teams,
- there is a reasonable agreement between discrete and continuum approaches on the overall displacements and stress fields. The continuum method leads to lower stress variations. The patterns of local deformation in the zone located above the repository, are rather different for two identical approaches. This causes significant differences on joints openings profiles. For the continuum approaches, this can be partly explained by the non-linear material models used for the equivalent medium,
- the hydraulic results are very different, both qualitatively and quantitatively. This raises the question of the predictability of reliable flow rates, when using highly non-linear flow laws.

In the frame of this benchmark, the discrete approach is certainly more accurate than the continuum approach, for which the chosen distance between fractures (100 m) is obviously an upper limit. However, the discrete approach is much more expensive to generate and to run and it can even become unfeasible if the number of fractures increases significantly.

Beyond this benchmark test, some further work is needed to validate these approaches (discrete or continuum). It can be achieved through comparisons with well-instrumented laboratory tests and in-situ tests. This is precisely the objective of the Test Cases considered in the DECOVALEX project.

7. REFERENCES

1   L. Jing, J. Rutqvist, O. Stephansson, C. Tsang, F. Kautsky, "DECOVALEX - Mathematical models of coupled THM processes for nuclear Waste repositories, Report of Phase 1'", SKI, Technical Report 93: 31, 1993.
2   P. Cundall, "A computer model for simulating progressive, large scale movements in blocky rock systems", Proc. Int. Symp. Rock Fractures, Nancy.
3   S.M. Tijani, "Viplef, Chef, Hydref - User's guide", ENSMP, Paris, 1991.

4   M. Oda, "An equivalent model for coupled stress and fluid flow analysis in jointed rock masses", Water Resour. Res., 22 (13), 1986.

5   A. Reuss, "Berechnung der Fließgrenze von Mischkristallen auf Grund der Plastizitäts-bedingung für Feinkristalle", ZAMM, 9, 1929.

6   D.T. Snow, "Anisotropic permeability of fractured media", Water Resour. Res., 5 (6), 1969.

7   P. Verpeaux, A. Millard, T. Charras, A. Combescure, "A modern approach of large computer codes for structural analysis", Proc. Structural Mechanics In Reactor Technology 10, Los Angeles, 1989.

8   J.P. Magnaud, S. Goldstein, "The finite element version of TRIO Code", 7th International Conf. on Finite Element in flow problems", Huntsville, USA, 1982.

9   Y. Ohnishi, H. Shibata, A. Kobayashi, "Development of finite element code for the analysis of coupled thermo-hydro-mechanical behavior of saturated-unsaturated medium", Proc. of Int. Symp. on coupled process affecting the performance of a nuclear waste repository, Berkeley, 1985.

10 O.C. Zienkiewicz, "The finite element method", Mc Graw-Hill, 3rd Edition, 1977.

11 A. Millard, A. Stietel, "DECOVALEX - Benchmark BMT1 - Comparaison des calculs couplés THM et TM", CEA/DMT Report N° 94-508, 1994.

12 A. Thoraval, E. Vuillod, "Modèle à l'échelle d'un stockage. Calculs complémentaires relatifs au BMT1 du projet DECOVALEX", INERIS Report N° 694 RP INE 94001, 1994.

O. Stephansson, L. Jing and C.-F. Tsang (Editors)
Coupled Thermo-Hydro-Mechanical Processes of Fractured Media
Developments in Geotechnical Engineering, vol. 79
© 1996 Elsevier Science B.V. All rights reserved.

Generic study of coupled T-H-M processes of nuclear waste repositories as near-field initial boundary value problems (BMT2)

T. Chan[a,b], K. Khair[a] and E.Vuillod[c]

[a] AECL Whiteshell Laboratories, Pinawa, Manitoba, Canada R0E 1L0

[b] Current correspondence address: c/o Ontario Hydro H16 G27, 700 University Ave., Toronto, Ontario, Canada M5G 1X6

[c] INERIS, Laboratoire de Mecanique de terrains, Parc Saurupt, 54042 Nancy, Cedex, France

Abstract

A generic study of coupled THM processes of nuclear waste repositories has been performed under the auspices of DECOVALEX I as a near-field initial boundary value problem. Designated as Benchmark Test, BMT2, this problem involves a 0.75-m x 0.5-m system of nine blocks of intact hard rock separated by two pairs of relatively soft fractures. Some or all of the coupled processes of conductive, convective and dispersive heat transport; nonisothermal fluid flow; and thermo-mechanical stresses and displacements in both the intact rock and the fractures were simulated by five research teams from five different countries. All teams simulated the fractures as discrete features. Each team employed one of the two fundamentally different numerical models, i.e., finite-element models and distinct-element models, in conjunction with two types of constitutive models of fracture behaviour. Two teams subsequently simulated higher thermal loading cases. The modelling results are compared and analyzed to gain insight into the pertinent near-field processes and bench-marking strategy.

1. INTRODUCTION

In a geological repository of nuclear fuel wastes the in situ state of stress, temperature and hydraulic pressure in the host rock will be subjected to a number of perturbations including excavation/construction of the underground openings, thermal loading generated by the decay heat from the wastes and glacial loading/unloading in the future. The rock mass responses to these perturbations are coupled phenomena involving thermal (T), hydrological (H), mechanical (M), and chemical (C) processes. Coupled processes imply that one process affects another and that the rock mass response to waste storage cannot be predicted by considering each process independently. Although coupling the chemical process with the other processes is important, the focus of the DECOVALEX project is on modelling coupled thermo-hydro-mechanical (THM) processes only.

Coupled THM processes are potentially important both in the near field and in the far field of a repository. Far-field studies are covered elsewhere in this book. The current chapter focuses on the near field. Various coupled THM phenomena can be envisaged in the vicinity of the waste packages or the storage/disposal rooms that may impact the waste isolation performance of a nuclear fuel waste repository constructed in low-permeability rock. These include opening or closure of fractures with accompanying changes in permeability, thermally driven fluid flow, glacial loading induced hydraulic gradient, thermally and thermomechanically induced pore pressure build up which may increase the potential for cracking due to the concomitant changes in the effective stress and failure criteria.

Coupled THM models and computer codes have been developed for assessing the possible influence of the above-mentioned phenomena to nuclear waste isolation. Before these models and codes can be applied with confidence, they must be verified and validated. Analytical solutions are essentially non-existent for general nonlinear coupled problems. It is, therefore, necessary to design bench mark test problems for inter-code comparison. In DECOVALEX-I a generic near-field initial boundary value problem, designated BMT2 (Bench Mark Test 2), was formulated by Atomic Energy of Canada Limited (AECL) and simulated by five research teams from five countries using two fundamentally different types of numerical models--finite-element and distinct-element methods--and various constitutive models. The bench mark test BMT2 consists of a 0.75 m x 0.5 m system of nine blocks of intact hard rock separated by two pairs of relatively soft fractures. All teams simulated the fractures as discrete features. Coupled processes of conductive, convective and dispersive heat transport; nonisothermal fluid flow; and thermo-mechanical stresses and displacements in both the intact rock and the fractures were simulated.

Details of BMT2 and the inter-code comparison performed in DECOVALEX-I have been previously published [1]. Subsequently, additional sensitivity analyses were undertaken by AECL and INERIS (Laboratoire de Mecanique de terrains, France) primarily to simulated higher thermal loading. The additional analyses provided new insights into near-field coupled THM processes. In this chapter we discuss the model comparison from this new perspective.

2. DEFINITION OF BMT2: A MULTIPLE-FRACTURE MODEL

This bench-mark test is a two-dimensional plane-strain simplification of the three-dimensional model reported by Guvanasen and Chan [2,3]. The detailed problem definition has been documented by Chan and Khair [4] and by the DECOVALEX secretariat [5]. The aim of this model is to investigate the coupled behaviour of discontinuities and intact rock in a coupled THM process, including an optional consideration for natural and forced heat convection in the rock matrix as well as along the fractures. Conversion of strain energy to heat energy is assumed to be negligible since strains and strain rates in the intact rock blocks would be small because the model is intended to represent hard rock.

2.1. Geometry
As illustrated in Figure 1, this two-dimensional model measures 0.75 m x 0.50 m and consists of nine rock blocks, separated by two "vertical" and two "horizontal" discontinuities.

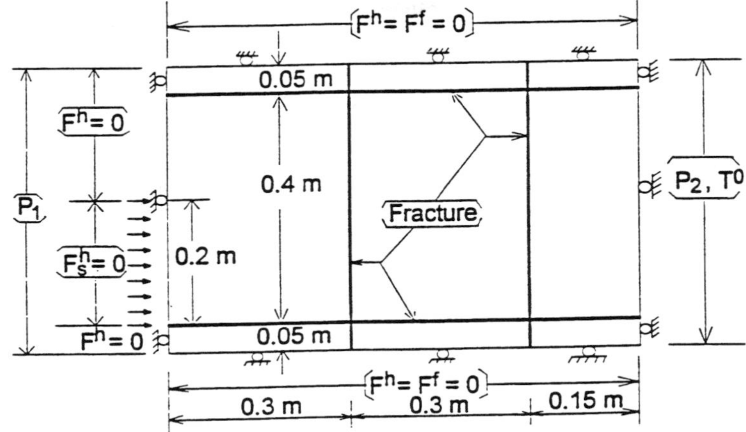

Figure 1.   Thermal, hydraulic and mechanical boundary conditions of the Muliple fracture
model, BMT2, DECOVALEX, Phase 1. $P_1 = 10$ KPa, $P_2 = 11$ KPa,
$F_s^h = 60$W/m^2 (heat source).

In order to make this bench-mark test amenable to simulation by a wider range of computer
codes used in DECOVALEX, the model has been simplified to a two-dimensional model in
the horizontal plane so that the terms "horizontal" and "vertical", as used to describe BMT2,
actually refer to the horizontal x and y coordinates, respectively.

2.2. Initial conditions

The initial conditions of the model include a constant temperature ($T^o = 15^\circ$C), a constant
fluid pressure ($P^o = 10$ *KPa*), constant initial (compressive) normal stresses ($\sigma^0_x = \sigma^0_y = -$
4 *MPa*) and a constant initial aperture of the discontinuities ($e^o = 300$ μm). The initial normal
and shear displacements are set to zero.

2.3. Boundary conditions

The normal displacements along all boundaries are set to zero. Zero fluid mass flux is
imposed on the top and bottom boundaries. A hydraulic pressure of $P = 10$ KPa is imposed
on the left vertical boundary and $P = 11$ KPa on the right vertical boundary to stimulate flow.
For thermal boundary conditions, zero thermal flux is imposed on the top and bottom

boundaries of the model. A constant temperature (T = $15^\circ C$) is imposed on the right vertical boundary. For the left vertical boundary, a heat flux Q = 60 W/m^2 is applied to the portion 0.05 m $\leq$ y$\leq$ 0.25 m for a period of 10^7 seconds after a steady-state flow condition is reached in the isothermal coupled hydromechanical calculation. The rest of the left vertical boundary has a zero thermal flux boundary condition. The heat flux of 60 W/m^2 corresponds to that at the wall of a 1.2-m diameter borehole into which is emplaced a 2.2-m high nuclear waste container with a heat generation rate of 500 W. Higher heat fluxes are subsequently considered, as discussed in Section 5 below.

2.4. Material properties

The material properties of the rock matrix and fluid are given in Table 1. The density (ρ_f) and dynamic viscosity (μ) of fluid are temperature-dependent properties according to the following relationships:

$$\rho_f = \rho_o [1 - \beta(T - T_o)] \tag{1}$$

$$\mu = A_I \exp[B_I / (273 + T)] \tag{2}$$

where ρ_o, β, T_o, A_I and B_I are given in Table 1.

It should be mentioned that Equation (1) does not accurately represent the nonlinear variation of water density with temperature in the range of T studied in this BMT2, but for this benchmark case, this type of linear variation is assumed.

2.5. Constitutive models for fractures

The constitutive model for the rock fractures may vary from one code to another. In BMT2 a modelling team can choose either (a) the Barton-Bandis (BB) model [6] or (b) the Coulomb-Friction (CF) model or a CF-like model. Details of these constitutive models are given in Chapter 3 to which the reader is referred. For the purpose of this BMT2, all teams employing CF-like models assumed a constant (normal and shear) fracture stiffness along with a smooth-walled, parallel-plate Snow type of fracture flow model [7]. On the other hand, the BB model is an empirical nonlinear model predicated on the premises that all the mechanical and hydraulic behaviors of a real fracture in rock can be related to the surface roughness and surface strength. In particular, the BB model involves a hyperbolic normal stress-normal closure relationship (stress-dependent stiffness), a normal stress-dependent shear strength, nonlinear shear stress-shear displacement relationship, shear dilatancy/contractancy behaviour and an empirical relationship between the mechanical and effective hydraulic apertures (see Section 3 below) that attempts to account for deviations from the Snow model caused by fracture roughness. Table 2 lists the different parameters needed for each model.

Admittedly the initial fracture aperture of 300 μm is on the extreme high end of in situ values which are frequently a few tens of μm. This large aperture was deliberately chosen for the fracture to produce some noticeable hydraulic and thermohydraulic effects for model verification.

Table 1
Rock matrix and fluid properties

Material	Parameter	Value
Rock Matrix	Young's modulus, E	50 GPa
	Poisson's ratio, u	0.25
	Density, ρ_s	2600 kg/m^3
	Uniaxial strength, σ_c	190 MPa
	Tensile strength, σ_T	50 MPa
	Thermal conductivity,λ	3.34 W/(m$^{2\circ}$C)
	Thermal expansion coefficient,α	10^{-5} C^{-1}
	Solid heat capacity, c_s	900 J/(kg$^\circ$C)
	Permeability tensor, $k_{xx} = k_{yy} = k_{zz}$	10^{-14} m^2
	Porosity, θ	0.35
	Biot's constant, M	6.5 GPa
Fluid	Density ρ_o, at reference, temperature T_0	997.1 kg/m^3
	Reference temperature, T_0	25°C
	Empirical constant, β(Equation 1)	$6\times10^{-4\circ}$C^{-1}
	Empirical constant, A_1 (Equation 2)	1.98404×10^{-6} Pa.s
	Empirical constant, B_1 (Equation 2)	1825.85°C
	Fluid heat capacity, c_f	4200 J/(kg$^\circ$C)

Table 2
Rock fracture parameters

Material	Parameter	Value
BB-model	Maximum closure of fracture, V_m	0.23×10^{-3}
	Initial normal stiffness, k_{ni}	10.3 GPa/m
	Joint roughness coefficient, JRC	7.5
	Initial permeability, $(e^\circ)^2/12$	7.5×10^{-9} m^2
	Joint compressive strength, JCS	80 MPa
	Residual friction angle, ϕ_r	30°
CF-model	Normal stiffness, k_n	10.3 GPa/m
	Shear stiffness, k_t	2.43 GPa/m
	Friction angle, ϕ	30°
	Dilation angle, i	1.3°

3. MODELLING APPROACHES

BMT2 was modelled by five teams: AECL, CNWRA (Centre for Nuclear Waste Regulatory Analysis, USA), INERIS, LBL (Lawrence Berkeley Laboratory, USA, sponsored by SKI, Swedish Nuclear Power Inspectorate); and, VTT (Technical Research Centre, Finland) with different codes [4, 8-11]. All the participating research teams represented the fractures explicitly as discrete features, with the exception of part of the VTT team calculations, as noted below. With the exception of the AECL team, all teams modelled BMT2 as a two-dimensional plane strain problem. Among the five teams three (AECL, LBL and VTT) employed the finite-element method (FEM) while the other two(CNWRA and INERIS) employed the distinct-element method (DEM). AECL and LBL assumed the rock matrix to be permeable whereas CNWRA and INERIS assumed the rock matrix to be impermeable. VTT did not model the fluid flow aspect of BMT2. The modelling methodology of each team is briefly summarized below.

Both AECL and LBL modelled BMT2 as a fully coupled THM problem except for the omission of the conversion of strain energy to heat energy, as noted in section 2 above. In fact, these two teams solved essentially the same governing equations with somewhat different numerical and programming implementations. The AECL team modelled BMT2 as a pseudo-plane strain problem with both natural and forced heat convection by using the 3-D MOTIF FEM code [2,3] (see also code summary near end of this book). The rock matrix was assumed to be thermo-poroelastic using a nonisothermal extension of Biot's theory of consolidation [12]. Fluid flow along discontinuities was modelled by a modified Snow's parallel-plate model [7]. Roughness of the surfaces bounding a fracture was modelled approximately using an empirical relationship between the mechanical aperture and effective hydraulic aperture given by the Barton-Bandis (BB) constitutive model [6]. In this model the effective hydraulic aperture, e, is related to the mechanical aperture, E, by the following equation:

$$e = \min{(E; E^2 \ / \ JRC^{2.5})} \tag{3}$$

where JRC is the joint roughness coefficient.

Two cases were simulated. In the main case heat transport by conduction, forced convection and natural convection were all modelled. Furthermore, a dispersive heat transport term was included to represent deviation of fluid velocity from idealized plug flow. Such dispersion can conceivably be caused by roughness of the fracture surfaces and Taylor dispersion. A no-convection (conduction only) case was also run to obtain results for comparison with other teams.

LBL modelled BMT2 as a plane-strain problem using the ROCMASS FEM code [13]. For the constitutive behaviour of the fracture a CF model with constant stiffness was employed in conjunction with a smooth-walled parallel-plate fluid-flow model. Dispersive thermal energy transport was not considered.

Both CNWRA and INERIS modelled BMT2 as a plane-strain problem using the UDEC DEM code [14] that does not consider heat convection in its mathematical formulation. Fluid flow along discontinuities were simulated with a smooth-walled parallel-plate model.

The rock matrix was assumed to be impermeable. The mechanical behaviour of discontinuities were represented by a Coulomb friction model with constant shear and normal stiffness and a dilation angle. CNWRA modelled the transient THM processes using the algorithm that has been incorporated in UDEC by the original code developers while INERIS employed an algorithm called the "Balloon Scheme" to simulate the hydraulic behaviour of the discontinuities [9]. A further difference between the approaches taken by these two teams was that CNWRA also studied the limiting case of forced convection along discontinuities by fixing the temperature along the two horizontal fractures at $15^\circ C$, i.e., assuming perfect cooling by convection.

The VTT modelled BMT2 by a combination of two codes, a standard continuum FEM code, ADINA-T [15], for heat conduction calculations and its own 2-D FEM code, JRTEMP[16], for mechanical and thermomechanical calculations without fluid flow. Only the JRTEMP calculation includes the discrete fractures explicitly. Therefore, only the thermal and mechanical response can be compared with other teams. For the constitutive relationship of a fracture Goodman's hyperbolic normal stress-normal displacement model [17] was combined with an early version of the Barton-Bandis model for the shear behaviour.

All the models submitted for comparison utilized a few hundred elements or zones except for the INERIS model which uses over four thousand triangular zones. Short descriptions of the physical basis, mathematical formulation and numerical methods for the computer codes used in modelling BMT2 are given in the Appendix.

4. COMPARISON OF THE MODELLING RESULTS

In this section modelling results for BMT2 obtained using different methods and codes are compared. Detailed results from individual research teams have been presented in the various team reports [4,8-11]. The temperature, displacement, principal stresses and hydraulic head calculated by the participating teams are plotted here for comparison as a function of distance or time at some selected monitoring points in the model. The locations of the monitoring points for BMT2 are shown in Figure 2. Spatial distributions of temperature and displacements are also presented at specific times. Fluid velocity comparison is deferred to Section 6.4 below where higher thermal loading causes more discernible departure from the initial velocity distribution.

4.1. Temperature distribution
Isothermal contours at time $t = 10^7$ seconds (at the end of the modelled period) for models without (designated NC) and with (designated C) consideration for heat convection are shown in Figures 3 and 4, respectively. For the sake of brevity the AECL (NC) and VTT isotherm plot have been omitted from Figure 3. Suffice it to say that the results from these two models agree very well with those shown. Within each of these two types of models temperature distributions from various teams are very similar. Noticeable difference in temperature can be seen between the two groups of models with (Figure 4) and without (Figure 3) convection. The temperature without convection is higher than that with convection and the temperature distribution for the latter is very much restricted to the area

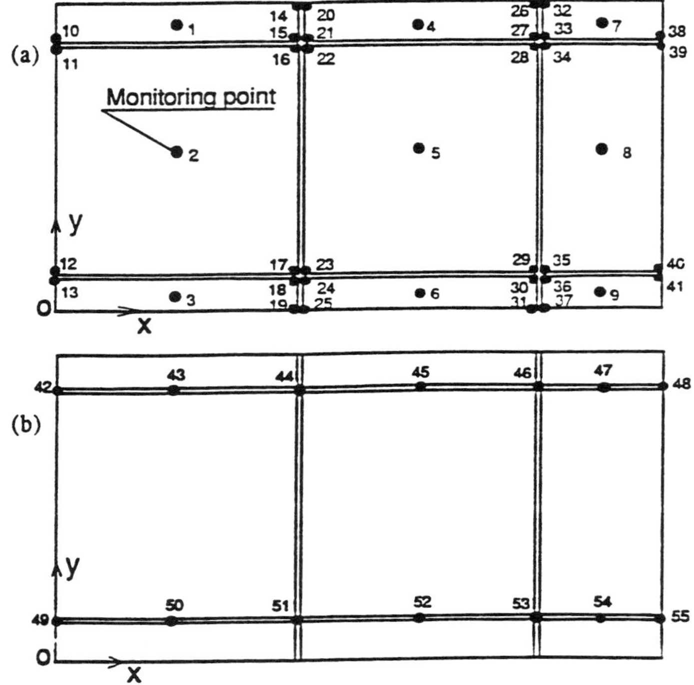

Figure 2.   Locations of monitoring points for BMT2 for: (a) stresses, displacements, temperature and hydraulic head; (b) fluid velocities.

defined by the two horizontal fractures, the left vertical boundary and the leftmost vertical fracture (cf. Figure 4). The AECL model has given a slightly higher temperature distribution than the CNWRA model, as expected from the consideration that the latter represents a limiting case.   It is not clear, however, why the LBL model predicts the lowest temperature distribution among the three models that simulated thermal convection.   Initially, it was surmised [1] that the LBL team's assumption of nondispersive heat convection had led to a sharp front and, consequently, oscillations in the numerical solution.   Subsequently, the AECL team repeated their BMT2 simulation by assuming zero thermal hydrodynamic dispersivity but did not observe any numerical oscillation or instability.   In contrast to the sharp front in solute transport caused by the dominance of advection over diffusion, for heat transport it appears that the thermal conductivity of the rock matrix is sufficiently high to smear out the temperature front in the fracture.   So the reason for the numerical oscillations in the LBL model remains unknown.   Presumably, the different numerical implementations

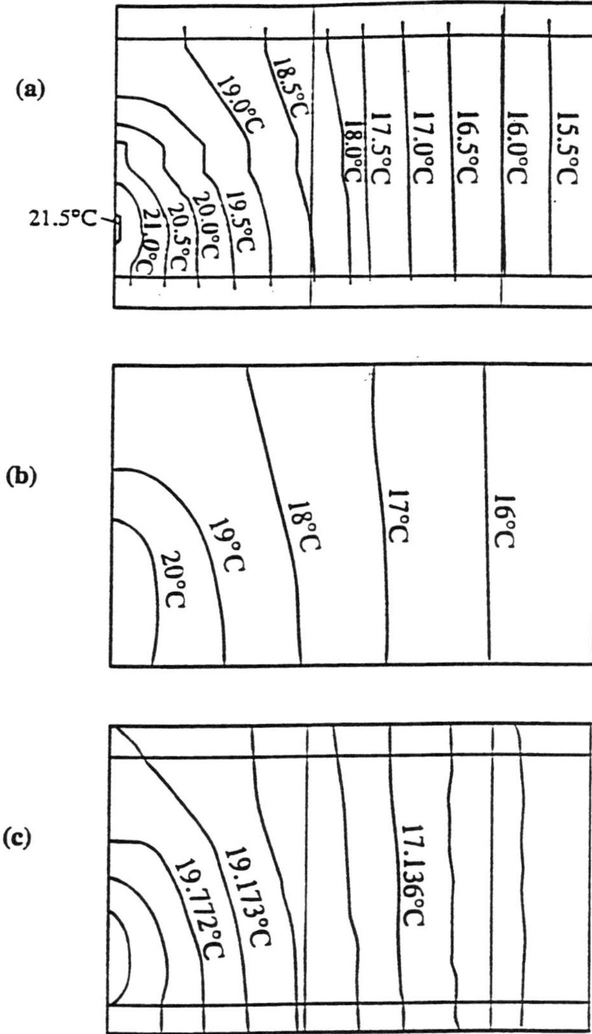

Figure 3.   Isotherms at time t = 10^7 second without heat convection: (a) CNWRA model; (b) INERIS model; (c) LBL model.

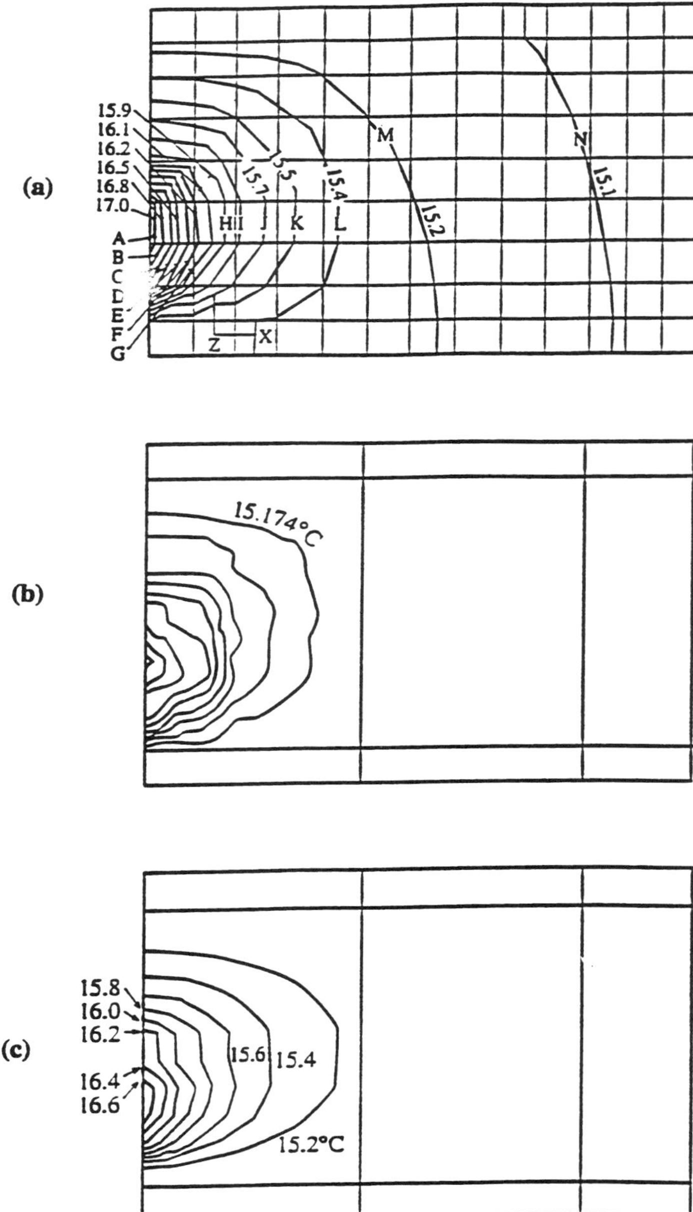

Figure 4. Isotherms at time t = 10^7 seconds with heat convection: (a) AECL model; (b) LBL model; (c) CNWRA model.

of the MOTIF and ROCMAS codes are responsible for their different behaviour. The maximum temperature is about $19 - 21\,^\circ C$ for the models without convection, but only $16 - 17\,^\circ C$ for the models considering convection.

The temperature at points 11 and 17 for cases with and without heat convection is plotted in Figure 5. Steady state has been reached for all models towards the end of the simulated period. For the INERIS model (using several thousand triangular zones) initially submitted for comparison the temperature is still rising at $t=10^6$ seconds but, from tabulated results submitted by the team, there is insignificant temperature rise after 3×10^6 seconds and steady state has been attained at 5×10^6 seconds. The other models start to approach steady state earlier at approximately 10^6 seconds. Subsequently, INERIS has repeated their calculation by means of a model consisting of several hundred zones and obtained results practically identical to those of CNWRA. Apparently, the slower approach to steady state of the fine-mesh INERIS model may reflect the transient behaviour of BMT2 more accurately than the other models.

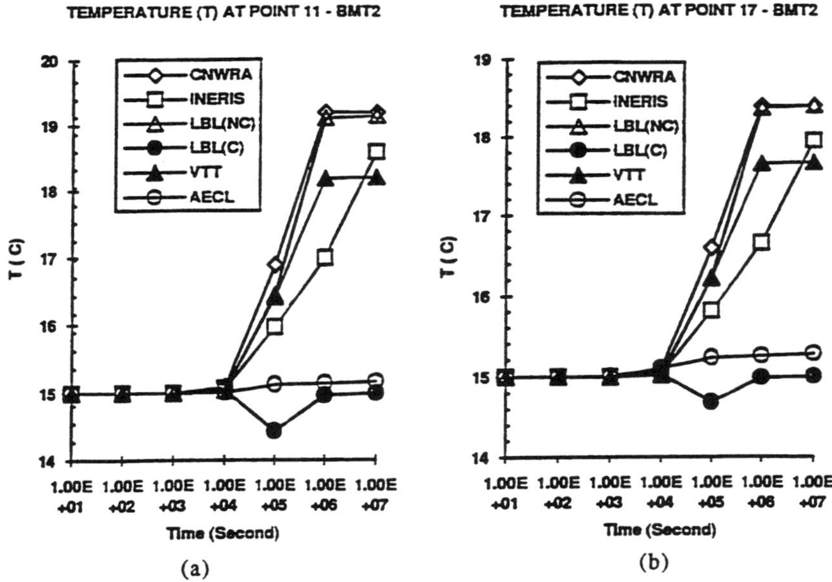

Figure 5. Temperature at points 11 and 17 versus time for BMT2. Heat convection is considered in the AECL and LBL(C) models.

Even without considering heat convection the maximum temperature rise (Figure 3) is only a meager $6\,^\circ C$ for this problem. This may appear surprising given a heat flux that is commensurate with a waste container with 500 W thermal power. In retrospect it is recognized that this is due to the combined effect of a modest input heat flux and a $15\,^\circ C$

constant temperature boundary on the right side of the model. In a nuclear fuel waste repository the rock will be heated by the thermal loading of many waste containers so that a constant temperature boundary condition at 0.75 m away from a particular container will not be a good approximation. Model results for higher thermal loading are discussed in Section 6.1 below.

4.2. Displacements

The displacement fields obtained by the three models that account for thermal convection (C), along with the corresponding results without convection (NC) are depicted as vector plots in Figure 6 and Figure 7 respectively. Computer readable data were not available from the LBL team for comparison. Plots found in the team report, however, display similar patterns to those shown in Figures 6 and 7. It can be observed that heat convection has significant impact on the model deformation. The deformation is more or less defined by the fractures into different zones when convection is considered, but not so obvious when heat convection is excluded. It should be noted that the displacement scales are quite different for the NC and C cases. Thus the thermally induced displacements are much larger in the NC models. Within each group of models (C or NC) there is remarkable similarity in the displacement fields obtained using different codes with different approaches. The area of maximum displacement corresponds to the area of maximum temperature rise near the heat flux boundary at the lower portion of the left vertical edge of the model. For this problem there is no evidence that DEM models lead to higher displacements. No sign of translational or rotational rigid-body movements, which are simulated in the DEM models but not in the FEM models, can be discerned from any of the numerical solutions. This is to be expected. In view of the boundary constraints rock deformation in BMT2 involves only the thermal expansion of the intact rock blocks and the closure of fractures (cf. Figure 10 below).

Figure 8 plots the time histories of horizontal displacement at point 17 at the lower right corner of rock block 2 (see Figure 2) for both cases with (C) and without heat convection (NC). For the NC case the AECL and CNWRA results are very similar. The INERIS displacement is comparable to those from the above two models at the end of the simulation but smaller at earlier time. The values of maximum displacement of 17-20 μm is comparable to, but lower than, free thermal expansion values, as expected. The VTT and LBL (NC) results are much lower than the other three teams. For the C case the three teams obtained similar displacements with the AECL values somewhat higher, and the LBL values much lower, than the CNWRA results. The slightly higher displacement in the AECL model is consistent with the slightly higher temperature. In fact, a back-of-the envelope calculation shows that a temperature difference of $0.5\,^{\circ}C$ would lead to a 1.5 μm difference in displacement, assuming free thermal expansion. Inherent numerical oscillations have caused the displacement to be lower in the LBL (C) solution. Generally, there is no obvious difference in the deformation of the rock mass between the distinct element approach and finite element approach.

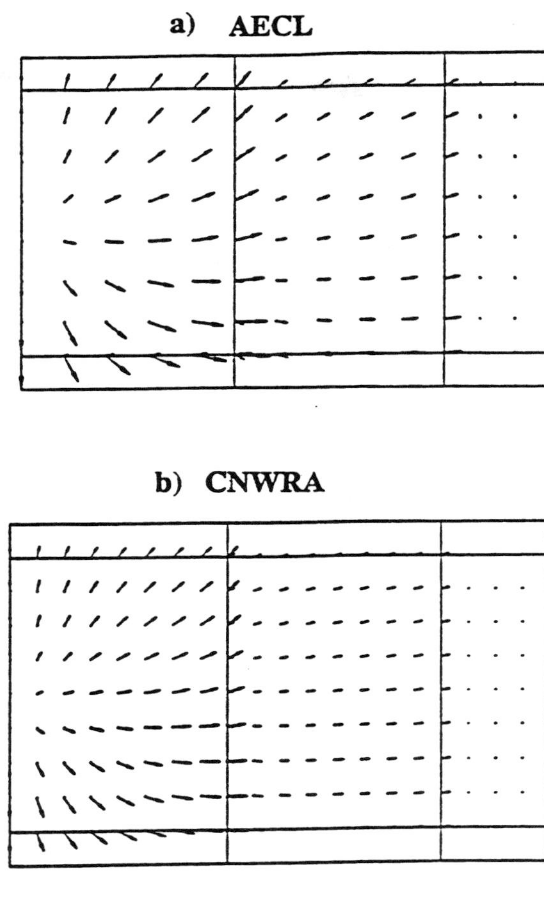

Figure 6.   Displacement vector plots at time t = 10^7 seconds with (c) heat convection:
(a) AECL; (b) CNWRA (limiting case)

4.3. Stresses

In BMT2 there is no mechanical loading and the hydraulic pressures are insignificant compared with the in situ and thermal stresses. Consequently, all stress changes in the model are caused by the temperature rise. Owing to the relatively small heat flux (60 W/m^2) from the heat source and the $15°C$ isothermal boundary condition on the right vertical boundary, the thermally induced stress increments in the model should not be great. This

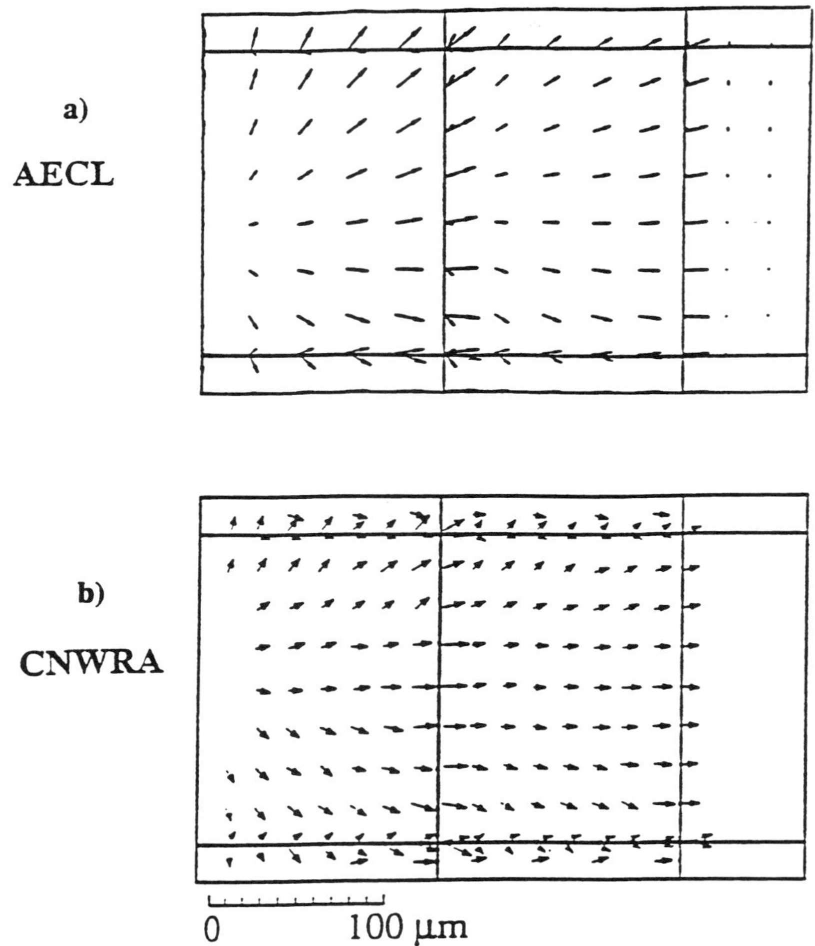

Figure 7. Displacement vector plot at time $t=10^7$ seconds without (NC) heat convection: (a) AECL; (b) CNWRA. Note that scale is different from Figure 6.

turns out to be the case in all model predictions for stress changes except for the VTT model, as illustrated in Figure 9, which shows the total normal stresses at monitoring point 2 (centre of the block with the heat flux boundary, Figure 2) as functions of time. The major stress disturbance occurs in the area close to the heat source located at the lower left corner of the model. With the exception of the conduction-only models of AECL (NC) and VTT, all models predicted maximum stress changes in the range 0.01-0.2 MPa. The AECL (NC) model without convection yielded stress increases slightly less than 1 MPa while the VTT model predicted stress increases of 1.5-2.5 MPa.

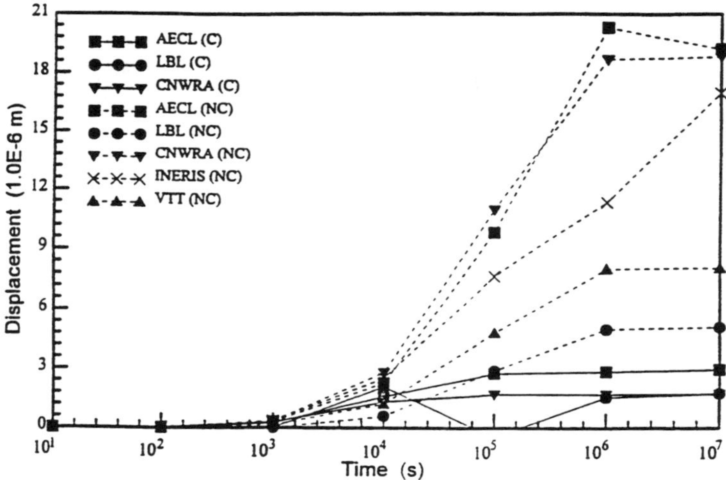

Figure 8. Horizontal displacement at point 17 (on upper surface of lower horizontal fracture) versus time.

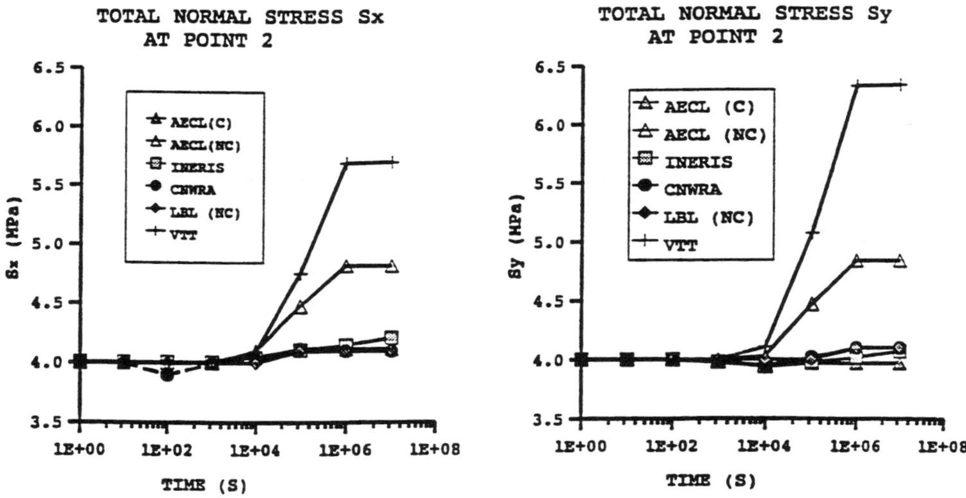

Figure 9. Normal stresses in x and y directions versus time at point 2, centre of block with heat flux boundary.

Detailed analysis of the various team reports reveals that most of the discrepancies between the modelling results can be attributed to the different values of normal stiffness employed by the various teams. The AECL team used the BB model with an initial (zero stress) normal stiffness of 10.3 GPa/m. Substituting the parameter values in Table 2 into the BB leads to a normal stiffness of 74.4 GPa/m at the in situ stress of 4 MPa. VTT used a Goodman hyperbolic model with a variable normal stiffness ranging from 880 to 2110 GPa/m. All the other teams utilized a constant normal stiffness of 10.3 GPa/m for their models, thus allowing the heated rock blocks to expand more easily into the softer fractures. This is responsible for their much lower thermal stresses. The apparent agreement of the AECL (C) model that includes heat convection with the conduction models of the teams that used a normal stiffness is just a fortuitous agreement caused by a higher fracture stiffness and a lower temperature rise that affect the thermal stress in opposite directions. A valuable lesson learned from this inter-code comparison is that sometimes apparent agreement of results can be deceptive.

4.4. Fracture aperture change

Figure 10 shows the aperture along the lower horizontal fracture at time $t = 10^7$ seconds as calculated by the models without (Figure 10a) or with (Figure 10b) heat convection. It can be observed that closure occurs at those regions of the fracture close to the heat source (close to the origin O of the coordinate system in Figure 10). For both the NC and C cases the AECL, CNWRA and LBL models all predicted similar fracture aperture. The maximum closure of the ranges from about 15 to 17 μm for the NC models and 2 to 3 μm for the C models. For the NC case the slightly lower fracture closure calculated by the AECL model is due to the use of a higher normal stiffness, which also leads to higher stresses, as discussed in section 4.3 above. For the C case (Figure 10b) the higher fracture closure predicted by the AECL model is consistent with the slightly higher temperatures in blocks 3 and 6 (Figure 2) below the lower horizontal fracture. The slightly smaller aperture change for LBL(C) model is also consistent with its lower temperature in comparison with the limiting case analysis of CNWRA. The model results indicate that taking heat convection into consideration reduces the calculated fracture closure by almost an order of magnitude. This is true both for the MOTIF and ROCMAS FEM models that have heat convection incorporated in their mathematical formulation and for the CNWRA limiting case analysis. With heat convection, the temperature rise and temperature gradient in the rock matrix becomes smaller resulting in lower thermal expansion and lower fracture closure.

4.5. Head distribution

Hydraulic head distributions along the upper horizontal fracture, as reported by various teams, are compared in Figure 11. Computed deviations from the initial hydraulic head distribution are rather insignificant for all teams. This is due to the prescribed fluid pressure boundary conditions and the modest temperature rise. In the case of the LBL model, it is almost identical to the initial value. Results from the other three teams agree well with each other. The AECL model predicted a somewhat nonuniform hydraulic gradient. In nonisothermal flow the fluid pressure is usually converted to a quantity known as reference head (or equivalent fresh water head). It is not clear whether the minor difference between the various model results may have arisen from the use of differing reference conditions for this conversion. In hindsight it would have been more preferable to require the modelling teams to submit their computed pressure distribution for comparison. As the head changes are so small, little general conclusion can be drawn from this comparison beyond the statement that in BMT2 both the total hydraulic pressure and the thermal and thermomech-

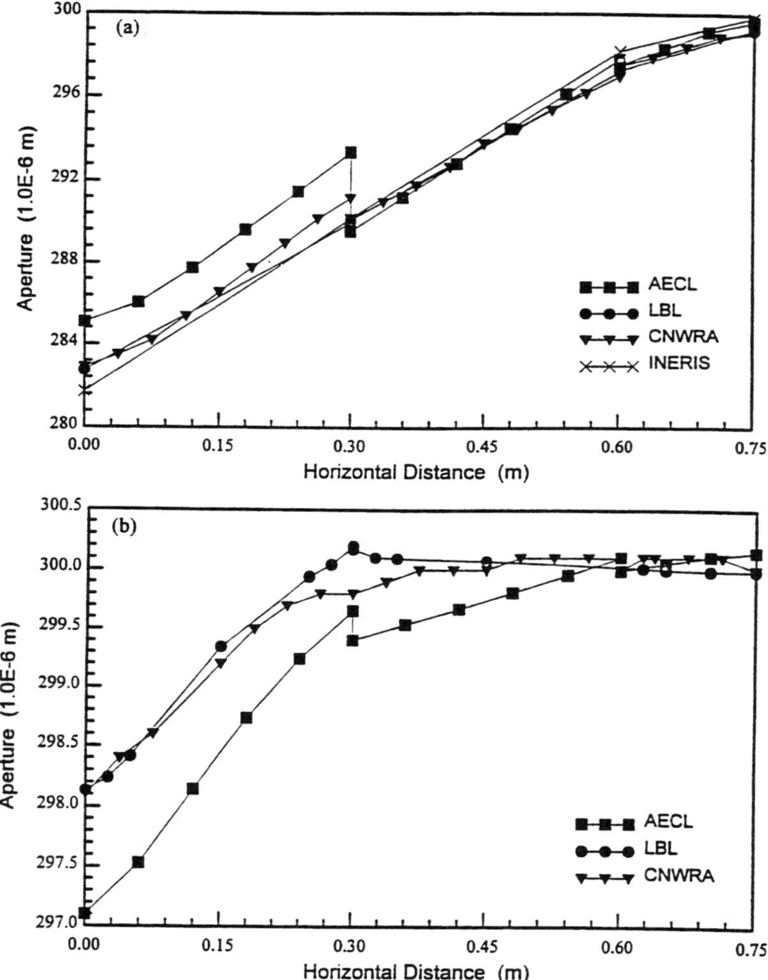

Figure 10. Aperture along lower horizontal fracture at time t = 10^7 seconds: (a) without heat convection; (b) with heat convection.

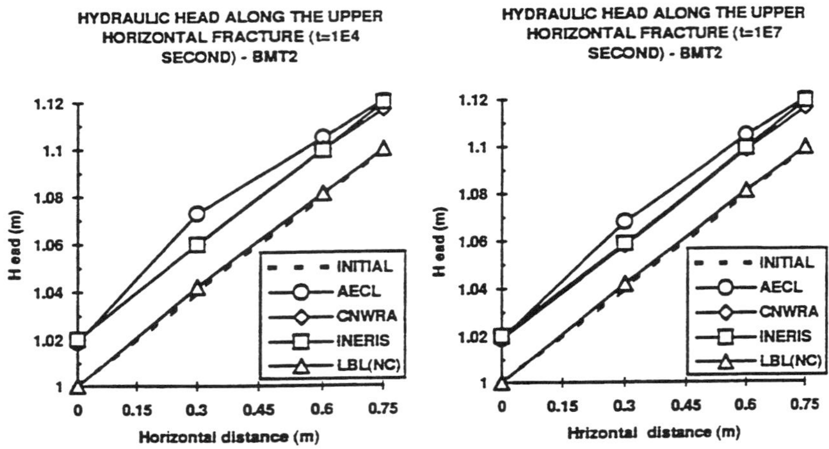

Figure 11.   Hydraulic head distribution along the upper fracture: (a) at time $t = 10^4$
seconds; (b) at time $t = 10^7$ seconds.

anically induced pore pressure changes are so insignificant that the flow field has little
influence on the mechanical stress and deformation.

4.6. Velocity distribution

The fluid velocity in a coupled THM problem like BMT2 is complicated by several
factors.   In light of the very minute head changes calculated for BMT2, as discussed in
Section 4.5 above, it was decided that this comparison might be more meaningful for higher
thermal loading. This is done in Section 6.4 below.

5. ADDITIONAL SENSITIVITY ANALYSES

As well as AECL's sensitivity analysis on the influence of the thermal hydrodynamic
dispersitivity on numerical stability and numerical oscillations and INERIS's study of the
effect of mesh density described in section 4.1 above, the two teams also modelled an
extension of BMT2 that considers higher thermal loading. The INERIS team simulated the
extended BMT2 for boundary thermal fluxes of 120, 180, 240 and 300 W/m^2 using UDEC
without heat convection while AECL simulated this extended bench mark test for thermal
fluxes of 180 and 300 W/m^2 using MOTIF both with and without heat convection. INERIS
did simulate forced convection (private communication, A. Thoraval) by means of an
enhanced version of the UDEC code that they developed [18]. Unfortunately, their model
assumed constant fluid viscosity and density so that the results are not suitable for
comparison with AECL's MOTIF model. Modelling results without heat convection for
various thermal loading from these two teams are compared in the next section.

6. COMPARISON OF RESULTS OF HIGHER THERMAL LOADING

Modelling results for the extended BMT2 with thermal loading in the range 60-300 W/m^2 from the AECL and INERIS teams are compared for temperature, stresses, aperture closure and fluid velocity along a discrete fracture. The INERIS results submitted for comparison were computed with their coarse-grid model that used a grid size of 7 cm. This grid is comparable to those utilized by other teams and has been shown to produce a similar transient behaviour for BMT2 to that predicted by other teams (see Section 4.1).

6.1. Temperature

Figure 12 shows the temperature profile along the horizontal line y = 0.15 m, which passes through the hottest point in the model, for a boundary heat flux of 300 W/m^2 at t = 10^7 sec. Isothermal contours at t = 10^7 sec are plotted in Figure 13 for AECL's model with and without heat convection for thermal loading of 180 and 300 W/m^2 . When heat convection is not considered, both the AECL and INERIS models have yielded almost identical temperature distributions. In fact the temperature field was found to be just directly proportional to the input heat flux. Thus, for instance, with a thermal loading of 300 W/m^2 the maximum temperature in the model is 47.5 °C. This corresponds to a temperature rise of 32.5 °C which is five times the temperature rise of 6.5 °C at the same location for a 60 W/m^2 thermal loading. This linear variation of temperature with the strength of the heat source is, of course, a manifestation of the linearity of the heat conduction equation.

The lowest curve in Figure 12 was calculated by AECL using a fully coupled THM MOTIF model that includes both natural and forced convection. Clearly, at higher thermal loading heat convection leads to a significantly lower temperature field for BMT2 than predicted from a heat conduction model. In contrast with the conduction model (Figure 3) where heat flux enters the left vertical boundary and exits the right, constant-temperature, vertical boundary, in the model with heat convection there is substantial heat loss (down the thermal gradient) to the two horizontal fractures, as well as to the right vertical boundary. The amount of heat energy, Q_{oT}, leaving the model domain by forced convection along the two horizontal fractures can be calculated as

$$Q_{oT} = ew\rho_f c_f v \Delta T \qquad (4)$$

where w is the width of the fracture (normal to the plane of the paper in Figure 1), v is the fluid velocity in the fracture, ΔT is the temperature rise at the exit point (leftmost point) of the fracture and the rest of the symbols have been defined in section 2.4. This convective heat energy loss works out to be about 11% and 30-31% of the thermal power of the heat source for the upper and lower fractures, respectively. From Figure 13 (d) it can be seen that, with 300 W/m^2 thermal loading the temperature rise in the lower fracture is not negligible. Thus assuming a 15 °C constant temperature at the horizontal fractures as in the CNWRA model discussed in Section 3 above would no longer be a good approximation.

Obviously, in BMT2 heat convection plays a very significant role in determining the temperature field, and consequently displacements, stresses, fracture aperture change and fluid velocities. The reader is cautioned not to generalize this conclusion to the near field of an actual repository. It should be recognized that some of the model parameters have been

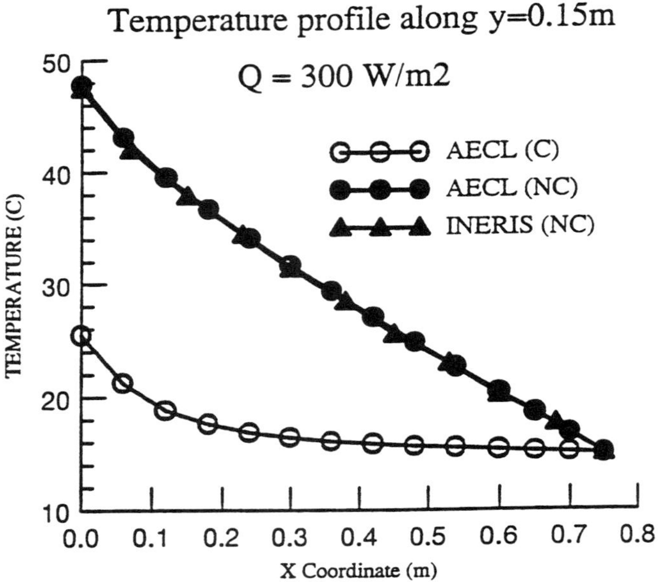

Figure 12. Temperature profile at time t = 10^7 seconds for thermal loading 300 W/m^2 along horizontal line y = 0.15 passing through the hottest point.

deliberately chosen to yield nontrivial effects to facilitate comparison between various modelling approaches. For example, the fluid pressure boundary conditions and fracture properties specified for BMT2 in Section 2 would lead to a fluid velocity of 8.7 mm/s (or over 750 m/day) in the horizontal fractures. In a properly characterized and selected site, it is most unlikely that this type of flow rate would exist except perhaps as a short transient.

6.2. Stresses

Time evolution of the normal stress, σ_{xx}, at point 2 at the centre of the lower left block of intact rock (Figure 2) is plotted in Figure 14 for a thermal loading of 300 W/m^2. As discussed in Section 4.3, owing to the use of a higher fracture normal stiffness the AECL model without convection yields higher stresses than the corresponding INERIS model. The AECL model that takes convection into consideration results in lower temperature and, therefore, correspondingly lower thermal stresses than the conduction only model.

6.3. Fracture aperture

The mechanical aperture at t = 10^7 sec is plotted in Figure 15 along the lower horizontal fracture for a thermal loading of 300 W/m^2 . It is now even more conspicuous that including heat convection in the model dramatically reduces the thermal expansion of the intact rock into the fracture. The fracture closure profile predicted by both models without convection are qualitatively very similar, including the discontinuity in the fracture aperture at the

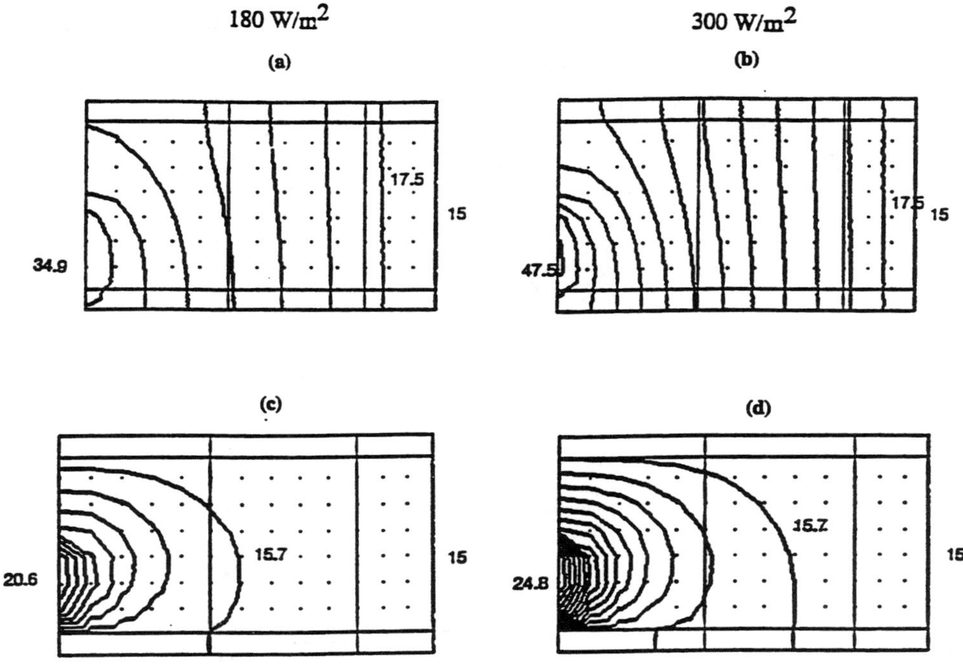

Figure 13. Isotherm contours at time t = 10^7 for two thermal loading according to AECL models: (a) and (b) without heat convection; (c) and (d) with convection.

intersection with the left vertical fracture at y = 0.3 m. There is less fracture closing in the AECL (NC) model, again because of the higher normal stiffness assumed.

6.4. Fluid Velocity

To compare fluid velocities, the horizontal component of the fluid velocity along the upper and lower horizontal fractures are plotted for the 300 W/m² thermal loading case at time t = 10^7 seconds in Figure 16. The general trend of the velocity distribution, in as much as the magnitude of the fluid velocity is greater closer to the heat source, is similar for the two models without considering heat convection. Nevertheless, the slopes of the curves from the two models are different. In both models the velocity was calculated as a postprocessing step using temperature-dependent fluid density and viscosity in an attempt to more closely approximate the real situation. INERIS assumed the linear relationship in Equation (1) whereas AECL used a quadratic equation which more accurately represents the variation of water density over the temperature range 15-50 °C. Apart from that the two teams also used different methods to calculate the fluid velocity. INERIS calculated the volumetric fluid flux along a fracture assuming the cubic law for a smooth-walled parallel-

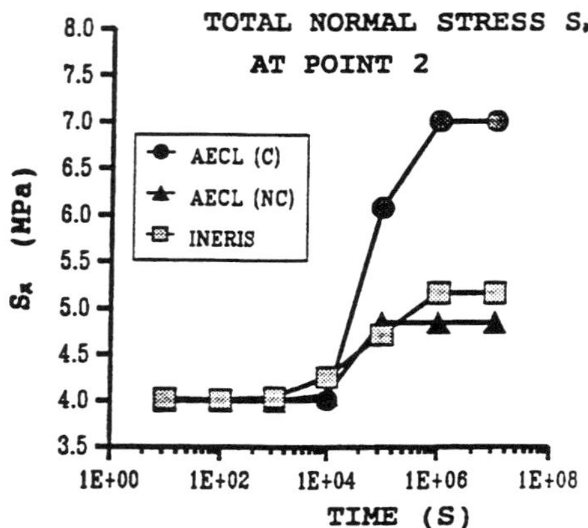

Figure 14.   Total normal stress in x direction at point 2 at time t = 10^7 seconds for thermal loading 300 W/m^2.

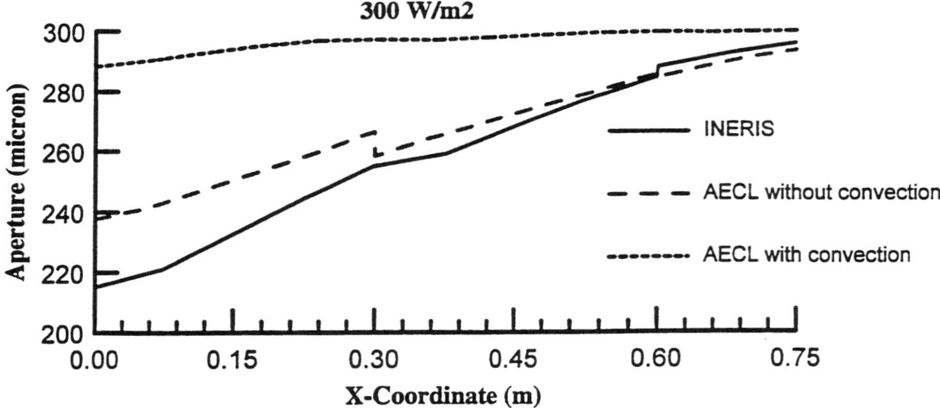

Figure 15.   Aperture along lower horizontal fracture at time t = 10^7 seconds for thermal loading 300 W/m^2. INERIS model does not consider heat convection.

plate fracture, in conjunction with fluid mass conservation in the fracture, which is consistent with UDEC's assumption of impermeable rock matrix. On the other hand, AECL employed a generalized Darcy's law:

$$q_i = -(k_{ij} / \mu)(\partial / \partial x_j)(p + \rho g x_3) \tag{5}$$

where q_i is the i-th component of the Darcy velocity or specific discharge and the other symbols have their usual meaning or are defined in Section 2.4 above. For fracture flow, since porosity is assumed to be unity, the fluid velocity $v_i = q_i$. For a slightly compressible fluid such as water in a horizontal fracture the last term in Equation (5) vanishes.

In a coupled THM problem such as BMT2 the fluid velocity, v, in a fracture is directly proportional to (a) fracture permeability which varies as the square of fracture aperture, (b) hydraulic head gradient and (c) fluid density which decreases weakly with rising temperature. In addition, v varies inversely with fluid viscosity which decreases strongly with rising temperature. Inspection of the equations of state reveals that water density decreases by only about 2 % over the temperature range of 15-50 $^\circ C$ which is negligible in comparison with the nearly 50 % reduction in water viscosity over the same temperature range. As the rock temperature increases, fracture closure would lead to decreased fluid velocity, while reduced viscosity would lead to increased fluid velocity. Which of these two competing factors dominates would depend on the equation of state and the constitutive relationship employed in the model, as well as the calculated temperature distribution. Since both the AECL and INERIS heat conduction models predicted the same temperature distribution, the water viscosity along a fracture would also be the same for both models. In this case, the higher magnitude of the velocity predicted by the AECL (no convection) model near the hotter end of the fractures is qualitatively consistent with a larger aperture (less closure) than the INERIS model. However, fluid velocity is also directly proportional to the hydraulic gradient which, unfortunately, is not available from INERIS for quantitative analysis of the discrepancies depicted in Figure 16.

For the AECL model that accounts for heat convection the velocity distribution is very similar to, but slightly lower than the initial velocity. Since the temperature rise is small, one would expect small changes in viscosity and fracture aperture (Figure 15) which both lead to small velocity changes. It is not clear, though, why the opposing effects of viscosity and fracture closure appear to cancel each other almost completely along both fractures. As no other modelling team has modelled BMT2 for the higher thermal loading using a fully coupled THM model, it must be concluded that fluid velocity calculation by any fully coupled THM has not been verified by this inter-code comparison exercise in DECOVALEX I.

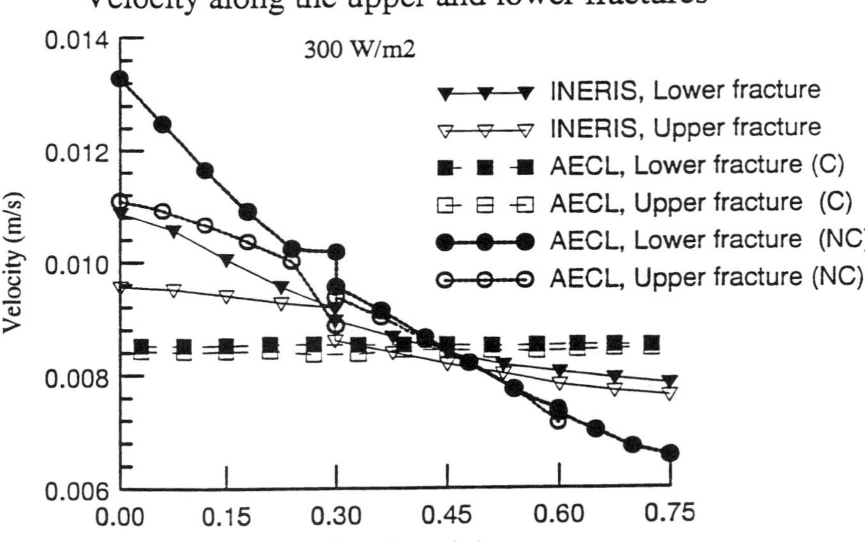

Figure 16.   Horizontal component of fluid velocity along the upper: (a) and lower (b) horizontal fractures at time t = 10^7 seconds.

7. SUMMARY AND CONCLUSIONS

The generic near field bench mark problem BMT2 has provided a rather complete spectrum of the coupled thermohydromechanical processes in a jointed rock mass for inter-code comparison. Among the coupled THM processes illustrated by BMT2, fracture closure caused by thermal expansion of the rock blocks and thermal energy dissipation by forced convection along fractures are the most prominent. Two fundamentally different methods-the finite-element method (FEM) and discrete-element method (DEM)- have been employed for simulating BMT2. In general, simulation results from these two categories of models agree with each other to about the same degree as among different models of the same category.

The bench mark study in this chapter suggest the following: (1) When only heat conduction is involved the temperature field can be predicted very well by all the participating models. (2) At present only the two FEM codes participating in this bench marking exercise have the capability of modelling convective heat transfer in the rock mass, although a team using a DEM model did perform a limiting case analysis with results in good agreement with one of the FEM models for a low thermal loading. (3) Displacements, stresses and fracture closure predicted by different models without considering heat convection generally agree well, with most discrepancies attributable to different choices of constitutive relationships or material parameters. (4) Among models taking heat convection into account the predicted temperature, displacement and fracture closure show reasonable agreement for the original BMT2. Unfortunately, owing to low thermal loading the stress changes for the original BMT2 are too small to allow a definitive conclusion to be drawn. No comparison can be made at higher thermal loading as only one team has undertaken these simulations taking heat convection into account. (5) The comparison of fluid velocity is inconclusive although results from the two teams attempting the higher thermal lading cases show generally similar trend for models without considering heat convection. Again, unavailability of other model results for comparison precludes the verification of fluid velocity calculated by the FEM model that includes convection. Consequently, the objective of code verification has only been partially achieved. (6) Sensitivity analysis at higher thermal loading has indicated that heat convection in rock fractures has significant effects on the distribution of temperature, deformation, stress and fluid velocity for this problem. Of course, in a real repository the significance - or otherwise- of heat convection would dependent on site specific parameters.

An important lesson learnt is that a bench mark problem should be simulated in at least two phases. First, all models must simulate identical geometry, initial and boundary conditions, constitutive relationships and material properties. Then, and only then, can the modelling teams be permitted to vary some of the problem specifications to study the effects of these variations. Another lesson learnt is that a modeller must be wary of the possibility of apparent agreement due to counteracting discrepancies annulling each other.

In future studies several changes to BMT2 can be made to better approximate in situ thermal, mechanical and hydraulic conditions in the near field of a nuclear waste repository in rock. These may include: (1) increasing the thermal loading by another factor of two; (2) using a less restrictive mechanical boundary constraint to enable a comparison between different modelling approaches when the assemblage of rock blocks are under less confinement such as near an excavation face; (3) using a vertical section instead of the horizontal section in BMT2 to facilitate inter-code comparison of modelled effects of natural thermal convection and gravity on fluid flow in fractures; (4) moving the right boundary further away from the heat source; (5) reducing the fracture aperture and increasing the fracture roughness and stiffness; and, (6) increasing the initial pore pressure and in situ stress.

8. ACKNOWLEDGMENT

The authors gratefully acknowledge members of the participating teams for their contribution of model results for comparison. They also thank Dr. C.F. Tsang and Dr. S. Nguyen for reviewing the manuscript and other participants in the DECOVALEX project for many stimulating discussions. The Canadian Nuclear Fuel Waste Management Program is funded jointly by AECL and Ontario Hydro under the auspices of the CANDU Owners Group (COG).

9. REFERENCES

1   T. Chan, K. Khair, L. Jing, M. Ahola, J. Noorishad and E. Vuillod. International Comparison of Coupled Thermo-Hydro-Mechanical Models of a Multiple-Fracture Bench Mark Problem: DECOVALEX Phase I, Bench Mark Test 2. Int. J. Rock Mech. Min. Sci. & Geomech. Abstr. Vol. 32, No. 5, pp.435-452 (1995).

2   V. Guvanasen and Chan T. Three-Dimensional Finite Element Solution for Heat and Fluid Transport in Deformable Rock Mass with Discrete Fractures. Proc. Int. Conf. on Computer Methods and Advances in Geomechanics, Cairns, pp. 1547-1552, Balkema Press (1991).

3   V. Guvanasen and T. Chan. A New Three-Dimensional Analysis of Hysteretic Thermohydromechanical Deformation of Fractured Rock Mass with Dilatancy in Fractures. Mechanics of Jointed and Faulted Rock (ed. H-P. Rossmanith), pp.437-442, Balkema, Rotterdam (1995).

4   T. Chan and K. Khair. Simulation and Results of the Multiple Fracture Model, DECOVALEX Project - Phase I, Bench-Mark Test 2. AECL-10780, AECL Research, Applied Geoscience Branch, Whiteshell Laboratories (1992).

5   Definition of Bench-Mark-Test 2, Multiple Fracture Model, DECOVALEX Series Document, Doc. 91/104, DECOVALEX Secretariat, Engineering Geology, Royal Institute of Technology, Stockholm, Sweden (1991).

6   N. Barton, M. Bandis and K. Bakhtar. Strength, Deformation and Conductivity Coupling of Rock Joints. Int. J. Rock Mech. Min. Sci. & Geomech. Abstr., 22, pp. 121-140 (1985).

7   D.T. Snow. A parallel plate model of fractured permeable media. Ph.D. Thesis, University of California, Berkeley (1965).

8   M.P. Ahola, S-M. Hsiung, L.J. Lorig and A.H. Chowdhury. Thermo-Hydro-Mechanical Coupled Modelling: Multiple Fracture Model, BMT2, Coupled Stress-Flow Model, TC1, DECOVALEX-Phase I. CNWRA 92-005, Center for Nuclear Waste Regulatory Analysis, Prepared for Nuclear Regulatory Commission, Contract NRC-02-88-005 (1992).

9   E. Vuillod, A. Thoraval and H. Baroudi. DECOVALEX - BMT2 - Analysis and Results: H-T-M Modeling. ANDRA Report 694 RP CER 92007, ANDRA (Agence Nationale Pour la Gestion des Déchets Radioactifs) (1992).

10 J. Noorishad and C-F. Tsang. Coupled Thermohydromechanical Modelling, Bench-Mark Test 2 (BMT2), Test Case 1 (TC1), DECOVALEX - Phase 1. Earth Science Division, Lawrence Berkeley Laboratory, Berkeley, California (1992).

11 O. Halonen. Bench-Mark Test 2 Without Fluid Flow. Report of DECOVALEX, Phase I, BMT2. Technical Research Center of Finland, Road, Traffic and Geotechnical Laboratory (1992).

12 M.A. Biot. Theory of Elasticity and Consolidation for a Porous Anisotropic Solid. J. App. Phys., **26**, pp.182-185 (1955).

13 J. Noorishad, C-F. Tsang and P.A. Witherspoon. A Coupled Thermal Hydraulic-Mechanical Finite Element Model for Saturated Fractured Porous Rocks. J. Geophys. Res., 89, pp. 10365-10373 (1984).

14   M. Board. UDEC (Universal Distinct Element Code) Version ICG1.5. Vols. 1-3. NUREG/CR-5429. Washington, D.C.: Nuclear Regulatory Commission (NRC) (1989).

15   K.J. Bathe. ADINA-T, A Finite Element Program for Automatic Dynamic Incremental Nonlinear Analysis of Temperatures. ADINA Engineering, Report AE81-2 (1981).

16   O. Halonen. Stability Calculations in Jointed Rock Mass. Technical Research Center of Finland, Publications 57, Espoo (1989).

17   R.E. Goodman. Methods of Geological Engineering in Discontinuous Rocks. West Publishing Company, St. Paul, (1976).

18   G. Abdallah, A. Thoraval, A. Sfeir and J.P. Piguet. Thermal Convection of Fluid in Fractured Media. Int. J. Rock Mech. Min. Sci. & Geomech. Abstr. Vol. 32, No. 5, pp.481-490 (1995).

10. APPENDIX: BRIEF SUMMARY OF CODES USED FOR BMT2

For the convenience of the reader the mathematical and numerical formulations, as well as the major assumptions and constitutive relationships are briefly summarized here for the computer codes employed to model the near-field coupled THM processes embodied in the generic initial boundary value bench mark problem, BMT2. Readers desiring further details should consult [1], Chapters 2-4 of this book, the code descriptions near the end of this book and, above all, the original references listed in Section 9 above.

10.1. MOTIF

MOTIF [2,3] is a three-dimensional finite-element code developed by AECL to model coupled fluid flow, heat transport, solute transport, and mechanical deformation in fractured porous media. The formulation is fully coupled except for the assumption that conversion of mechanical energy to heat energy is negligible. Biot's consolidation theory [13] generalized to the nonisothermal case forms the theoretical basis. Four governing equations for the thermoporoelastic media are solved. These are the generalized Biot's equilibrium equation, fluid mass balance equation in conjunction with the generalized Darcy's law, heat energy balance equation in conjunction with the generalized Fourier's law and solute mass balance equation. Heat transport by conduction, convection and hydrodynamic dispersion are all accounted for. Governing equations are derived for planar fractures by integrating the above four equations over the thickness of the fracture. These governing equations are supplemented by the equations of state for fluid density and viscosity as functions of temperature, pressure and salinity. The rock matrix is assumed to be linearly thermoporoelastic and transversely isotropic. The fractures are modelled using the Barton-Bandis constitutive relationship. Spatial discretization is achieved by means of the Galerkin finite-element method. A weighted finite-difference scheme is utilized in the time domain for solving transient problems. MOTIF employs the Picard iterative technique for solving the nonlinear coupled equations.

The rock matrix is represented by 8-noded solid elements. Fractures are represented by 4-noded planar elements for fluid flow, heat transport and solute transport analyses and by 8-noded planar joint elements for stress analysis. Arbitrary 3-D geometry and time - dependent boundary conditions can be simulated in a MOTIF model. A special feature of the code is that the planar element can represent a rock fracture or fracture zone that is arbitrarily oriented in 3-D space.

10.2. ROCMAS

This 3D simulator is designed by LBL [13] to address coupled occurrences of heat flow, fluid flow and deformation in geologic medium. The coupled thermohydroelasticity formalism of the code is based on a generalization of Biot's 3D consolidation theory for inelastic material and its extension for consideration of coupled energy transport. The constitutive material models include: associated and non-associated strain softening/hardening continuum; dilating and strain softening elastoplastic fractures; ubiquitous elastoplastic fractures; no-tension continuum and hydraulic nonlinearity. Linearization schemes consist of: direct iteration; Newton-Raphson and mixed and modified Newton-Raphson method. In spite of all these general features available in the code the only type of coupled THM simulations based on BMT2 submitted by LBL for inter-code comparison were based on a plane-strain model using a Coulomb friction constitutive relationship with constant stiffness for the initial low thermal loading case.

10.3. UDEC

UDEC (Universal Distinct Element Code) is a two-dimensional distinct element code [14] for coupled thermomechanical analysis for discrete block systems and coupled hydromechanical analysis for flow through fractures. The rock mass is assumed to be an assemblage of discrete (rigid or deformable) blocks interfaced by discontinuities (fractures). For deformable blocks, an internal discretization with constant-strain triangle zones (finite-difference elements) is used for solid block deformation. The governing equations are based on Newton's second law applied to the translational and rotational motion of the blocks. These equations are solved by a central difference time marching scheme. Fluid flow is conducted only through the fractures which are assumed to obey the cubic law for parallel-plate fractures. A fracture is divided into fluid volume domains for analysis. No poroelasticity is considered for the solid matrix. Heat conduction through the solid matrix is the only mode of heat energy transport modelled. Coupled TH and coupled HM processes can be simulated. Constant fluid properties are assumed. The solid block are assumed to be isotropic and linearly thermoelastic. A Coulomb friction joint model with constant stiffness is employed. UDEC calculates total stress for the blocks and effective stress for the fractures.

INERIS, one of the two teams using UDEC for BMT2, has developed a new algorithm [9] known as the "Balloon Scheme" for coupled HM calculations with UDEC. This scheme entails two steps: (1) a hydraulic step in which the flow rate in a (assumed rigid) fracture is calculated from the pressure gradient using the cubic law and (2) a mechanical force balance step for calculating fracture deformation. During step (1) excess (out of balance) fluid volume is stored temporarily in an imagined balloon. During step (2) this stored excess fluid is transferred back to the fracture domain and the fracture will be allowed to contract or dilate according to the calculated effective stress. The code cycles iteratively between these two steps until convergence is attained.

10.4. ADINA-T

ADINA-T [15] is a standard FEM code for heat conduction analysis in 1-, 2- or 3-D space with prescribed temperature or heat flux; or (forced) convective and radiative heat transfer coefficient boundary conditions.

10.5. JRTEMP

JRTEMP is a 2-D FEM code developed by VTT [16] for thermomechanical simulations in an impermeable rock mass with discrete fracture elements. It is a generalization of a similar mechanical equilibrium analysis code described in Goodman's book [11].

O. Stephansson, L. Jing and C.-F. Tsang (Editors)
Coupled Thermo-Hydro-Mechanical Processes of Fractured Media
Developments in Geotechnical Engineering, vol. 79
© 1996 Elsevier Science B.V. All rights reserved.

311

Generic study of coupled T-H-M processes in the near field (BMT3).

P. Wilcock.

AEA Technology, 424.4 Harwell, Didcot, Oxfordshire, OX11 0RA, United Kingdom

Abstract

Groundwater flow through the fractured rock surrounding underground radioactive waste repositories is coupled to mechanical changes resulting from both in-situ and repository induced stresses and by thermal loading resulting from heat generated by the radioactive waste. It is important to model these effects carefully to provide a full picture of the flow through the surrounding rock mass. The DECOVALEX Bench-Mark Test 3, the Near-Field Test Case, has been designed to test the capability of different models to address complex problems on a realistic scale. The test case addresses the coupling between the thermal, hydraulic and mechanical (T-H-M) processes that might affect flow in a 50m region around a repository tunnel. The rock mass is initially in hydro-mechanical equilibrium, at time $t = 0$ a tunnel is excavated and the flow is modelled until time $t = t^*$ when the network has reached a new equilibrium, which must be defined by each research team. After this, thermal loading is started and the thermal effects on the network are modelled for a further 100 years. Eight research teams modelled this test case using either fully coupled models, discrete flow models with approximations for the stress field or continuum models. The results of modelling the test case were pressure, temperature, water flux and stresses at specified output points and lines. A selection of these results are presented and compared. The results show remarkable agreement considering the range of different approaches. An improved understanding of the coupled processes affecting flow around a repository tunnel has been obtained.

1. INTRODUCTION

Many countries are considering the disposal of radioactive waste in deep underground repositories. In such repositories, the surrounding rock forms a natural barrier limiting the release of radionuclides into the human environment. The principal way in which radionuclides may escape is by the dissolution and transport in groundwater flowing through the rock. It is vital, therefore, to understand the flow of groundwater when considering the safety of these repositories. Groundwater flow through rock is coupled to the mechanical changes induced both by in-situ and repository induced stresses and by thermal loading resulting from heat generated by the radioactive waste. It is important to model these effects carefully to provide a full picture of the flow through the surrounding rock mass.

In this chapter the modelling of these coupled processes is discussed. Different approaches to the modelling of an idealised benchmark test are compared.

The DECOVALEX Bench-Mark Test 3, the Near-Field Test Case [1], has been designed to test the capability of different models to address complex problems on a realistic scale. The test case addresses the thermal, hydraulic and mechanical (T-H-M) processes that might affect flow in a 50m region around a repository tunnel. The rock is initially in hydro-mechanical equilibrium, at time t = 0 a tunnel is excavated and flow is modelled until time t = t* when the network reaches a new equilibrium. After this, thermal loading is started. Section 2 provides a complete description of the test case.

Eight research teams modelled the test case using a number of different approaches. The models used were fully coupled models, discrete flow models with approximations for the stress field or continuum models. When modelling coupled processes, the complexity of the numerical models is greatly increased and it is currently not computationally feasible to solve the fully coupled problem on large three-dimensional regions with realistic fracture densities. The test case itself was therefore simplified by specifying a two-dimensional problem but other simplifications also had to be made. Section 3 looks at the various simplifications made by the modelling teams and the impact on the modelling of the problem. An introduction to the methods used by each team is given followed by a general discussion of approaches to modelling coupled processes in the near-field of a waste repository.

The results of modelling the test case were pressure, temperature, water flux and stresses at specified output points and lines. It is not possible to present all the results from the research teams and so a selection of results are presented and compared in Section 4. With the diverse modelling approaches, the results are bound to be different. Reasons for discrepancies between the results are also discussed in this section.

The final section discusses the significant results of this test case and identifies areas for further work.

2. DEFINITION OF AN IDEALISED BENCHMARK TEST

To test the capability of different models to simulate T-H-M coupled processes in a near-field repository environment, an idealised benchmark test, Bench-Mark Test 3 (BMT3) [1], has been set up. In this test case, a repository tunnel is located at a depth of 500m and nuclear waste, which is a source of heat, is disposed in a borehole below this tunnel. The test is defined as a two-dimensional plane strain problem in which the rock mass surrounding the tunnel is assumed to be highly fractured with mechanical properties independent of temperature. The fractures themselves are defined by two parallel, planar surfaces, with an effective hydraulic aperture through which fluid may be conducted. The thermal conductivity and expansion of the rock matrix are assumed to be isotropic and the heat source has the same properties as the rock mass.

2.1. Model geometry
A vertical plane 50m × 50m, located at a depth of 500m below the ground surface has been considered. The geometry, dimensions and coordinate system of this test case are presented in Figure 1. The two-dimensional fracture pattern is generated from a three-

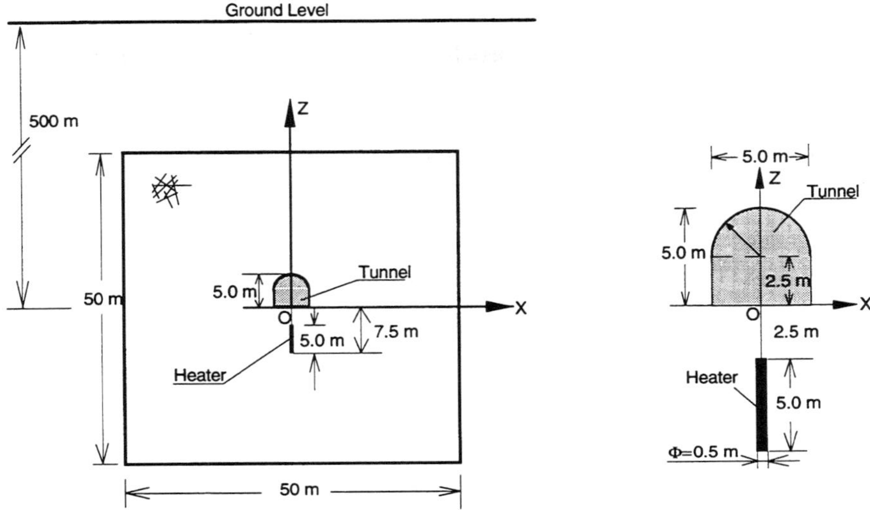

Figure 1: The geometry, dimensions and co-ordinate system of BMT3.

dimensional realisation of a model of the fracture network in the SCV region of the Stripa Mine (Sweden). The traces of fractures intersecting a cross-section through the three-dimensional network were used to define a two-dimensional network. The fractures were then lengthened to maintain the same connectivity as for the original network. Finally a number of modifications were made to the two-dimensional network geometry to remove unphysical features that, in particular would cause difficulties for fully coupled hydro-mechanical numerical models. These modifications are summarised below:

1 Eliminate triangular blocks bounded by one or more joints with length < 2.0m;
2 Remove all joints with length < 0.5m;
3 Remove all isolated joints with length < 1.0m;
4 Terminate end of joint at the point of intersection with another joint if the length of the end is less than 0.5m;
5 Extend joints that approach another joint with separation < 0.5m;
6 Eliminate blocks bound by crossing joints with an angle < 5.0°.

The resulting reference network is shown in Figure 2. This dense network is well connected with a hydraulic conductivity of 1.297×10^{-10}m/s which is consistent with values obtained at Stripa. There are 6581 fractures with apertures ranging between 0.2μm and 45μm.

The problem specification for BMT3 gives each individual fracture location, length and hydraulic aperture for use in a discontinuum analysis of the problem. Conversely, equivalent permeabilities for different square cell sizes are supplied for use in continuum formulations.

2.2. Boundary and initial condiditons

The assumed mechanical boundary condition is that all outward boundary surfaces of

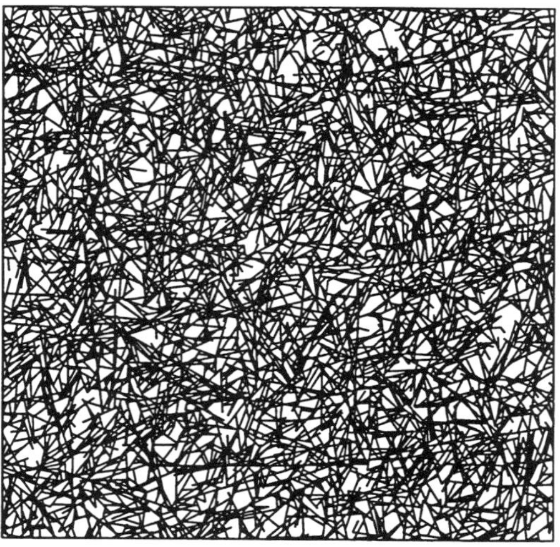

Figure 2: The reference network.

the model are fixed in the normal direction, except the top surface at which the weight of the overburden rocks is applied. The displacements are free at the surface of the tunnel. The mechanical initial condition is defined by the initial isotropic state of stress given by the principal components σ_{xx}^o, σ_{yy}^o and σ_{zz}^o (equation 1) and displacement $(U_x, U_y) = (0, 0)$:

$$\sigma_{xx}^o = \sigma_{yy}^o = \sigma_{zz}^o = -\rho_r g(500 - z) \tag{1}$$

where ρ_r is the density of the rock, z is measured vertically upwards with the coordinate origin at the base of the tunnel, and compressive stress is taken to be negative.

The hydraulic boundary conditions consist of no flux along the two vertical surfaces and the bottom surface, as well as a constant water pressure, corresponding to a depth of 475m, applied along the top surface. Initially there is no flow into the region and hydrostatic pressure everywhere. When the tunnel is excavated, a constant atmospheric pressure is applied around the tunnel surface. All fractures are assumed to be fully saturated.

The initial thermal conditions consist of a constant initial temperature (T_0) of 27°C throughout the model. This initial temperature is held fixed along the top surface of the model, while both vertical surfaces and the lower boundary surface are assumed adiabatic. The tunnel surface is assumed to behave as a convective boundary upon excavation in which the flux condition is given as

$$\phi_t = H \cdot (T_{wall} - T_g) \tag{2}$$

where ϕ_t is the thermal flux across the tunnel surface (W/m^2), $H = 7W/m^2 \cdot °C$ is the coefficient of surface heat transfer, T_{wall} is the wall temperature (°C) and $T_g = 27°C$ is the constant tunnel temperature.

The heat source simulating the waste canister is assumed to decay exponentially with time according to the following relation

$$Q(t) = Q_0 \cdot \exp(-\beta \cdot t) \tag{3}$$

where $Q(t)$ is the heat flux at time t (W/m^3), $Q_0 = 0.47 \times 10^4$ W/m^3 is the initial heat flux and $\beta = 0.02$ 1/year is the heat decay coefficient. The total power output of the heat source, $Q_v(t)$ over the borehole is given by

$$Q_v(t) = \int\int_{source} Q(t)dv = \int_{-0.25}^{0.25}\int_{-7.5}^{-2.5} Q(t)dzdx \tag{4}$$

2.3. Material properties and parameters

Material properties and other data for the thermal, hydrological and mechanical behaviour, as given in the problem specification are given below

Rock matrix

Young's Modulus, E	$= 60$ GPa
Poisson's ratio, ν	$= 0.23$
Density, ρ_r	$= 2670 Kg/m^3$
Porosity, θ	$= 10^{-4}$
Uniaxial Compressive Strength, σ_c	$= 200$ MPa
Tensile Strength, σ_t	$= 10$ MPa
Thermal Conductivity, λ	$= 3$ W/m·°C
Specific Heat Capacity, C_s	$= 900$ J/Kg·°C
Thermal Expansion Coefficient, α	$= 9.0 \times 10^{-6}/°C$

Fractures

Normal Stiffness, K_n	$= 100$ GPa/m
Shear Stiffness, K_s	$= 10$ GPa/m
Cohesion, C	$= 0.1$ MPa
Friction Angle, Θ	$= 30°$
Dilatancy Angle, i	$= 0°$
Tensile Strength, σ_c^t	$= 0$ MPa

Fluid

Both the fluid density, ρ_f, and dynamic viscosity, μ, are assumed to be temperature dependent as follows.

$$\rho_f = \rho_f^0 \left[1 - \epsilon(T - T^0)\right] \tag{5}$$

where

ρ_f^0	$= 1000$ Kg/m^3
ϵ	$= 6 \times 10^{-4}$ 1/°C
T^0	$= 20$ °C.

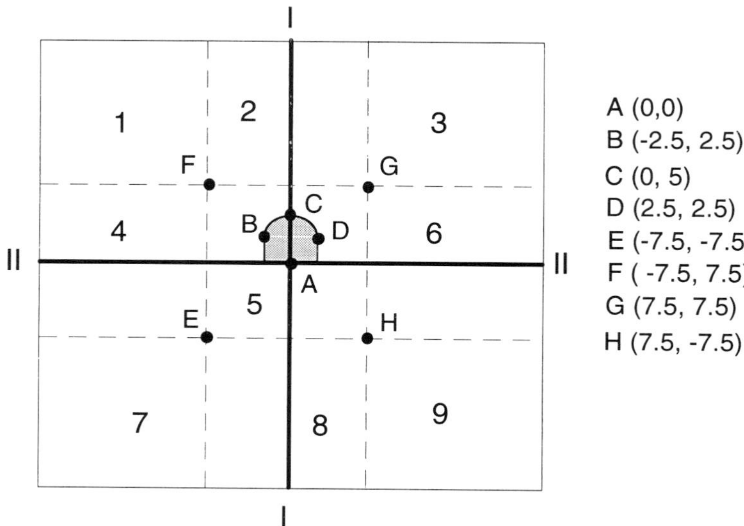

Figure 3: Output specifications for BMT3.

and

$$\frac{1}{\mu} = \frac{1}{\mu_0}\left[1 + \gamma(T - T^0)\right] \tag{6}$$

where
$$\mu_0 = 1 \times 10^{-3} \text{ Ns/m}^2$$
$$\gamma = 3.2 \times 10^{-2} \text{ 1/K}$$

Fluid heat capacity, $C_f = 4200$ J/Kg·°C

2.4. Loading sequence

There are three phases to the test case as follows:

I   The system is in initial hydromechanical equilibrium

II   The tunnel is opened at time t=0 and time is taken for a a new hydromechanical equilibrium to be reached.

III   Thermal loading is started and continues for 100 years

2.5. Results and output specification

To compare the results obtained from different research teams, plots of temperatures, water flux, displacement components, stresses and porosities were requested at different times throughout the three loading sequences. Figure 3 shows the monitoring points, lines and the regions where the specified output is required. These will be referred to when describing the results.

3. COMPARISON OF DIFFERENT MODELLING APPROACHES TO BMT3

Bench-Mark Test 3 was modelled by eight research teams (INERIS, CNWRA, NGI, VTT, AEAT, ITASCA, CEA/DMT and KPH, see Chapter 2 for their full names) using a number of different approaches. This test case posed a significant challenge to the modelling of coupled processes as a realistically sized fracture network with a realistic density was considered. The problem was simplified, in the first place, by reducing it to two dimensions and research teams were requested to make any further simplifications that were necessary to model the test case.

Both discrete and continuum modelling approaches were adopted to analyse the test case. Three research teams used three different continuum based computer codes which applied either finite-element or finite-difference methods to discretise the models. The finite-element codes THAMES and CASTEM 2000 were used by KPH and CEA/DMT, respectively, and the finite-difference code FLAC was used by ITASCA. Four research teams, INERIS, CNWRA, NGI and VTT, employed a discontinuum approach in which the discrete element code UDEC was used. The other team, AEAT, adopted a discrete fracture network approach using the fracture network code NAPSAC.

Different methods were used by each team to represent the problem, so that even teams using the same code have different representations of the problem. With this in mind, this section first gives an introduction to the methods used by each team to represent this test case. Detailed descriptions of the computer codes used are given as appendices to this book and are not included here. The section concludes with a general discussion of the approaches to modelling coupled processes in the near-field.

3.1. Discrete element models

INERIS research team

The INERIS research team used a discontinuum approach in which the discrete element code UDEC was used [2-4]. An attempt at generating the entire reference network using UDEC took around $2\frac{1}{2}$ days on a HP750 PC. The size of files produced and the memory required to model BMT3 using a sensible UDEC mesh were too great if the original network was used and so the network had to be simplified.

The statistical distributions of fracture length, orientation and hydraulic aperture in the reference network were investigated. A study of the distributions of these parameters on smaller regions of the network showed that distributions were homogeneous in space. Further studies showed no correlation between length, orientation or aperture.

A simplified model consisting of three nested regions with different fracture patterns was constructed (Figure 4). The three regions were defined as

- an inner region with area defined by $-10\text{m} \leq \text{x} \leq 10\text{m}$ and $-10\text{m} \leq \text{z} \leq 10\text{m}$,

- a transitional region with area defined by $-15\text{m} \leq \text{x} \leq -10\text{m}$, $10\text{m} \leq \text{x} \leq 15\text{m}$ and $-15\text{m} \leq \text{z} \leq -10\text{m}$, $10\text{m} \leq \text{z} \leq 15\text{m}$ and

- an outer region which is the remaining part of the model domain.

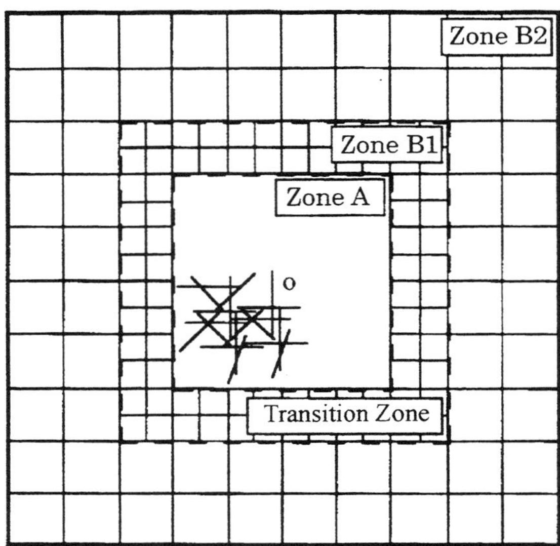

Figure 4: Representation of the simplified fracture network used by the INERIS team.

The fracture network in the inner region was taken directly from the reference network. Two options were considered. In model 1, the inner region consisted of all the fractures from the reference network and in model 2 only fractures with apertures greater than 3μm were included. Both these models were used in the analysis of the test case to investigate to what degree the connectivity of the network would be affected by deleting small aperture fractures.

As there was no preferred orientation of fractures in the reference network, two orthogonal sets of fractures, with horizontal and vertical directions respectively, were introduced into the transitional region, with spacing 2.5m, and into the outer region, with spacing 5m. The equivalent properties for the rock matrix were calculated by performing a numerical test on a 5×5m region of the reference network, using UDEC. This gave an equivalent Young's modulus of intact rock to be $E_e = 21.5$ GPa. Using Singh's relations [5]

$$\frac{1}{E_e} = \frac{1}{E_i} + \frac{1}{SK_n}, \qquad \nu_e = \nu_i \frac{SK_n}{E_i + SK_n}, \qquad (7)$$

where E_i is the given Young's modulus of the intact rock (MPa), S is the fracture spacing (m) and K_n is the fracture normal stiffness (MPa/m) and taking $E_i = 60 \times 10^3$MPa, and $K_n = 10^5$ MPa/m as given in the problem specification, the spacing of the two sets of fractures considered as the mechanical equivalent of the reference network was found to be 0.337m. This gives an equivalent Poission's ratio ν_e of 0.083, using $\nu_i = 0.23$, as given in the specification. Equations 7 are used to derive equivalent Young's moduli and Poisson's ratios for the transitional region and outer region, using the respective fracture spacings in this region. Similar methods were used to find the equivalent rock properties for the inner region in model 2 (where fractures with aperture of 3μm were discarded). The values of these properties are given in Table 1.

Table 1
Equivalent rock properties for simplified fracture network.

	Inner region (model 2)	Transitional region	Outer region
Young's modulus, E (GPa)	30.346	23.622	22.557
Poisson's ratio,ν	0.116	0.091	0.087

The apertures of the fractures in the transitional and outer regions were calculated so that the resulting network would have permeability equivalent to the reference network. The average aperture of the reference network is 4.576μm. For a system of fractures with spacing S and aperture e, the permeability for an equivalent continuum, k, would be,

$$k = \frac{\rho_f g}{12\mu} \frac{e^3}{S} \tag{8}$$

where ρ_f is the density of fluid, g is gravitational acceleration and μ is the dynamic viscosity. Assuming the viscosity remains constant, permeability is proportional to e^3/S and this leads to the aperture of fractures in the transitional region, being 8.91μm and in the outer region being 11.23μm.

To enable UDEC to cope with the specified test case, a modification was made so that the number of fracture sets for which individual properties could be assigned was increased from 10 to over 500. This enabled each fracture in the inner region of both models to be assigned the aperture specified in the reference network.

The resulting models had 1047 fractures, 1950 blocks and 24613 finite-difference zones for model 1 and 608 fractures, 677 blocks and 24391 finite-difference zones for model 2.

CNWRA research team

The CNWRA research team adopted a discrete element approach in which the UDEC code was used [6]. Due to the large number of fractures, several assumptions and simplifications were made for the distinct element model using UDEC. The first simplification was to assume a vertical plane of symmetry through the tunnel axis based on the symmetrical boundary conditions, the random nature of the fracture distribution and the observation that, on a large scale, the equivalent hydraulic data provided for the problem did not vary much in the x and z directions. The second simplification was to reduce further the number or density of the fractures to make the problem computationally feasible. Two separate cases were run with the UDEC code.

The first case (Case A) included a representative fracture distribution only for the immediate region around the tunnel, while the outer region was approximated using equivalent matrix and fracture properties. The inner region around the tunnel contained a fairly dense set of fractures as simplified from the original problem by including only those fractures longer than 2m. The actual fracture stiffnesses and strength properties given in the problem specification were used, along with the intact rock properties for the blocks in this region. As UDEC cannot normally assign apertures and other fracture properties to individual fractures, representative fracture apertures for this inner model had to be defined. It was decided to divide the fractures into sets according to their orientation and assign each fracture set an average aperture which was calculated as the median of the specified apertures within each fracture set.

The outer region extending to the model boundaries contained a uniform set of evenly spaced horizontal and vertical fractures with an assumed spacing of 5m. These outer fractures were necessary primarily to allow connectivity of the flow from the top boundary where a constant fluid pressure was applied to the inner region containing the tunnel, since UDEC considers flow through fractures only. Appropriate mechanical and hydrologic properties were developed for the rock mass in the outer region to represent the original problem. The fracture stiffness for the horizontal and vertical fractures were set to be the same as for the fractures within the inner region, however, the intact elastic rock properties of the square blocks were reduced from the values specified in the problem to make the overall mechanical stiffness of this outer region equivalent to that of the original problem.

A value for the reduced intact rock modulus was determined from Singh's relations (equations 7). A simple numerical biaxial test on a small portion of the reference network was conducted, resulting in a value of 1.2×10^4 MPa for E_e. Taking $K_n = 10^5$ MPa/m, as given in the problem specification and $S = 5$m, E_i was found to be 1.23×10^4 MPa. In addition to reducing the intact elastic modulus within this outer region, the fracture hydraulic apertures were also modified so as to realistically represent the equivalent permeability. Using equation 8 and the average equivalent permeability of 0.13×10^{-9} m/s given in the problem definition, an equivalent aperture e of 9.26μm was obtained.

To reduce the complexity of the blocks for corner rounding and zoning, the horizontal and vertical fractures of the outer region were allowed to extend through the inner region. In the inner region, these fractures were assigned artificial properties to ensure that they were hydraulically and mechanically inactive in this region. This model contained 211 fractures, forming 337 distinct element blocks and 1540 finite-difference zones.

The second case again assumed vertical symmetry but this time modelled a representative fracture distribution throughout the whole 25×50m symmetric region. This model still did not incorporate the fractures specified in the original reference pattern. Only those fractures longer than 1.5m were included in the model. Again, a uniform set of hydraulically and mechanically inactive artificial fractures were put into the model to facilitate zoning. Thus, the second model contained 1799 fractures forming 2497 discrete element blocks. These blocks were discretised into 6001 finite-difference zones.

NGI research team

A discontinuum approach was applied in which the UDEC discrete element code was used [7]. To reduce the complexity of the problem to one which was feasible for a UDEC simulation, symmetry was assumed about the vertical axis. It was acknowledged that this property of symmetry was not actually correct but allowed for a more detailed representation in the models being used. Analyses were performed for two symmetric models, one based on the fracture geometry of the right half and the other one the fracture geometry of the left half. It was decided to distinguish between areas close to the excavation, using a relatively fine fracture representation, and areas remote from the excavation where a coarser system is used. The area of the inner region was centered on the tunnel and is defined using x-z coordinates by -10m $\leq$ x ≤ 10m and -7.5m $\leq$ z ≤ 12.5m, which is a shift of 2.5m upwards from the inner region defined by CNWRA. The approach for extracting a representation of a network for the inner region was identical to the approach adopted

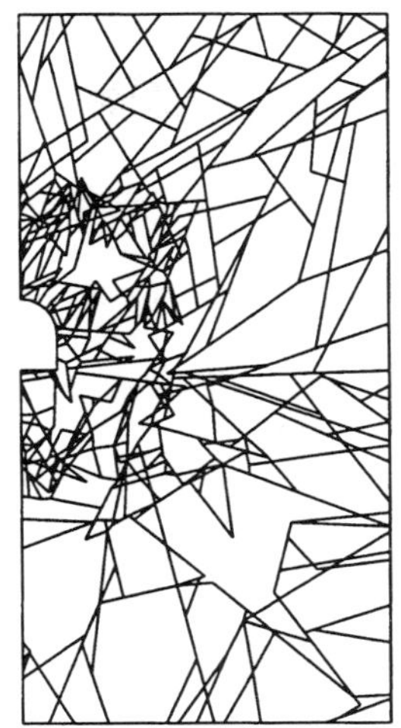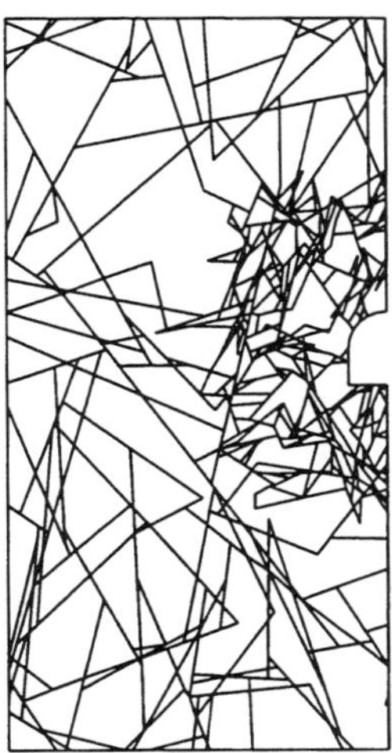

Figure 5: Fractures and blocks for the right half and left half NGI models.

by CNWRA, described above. The network used in the outer region was obtained in the following way:

- All fractures which have an end point inside the area defined by $-25m \leq x \leq -14m$ and $-25m \leq z \leq -14m$ were extracted from the reference network.

- The extracted set was mapped onto the total area of the test model ($-25m \leq x \leq 25m$ and $-25m \leq z \leq 25m$), which results in a sparser network of fractures over the domain.

- From this fracture set, all fractures having an end point inside $-8m \leq x \leq 8m$ and $-8m \leq z \leq 8m$ were excluded.

The resulting models consist of 514 fractures and 519 discrete blocks for the right-half model and 510 fractures and 505 discrete blocks for the left-half model. The fractures and blocks for the models are shown in Figure 5. The total number of finite-difference zones in the right-half model was 1587 and in the left-half model 1581.

VTT research team

The VTT research team used the discrete element code UDEC to model BMT3 [8]. Vertical symmetry was assumed, however this did not sufficiently reduce the number of fractures and so further simplifications had to be sought. A simplification based on eliminating small aperture fractures was considered but rejected because this could result in the change of continuous flow pathways and therefore a change in the conductivity of the network. The simplification was accomplished by modifying orientation angles of longer fractures of larger apertures in an attempt to preserve the continuous flow paths through the network. The number of fractures in the close vicinity of the tunnel was increased, as described below, to more accurately represent the water flow into the tunnel.

The left-hand side of the network was modelled and this was divided into an inner area (defined by $0m \leq x \leq 12.5m$ and $-10m \leq z \leq 15m$) and an outer area. Six fracture sets were defined by their orientation angles with the x-axis which were 75°, 45°, 15°, -15°, -45° and -75°, respectively. First, the six fracture sets were generated over the whole area with an average spacing of 2.1m. In the inner region, the six sets were generated again giving a final average spacing on that region of 0.8m. The resulting network has 814 fractures and 496 discrete blocks, which was modelled with a discretisation of 1308 finite-difference zones.

The apertures of the fractures in the inner and outer zones were calculated from an equivalent permeability for the network of $k = 0.1322\times10^{-9}$m/s, using equation 8, and were found to be 6.99×10^{-6}m for the outer region and 5.06×10^{-6}m for the inner region. The residual aperture (the minimum aperture that a fracture can have) was set to 3×10^{-6}m. All material properties for the rock matrix and fracture stiffness remained the same as given in the specification.

3.2. Discrete fracture network models

AEAT research team

A discrete fracture network approach was adopted by the AEAT team [9]. The NAP-SAC fracture network code was used to model flow after excavation of the tunnel up to equilibrium. NAPSAC has been designed to simulate flow through large complicated networks and so the 6580 fractures before tunnel excavation and 6559 fractures after tunnel excavation were represented explicitly. The network was discretised by assigning a pressure node at each fracture intersection, so that pressure was calculated at a total of 23118 nodes.

This was the only modelling approach that did not require a simplification of the geometry. This was made possible at the expense of not solving the fully coupled hydro-mechanical problem. Instead, a one-way coupling, relating the aperture of the fractures to the stress field around the tunnel, was assumed. Further approximations were made by assuming that the stress field can be represented by an idealised continuum analytical solution. (This has been shown to be a reasonable approximation to the discrete stress field in studies for the OECD Stripa project [10-11].) Three different stress-aperture couplings were used to model the effect of stress on the fracture apertures and the subsequent effect on the flow field.

The specified apertures of the reference network are those that apply to the in-situ network under the equilibrium stress field before the tunnel has been drilled. To calculate the analytical stress field, the tunnel is taken to be circular, with radius a = 2.5m centred at x = 0.0 and at a depth of 497.5m. The Kirsch 2-D linear elastic solution (equation 9), giving components of the stress field in plane polar co-ordinates, (r, τ), is used:

$$\sigma_{rr} = \sigma_0 \left(1 - a^2/r^2\right),$$
$$\sigma_{r\tau} = 0,$$
$$\sigma_{\tau\tau} = \sigma_0 \left(1 + a^2/r^2\right). \tag{9}$$

where $\sigma_0 = \rho_r gd$ is the magnitude of the initial stress field at depth d. The tangential component of stress, $\sigma_{\tau\tau}$, has maximum $2\sigma_0$ on the surface of the tunnel and tends to σ_0 as the distance from the tunnel increases and the radial component of stress, σ_{rr}, has minimum zero on the surface of the tunnel and also tends to σ_0 as the distance from the tunnel increases. It should be noted that as this circular tunnel lies inside the tunnel defined in the specification, the effect of stress changes on the fractures close to the bottom corners of the tunnel will not be as well represented as that on the fractures in the rest of the network.

For each section of fracture between intersections, the change in normal stress (from σ_{n_0} to σ_n) across the fracture due to the tunnel excavation is calculated at the centre point of the section, using the above solution. The change in stress is converted to a change in aperture using one of following three stress-aperture couplings:

$$
\begin{aligned}
&T = \text{constant}, \\
&T/T_0 = (\sigma_n/\sigma_{n0})^{-0.5}, \\
&e - e_0 = \max\left(-(\sigma_n - \sigma_{n_0})/\text{RKN}, e_{\min}\right).
\end{aligned}
\tag{10}
$$

In the first coupling the transmissivity, T, of each fracture is assumed to be unchanged by the excavation of the tunnel and so the effect on the fracture network of the change in stress is neglected. The second coupling is a compliance law that has been used to model the effects of tunnel excavation in experiments carried out at the Stripa Mine [12]. As the transmissivity of a fracture is dependent on the cube of the aperture and the power law exponent is 0.5, as measured at Stripa, it is seen that a very large change in stress is required for this model to produce a significant change in aperture. The third law relates the change in aperture (from e_0 to e) directly to the change in stress and the fracture normal stiffness, RKN. This can produce quite a significant change in aperture for the range of stress changes being considered. It was necessary to define a minimum aperture to ensure that all apertures remain positive. A minimum aperture of 1μm was chosen, as there would be very little flow through fractures with apertures below this.

3.3. Finite-element continuum models

CEA/DMT research team

The CEA/DMT research team used a continuum approach to model this test case [13]. First, the governing equations were defined. The rock matrix was assumed to be a linearly isotropic elastic material obeying Hooke's law:

$$
\sigma_{ij} = \lambda\epsilon_{kk}\delta_{ij} + 2\mu\epsilon_{ij}
\tag{11}
$$

where σ_{ij} and ϵ_{ij} are stress and strain tensors and λ and μ are Lamé's coefficients. Poiseuille's law was used to describe fluid flow through a single fracture, written

$$
\mathbf{v} = -\frac{e^2 g}{12\nu}\nabla(p/\rho_f g + z)
\tag{12}
$$

where $\mathbf{v}$ is the average velocity vector of fluid flow inside the fracture, e is the aperture, g is the gravitational acceleration, ν is the kinematic viscosity and p is the fluid pressure.

From simple statistical analysis of the reference network, the fracture orientation distribution was found to be roughly uniform but the length and aperture distributions were

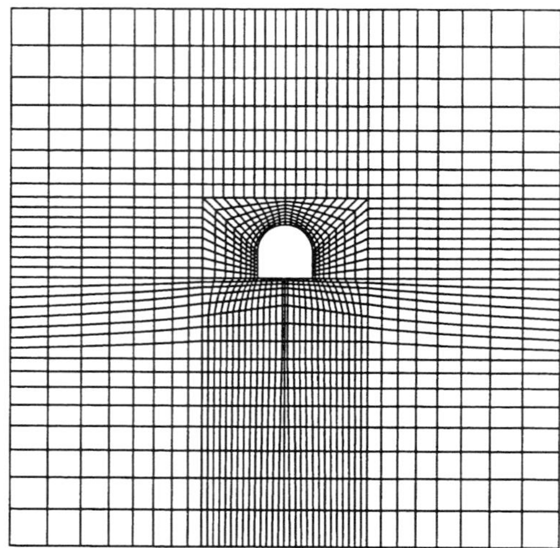

Figure 6: Finite-element mesh used for the CEA/DMT models.

skewed. The fractured rock mass was taken as a continuous material governed by Biot's anisotropic poroelasticity equations [14]. A homogenisation process using Oda's crack tensor concept [15] was applied. An equivalence relation between Oda's crack tensor theory approach and Biot's poroelasticity concept was established so that the permeability, porosity and strain tensors in the governing hydro-mechanical equations were homogenised for the fractured media giving equivalent properties to be used in the model.

Scale dependence and spatial variability of the equivalent permeability were investigated. Using both nested domains and spatial partition techniques, the variability of the equivalent principal permeability tensor with respect to the homogenisation scale was investigated. It was found that it was only when the homogenisation scale was larger than 25m that the two principal values of the permeability tensor and the inclination angle γ of the principal axes to the x-z axes reached roughly stabilised values. It was therefore established that the Representative Elementary Volume (REV) should be about 25×25m. A 15×15m scale may be taken as a lower bound of the REV.

The anisotropy in the equivalent permeability and compliance tensors eliminated any possibility of using symmetry to reduce the size of the geometric model. The temperature field was assumed to be independent of the hydro-mechanical response of the rock mass and was first determined. The pre-determined temperature field was then taken as input for the coupled hydro-mechanical analysis. Two cases were considered. An isotropic case in which the material data were selected empirically and an anisotropic case in which the material data for the equivalent properties were derived by homogenisation on a scale of 50m. A finite-element mesh was established with 1344 quadrangular elements and 4216 nodes (see Figure 6) and used for both cases.

KPH research team

To analyse BMT3, a continuum model was assumed in which the code THAMES was used [16]. THAMES is a finite-element code which can simulate fully coupled T-H-M problems in a continuum. A homogenisation process must be taken to replace the fractured medium with a continuum of equivalent properties representing the T-H-M behaviour of the fractured medium. As well as the formulations for the equivalent properties, the size of the REV over which the homogenisation is valid must be determined.

According to the specification of BMT3, the flow takes place only through fractures and the rock matrix is assumed impermeable. The cubic law for flow through parallel plates (equation 12) was used to model the flow through the fractures. Using the Oda's Crack tensor theory approach, the equivalent permeability tensor and equivalent strain and compliance tensors were derived.

To determine an REV, the fracture data from the test case specification was used to provide statistical information on the fracture network. This information was then used to randomly generate an equivalent network for different volumes and thus compute the components of the crack tensor for different volumes. The smallest volume giving a stable value was defined to be the REV. In this test case, the REV value was found to be a 10m square.

The equivalent properties for each finite-element were derived from the fracture geometry of a surrounding area of the REV size with the same centre as the element. No symmetry was assumed for this model because of the apparent anisotropy in the equivalent properties of the medium.

3.4. Finite-difference continuum model

ITASCA research team

ITASCA considered the problem of 6580 fractures to be too large to model explicitly. By plotting the frequency of fracture orientations, it was found that there were no dominating orientations among the specified fractures. Further, it was noted that all orientations were represented over the entire problem area. Based on these observations, it was decided to analyse the problem as a plane strain continuum using the two-dimensional finite-difference continuum code FLAC [17]. Vertical symmetry was assumed as the difference in hydraulic conductivities for both sides of the network was low, and the FLAC grid, consisting of 735 finite-difference zones, constructed to model the problem is shown in Figure 7. This modelling approach required the continuum mechanical, hydraulic and thermal properties of the rock mass to be derived from those specified properties.

Two different approaches were used to obtain the continuum mechanical properties of the rock mass. A numerical biaxial test was performed on a block of the reference network, 10m high and 20m wide, using the UDEC code. The rock mass strength obtained by this method turned out to be very low. The second method was a Rock Mass Rating (RMR) [18] based on fracture observations in a $20 \times 20\text{m}^2$ block of referenced network, centered on the tunnel. As it was felt that the strength parameters obtained by this method were not applicable to the type of rock modelled in this test case, the RMR-rating was assumed equal to the RMS (Rock Mass Strength) [19] system. This approach gave results comparable to those obtained from the ultra large Stripa granite core [20].

Horizontal and vertical hydraulic conductivities for different levels of discretisation, with length scales ranging from 5 to 50m were given in the problem specification. The

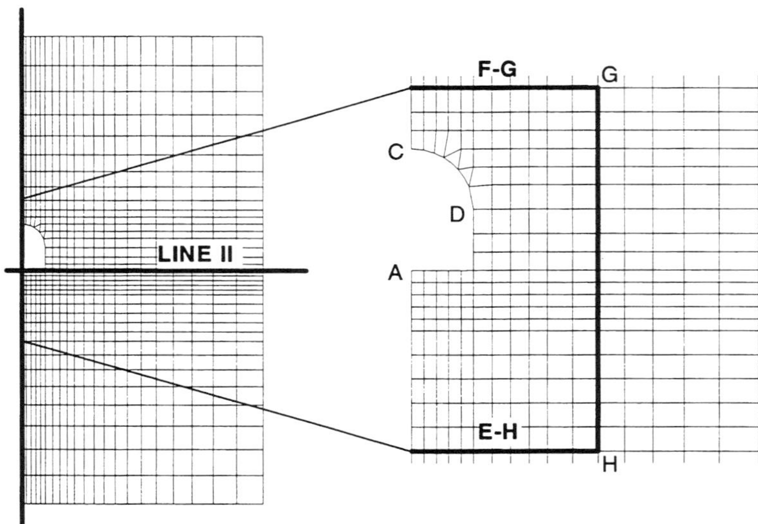

Figure 7: Finite-difference grid used for the ITASCA model.

conductivities calculated on each quarter of the network were used in the continuum analysis of the problem. These conductivities, K, were converted to 'permeabilities', Fk, used by FLAC, by the following equation

$$Fk = \frac{K}{\rho_f g}. \qquad (13)$$

This gave some distinction between the hydraulic properties of the upper and lower half of the region. The given porosity of the solid intact rock was converted to an equivalent porosity of a continuum by adding the void space from the fractures to the pore volume. An average horizontal and vertical spacing (S_x and S_z respectively) between fractures was calculated from the hydraulic conductivity using equation 8. The void space from the fractures, V, was calculated using

$$V = a \left(\frac{1}{S_x} + \frac{1}{S_z} \right) \qquad (14)$$

where $a = 2.2 \mu m$ is the most frequent aperture. This resulted in porosities of 1.5×10^{-4} for the upper half of the region and 2×10^{-4} for the lower half of the region.

The specified thermal properties were used for the continuum rock mass.

The mechanical influence on the rock mass hydraulic properties were considered by assuming a linear relation between the change in aperture and the change in normal stress. This was converted to the following relationship between the conductivity of the rock mass and the change in normal stress

$$K_{xx} = \left(\frac{\rho_f g}{12\mu} \right) \frac{(a_0 + (\sigma_{zz} - \sigma_{zz}^0)/K_N)^3}{S_x}, \qquad K_{zz} = \left(\frac{\rho_f g}{12\mu} \right) \frac{(a_0 + (\sigma_{xx} - \sigma_{xx}^0)/K_N)^3}{S_z} \qquad (15)$$

where $a_0 = 2.2\mu m$ is the initial aperture. This value was also taken as the residual aperture, the value below which no aperture is allowed to decrease. K_N is the fracture normal stiffness, σ^0_{xx} and σ^0_{zz} are the initial stresses in the x and z directions, respectively. To avoid large conductivity contrasts, the change in conductivity was limited to 10 times the initial conductivity. This restriction was believed to have only a small effect on the overall flow rates.

Due to the symmetry assumption, the total heat flux was reduced by half. The symmetry line made it possible to apply the heat source as a boundary flux with a height of 5m. In the problem statement, the water density and the dynamic viscosity were supposed to be temperature dependent. Therefore, the conductivity of the rock mass will be temperature dependent. Substituting equation 13 into equation 15 and using the relation between viscosity and temperature defined in the previous section, the following relation is obtained

$$Fk_{xx} = \frac{\left(1 + \gamma(T - T^0)\right)\left(a_0 + (\sigma_{zz} - \sigma^0_{zz})/K_N\right)^3}{12\mu^0 S_x},$$

$$Fk_{zz} = \frac{\left(1 + \gamma(T - T^0)\right)\left(a_0 + (\sigma_{xx} - \sigma^0_{xx})/K_N\right)^3}{12\mu^0 S_z} \tag{16}$$

showing that the 'permeability' will be both stress-dependent and temperature-dependent. Transportation of thermal energy with water was not modelled, but the temperature dependency of the fluid density and dynamic viscosity was modelled as specified. A bulk modulus of the water of 2×10^8 Pa corresponding to 0.45% air content was used for the analysis.

3.5. Discussion of modelling approaches

The continuum, discontinuum and discrete fracture network approaches presented above all address the same basic issues, namely, i) how to represent the fractured medium geometrically and ii) how to represent the thermo-hydro-mechanical properties of the fractured medium and satisfy all the conservation laws at the same time. Geometrically, the three approaches give three levels of representations for a fractured rock mass. The continuum approach approximates the whole system as an intact material continuous everywhere. The discrete element approach represents the rock blocks and fractures as independent structural units of a fractured rock explicitly. The discrete fracture network model considers the connected fracture space only. The physical models must then be formulated according to these geometrical assumptions.

For the continuum approach, the fractured medium is modelled as an equivalent porous medium with properties representing the thermo-hydro-mechanical behaviour of a fractured medium. One approach to getting the equivalent properties of the continuum is a homogenisation process in which an REV, over which the homogenisation is valid, must be determined. Computationally, the size of the REV depends on the governing equations, especially the constitutive equations used in the homogenisation. For fractured hard rocks, there may exist fractures at a range of different length scales and a REV may become very large or might not exist at all, in which case this process may not be used. An alternative approach is to derive equivalent properties by taking an average over the ensemble of possible realisations, which does not require an REV to be defined.

Another problem with the REV concept is that for the same problem, the REV for hydraulic analysis and the REV for deformation analysis may differ. For example, if Oda's crack tensor approach is used, then the equivalent compliance tensor is obtained by using all the fractures, whether they contribute to the flow field or not, since they all contribute to the deformability of the idealised continuum. This may result in the REV being smaller than it should be for flow or the estimated hydraulic permeability being higher than it should be, since no account is taken of the connectivity of the fractures. In addition to this, REVs are not only scale dependent but may also depend on the deformation of the rock mass or may not be constant with time. The mathematical models would be more complex if this was taken into account.

The discrete fracture network approach considers the void space between connected fractures which is occupied by the fluid and neglects the permeability of the rock matrix. Stresses are estimated by an analytical solution and the main focus is in the calculation of fluid flow through the fractures. The effects of the flow field on the rock deformation and temperature were not considered. This was compensated by a detailed flow analysis through a fully represented fracture network, often the dominant apsect of the T-H-M processes in hard rocks. The 'one-way' coupling limitation may present some problems if the effect of thermal gradients and rock deformation on the flow field is significant. These physical processes could be represented more directly within this approach, although some simplifications are implicit in neglecting the explicit representation of the rock matrix between fractures.

For the four UDEC models, the main problem was how to simplify the fracture network to enable the numerical implementation of the discrete element approach. The approaches to network simplification were similar in that an outer region was used with the fracture pattern in this region being either replaced by: i) two idealised sets of orthogonal persistent fractures, or ii) a simplified fracture network with a smaller fracture population. The hydro-mechanical properties of the idealised or simplified regions were obtained by either numerical testing small test models or empirical means to estimate the properties of the equivalent rock matrix in the outer regions. Two problems may arise with this kind of simplification. Firstly, the conservation laws for the simplified outer regions and across boundaries between inner and outer regions may not be satisfied after simplification. Also, the empirically estimated equivalent properties of the rock matrix and artificial fractures for the outer region may not represent the equivalent properties for the simplified area.

The nature of the simplification is important. Vertical symmetry was used by the ITASCA, CNWRA, NGI and VTT models. The homogenisation carried out by CEA/DMT and KPH and the discrete fracture network calculations on the original fracture network carried out by AEAT showed that no vertical symmetry exists, either hydraulically or mechanically. The NGI, CNWRA, INERIS and VTT models all simplified networks by discarding fractures. If the eliminated fractures are all isolated fractures, then their elimination will only affect the deformability of the equivalent rock mass, not the permeability. If, however these eliminated fractures form part of the connected network of fractures, their elimination may change the flow pattern and the equivalent permeability of the model.

4. THE RESULTS OF MODELLING BMT3

As described in section 2, BMT3 has three loading phases: initial hydro-mechanical equilibrium; tunnel excavation plus time until new hydro-mechanical equilibrium and thermal loading. The approaches described above were used to model these sequences with the exception of the NGI and AEAT teams, who did not model the thermal loading sequence. In this section, a selection of the results for each stage are compared in turn and the following abbreviations have been used for the different cases of each model.

CEA	anisotropic case of the CEA/DMT model
KPH	the KPH model
ITASCA	the ITASCA model
CNWa	coarse model of the CNWRA model
CNWb	detailed model of the CNWRA model
NGIa	right half of the NGI model
NGIb	left half of the NGI model
INEa	the INERIS model with aperture not less than $3\mu m$ in region A
INEb	the INERIS model with all fractures in region A
VTT	the VTT model
AEAa	the AEAT model with the effect of stress neglected
AEAb	the AEAT model with second stress-aperture coupling
AEAc	the AEAT model with third stress-aperture coupling

4.1. Initial hydro-mechanical equilibrium

The first loading sequence consists of establishing initial hydro-mechanical equilibrium in the model without the tunnel. With some approaches this was a case of specifying the appropriate input parameters to the numerical models. For example, in the AEAT approach which uses the reference network without simplification, this initial condition was imposed by specifying the initial pressure at all pressure nodes and inputing the apertures of each of the fractures to be the in-situ apertures specified in the model. For the continuum models, the initial properties of the network were derived as described in the previous section. The KPH team used THAMES to calculate the initial steady state for their model.

For approaches using the UDEC code it was necessary to consolidate the numerical models used under in-situ conditions. The initial in-situ stress varies linearly from 12.44 MPa at the top of the model to 13.75 MPa at the bottom of the model. Similarly, the fluid pressure within fractures varies linearly from 4.66 MPa at the top to 5.15 MPa at the bottom of the model. The effective stress across the joints is the total mechanical stress minus the fluid pressure. No water flow through the joints was considered at this stage of the analysis.

This loading process changes the original apertures input to UDEC. The CNWRA team applied the in-situ stress field so that joint normal displacements were inhibited which meant that at hydro-mechanical equilibrium the fracture apertures remained the same as the input values. The INERIS team made a new development to UDEC which reset the apertures to the specified ones at the end of loading. The NGI team made a series of trial and error analyses by varying the input apertures so that the apertures obtained at the end of loading were in good agreement with the reference data. All displacements were initially set to zero.

4.2. Hydro-mechanical equilibrium after tunnel excavation

The second loading sequence corresponds to excavating the tunnel at time t = 0 and allowing the model to reach a new hydro-mechanical equilibrium as a result of the excavation-induced changes. Prior to tunnel excavation, the NGI team reduced the pore pressure in the non-excavated tunnel to zero and to use the steady-state flow option in UDEC to perform calculations until mechanical and hydraulic equilibrium was obtained. Similarly, CNWRA set the fluid pressure equal to zero in the region of the tunnel and conducted some explicit fluid flow timestepping, using UDEC, before the tunnel blocks were removed. The purpose of this was to model the increase in effective stresses in the joints ahead of the tunnel face caused by drainage through an advancing excavation. If this effect was not considered, the UDEC analyses would predict rock wall stability problems due excavation.

Output during this loading sequence was requested at times t = 1/12 and t* years where t* is the time after tunnel excavation for a hydro-mechanical equilibrium to be reached. It was widely agreed that mechanical changes occurred instantaneously compared to the fluid response and so t* was defined to be the time at which the changes of water flux into the tunnel is less than 1% over a period of 1 week after tunnel excavation. Different modelling techniques implied a range of different t* as shown in Table 2 below. Although a transient option is available in UDEC, the complexity of each of the different UDEC models meant that a very small timestep needed to be taken, resulting in very slow convergence of the fluid flow. Thus, the four teams using UDEC models assumed a steady-state flow and so t* could not be calculated.

Table 2
Time to reach hydro-mechanical equilibrium in years

	AEAT	KPH	CEA/DMT	ITASCA
t*	0.2	0.52	0.05	0.11

Figure 8 shows the horizontal stress along line I for time t = t* and Figure 9 shows the vertical stress along line II at the same time. The first thing to observe is that the stresses calculated by the different models are in fairly good agreement with each other, the main discrepancies occurring close to the tunnel. The AEAT models used an analytical solution to obtain the stress field caused by tunnel excavation and these figures show this to be a valid approximation. The results also show that the effect of tunnel excavation is limited to a small zone around the excavation.

By considering the stress field for the entire models, it was observed by most teams that the maximum compressive stresses occurred close to the tunnel. The NGI team noted that while the maximum compressive stress were approximately parallel to the tunnel wall, the smallest compressive stress were directed normal to the tunnel wall. As a result, the joints approximately normal to the tunnel tended to close while joints approximately

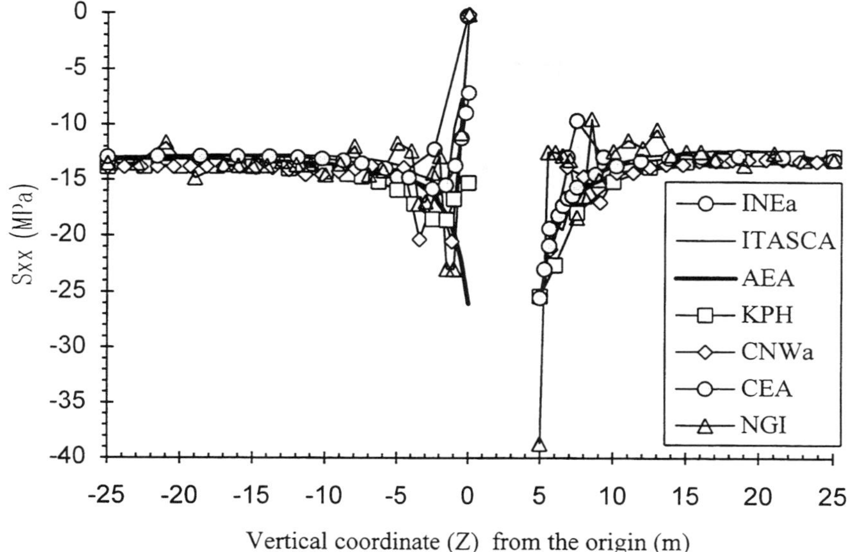

Figure 8: Horizontal stress (S_{xx}) along line I for time t = t*.

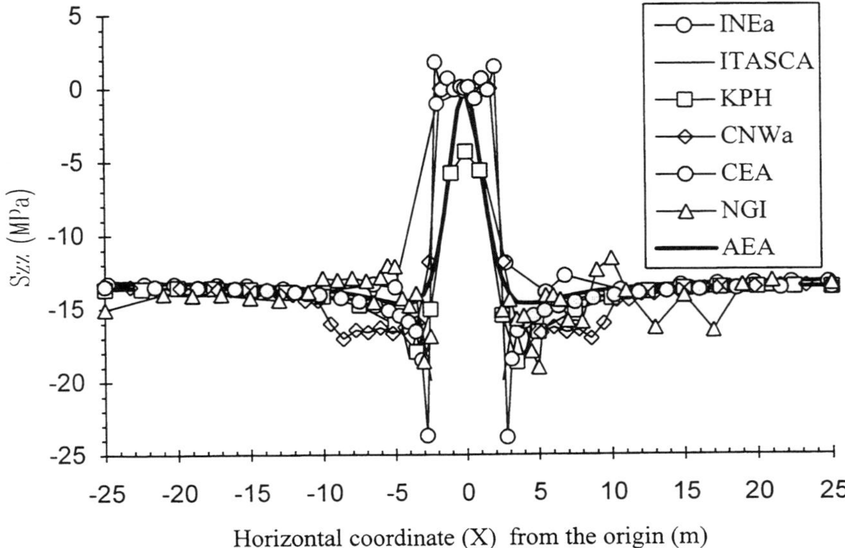

Figure 9: Vertical stress (S_{zz}) along line I for time t = t*.

parallel to the tunnel tended to open up. This effect was also observed in the AEAT models.

In the ITASCA model, the vertical and horizontal permeability are dependent on the stress field. It was observed in this model that the horizontal permeability was increased below and above the tunnel due to reduced vertical stress in those areas and the vertical permeability increased gradually from the tunnel wall due to reduced horizontal stresses there. No decrease in permeability occurred because the model assumed that the initial aperture was the residual aperture.

Table 3 shows the horizontal and vertical displacements at output points A-H for time $t = t^*$. The AEAT models have no deformation analysis and so are not present here. The NGI models showed a small asymmetry in the results from the two models showing the effect of the different joint geometries in the right and left half of the models. Maximum displacements in the model range from about 2mm to 411.0mm in the case of the INEb model. Again the largest displacements occur close to the tunnel.

Table 3
Horizontal (U_x) and vertical (U_z) displacements (mm) at output points A-H for time $t = t^*$.

		A	B	C	D	E	F	G	H
CEA	U_x	0.0	6.85	-0.36	-7.28	0.77	1.39	-1.84	-1.01
	U_z	2.83	-8.05	-14.55	-8.40	-3.30	-9.12	-9.37	-3.41
KPH	U_x	0.23	6.31	0.10	-4.82	0.92	1.69	-1.37	-0.87
	U_z	5.55	-2.31	-6.06	-2.07	0.43	-2.87	-2.57	0.27
ITASCA	U_x	0.0	*	0.0	-1.9	*	*	-0.6	-0.4
	U_z	0.6	*	-3.9	-2.4	*	*	-2.6	-0.9
CNWa	U_x	0.0	*	0.0	-11.94	*	*	-0.59	-0.56
	U_z	3.06	*	-4.47	-11.68	*	*	-1.86	0.13
CNWb	U_x	0.0	*	0.0	-7.24	*	*	-0.61	-0.57
	U_z	2.91	*	-4.45	-6.95	*	*	-2.14	-0.09
NGI[1]	U_x	0.0	-1.59	0.0	-2.07	-0.20	-0.45	-0.45	-0.25
	U_z	2.04	-0.30	-2.03	-1.07	0.26	-0.38	-0.91	-0.06
INEa	U_x	0.89	2.43	0.41	-4.46	0.21	0.60	-0.25	-0.30
	U_z	4.43	-0.65	-1.76	-2.78	0.09	-0.56	-0.48	0.17
INEb	U_x	2.31	193.9	101.5	-411.0	1.49	5.62	-5.91	-1.76
	U_z	72.44	-16.38	-164.0	-134.6	1.78	-4.55	-6.18	1.52
VTT	U_x	0.0	*	0.0	-1.80	*	*	-4.43	2.88
	U_z	1.75	*	-2.93	-1.33	*	*	-1.60	-1.22

*values not calculated as symmetry is assumed.
[1]results from two NGI models have been combined.

The results show a big difference in the displacements for the two INERIS models. This is an indication that the deformation of the surrounding rock mass depends a lot on the number of joints intersecting the tunnel boundaries. Displacements are larger in the case of INEb where there are 48 intersections with the tunnel boundary than INEa which has 21 intersections with the tunnel boundary. Although the mechanical properties of the rock were different in the two models, these results show that the complexity of the fracture network is an important factor in the stability of the tunnel wall.

Table 4 gives the water flux across the tunnel surface (segment ABCD) and monitoring segments EF, FG, GH, HE. There are negligible differences between the AEAa and AEAb models. By comparing the resulting aperture distributions of applying the three different stress-aperture couplings to the model, it was concluded that the second stress-aperture coupling had little effect on the fracture network and so the AEAb model was discarded. The AEAc results seem more realistic and indicate the significance of the stress field to the coupled hydro-mechanical behaviour of the jointed rock mass. This model also reveals that symmetry about the vertical central line does not exist from a hydraulic point of view, by comparing the water flux across segment EF ($0.74 \times 10^{-8} \mathrm{m}^3/\mathrm{s}$) and GH ($2.55 \times 10^{-8} \mathrm{m}^3/\mathrm{s}$).

Table 4

Water flux (m^3/s) across the tunnel surface (segment ABCD) and monitoring segments EF, FG, GH, HE at time t = t*.

	ABCD	EF	FG	GH	HE
CEA	0.76e-6	-0.78e-15	0.34e-14	0.47e-15	-0.4e-15
KPH	1.22e-8	1.54e-10	-2.72e-9	-2.75e-8	1.02e-9
ITASCA[1]	2.36e-7	*	8.5e-8	5.8e-8	2.4e-8
CNWa[1]	4.248e-8	*			
CNWb[1]	6.586e-8	*			
NGIa[1]	1.718e-8				
NGIb[1]	1.894e-8				
VTT[1]	5.78e-8				
AEAa	1.445e-7	0.122e-7	0.890e-7	0.278e-7	0.143e-7
AEAb	1.402e-7	0.119e-7	0.870e-7	0.278e-7	0.130e-7
AEAc	7.178e-8	0.743e-8	2.929e-8	2.551e-8	0.918e-8

*values not calculated as symmetry is assumed.

[1]values are corrected for the symmetry plane, ie fluxes into tunnel and across segments FG and EH (where applicable) have been doubled.

Significant differences exist between the CEA model and the other models for the tunnel inflow. This may indicate the limitation of Oda's crack tensor approach for homogenisation without a preliminary screening process to eliminate the isolated fractures, as discussed in the previous section. The difference between the water flux at the monitoring segments between the CEA model and the KPH model may also be attributed to the different treatments of aperture changes due to deformation. The apertures in the KPH model depend on the porosity and strains but remain constant in the CEA model.

The discrete models showed better agreement with each other, although the differences are still non-negligible, suggesting that the modifications to the reference network may affect the fluid flow. One illustration of this can be given by comparing the two CNWRA models. The tunnel inflow for the CNWb was higher than for the CNWa model. This discrepancy may be due to the fact that CNWb model has a higher fracture density around the tunnel than the CNWa model. The ratio of flow rate to fracture density was calculated for the two models and gave values of 2.013×10^{-8} and 2.288×10^{-8} for CNWa and CNWb, respectively, giving better agreement.

4.3. Thermal loading

At time t^*, the thermal loading is started. Output was required for specified time instants $t = t^*+1$, t^*+t_{max}, t^*+30 and t^*+100 years, where t_{max} is the time after heating starts at which the temperature at source point reaches its maximum. Table 5 gives the calculated values for t_{max}, showing this to be about 4 years for all models.

Table 5
Time at which temperature at source point reaches its maximum in years.

	INERIS	CNWRA	VTT	KPH	CEA/DMT	ITASCA
t_{max}	4.25	4	4.2	4.5	4.6	4.08

Figure 10 shows the distribution of temperature along line I and line II for time t^*+t_{max}. There is very close agreement between the distributions, providing verification of the different codes. The temperatures calculated by UDEC assume a continuum, so the simplifications made in the problem have no effect on the resulting temperatures calculated. It can be concluded that the overall thermal behaviour of BMT3, predicted by all teams, is reasonable. The KPH model considered the heat convection for this problem and by noting that the temperature produced by this model is not much different to the that produced by the other models, it is reasonable to suggest that the heat convection does not have a great effect on the thermal behaviour. When significant fluid flow rates occur, convective heat flow through the fluid phase can transfer heat faster and more extensively than conductive heat flow through the solid phase. Therefore, as the flow rates in this problem are small, this is essentially a thermal conduction problem.

Figures 11 and 12 show the horizontal stress along line I and the vertical stress along line II for time t^*+t_{max} years, respectively. The analytical solution for stress used by the AEAT team is also shown. Disagreements in stresses between the KPH model and others close to the tunnel are relatively significant, but diminish considerably away from the tunnel and heat source. In the CNWa model, the horizontal stress within the immediate tunnel flow was seen to increase to approximately 40 MPa, compared with only 20 MPa after excavation. The INERIS team also noted an increased stress at the start of thermal loading.

It was widely agreed that the introduction of a heat source deformed the fracture network. Thermal expansion of the rock mass was seen as the main reason for increased displacement magnitudes, which were observed to increase even after the temperature reached its maximum. It was also seen that the vertical displacements were much larger in magnitude than the horizontal displacements, due perhaps to the effects of the tunnel

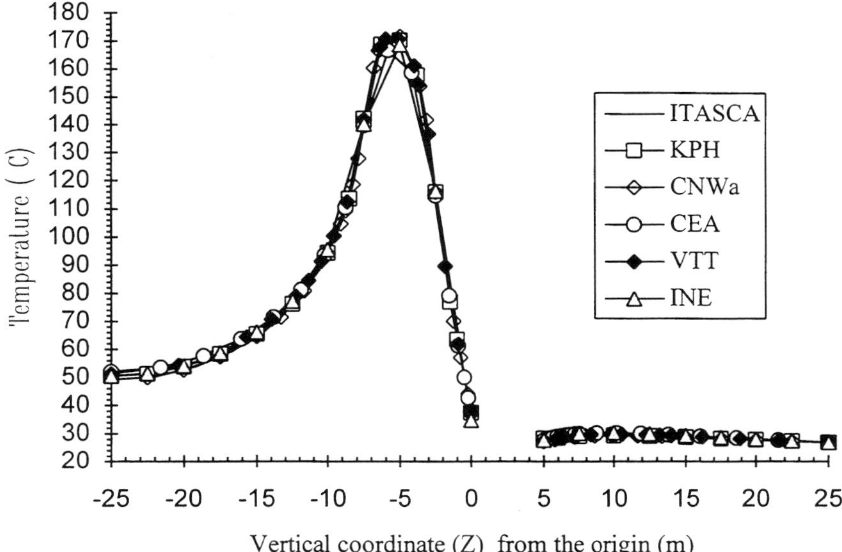

Figure 10a: Temperature along line I for time $t = t^* + t_{max}$.

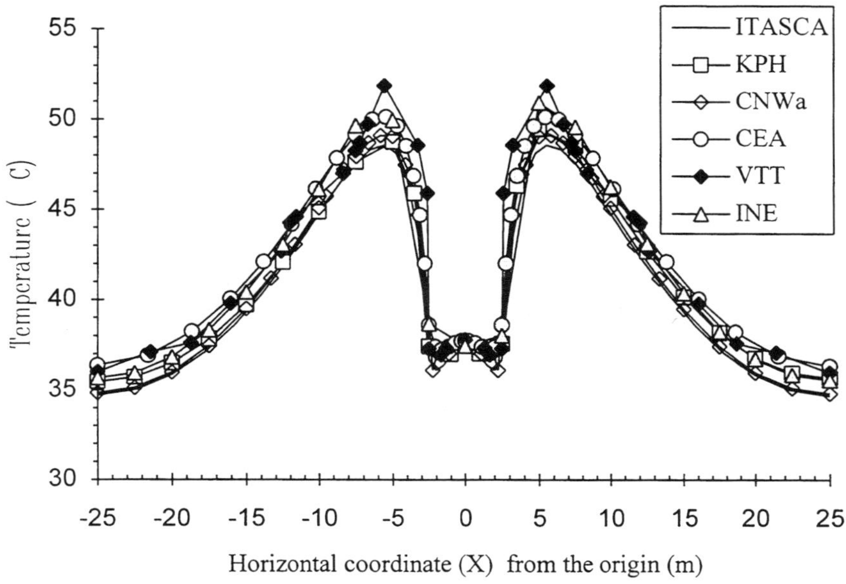

Figure 10b: Temperature line II for time $t = t^* + t_{max}$.

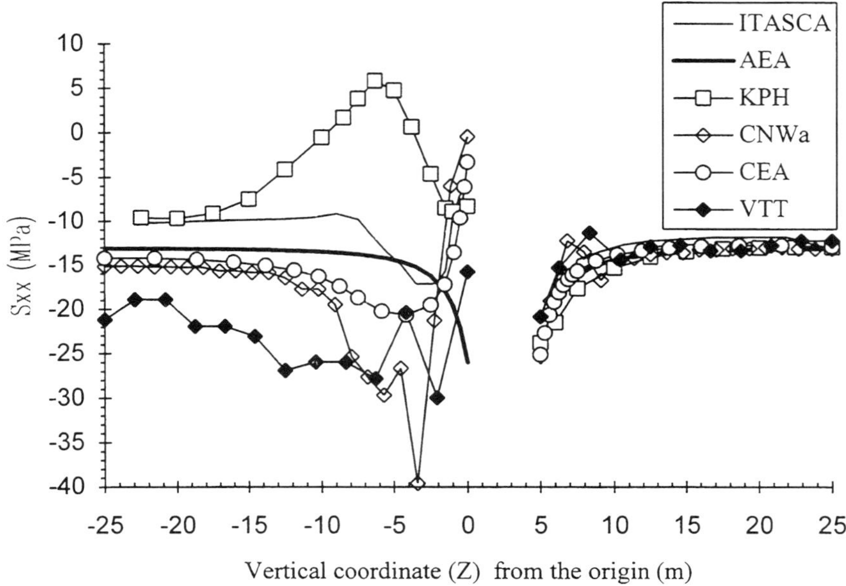

Figure 11: Horizontal stress (S_{xx}) along line I for time $t = t^* + t_{max}$.

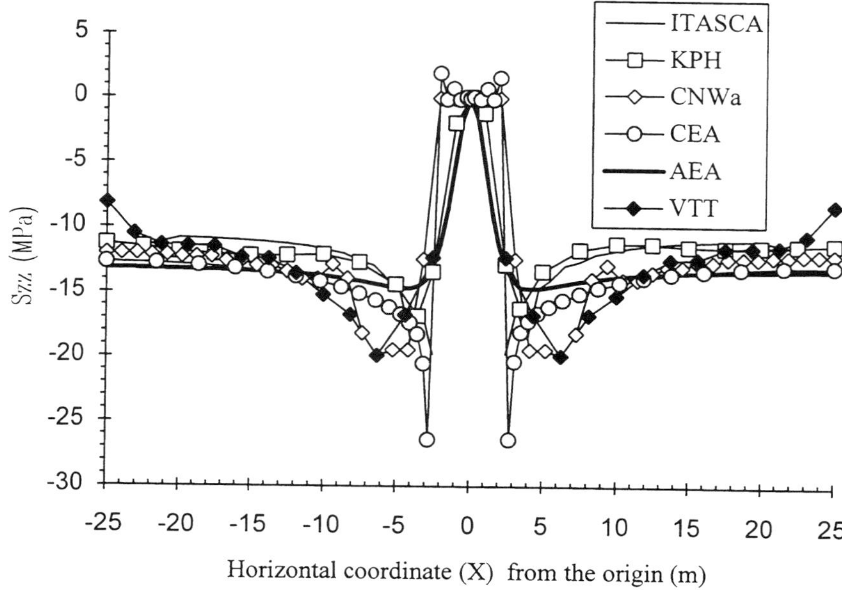

Figure 12: Vertical stress (S_{zz}) line II for time $t = t^* + t_{max}$

excavation and boundary conditions in the horizontal direction. In the CNWa model, upward displacement of the tunnel floor on the order of 15mm was shown to occur as a result of thermal expansion. The maximum inward displacement along the tunnel wall increased to a value of 47.3mm, causing some slight separation of the blocks in this area.

Table 6
Water flux (m^3s^{-1}) across the tunnel surface (segment ABCD) and monitoring segments EF, FG, GH, HE at times $t = t^*+1$, t^*+t_{max}, t^*+30 and t^*+100 years.

	$t = t^*+1$	$t = t^*+t_{max}$	$t = t^*+30$	$t = t^*+100$
CEA	0.82e-6	0.80e-6	0.76e-6	0.76e-6
KPH	1.01e-8	8.27e-9	5.24e-9	3.88e-9
ITASCA[1]	27.1e-8	21.5e-8		
CNWa[1]	3.892e-8	1.685e-8	3.528e-8	3.616e-8
CNWb[1]	5.054e-8			
VTT[1]	5.56e-8	5.48e-8		

[1]values are corrected for the symmetry plane, ie fluxes into tunnel have been doubled.

Thermal expansion of the rock mass also affected the flow around the tunnel. Table 6 shows the water flux across the tunnel surface at times $t = t^*+1$, t^*+t_{max}, t^*+30 and t^*+100 years. A decrease in flow rate after 4 years of heating was observed on the tunnel in the CNWa model. Neglected in the CEA model, but considered in the KPH model, are the change of fluid density due to the temperature and pressure, strain-induced heat production, heat convection and changes of heat capacity due to porosity changes. The combined influence of this difference is reflected in the different fluxes between the two models. The ITASCA team noted that the thermal load caused the permeability of the rock mass to change due to the induced stress and changed properties of the fluid. The results in Table 6 confirmed that this has some effect on the flow.

5. CONCLUSIONS

The results from eight research teams modelling BMT3 have been presented and compared. This test case is a comprehensive bench mark test problem for coupled T-H-M processes in fractured rocks and is well suited to test the validity of the mathematical models, algorithms and computer implementations. The geometrical model comes from a two-dimensional realisation of a realistic three-dimensional fracture system in the Stripa mine, and the boundary conditions are properly defined to test the code performances. One possible shortcoming is that the size of the model is too small. Research teams noted that for both thermal and hydraulic analysis, the boundary conditions affected the results as time increased.

With the diverse modelling approaches, the results are bound to be different. However, there is good agreement on the temperature distribution and also fairly good agreement for displacement and stress. The major difference is that the models predict different groundwater fluxes. This may be attributed to the fundamental difference between the continuum and discontinuum approaches. The discontinuum approach is more sensitive to the change of fracture apertures than the continuum approach. Fluid flow may take place through a number of significant pathways of connected fractures and therefore the change in fracture apertures may have a direct effect on the flow rate. This effect is better modelled by a discontinuum approach which models the fractures explicitly.

The effect of heat convection by fluid flow on temperature, as considered by the KPH team, may be said to not have a significant impact on the results for this problem. The difference in temperature between the KPH model and the other models is not significant and should not, in turn, induce significant thermal stress increments. The difference in stresses between the KPH model and those of other teams must be caused by the difference in homogenisation approaches, constitutive relations and parameter values, discretisation and other numerical techniques.

An important issue is how to characterise a fractured medium by the continuum approach. Two teams used homogenisation schemes which resulted in different REVs and tensors of strain and permeability. The existence of a valid REV for this problem and its relation with the size of finite elements has not been resolved and clearly more work needs to be done. The CEA/DMT team used constant properties calculated with a homogenisation scale equal to the domain size, but are carrying out work to investigate the use of smaller homogenisation scales, allowing for better spatial resolution of the hydro-mechanical processes in this test case.

A possible representation of the test case is to use a mixed scheme in which an inner, small scale, region containing the tunnel and heat sources is represented by a discrete model and this is surrounded by a continuum model extending to the problem boundary. This approach was tried by some teams using the UDEC code. The key problem here is how to ensure continuity of pressure and flux at the interfaces of the two regions and obedience to all physical laws both in the outer continuum region and at the interfaces.

The capability of the discrete element modelling is limited by the required computer time and storage demands for practical problems since they increase dramatically with the increase of the fracture numbers and domain sizes. The discrete fracture network model used by the AEAT team was the only model to consider the reference network without simplification. This approach was useful in identifying how significant a role the discrete fracture network played in shaping the flow field as opposed to the average flow properties of the rock.

To summarise, this test case has provided a successful benchmark to test the capability of different models to address realistic problems on a realistic scale. The problem itself was simplified by specifying a two-dimensional system and a wide range of modelling techniques were employed which required further simplifications to the problem. The results have enabled a greater understanding of the coupled processes affecting groundwater flow around a repository tunnel although the problems are difficult and there is scope for further work. It should be noted that modelling flow in a two-dimensional system may not be appropriate as the geometry and connectivity of a three-dimensional system is much more significant for groundwater predictions. An alternative would be to allow teams to start from raw data. The approaches developed here could then be used to model the coupled processes associated with real field experiments.

6. ACKNOWLEDGEMENT

Funding for the production of this chapter was provided under a contract from UK Nirex Ltd. to AEA Technology.

7. REFERENCES

1   Baroudi, H., L. Dewiere, M. Durin, A. Herbert, A. Makurat and F. Plas, Specifications of a Bench-Mark Test for DECOVALEX Project: Near-Field Repository Model. DECOVALEX Document, 1992.

2   Thoraval, A. and E. Vuillod, Thermo-Hydro-Mechanical coupled modelling in the near field. ANDRA Report 694 RP INE 93011, 1993.

3   Vuillod, E., Additional results about the T-H-M coupled modelling, BMT3. ANDRA Report 694 RP INE 94002, 1994.

4   Thoraval, A. and E. Vuillod, DECOVALEX BMT3 Eleménts de synthèse. ANDRA Report 694 RP INE 94004, 1994.

5   Singh, B., Continuum characterisation of jointed rock masses, Part I - the constitutive equations. Int. J. Rock Mech. Min. Sci. & Geomech. Abstr., Vol. 10, pp. 337-345, 1973.

6   Ahola, M., L. Lorig, A. Chowdhury and S-M. Hsiung, S-M., T-H-M coupled modelling: near field repository model, BMT3, DECOVALEX Phase II, Centre for Nuclear Waste Regulatory Analysis, San Antonio, Texas, USA, 1993.

7   Hansteen, H., DECOVALEX - Phase II Bench-Mark Test 3, UDEC discontinuum modelling of near field Test Case BMT3, Report 921061.4, Norwegian Geotechnical Institute, 1994.

8   Pöllä, J., DECOVALEX - phase II, bench mark test 3, near-field repository model, Technical Research Center of Finland (VTT), Communities and Infrastructure, 1994.

9   Wilcock, P., The results of applying the NAPSAC fracture network code to model BMT3: the DECOVALEX Near-Field Test Case, UK Nirex Ltd Report, NSS/R365, (in preparation) 1994.

10   McKinnon, S and P. Carr, Site characterisation and validation - stress field around the validation drift. Stripa Project Technical Report 90-09, SKB, Stockholm, 1990.

11   Monsen, K., A. Makurat and N. Barton, Disturbed zone modelling of SCV validation using UDEC-BB, models 1 to 8 - Stripa Phase 3. Stripa Project Technical Report 91-05, SKB, Stockholm, 1991.

12   Herbert, A., J. Gale, G. Lanyon and R. MacLeod, 1991. Modelling for the Stripa site characterisation and validation drift inflow: Prediction of flow through fractured rock. Stripa Project Technical Report 91-35, SKB, Stockholm, 1992.

13   Ababou, R., A. Millard, T. Treille and M. During, Coupled
     thermo-hydromechanical modelling for the near field benchmark test 3 (BMT3)
     of DECOVALEX Phase 2, Progress report, Rapport DMT/93/488, 1993.

14   Biot, M., General theory of three-dimensional consolidation, J. Appl. Phys.,
     Vol. 12, pp. 155-164, 1941.

15   Oda, M., An equivalent continuum model for coupled stress and fluid flow
     analysis in jointed rock masses, Water Resources Research, Vol. 22, No. 13,
     pp. 1845-56, 1986.

16   Kobayashi, A., K. Hara, T. Fujita, and Y. Ohnishi, Analyses of BMT3 with
     THAMES, Hazama Corporation, Power Reactor & Nuclear Fuel Development
     Corporation, and Kyoto University, 1993.

17   Israelsson, J., DECOVALEX, Bench-Mark test 3, Thermo-hydro-mechanical
     modelling, SKB Arbetsrapport 93-31, 1993.

18   Bieniawski, Z., The geomechanics classification in rock engineering applications.
     Int. Congress of Rock Mechanics, Montreux Vol. 2, 1979.

19   Stille, H., T. Groth and A.Fredriksson, FEM-analys av Bergmekaniska Problem
     Med JOBFEM. BEFO Report No. 307:1/82. Stockholm 1982.

20   Thorpe, R., D. Watkins, W. Ralph, R. Hsu and S. Flexser, Strength and
     permeability test on ultra-large Stripa granite core. Technical Information Report
     No. 31, LBL-11203, SAC-31, UC-70. Lawrence Berkeley Laboratory, University
     of California, 1980.

O. Stephansson, L. Jing and C.-F. Tsang (Editors)
Coupled Thermo-Hydro-Mechanical Processes of Fractured Media
Developments in Geotechnical Engineering, vol. 79
© 1996 Elsevier Science B.V. All rights reserved.

Mathematical simulations of coupled THM processes of Fanay-Augères field test by distinct element and discrete finite element methods

A. REJEB

Institut de Protection et de Sûreté Nucléaire
Département de Protection de l'Environnement et des Installations
60-68, avenue du Général Leclerc, B.P. 6
92265 Fontenay-aux-Roses Cedex, FRANCE

Abstract

This chapter presents Test Case 2, designed as TC2 of the DECOVALEX Project, concerning the three-dimensional modeling of the thermomechanical behavior of a granitic mass subjected to controlled heating during the Fanay-Augères THM experiment in France. This modeling considers the main fractures of the mass, which is an advance over previous models for interpreting the results of this experiment in a continuum medium. The discrete approach was adopted by representing all the fractures explicitly with their properties and specific behavior. Two numerical methods were used by the research teams : the Distinct Element Method and the Finite Element Method, with special joint elements. A comparative analysis is made here of the numerical results generated by each of these methods. Moreover, the results of calculations for continuous and discontinuous media are compared with the various measurements made in the Fanay-Augères site, to bring out the effect of the modeled discontinuities.

1. INTRODUCTION

From the geotechnical point of view, the main characteristic of highly active radioactive waste is the heat generation. Storing it in a deep geological medium such as granite will produce a temperature rise in the rock, along with strains, and a change in the existing stresses. It is important to quantify these changes, on the basis of a thermomechanical analysis, to make sure the disposal galleries will stand up mechanically and to evaluate the risks of fracturing the rock or of opening up existing discontinuities.

According to the long duration of the phenomena and the scale of the medium considered, the only way to quantify these changes experimentally is to implement on a reduced scale. Approaching the problem this way, the need for *in situ* heating experiments appeared. The measurements made during these experiments should make it possible to refine the numerical models and validate the computer codes for subsequent use in simulating real disposal situation configurations. This was the context in which the IPSN (Institut de Protection et de Sûreté Nucléaire) proposed to the organizations involved in the international DECOVALEX project to use the Fanay-Augères THM experiment as a test case (Jing *et al.*, 1995).

2. PRESENTATION OF TEST CASE 2 (TC2)

2.1. Objective
A great deal of modeling work has already been done on the Fanay-Augères THM experiment by the Ecole des Mines de Paris, on behalf of the IPSN, in the framework of European project n° FI 1W/0246 (Rejeb *et al.*, 1994). The conclusions drawn from the various thermomechanical models for continuum medium, and the experimental validation of these models, showed the need to take the main fractures of the rock mass into account (Rejeb, 1992). The objective of the proposed test case 2 is thus twofold :

-   to continue the modeling of this experiment under the assumption that the medium is discontinuous ;
-   to compare the various approaches used by the research teams participating in this test.

2.2. TC2 terms
This is a simulation of the Fanay-Augères THM experiment using a three-dimensional model including the six most conductive fractures of the granitic mass. The first part of the work consists in performing and analyzing coupled thermal and mechanical computations using an exact reproduction of the laboratory geometry and the real experimental conditions. The results of this computation were then compared with the temperature and strain measurements recorded during the experiment, to judge the validity of the model.

It should be noted that, despite the common name "THM experiment", the hydraulic aspect has been investigated very little because of the unsaturated environment, and the few measurements that have been made of the permeability of certain fractures are not very significant. So the TC2 exercise deals only with the thermo-mechanical aspect, for which a solid base of reliable measurements is available.

2.3. TC2 data
The main data needed to compute TC2 were supplied to the participants (Gros, 1993), and will be described only briefly here.

2.3.1. Laboratory geometry and heating data

The experimental site is located at a depth of 100 m and can be accessed through a pre-existing drift. As figure 1 shows, the site consists of a test room, an access gallery, and a side gallery. The access gallery was used to dig the test room, which is a rectangular box measuring 12 m x 10 m x 5 m. The side gallery was used to drill the holes for precutting the test room floor, and to drill five other holes for the heat source elements. The heat source, buried 3 m under the floor, consists of five cylindrical heaters, 1.5 m long and 0.15 m in diameter, spaced 0.30 m apart between axes.

To simplify the TC2 input data, the modeled area was limited to the test room and the instrumented rock mass with the heat source (figure 2). Not all the environment consisting of the three galleries is taken into consideration in the modeling. The model geometry is defined in the local (X, Y, Z) coordinate system of reference used in positioning the measuring instruments in the granitic mass.

As far as the heating characteristics are concerned, the total power of the source is 1 kW, or 200 W per heating element. The exact duration of the heating is 51 days, 21 hours, and 30 minutes ; and that of the cooling is 73 days, 14 hours, and 30 minutes. The heating phase was interrupted by four accidental electric power failures lasting 3, 22, 0.5, and 11 hours.

2.3.2. Thermomechanical characteristics of the rock mass

The thermal characteristics of the Fanay-Augères granite, as measured on laboratory samples, are the following :

- thermal conductivity $\lambda = 2.13$ W/m/K at 20 °C
  $\lambda = 1.95$ W/m/K at 84 °C
- specific heat $C = 2.1089 \times 10^{-6}$ J/m^3/K.

The mechanical parameters, also measured in the laboratory, are :

- Young's modulus $E = 55650$ MPa
- Poisson's ratio $\upsilon = 0.22$
- bulk density $\gamma = 26.2$ kN/m^3

The coefficient of thermal expansion (α) is related linearly to the temperature (T) by the equation (in K^{-1}) : $\alpha\,(T) = 10^{-6}\,[5 + 0.1 \times (T - 295)]$.

The expansion measurements made in the survey holes yielded Young's moduli of 32000 MPa for the unfractured areas, and 28200 MPa for fractured areas.

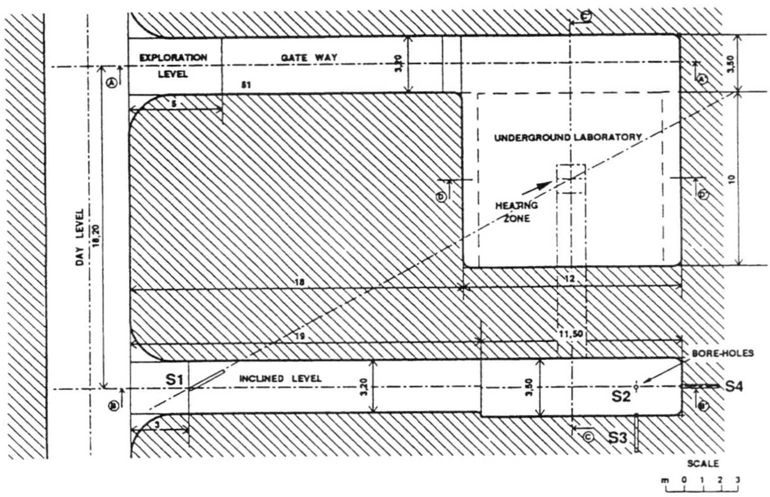

Figure 1. Layout of the experimental arrangement.

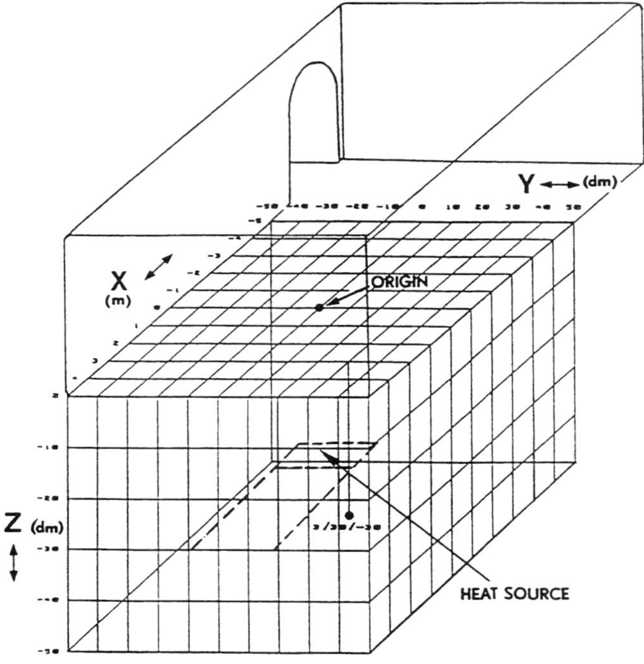

Figure 2. Spatial diagram of the Fanay-Augères test site.

2.3.3. Characteristics of main fractures

The rock at the Fanay-Augères test site is moderately fractured granite. The fractures in the test floor were carefully mapped and characterized. Figure 3 shows the network of fractures in the 10 x 10 m² test area. All the fractures were numbered, and the coordinates and description of each fracture were made available on diskette. For the purposes of the TC2 exercise modeling, only the six most conductive fractures were considered. These are represented by thick lines in figure 3. Table 1 gives the mechanical characteristics of two the families of fractures retained for the simulations.

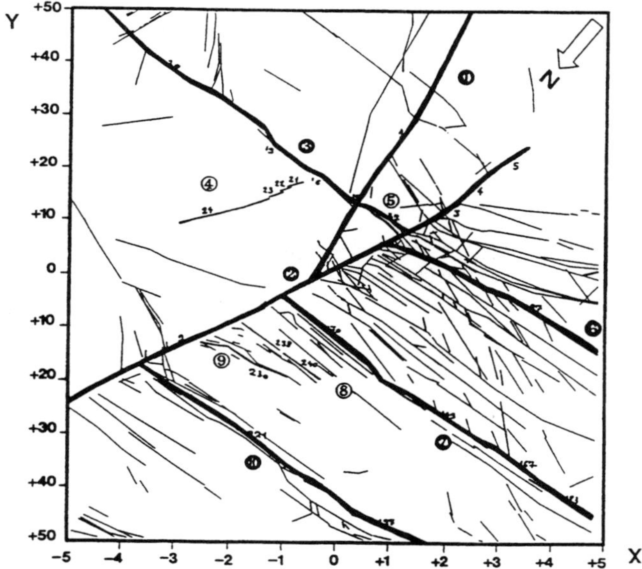

Figure 3. Network of fractures at the experimental floor, and the six fractures modeled in TC2.

Table 1
Mechanical properties of fractures

	Normal stiffness	Shear stiffness	Cohesion	Friction angle	Dilatancy angle
	K_n [MPa/m]	K_s [MPa/m]	C [MPa]	φ [°]	i [°]
Family 1	638 000	577	0.03	34	47
Family 2	793 000	560.6	0.15	37.2	35.4

2.3.4. Initial conditions and loading

The initial state for the computation is defined as a uniform, isotropic temperature field of 13°C throughout the rock mass, which is the mean value of initial temperatures measured in situ.

The initial stress state, determined in situ by hydraulic fracturing, is defined by :

$\sigma_v = 0.001\,\gamma\,z$ vertical stress ;

$\sigma_h = 0.92\,\sigma_v$ minor lateral stress, 60° from north ;

$\sigma_H = 1.35\,\sigma_v$ major lateral stress, 150° from north.

z is the depth below the surface, which is 100 m.

The rock mass is subject only to the thermal stress due to heating and cooling, the characteristics of which are given in section 2.3.1 above.

2.3.5. Thermomechanical experimental data base

A major experimental program and high-performance instrumentation made it possible to monitor the rock mass temperature and strain variations during the heating and cooling process. A total of 153185 measurements were made of :

-    the temperatures, with sensors (coded TE-1 to TE-79) placed at the source, in the rock mass, and on the floor surface ;
-    the vertical displacement of the rock mass, using extensometers installed in three boreholes (F1, F2, F3) ;
-    the heaving and lowering of the floor surface, with eight V-shaped photoelectric cells (V1 to V8) ;
-    the longitudinal and transversal strains in the floor, using electric extensometers (EL/ET-1 to EL/ET-18) ;
-    the opening and closing of the main fractures, 1 and 2 (EF1, EF2) ;
-    the vertical shear of fractures 1 and 2, using electric extensometers (EV1 to EV4).

The experiment proceeded successfully both as concerns the operation of the heating arrangement and the performance of the acquisition system. The measurement data are excellent, from the point of view both of the instrument operation and of the quality and accuracy of the readings, which thereby constituted a sound experimental data base, which was provided to the participants in the form of diskettes.

A survey of the measurement results and interpretations was published by Rejeb *et al.* in 1990. A few examples of this data will be presented in the comparison of the numerical results and measurements in section 4 below.

3. MODELING OF TEST CASE 2 (TC2)

In addition to the data supplied, as described above in section 2, the modeling also required other specifications that the researcher should adopt in developing his model and performing his computations in three dimensions. The third part of this chapter presents the specifications of the different models developed, with an emphasis on the common points and differences, to facilitate the comparison of the results later. These specifications concern particularly the approaches and methods used, along with the models and meshes, and the principles of the thermal and mechanical computations.

3.1.   Research teams, codes, and approaches

Three funding organizations decided to analyze the TC2 case - ANDRA and IPSN in France, and SKB in Sweden - with collaboration of the following research teams, respectively : INERIS, Ecole des Mines de Paris (EMP) and Clay Technology AB (CT). The specifications of the different modelings were taken from the reports of these research teams, respectively Thoraval and Hosni (1994), Vouille *et al.* (1995) and Börgesson *et al.* (1994).

The approach ordinarily used in modeling a rock mass assumed to have a limited number of fractures is the discrete approach, where the bulk is represented by an assembly of blocks separated by discontinuities. The two numerical methods used for solving the problem by this approach are :

-   the Distinct Elements Method (DEM), in which the blocks may be rigid or deformable, and the discontinuities are represented by block to block contact. The mechanical computation is of the explicit, dynamic type, and the differential equations of motion are solved by finite differences. This method was used by the INERIS team, with the 3DEC numerical code ;
-   the Finite Element Method (FEM), in which the blocks are discretized by standard finite elements and the discontinuities by joint elements, as in the DEM. The equations of quasi-static equilibrium are written for the mechanical aspect. This was the method used by the EMP, with the VIPLEF code, and by CT, with the ABAQUS code.

These methods and numerical codes will not be described in details here, but are presented in other chapters of this book.

In addition to the models developed by these three research teams as part of the DECOVALEX project, the comparison will also include results of a three-dimensional modeling previously developed at the EMP using the continuum approach, in which the rock mass is assumed to be homogeneous (Rejeb, 1992), so that the results of the discrete and continuum approaches can also be compared. Table 2 recapitulates the thermomechanical models with their approaches, methods, and the numerical codes used to date, to simulate the Fanay-Augères THM experiment.

Table 2
Summary of TC2 models

	Funding organization	Research team	Approach	Method	Code
	ANDRA	INERIS	Discrete	DEM	3DEC
DECOVALEX	IPSN	EMP	Discrete	FEM	VIPLEF
	SKB	CT	Discrete	FEM	ABAQUS
IPSN-CEC	IPSN	EMP	Continuum	FEM	VIPLEF

3.2. Description of models

The geometry of the structure to be modeled in three dimensions consists of a large parallelepiped representing the surrounding rock, containing :

- a parallelepipedic void representing the test room, including the service corridor ;
- six fracture planes intercepting the heated rock block located under the test room ;
- five hollow cylinders representing the heating source placed under the test room floor.

Each engineer used his own judgment in choosing the dimensions of the model boundaries. Figure 4 and table 3 give a simplified diagram of the structure geometry and the dimensions adopted by each team, respectively.

Note : To distinguish the two models developed at the EMP, we use hereafter EMP$_c$ to denote the model with continuum approach, and EMP$_d$ for the model with the discrete approach.

The three teams used the TC2 data to determine and represent the planes of the fracture systems retained (1, 2, 3, 6, 7, and 10). However, since the extent of these fractures was not defined, the INERIS and CT teams stopped the fractures at the boundary of the instrumented rock mass, while the EMP team extended them to the outer limit of the surrounding rock. Towards the center of the floor, the fracture plans were truncated in the same way, stopping them at the plane of fracture 2, which passes through the entire mass.

Only the EMP models took into account the real geometry of the heat source using five hollow cylinders. In the other models, these were represented by set of points, in order to simplify the structure geometry.

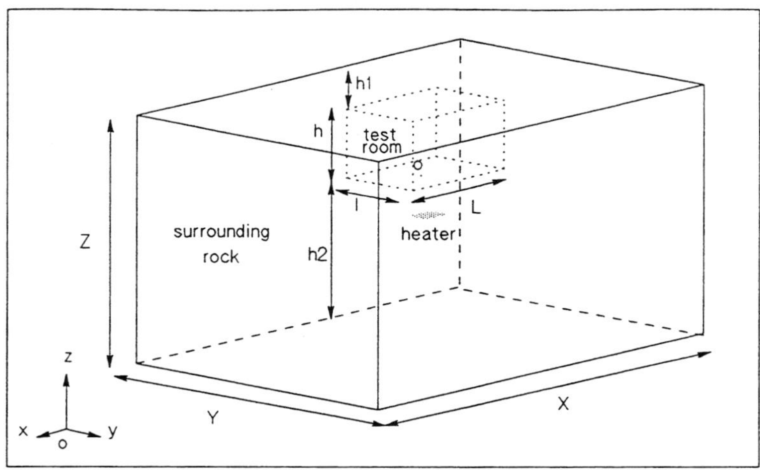

Figure 4. Model geometry.

Table 3
Model dimensions for each team

Dimension [m]	CT	EMP$_c$	EMP$_d$	INERIS
X	40	30	23.5	49
Y	40	30	22	44
Z	40	15	29	39
L	13	13	13.5	13
l	13	10	12	10
h	5	5	5	5
h1	15	0	9.5	15
h2	15	10	14.5	19

The mesh used for this structure is not so simple as its geometry might lead one to suspect. Generating a three-dimensional mesh of the large parallelepiped with a parallelepipedic void and five cylindrical holes is a tedious work, considering the opposing constraints of constructing a fine mesh in the areas where the gradients are high, while limiting the total number of nodes. Including the six fracture planes with the inclinations and intersections makes the process of mesh generation even more complex, and explains why some flexibility is needed when evaluating the quality of the meshes obtained by the various codes. Figure 5 gives some views of these meshes and table 4 indicates the number of elements, their types, and the number of nodes for each model.

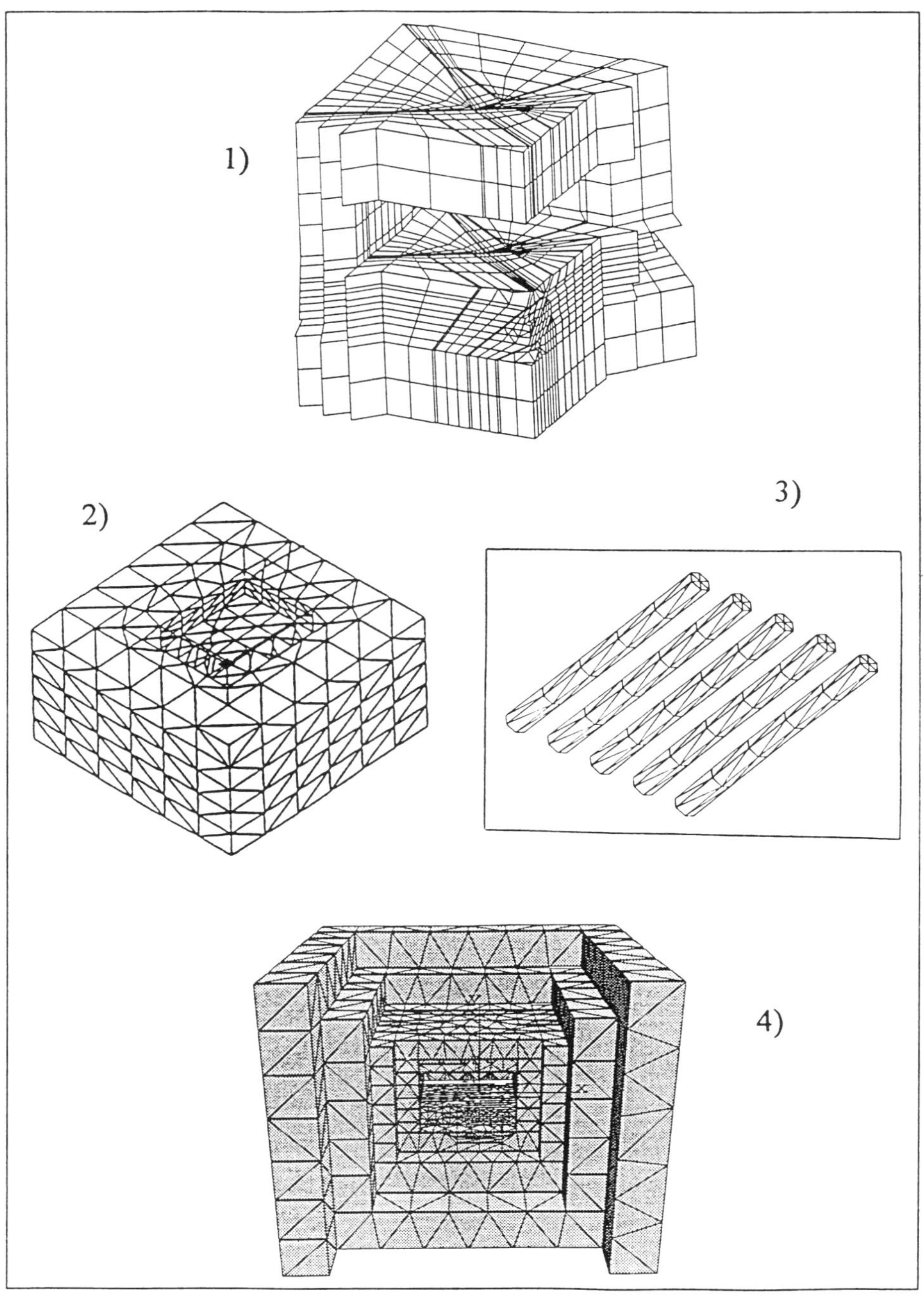

Figure 5. Views of the different meshes : CT(1), EMP_c(2), EMP_d(3), INERIS (4).

Table 4
Characteristics of the different meshes

Mesh	CT	EMP_c	EMP_d	INERIS
Number of nodes	6283	7307	4739	6924
Element types	trilinear hexaedra 8 nodes	quadratic tetrahedral 10 nodes	linear tetrahedral 4 nodes	linear tetrahedral 4 nodes
Number of elements	7180	4794	16771	19461

3.3. Thermal modeling principles

The usual method for solving a rock thermomechanical problem starts by computing the temperature field, which is then used as input data for computing the mechanical effects. To perform this computation on the TC2 configuration, all the teams adopted the assumption that the temperature is at each time continuous throughout the structure, even across a newly opened fracture.

The principle of this modeling is based on the transient-state solution of the heat conduction equation by FEM to compute the temperature at each point in time. This is the principle adopted by the EMP and CT teams.

However, to compute the temperature with 3DEC, the INERIS team uses an analytical solution yielding the temperature at all points in space and time due to a punctual heat source. To do this, each source cylinder is discretized into 50 points. The temperature at a given point is simply the sum of temperatures due to the contribution of the infinitesimal sources.

The boundary conditions adopted by the various teams for the thermal calculations are reported in table 5, where :

- Γ_{outer} means the outer boundary of the surrounding rock ;

- Γ_{inner} the boundary consisting of the roof, walls, and floor of the test room ;

- Γ_{heater} the boundary consisting of the heat source.

Despite the different ways adopted to represent the heat source, the same data (power, heating and cooling durations) are used by all the teams.

Only Clay Technology AB used a thermal conductivity, λ, that varies linearly with the temperature. All the other teams used the value of λ at atmospheric temperature.

Table 5
Thermal boundary conditions

Boundary	CT	EMP$_c$	EMP$_d$	INERIS
Γ_{outer}	adiabatic	adiabatic	adiabatic	adiabatic
Γ_{inner}	heat convection, $K = 10 W/m^2 K$, $\theta_i = 13°C$, $\theta_f = 9°C$	linear decrease of temperature during the test (13°C to 10°C)		adiabatic
Γ_{heater}	8 nodes (1.2 m²), no electric breakdowns	5 cylinders with electric breakdowns	5 cylinders with electric breakdowns	5 x (50 points) with electric breakdowns

3.4. Thermomechanical modeling principles

The computations using continuum or discrete approaches included the following five steps :

1- application of in situ stresses and computation of equilibrium,
2- excavation of test room,
3- temperature computation,
4- matching of computed temperatures with measurements,
5- thermomechanical computation of the temperature effect.

It should be noted that step 4 was used only in the modeling of the Fanay-Augères experiment in continuum medium by the EMP team, in order to avoid the repercussion of the difference between computed and measured temperatures on the mechanical results. In the framework of TC2, none of the teams used corrected temperatures in the thermomechanical computation.

As concerns the T-M coupling, only the thermal effect on the mechanical aspect is computed here. The temperatures are computed independently from the mechanical computation, considering the negligible mechanical effects on the thermal properties. However, there are two ways of computing the thermal effect on the mechanics, depending on whether or not the calculations are alternated :

1- either the computation sequences are independent, in which case a sequence consists of calculating the temperature at date t and, after adding the thermal stress field from time 0 to time t, the mechanical equilibrium state is computed ;

2- or the sequences may be consecutive other. A sequence in this case is a calculation of the temperature variation between times t_i and t_{i+1}, followed by a calculation of the mechanical equilibrium where the induced thermal stress variations are computed between times t_i and t_{i+1}.

Only the INERIS team performed the computations the first way, which is equivalent to assuming that the global behavior of the mass remains elastic throughout the modeling, and that the mechanical results do not depend on the thermal loading path applied. This assumption is especially incompatible with the discrete approach, in which the fractures are modeled by assigning them a nonlinear behavior. This is clearly confirmed by the experimental measurements during the cooling phase.

All the teams adopted the same modeling principle for the thermomechanical computations. They assumed that the entire mass is elastic, and introduced the six fractures with elasto-plastic behavior. In some models, in order to reflect the fracturing state of the rock mass, different values of Young's modulus, E, were assigned to three distinct zones. Figure 6 shows the location of these zones, and table 6 gives the values of E used by each team.

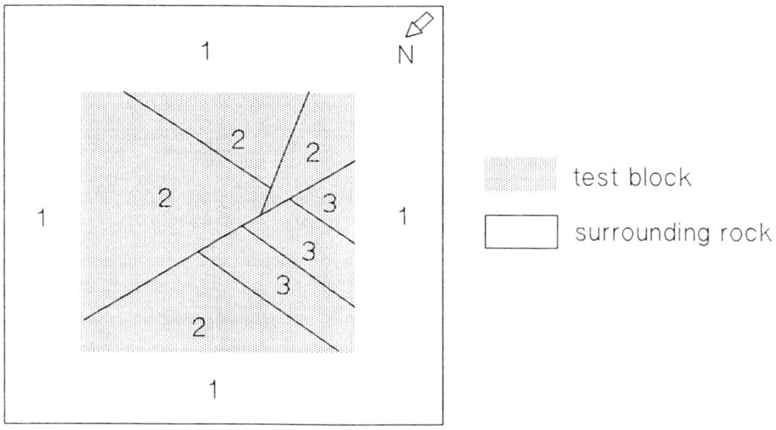

Figure 6. Location of different rock mass fracturing state zones.

Table 6
Different values of Young's modulus E [MPa] used in computations

Zone	CT	EMP$_c$	EMP$_d$	INERIS
1 no fractured zones	55 650	55 650	55 650	55 650
2 slightly fractured zones	30 000	55 650	55 650	32 000
3 highly fractured zones	30 000	55 650	55 650	28 000

Table 7 recapitulates the fracture constitutive laws and the mechanical parameters used by each team in performing the thermomechanical computations with the discrete approach. The EMP considered that the maximum closing of the joint would be 46×10^{-6} m in characterizing the normal behavior of the fracture. Moreover, the finite elements representing the fractures in the CT model are 13 cm wide, which is much larger than the opening of the real fractures.

Table 7
Fracture data modeling

Fracture	CT	EMP_d	INERIS
Behaviour law	Drucker-Prager	Mohr-Coulomb	Mohr-Coulomb
Mechanical parameters (see table 1)	Family 1	Family 2	Family 1

Table 8 uses the above notation to summarize the boundary conditions applied in the thermomechanical computations.

Table 8
Thermomechanical boundary conditions

Boundary	CT	EMP_c	EMP_d	INERIS
Γ_{outer}	fixed	Weight of the overburden on the roof Normal displacement is fixed on lateral surfaces		$\sigma = \sigma_i$
Γ_{inner}	free	free	free	8 corners of the chamber are fixed

4. COMPARISON OF NUMERICAL RESULTS WITH MEASUREMENTS

4.1.　General

We will not attempt here to review the whole of the simulation results supplied by the research teams, but rather to look at the computation results corresponding to the quantities measured during the experiment. To do this, an effort was made to collect the maximum amount of computed data from the various teams. These data were then subjected to graphic analysis to compare the results of the four models with the experiment for the six quantities measured at a few hundred points.

In the following sections, a comparative analysis and synthesis are made for each measured quantity, illustrated by an objective choice of the most significant curves : a time variation of the quantity measured at a given point. The approach used for this analysis consists in :

- comparing the results of the three computations using the discrete approach, to reveal the common points and the differences obtained with the two numerical methods used (FEM and DEM) ;

- comparing the results of the discrete and continuum approaches to bring out the effect of including the fractures in the models.

Of course, our reference for these two types of comparison is the set of experimental data, which is of major importance for assessing the validity of the models developed.

4.2.　Temperatures

Let us first of all recall that the thermal modeling principle developed by the various teams is based on the continuum medium assumption. So emphasis will be placed on the numerical methods used in analyzing any differences between the results. The simple notation of EMP_c and EMP_d models is used here just to distinguish between the thermal simulations corresponding to the mechanical computations of the continuum and discrete media.

The temperature histories predicted by the four models have been shown on the same graph for each of the 43 sensors. Figure 7 shows the data for four sensors. The (x, y, z) coordinates of these sensors are given in parenthesis in the title of each graph.

On the whole, most of the computations show good qualitative agreement with experiment, though the temperatures are overestimated near the source. This overestimation is attenuated farther away from the source. It is not as large in the case of the EMP models, which simulate the real geometry of the source.

Rejeb (1992) attributes this difference rather to the thermal parameters used in the calculations, which were obtained by measurements on small laboratory samples. By simply matching the numerical data generated by the EMP_C model with the temperature measurements, we deduced that there existed a 40 % difference between the true thermal parameters in the rock mass and those measured in the laboratory. This difference is mainly due to the scale effect.

At two-by-two comparison of the theoretical results shows a few further differences, which can be summarized as follows :

-    A good quantitative agreement is observed between the EMP results and those of INERIS near the heat source, where the maximum difference is 4° C in the case of sensor TE-43. This means that the source discretization in the INERIS analytical computations was good. On the other hand, the difference becomes greater as we approach the surface, as shown by the sensor TE-3. This difference is due to the poor choice of adiabatic boundary condition in the test room, when there is actually a lowering of the temperature. Considering the thermal computation principle in the 3DEC code (analytical solution when UDEC is using finite differences), we can draw no conclusion concerning the DEM and FEM methods for solving such a thermal problem, on the basis of this comparison.

-    Between the CT results and those of the EMP, using bothly the FEM, it is observed that the CT model is the one that overestimates the most the temperatures close to the source (in the case of sensor TE-43). This overestimate can be explained by the fact that, in this model, the electrical outages that occurred during the heating were not taken into account, and that the source area was taken to be equal to 1.2 m^2 rather than 3.61 m^2, which would of course favor a greater supply of heating energy into the mass. On the other hand, for points far from the source, the CT computations yield results closer to the experiment than those of the EMP.

-    The two EMP computations, performed under quite the same conditions and with the same code, differ slightly because of the fineness of the meshes (7307 nodes compared to 4739), particularly around the source.

Let us finally state that the case of sensor TE-71 shows that the test room wall surface convective transfer condition used in the CT calculations better reflects the measurements in the heating phase than the EMP's decreasing linear temperature condition.

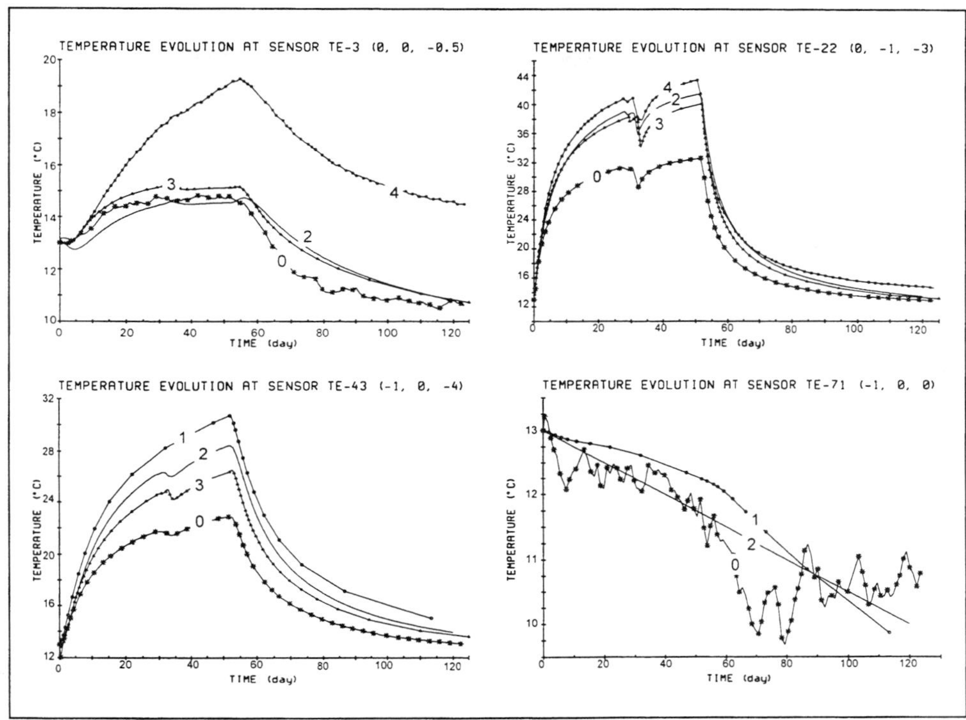

Figure 7. Comparison between measured temperature (curve 0) and temperatures computed by CT (1), EMP_c (2), EMP_d (3), and INERIS (4).

4.3. Vertical heave of the floor

The principle of this measurement is to record the vertical displacement of the floor, by means of the relative motion of photoelectric cells (V2, V4, V5, V6, and V7) in the center, with respect to reference cell V8, which is placed in one corner of the test room. This relative measurement suppresses any errors due to the instability of the laser source. It should be noted that the raw measurement data obtained by this prototype device required a large volume of processing, which could cast doubt on its reliability. It is also felt that the vertical displacements obtained, ranging from 425 to 170 µm, are rather high compared with those recorded by borehole extensometers.

The comparison between these measured plots and the curves generated by the simulations, in figure 9, is therefore purely qualitative. The photocell layout is given in figure 8, where the blocks and the fractures are also numbered, to make their motions easier to describe.

All the curves show that the computations for a discrete medium are consistent with the measurements of cells V4, V5, and V6. For cells V2 and V7, the computed values underestimate the measurements. This is explained by the fact that this computation shows that blocks 1 and 5 have heaved the most, which is just where cells V4, V5, and V6 are located (see figure 9). The major difference between this calculation and the experiment at cells V2 and V7 shows, though, that the computed heaving of blocks 2 and 6 is much less than what was actually observed.

The irregular heaving of the various blocks as computed for the discontinuous medium can be explained, for its part, by the location of the source with respect to the blocks. More precisely, the analysis of the EMP_d model results shows that two-third of the source is located within block 1 and the remaining third within block 5. The expansion therefore occurs preferentially in these blocks, whence the high heaving computed for cells V4, V5, and V6.

Despite the common features of the various curves obtained by the discrete approach, one notes that the INERIS computations yield greater heaving. This increased heaving results from the previously mentioned difference between the measured and computed temperatures on the mechanical phenomenon, whence the advantage of matching the theoretical temperatures with measurements before proceeding with the mechanical calculations.

The results of the continuum computations show a regular heave of the floor of no more than 100 µm. The smallest data occur in the case of cells V4, V5, and V6; but they are greater than those obtained by the EMP_d model for cells V2 and V7, and by the CT model for cell V2 during the heating phase.

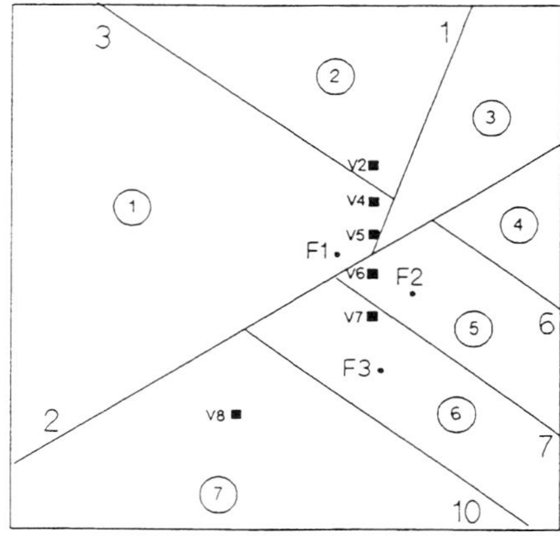

Figure 8. Location of laser cells (V2 to V8) and boreholes (F1 to F3).

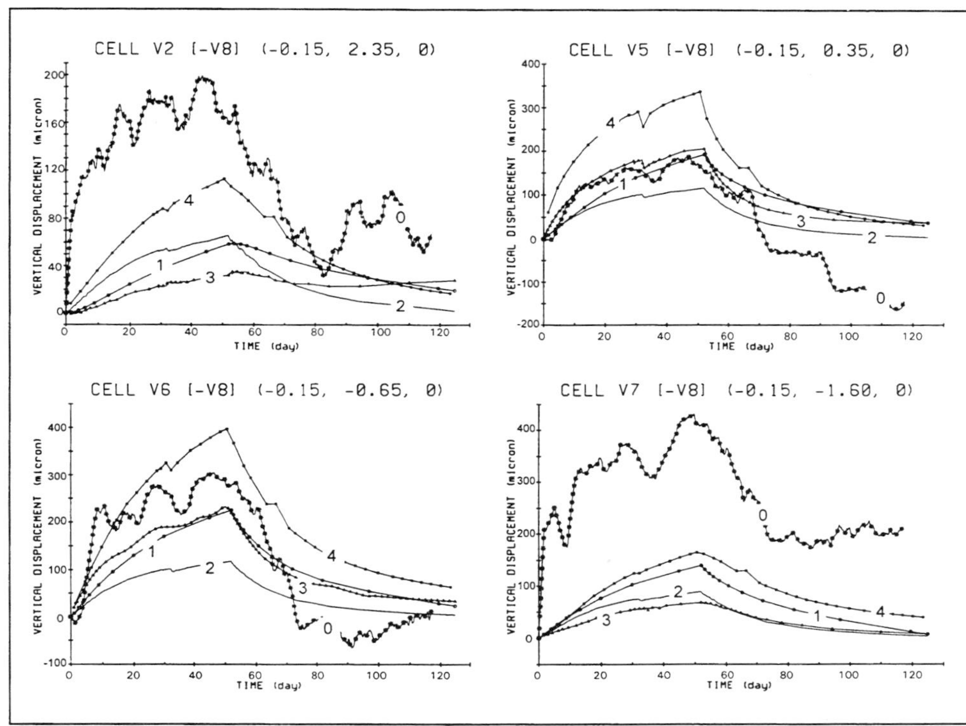

Figure 9. Comparison of vertical heave of the floor, as measured (curve 0) by laser cells, and as computed by CT (1), EMP_c (2), EMP_d (3), and INERIS (4).

4.4. Expansion of the rock

The vertical depthwise displacement is measured by three borehole extensometers (F1, F2, and F3), each including four displacement sensors. These sensors measure the time variations of the distance between the floor and the sensors' anchor points in the borehole. Figure 8 gives the layout of these boreholes on the test room floor, with respect to the modeled fractures.

For better analysis of the comparison of results, let us consider the three boreholes separately in the heating and cooling phases. Figure 10 illustrates the case of extensometer EF-12 in borehole F1, that of EF-24 (borehole F2), and EF-31 and EF-34 (F3). Each of these examples represents the results obtained in each borehole, and the order of the various results is always the same in the four anchoring points.

For borehole F1, the FEM computations for a discontinuous medium by the CT and EMP teams are in excellent agreement with the measurements, while those using the DEM (INERIS) overestimate the experimental data by at least 30 %, as is shown by the example of extensometer EF-12. The high strain values calculated by INERIS are due partly to the overestimation of the temperatures, and partly to the low Young's modulus used : that is, since the rock mass being less rigid tends to deform more easily under the effect of the heating.

Moreover, the discontinuous computations of floor heave presented in the previous section show that borehole F1 is located in a block which has been subjected to major heaving. These calculations therefore lead to greater extensions than the continuum calculations, and are closer to the experimental data, especially during the heating period. During the cooling, the results of the three models using the FEM, both in continuum and discontinuous medium, match the experimental data perfectly. At this borehole, the material does not seem to have undergone enough plastic strain, which might have distinguished the continuum and discrete results during the cooling.

In the case of borehole F2, the situation is the reverse during the heating phase, and the INERIS data matche the experiment more closely than those of the CT or EMP teams. The latter yield very similar results, which underestimate the measured extensions by about 35 %, as shown in the example of extensometer EF-24. It is also noted that the continuum results are practically the same as those generated by the FEM for a discrete medium.

During cooling, the contractions computed by DEM are marked by a very high degree of irreversibility, resulting in a general difference of the order of 60 % with respect to the measurements. The FEM results, regardless of the approach, are in better agreement with experiment. However, the interpretation of the EF-24 measurement to the effect that there is a fracture between the depths of - 3.5 m and - 4 m which opened up during the heating process but did not close during cooling is not confirmed by the computations, since no fracture intercepts this borehole section in the INERIS model.

For borehole F3, which is the farthest one from the source, the computation by the discrete approach shows that this hole is located in a block where the heaving is rather small. The strains, arranged in decreasing order, are those computed during the heating by the INERIS, CT, EMP_c, and EMP_d teams, with a major overestimation of the measurements by the INERIS model. The CT and EMP_c models envelope the experimental data well, as indicated in the examples of EF-31 and EF-34 in figure 10. These examples also show that the EMP continuum calculations are closer to experiment than the ones for the fractured medium.

Let us point out that the three elastic and elasto-plastic computations by FEM leading to practically identical results show that it is due to the external temperature variation imposed in the thermal calculation that the rock mass continues to deform during cooling, beyond its initial zero-strain state, which is not the case in the INERIS computations.

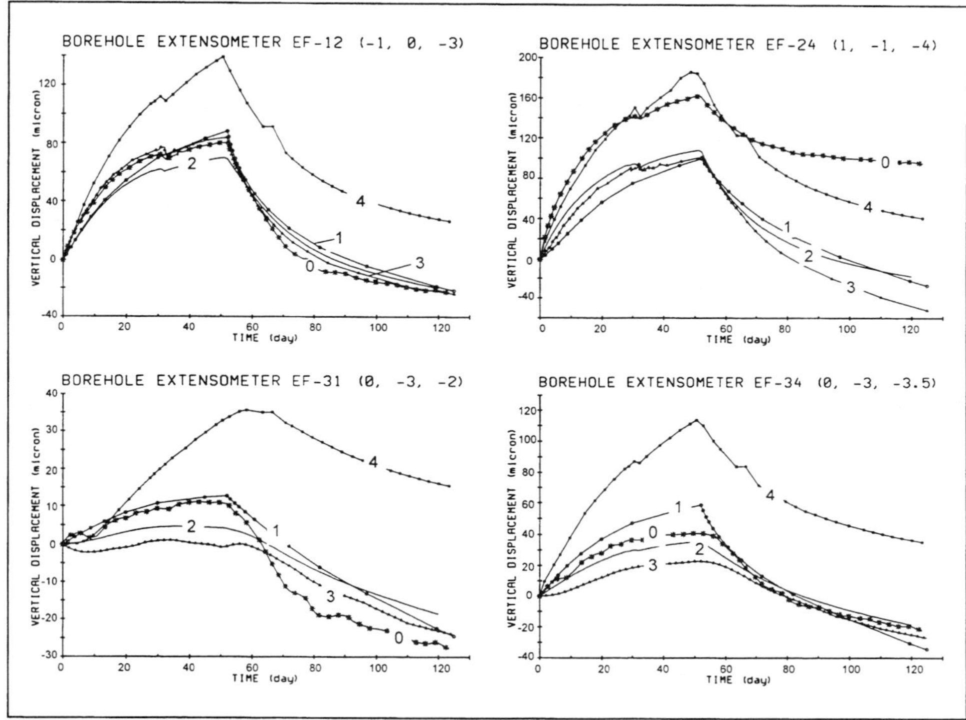

Figure 10. Comparison of vertical displacement, as measured (curve 0) by the borehole extensometers, and as computed by CT (1), EMP_c (2), EMP_d(3), and INERIS (4).

4.5.   Strains in the floor

The floor strains are measured by electric surface extensometers that record the relative displacement of two Invar studs embedded in the rock mass. The instruments denoted EL measure the ε_x strains and those denoted ET measure the ε_y strains. Figure 11 gives their position on the floor with respect to the simulated fractures. Figure 12 shows six graphs plotting the strains ε_x and ε_y at three different points.

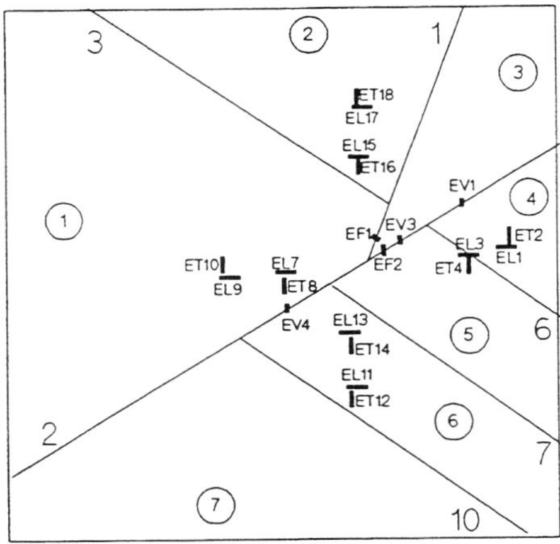

Figure 11. Location of electric extensometers on the simulated floor.

Generally, a large disagreement is observed between the experimental and computed data in all of the plots, which proves that the floor surface motions are more complex than predicted by the models. Moreover, the wide disparity in the comparisons provides us with no explanation for the differences by simple analysis of the curves. So it is difficult to draw any conclusions concerning the approaches and methods used.

However, it is clear that the CT results deviate the most from the real situation. The INERIS and EMP curves for continuum and discrete media usually follow the same tendencies as the experimental plots. These tendencies are of two types, depending on whether the measured strain is tangential (extensometers ET-2, ET-8, and EL-17) or radial (EL-1, EL-7, and ET-18), marked by a slight contraction at the beginning of the heating. It should be pointed out that these tangential and radial strains are defined in a polar axis system, referenced to the center of the floor. Quantitatively, the difference between the calculations and measurements is generally less for the points in slightly fractured zones, as shown in the cases of EL-7 and EL-17. This smaller difference is understandable in the case of the continuum model (EMP$_C$), but strange for the INERIS' discrete-medium computation.

In order to better analyze the floor movement, Vouille *et al.* (1995) plotted the two horizontal components of the displacement (U and V, respectively, along the X and Y axes) at different dates. These plots show that U is subject to greater discontinuities than V when passing through the fractures. In other words, the morphology of the simulated fracture network favors the opening of cracks in the X direction more than in the Y direction. This very interesting finding is in a good agrement with the real situation, in which the ε_x measurements (by EL-1, EL-7, and EL-17) are often smaller than the ε_y (ET-2, ET-8, and ET-18). Averaging the total of each of these measurements gives $\varepsilon_x = 26$ μm/m and $\varepsilon_y = 47$ μm/m.

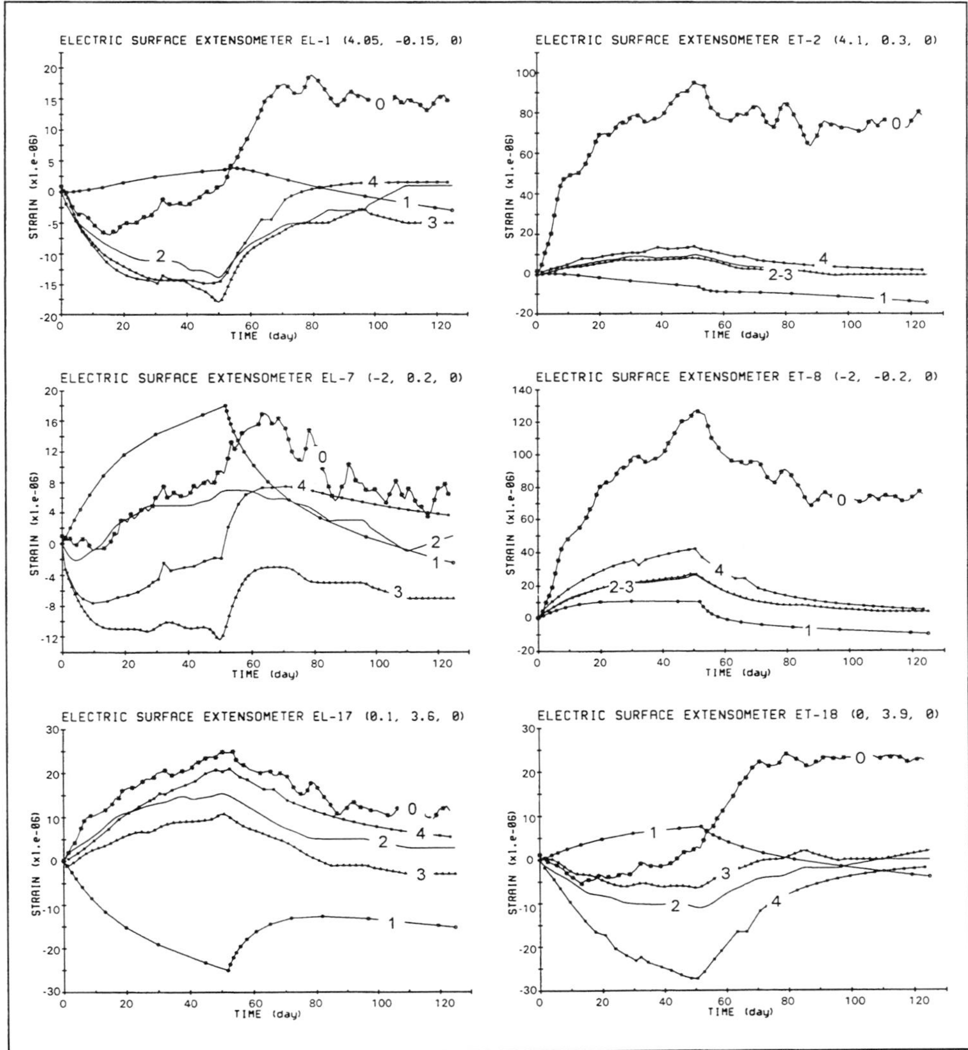

Figure 12. Comparison of strains in the floor, as measured (curve 0), and as computed by CT (1), EMP_c (2), EMP_d (3), and INERIS (4).

4.6.   Fracture openings

Here we compare the results of the various computations using the discrete approach with the openings of fractures 1 and 2 as measured by electrical extensometers EF1 and EF2, respectively, placed in the center of the floor.

The two graphs in figure 13 show that none of the three models reflects the real motions of these fractures during the heating and cooling processes. However, there is generally a qualitative consistency between the theory and experiment, except for the opening of fracture 2 as computed by the CT team.

The DEM used by INERIS yielded better estimations of the opening during the heating, and poorer ones during the cooling, where the two fractures close almost completely. This discrepancy is probably due to the oversimplified scheme of this mechanical computation (see section 3.4).

The EMP results by FEM underestimate the measurements of fracture 2 and overestimate those of fracture 1, showing some irreversibility of the motion in the cooling phase. This large difference between the opening predicted for the two fractures is confirmed by the computation which yields an extension of the floor after the heating that localizes at the level of fractures 1, 7, and 10, as shown in figure 14.

The major disagreement between the CT results and the measurements is probably due to the relatively thick joint elements used in the model.

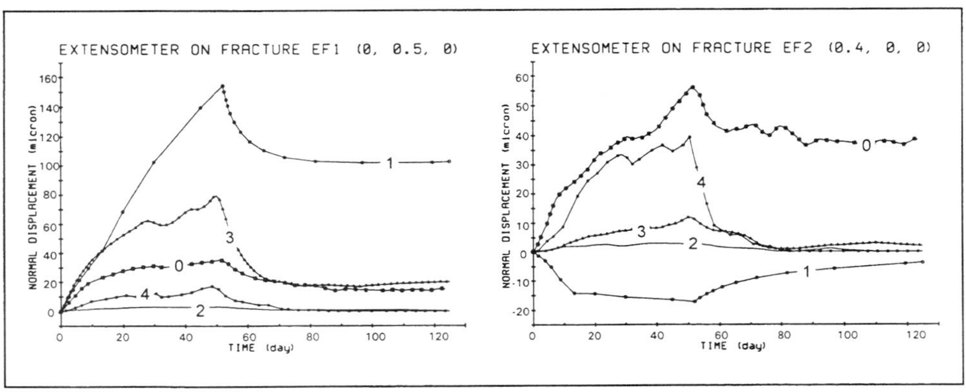

Figure 13. Opening of fractures 1 and 2. Comparison of values as measured (curve 0), and as computed by CT (1), EMP_c (2), EMP_d (3), and INERIS (4).

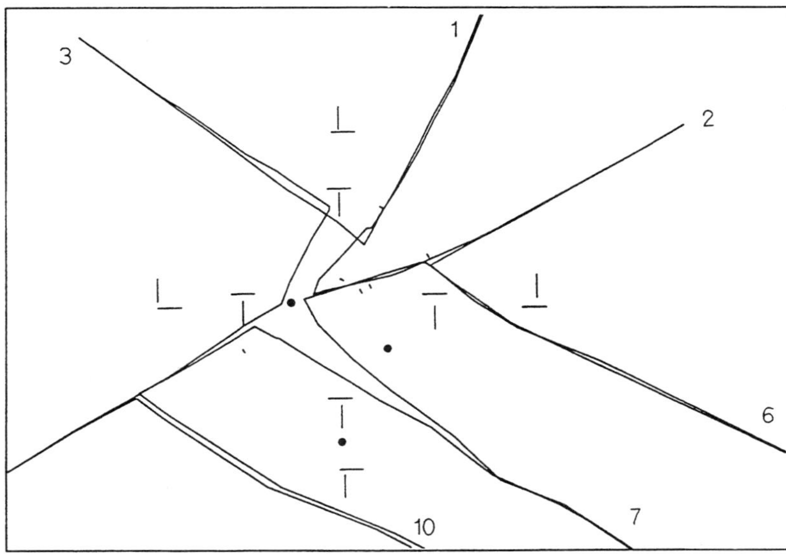

Figure 14. Deformed shape (x 10 000) of the six fractures at the end of heating as computed by $\mathrm{EMP_d}$.

4.7. Vertical shear displacement

Here we consider the differential displacement of the two edges of each main fracture, 1 and 2, measured by four EV-type electric extensometers. Their positions are given in figure 11.

In figure 15, it can be seen that a very large difference exists between the experimental and computed curves. The purely elastic computations are clearly of little interest for this measured quantity. The theoretical curves often have similar trends (as for EV1 and EV3). These trends are often related to the motions of the blocks with respect to each other, and which are specific to each computation. The measured vertical shear does not exceed 10 μm on fracture 2, and 3 μm on fracture 1, while the values computed by both methods range as high as 90 μm, which sheds doubt on the tangential stiffness value assigned to the simulated fractures.

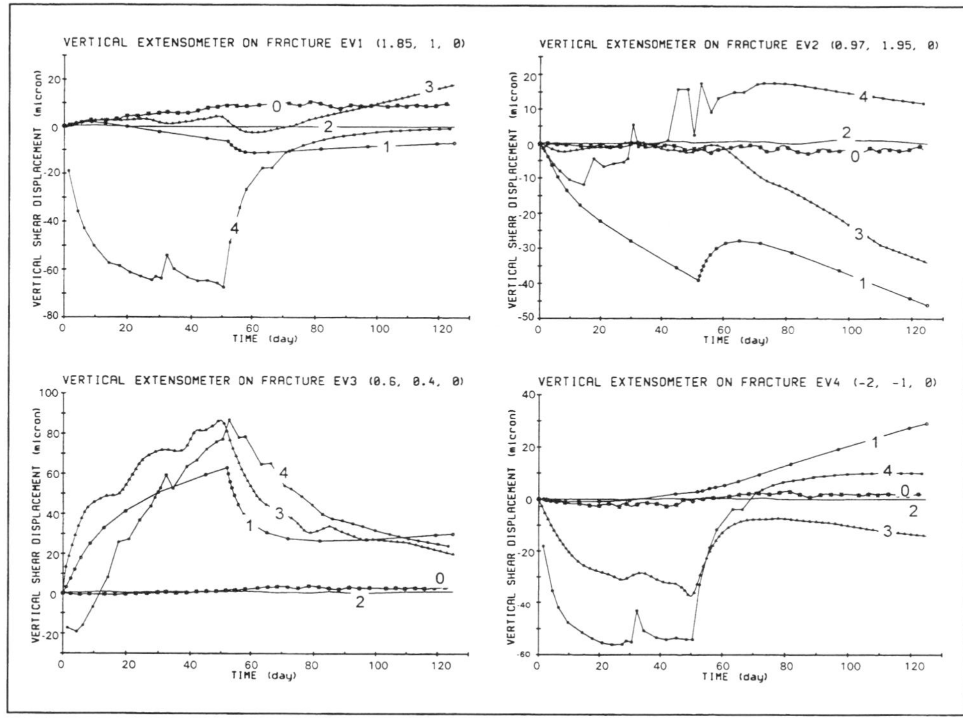

Figure 15. Vertical shear displacement on fractures 1 and 2 - Comparison of measurements (curve 0) with values computed by CT (1), EMP_c (2), EMP_d (3), and INERIS (4).

5. CONCLUSION

The DECOVALEX project TC2 test case is a very useful exercise for understanding the three-dimensional modeling complexity of fractured rock mass. This exercise achieved the objectives sought, in that the modeling of the Fanay-Augères experiment made it possible to compare the response of a continuum to that of a discontinuous medium, and furthermore to validate the models developed by comparing the simulation results with *in situ* measurements. Such comparisons give a better appreciation of the possibilities and limitations of the three-dimensional computer codes used.

Generally speaking, the TC2 example simulated by the discrete approach shows that the Finite Element and Discrete Element Methods are both entirely adequate for modeling the behavior of a set of blocks bounded by fractures having their own specific mechanical behavior. Quite aside from the method adopted, the main difficulty in studying this test case has to do with its three-dimensional aspect. The computer codes used all showed drawbacks in the still-fastidious mesh generation process, and also in the lack of any mechanical law reflecting the real behavior of a fracture.

For the comparative analysis, it is sometimes difficult to discern the reasons for certain differences occurring between computations, considering that the research teams did not always use the same parameters or assumptions. The different assumptions are particularly important for the fracture lines, which were in some cases extended to the outer limits of the model and sometimes restricted to the block of heated rock. With better coordination among the teams, the comparison would have been both easier and more accurate.

However, the following general conclusions can be drawn :

-    The temperature field calculations are in good qualitative agreement with experiment, and the use of thermal parameters specific to the rock mass will yield a better quantitative agreement between simulations and measurements.

-    The thermomechanical computations for a discontinuous medium showed a more or less heterogeneous heaving of the floor, which was closer to the real situation than that obtained by the continuum calculations.

-    As concerns the deformations at the floor surface, the fracture modeling is less beneficial. The results obtained are quite varied, and show that the real movement is much more complex.

-    The discrete approach can nevertheless serve to calculate the opening and vertical shear of the fractures. The qualitative and quantitative agreements with the experiment are less evident, which may be explained by the inability of the used models to reflect the irreversibility of the crack behavior as observed in situ.

Finally , it should be emphasized that the mechanical aspect requires further developments to improve the three-dimensional modeling of fractured rock masses, considering the crucial importance that fracturing can have on the variations of the hydraulic conditions, particularly where any tightness failure in the geological barrier offers a means of escape for radionuclides. While downscale experiments may not represent all the complexity of a real storage site, they can be of use in mastering the modeling of the physical phenomena involved, and in validating the numerical tools.

6. ACKNOWLEDGEMENTS

The author thanks the research teams and their funding organizations for supplying the results of their modeling work on the Fanay-Augères experiment.

7. REFERENCES

L. Börgesson, J. Hernelind and H. Lind, DECOVALEX - Calculation of Test-Case 2 in phase 3 - Fanay-Augères THM test - Thermo-mechanical modeling of a fractured rock volume, Clay Technology AB and Fem-Tech AB Report, (1994).

J. Cl. Gros, DECOVALEX - Phase 3 - Test Case 2 - Fanay-Augères THM test, DPEI/SERGD Report No. 92/43, (1993).

L. Jing, C. - F. Tsang and O. Stephansson, DECOVALEX - An International Co-Operative Research Project on Mathematical Models of Coupled THM Processes for Safety Analysis of Radioactive Waste Repositories, Int. J. Rock Mech. Min. Sci. & Geomech. Abstr. Vol. 32, No. 5, pp. 389-398, (1995).

A. Rejeb, G. Vouille and S. Derlich, Numerical modeling of the thermomechanical behavior of a granitic mass - Application to the simulation of Fanay-Augères THM experiment, Rev. Franç. Géotech. No. 53, pp. 21-31, (1990).

A. Rejeb, Comportement thermomécanique du granite - Application au stockage des déchets radioactifs, Doctor thesis, Ecole des Mines de Paris, (1992).

A. Rejeb, G. Vouille and J. C. Gros, Expérience thermo-hydro-mécanique de Fanay-Tenelles, Interprétation des résultats expérimentaux et validation de modèles thermomécaniques, EUR 14962 FR Report, (1994).

A. Thoraval and A. Hosni, Test Case 2 - Fanay-Augères THM test, ANDRA Report No. 694 RP INE 94 003, (1994).

G. Vouille, S.M. Tijani and B. Humbert, Projet DECOVALEX - Test Case 2 - Modélisation thermomécanique tridimensionnelle de l'expérience THM de Fanay-Augères, EMP Report No. 10.02.95.RR.01, (1995).

O. Stephansson, L. Jing and C.-F. Tsang (Editors)
Coupled Thermo-Hydro-Mechanical Processes of Fractured Media
Developments in Geotechnical Engineering, vol. 79
© 1996 Elsevier Science B.V. All rights reserved.

Experimental investigation and mathematical simulation of coupled T-H-M processes of the engineered buffer materials, the TC3 problem.

T.Fujita[a], A.Kobayashi[b] and L.Börgesson[c]

[a]Power Reactor and Nuclear Fuel Development Corporation, Tokai-mura, Naka-gun, Ibaraki-ken, Japan

[b]Iwate university, Morioka-shi, Iwate-ken, Japan

[c]Clay Technology AB, Ideon Research Centre, Sweden

Abstract

Evaluation of coupled heat and water transfer and stress changes due to swelling / shrinkage in engineered clay barriers is an important issue in the safety assessment of engineered clay barriers. Comparison of the observed results from experimental investigation using a large scale laboratory test and simulated values of coupled thermo-hydro-mechanical (T-H-M) processes within engineered buffer material using numerical models was performed in Phase III of the DECOVALEX project. This problem has been identified within the project as the Test Case 3 (TC3).

The large scale experiment was conducted using the Big-Bentonite facility (BIG-BEN) at Tokai Works, Power Reactor and Nuclear Fuel Development Corporation (PNC). The BIG-BEN is a physical model of a disposal borehole (pit) included in the current design of a repository concept considered by PNC. The experiment consists of an electric heater surrounded by glass beads, a carbon steel overpack, buffer material and man-made rock. The buffer is a mixture of bentonite and sand. The heater was operated at 0.8 kW. Water was injected at the buffer / man-made rock interface at a pressure of 0.05 MPa. The duration of the experiment was five months.

Four variables were of interest in this study: the temperature, moisture content, vertical stress and horizontal stress. The temperature changes and swelling pressures were continuously monitored and moisture content measured gravimetrically at the end of experiment (five months).

The coupled T-H-M processes were simulated by three Research Teams: Clay Technology (CLAY TECH.)/SKB, Center for Nuclear Waste Regulatory Analysis (CNWRA)/NRC and Kyoto University-PNC-Hazama Corporation (KPH)/PNC using different models. The results of the validation suggested that the changes in temperature and moisture content within the buffer can be simulated well; simulated values by all three teams matched the observed data. Observed values for vertical and horizontal stresses were not available and comparison was made between the values simulated by CLAY TECH. and KPH. The results showed a discrepancy in the simulated stress values. These discrepancies may be explained by the different parameters used in the two models although problems associated with the conceptual models can not be ruled out. It appears that further investigations are needed to improve numerical models to simulate adequately the coupled T-H-M processes in the buffer.

1. INTRODUCTION

Geological disposal of heat generating radioactive waste in many countries is based on multibarrier concept of engineered and natural barriers [1]. The engineered barriers within one of the repository concept of Power Reactor and Nuclear Fuel Development Corporation (PNC) include vitrified radioactive waste, overpack and buffer material [2]. The major roles of the buffer material are to reduce the groundwater flow, to protect the overpack from degradation and to minimize the migration of radionuclides.

Estimation of heat transfer, water flow and stress changes in the engineered clay barrier system and the evaluation of the impact of these changes on the sealing characteristics of clay barriers are important issues in the safety assessment of vault engineering. The processes fiat take place within in the clay barrier system am coupled and generally they are numerically evaluated using simulation models. In order to increase our confidence in the ability of our models to describe thermo-hydro-mechanical (T-H-M) processes comparison are made between calculated values and experimental observed values. To investigate the coupled T-H-M processes at a borehole (pit) level and to evaluate the existing technology for the emplacement of buffer material, a full-scale experimental facility for engineered barrier system, Big Bentonite facility (BIG-BEN) based on PNC current design concept was constructed at Tokai Works, PNC in 1989.

A series of experiments were performed using the BIG-BEN using different boundary conditions regarding the moisture flux at the interface of the man-made rock and buffer. The first experiment, ran in 1990, was performed with zero moisture flux in the buffer. The results from this experiment showed that moisture content in the buffer redistributes in response of thermal gradient. This process was successfully simulated by a model that described the water vapor movement under thermal gradient [3]. The second experiment was ran with partial water supply to the buffer. The third experiment was ran with uniform water injection at the boundaries of the buffer, i.e. the interface of the man-made rock and buffer, and water seepage from the surface of the pit. This experiment was intended to mimic the conditions in the vault where water flow from the joints of the surrounding rock into the buffer. The temperature, moisture content and swelling pressure changes were either continuous monitored or measured at the end of the experiment. This section present the results of the third test and comparison between observed values and simulated (calculated) values using different numerical models.

2. EXPERIMENTAL FACILITY AND BUFFER INITIAL AND BOUNDARY CONDITIONS

Figure 1 shows the general layout of the BIG-BEN. A cylindrical man-made rock (reinforced concrete) with a diameter of 6.0 m and a height of 5.0 m is located underground, below the floor of the laboratory room. The man-made rock has a central pit with a diameter of 1.7 m and a depth of 4.5 m. The dimensions of the pit are approximately equal to that of the boreholes proposed to be drilled in a disposal vault within one of the current design of PNC repository concept. An elected heater in a cylindrical overpack with diameter of about 1.0 m and height of 2.0 m was placed in the central cavity of the reinforced concrete and surrounded by compacted buffer. The radial gap between the overpack and the buffer, approximately 0.02 m, and the gap between buffer and the man-made concrete, approximately 0.03 m, had been filled with quartz sand.

The man-made rock and the buffer were installed to monitor the temperature, heat flux, water content, displacement of the overpack and swelling pressure. Specification of sensors used in the experiment are listed in Table 1. The instruments were located within three horizontal planes normal to the longitudinal axis of the pit; in the region above the heater, at the central plane level of the heater and in the region beneath the heater at the level of the bottom of the overpack.

The buffer material, a mixture of bentonite and quartz sand, was packed with a temper to the dry density of 1.6 g/m^3 and at an average gravimetric water content of 16.5 % with standard deviation of the mean of ±1.0%. The heater was operated at a constant power output of 0.8 kW to limit the temperature of the buffer material to a value < 100 °C. Water was injected at the buffer - man-made rock interface at a pressure of 0.05 MPa.

3. FUNDAMENTAL PROPERTIES AND PARAMETERS

3.1 BUFFER MATERIAL

The buffer material used is a mixture of sodium bentonite (KUNIGEL-V1 [4]) of 70 % and quartz sand of 30 %. The basic properties of the buffer material are shown in Table 2.

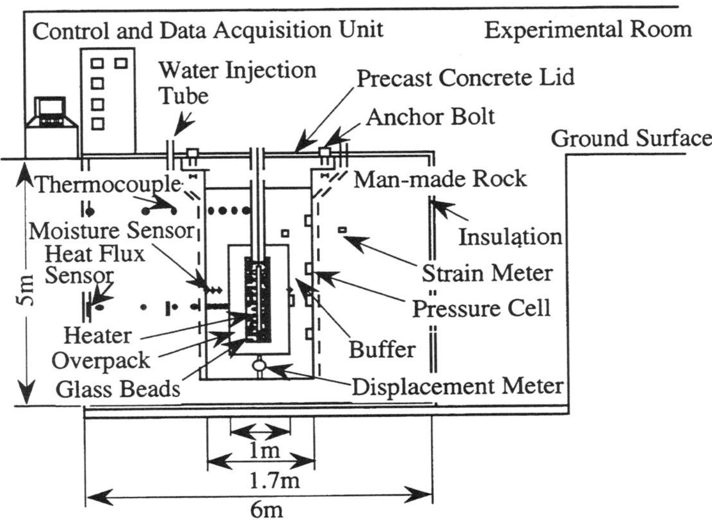

Figure 1. Schematic view of the layout of the experiment using the BIG-BEN

Table 1. Specification of sensors used in the experiment

Variable measured	Sensors	Number
Temperature	K-type sheathed thermocouples [JIS C 1602] Sheath diameter 1.6 mm Heater, Overpack, Buffer material Sheath diameter 3.2 mm Man-made rock	139
Heat flux	Heat flow meter MF-81,Temperature range -250 to +280°C Overpack MF-81, CN-130, Temperature range -50 to +120°C Man-made rock	8
Water content of Buffer material	Gypsum moisture block (RM-8) Water content range 5 to 25% Heat probe moisture (IDL-1600) Water content range 5 to 25% Humidity meter (HN-L20, HN-T21) Water content range below 13% Psychrometer (PCT-55) Water content range above 17%	38
Displacement of Overpack	Strain gauge type of displacement meter (BRD-50A) Capacity 50mm	1
Strain of Man-made rock	Strain gauge (BS-8F) Capacity ± 1000E-6 (strain)	12
Swelling pressure of buffer material	Pressure cell (F7/14QM200AZ4) Maximum pressure 20 MPa	20
Stress of buffer material	Pressure cell (S6 KF 50) Maximum pressure 5 MPa	4
Pore pressure	Pressure cell (P4KF50) Maximum pressure 5 MPa Pore pressure meter (BP20KC) Maximum pressure 2 MPa	4
Water pressure of water tube	Pressure meter (PG-50KU) Maximum pressure 5 MPa	1

Table 2. Basic properties of buffer material

Parameters	Value
Initial wet density ρ_t [g/cm^3]	1.86
Initial dry density ρ_d [g/cm^3]	1.6
Initial gravimetric water content ω_0 [%]	16.5±1.0
Initial porosity n_0 [%]	41
Young's modulus E [MPa]	27 (w =16.8) 10 (w =25.9)
Uniaxial compressive strength q_u [MPa]	15 (w =16.8) 6.8 (w =25.9)
Swelling pressure P_s [MPa]	derived from Figure 2
Saturated permeability k [m/s]	4.0×10^{-13}
Thermal conductivity λ [W/m°K]	$0.33 + 0.031\theta$
Specific heat C_p [kJ/kgK]	$(100 + 4.2\omega)/(100 + \omega)$
Isothermal water diffusivity D_θ [m^2/s]	derived from Figure 3
Relation between θ and Suction ψ [MPa]	derived from Figure 4
Coefficient of thermal expansion α [/°C]	1.0×10^{-4}

Volumetric water content ; θ [%] $= \omega\rho_d/\rho_w$,   Density of water ; ρ_w[g/cm^3]

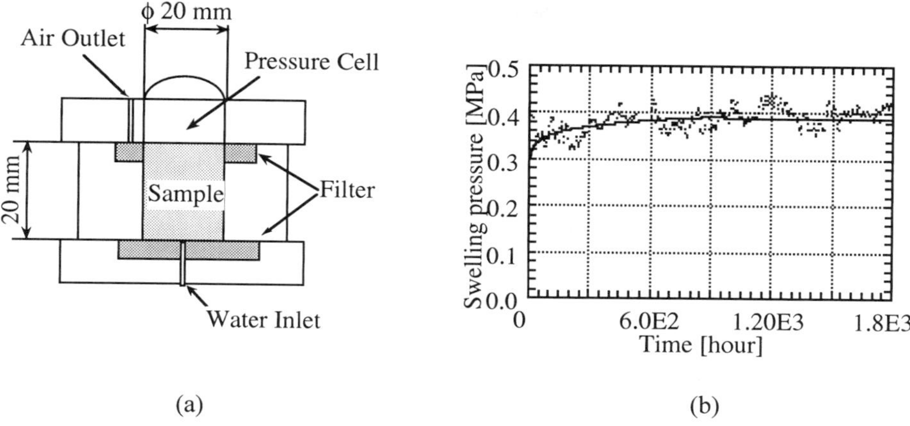

Figure 2. Equipment for swelling pressure experiment (a) and time history of swelling
pressure (b)

The result of one swelling pressure experiment is presented in Figure 2 (b). The curve shows
a stabilization point at around 0.4 MPa.

The governing differential equation for one dimension unsaturated water flow under
isothermal condition [5] is

$$\frac{\partial \theta}{\partial t} = D_\theta \frac{\partial^2 \theta}{\partial x^2} \tag{1}$$

where q is the volumetric water content, t is time, D_q is the isothermal water diffusivity, and x is
the coordinate. Figure 3 shows the measured isothermal water diffusivity of the mixture
compacted to the dry density of 1.6g/cm^3. The mean values are 1.9x10^{-6} cm²/sec (T=25°C) and
8.3x10^{-6} cm²/sec(T=50°C).

The suction was measured by a thermocouple psychrometer under confined condition to keep
the dry density constant at 1.6 g/cm^3. Figure 4 shows the relations between the suction and the
volumetric water content.

By assuming that suction is zero at the saturation (the saturated volumetric water content =
41%), the relationships y vs. q presented in Figure 4 are expressed as follow:

$$\psi = 1.06E6 \times 10^{(-0.0325\theta)} \qquad at \ T=25°C \tag{2}$$

$$\psi = 8.81E5 \times 10^{(-0.0334\theta)} \qquad at \ T=50°C \tag{3}$$

The properties of the man-made rock, the glass beads including heater and the overpack are
shown in Table 3.

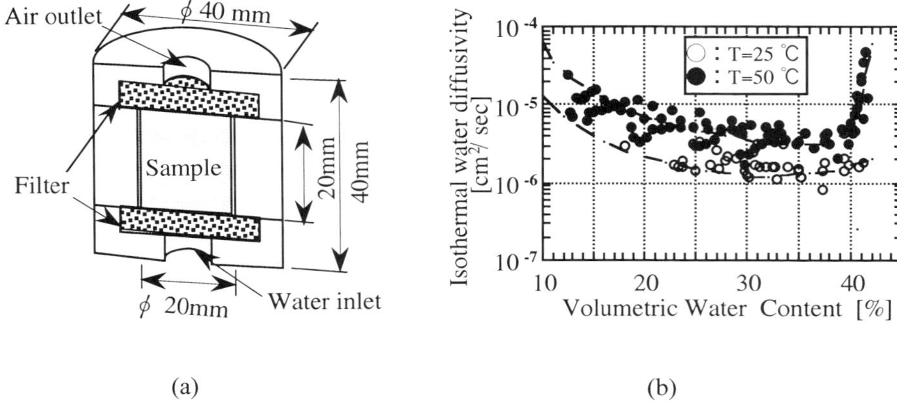

Figure 3 Equipment for water diffusion experiment (a) and isothermal water diffusivity vs. volumetric water content relationship (b)

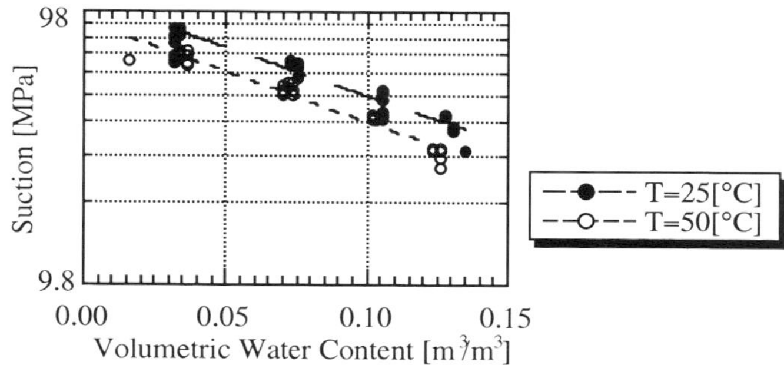

Figure 4. Relations between the measured suction and the volumetric water content

Table 3 Properties of man-made rock, glass beads and overpack

	Man-made rock	Glass beads	Overpack
Initial density ρ_i [g/cm³]	2.3	1.6	7.8
Young's modulus E [MPa]	2.50x10⁴	8.20x10⁴	2.00x10⁵
Poisson's ratio n [-]	0.167	0.3	0.3
Thermal conductivity λ [W/m°K]	1.88	0.255	53.0
Specific heat C_p [kJ /kg°K]	0.75	0.84	0.46
Coefficient of thermal expansion α [/°C]	1.00x10⁻⁵	1.00x10⁻⁵	1.64x10⁻⁶

4. RESULT OF BIG BEN EXPERIMENT

4.1 TEMPERATURE
The thermocouples were used to monitor the change in temperature during the experiment. The change in temperature distribution in the overpack, the buffer material and the man-made rock as a function of the distance from the center at the level of the heater cross sections, is presented in Figure 5 and the temperature distribution at 5 months after start, in longitudinal direction, is presented in Figure 6.

4.2 WATER CONTENT
After 5 months, the buffer material was sampled using a split-spoon sampler, and the water content was measured gravimetrically in a laboratory. The initial water content was 16.5% and the theoretical saturated water content is 26.5%. The contours of water content at two sections in the buffer material after 5 months are presented in Figure 7.

4.3 SWELLING PRESSURE
The swelling pressure was measured using the pressure cells installed at the surface of the pit. Figure 8 (a) shows the position of pressure cells and Figure 8 (b) shows the time history of the swelling pressure at different depths below the ground surface.

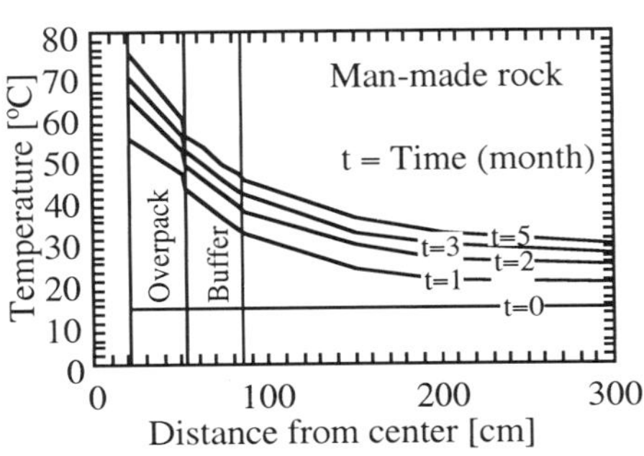

Figure 5. Time-dependent variation of the temperature distribution

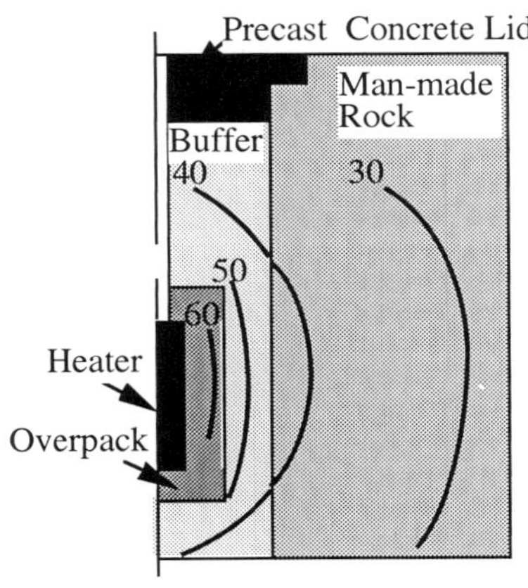

Figure 6. Contours of temperature after 5 months

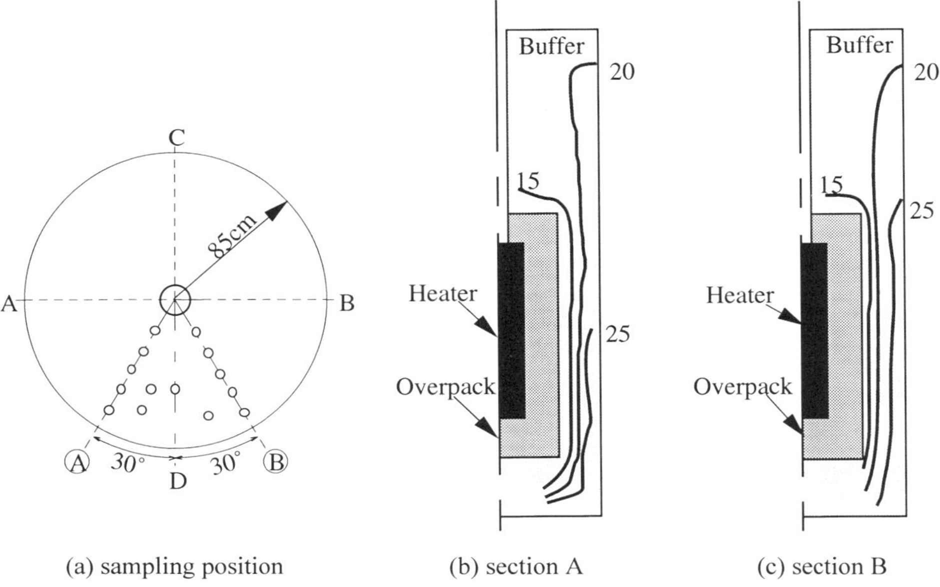

Figure 7. Contours of water content after 5 months in two different longitudinal sections.

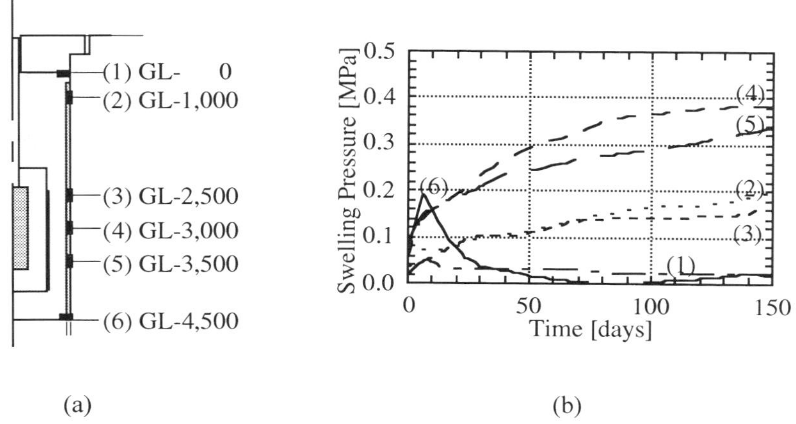

Figure 8. Cell position (a) and time history of swelling pressure of the buffer (b)

4. MODELING METHODOLOGIES FOR TC3

The experiment was simulated by CLAY TECHNOLOGY (CLAY TECH.)/SKB and Center for Nuclear Waste Regulatory Analysis (CNWRA)/NRC using ABAQUS and Kyoto university, PNC and Hazama corporation (KPH)/PNC using THAMES [6]. The modeling techniques used by different team are briefly described below. The details of the models used by the three teams are described several papers [7], [8], [9].

4.1 *CLAY TECH.* MODELING APPROACH

The heat transport is described using Fourier's law. The evaporation-condensation process is taken into account by changing the thermal conductivity with temperature and degree of saturation.

$$q_{Ti} = -\lambda_{ij}T_{,j} \tag{4}$$

$$C_p\rho\frac{\partial T}{\partial t} = \lambda_{ii}T_{,ii} \tag{5}$$

where q_{Ti} is the heat flux vector, λ_{ij} is the thermal conductivity tensor, T is temperature, C_p is the specific heat, ρ is the density and t is time.

For both the full water saturated and the unsaturated states, the water flow process is modeled by using the Darcy's law :

$$nSrV_w = \left\{k_{ij}h_{,i}\right\}_{,j} \tag{6}$$

where n is the porosity, Sr is the degree of saturation, V_w is the average velocity of water, k_{ij} is hydraulic conductivity tensor and h is the total head.

The vapor flow driven by the temperature gradient is described by :

$$q_{vi} = -D_{Tvij}T_{,j} \tag{7}$$

where q_{vi} is the vapor flux vector, D_{Tvij} is the thermal vapor diffusivity tensor.

The effective stress theory is applied to describe the stress in the buffer material

$$\sigma_{ij} = \sigma_{ij} - Sru \tag{8}$$

where σ_{ij} is the total stress, σ_{ij} is the effective stress, Sr is the degree of saturation and u is the pore pressure. While the buffer material is assumed to be elastic, the porous bulk modulus κ is given by

$$\kappa = \frac{\Delta e}{\Delta \ln P} \tag{9}$$

where P is the effective mean stress, e is the void ratio.

The swelling behavior of the buffer material is modeled by using a procedure called moisture swelling, which assumes that the amount of volumetric swelling of the solid skeleton is a function of the degree of saturation. The logarithmic measure of swelling strain $\varepsilon^{vs}{}_{ii}$ is calculated with reference to the initial saturation as follows :

$$\varepsilon_{ii}^{vs} = t_{ii}\frac{1}{3}\left\{\varepsilon^{vs}(Sr) - \varepsilon^{vs}(Sr_0)\right\} \text{ (no sum on } i) \tag{10}$$

where $\varepsilon^{vs}(Sr)$ and $\varepsilon^{vs}(Sr_0)$ are the volumetric swelling strains at the current and the initial degree of saturation and t_{ii} is anisotropic swelling ratios. The specific relationship between $\varepsilon^{vs}(Sr)$ and Sr is to be presented in Chapter 5.

The coupled T-H-M process is simulated by combining thermal calculation and coupled hydraulic-mechanical calculation. The temperature is calculated as a function of time assuming that the properties related to the water flow and mechanical behaviors are constant during a given time step. A coupled hydraulic-mechanical calculation is then made with the temperature data obtained from the temperature calculation. The temperature calculation is repeated using the results from the coupled hydraulic-mechanical calculation at each time step. The above calculations are repeated until the difference between the iterated calculations becomes smaller than a given small value. Figure 9 illustrates the procedure of the coupled T-H-M analysis performed by CLAY TECH.

4.2 *CNWRA* MODELING APPROACH

CNWRA also used ABAQUS in order to solve TC-3. The basic equations used by CNWRA to describe the T-H-M processes are the same with CLAY TECH. except for thermally driven vapor flow in equation (7) and porous elasticity in equation (9) which were not used ; the gas flow and vapor phase are not considered.

Figure 10 illustrates the procedure of the coupled T-H-M process performed by CNWRA.

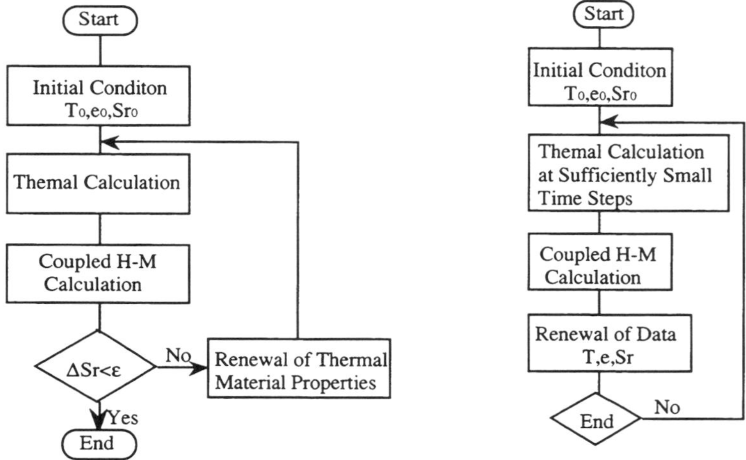

Figure 9   Procedure of coupled T-H-M analysis by CLAY TECH.

Figure 10   Procedure of coupled T-H-M analysis by CNWRA

4.3 *KPH* MODELING APPROACH

The KPH team used THAMES model, a finite element code that describes fully coupled T-H-M process in the saturated-unsaturated geological medium.

Considering the energy change by evaporation, the energy conservation equation is given as :

$$
\left(\rho C_v\right)_m \frac{\partial T}{\partial t} + nS_r\rho_f C_{vf} V_{fi} T_{,i} - \lambda_{Tm} T_{,ii} + \rho_f HD_{\theta ij} \frac{\partial \theta}{\partial \psi}\left(h_{,j} - z_{,j}\right)
$$
$$
+ nS_r T \frac{\beta_T}{\beta_p}\left\{ D_{\theta ij} \frac{\partial \theta}{\partial \psi}\left(h_{,j} - z_{,j}\right)\right\}_{,i} + \frac{1}{2}(1-n)\beta_T \frac{\partial}{\partial T}\left(u_{i,j} + u_{j,i}\right) = 0
\tag{11}
$$

where $(\rho C_v)_m$ is the volumetric specific heat of the material consisting of liquid and solid particles, n is the porosity, Sr is the degree of saturation, ρ_f is the density of water, C_{vf} is the volumetric heat capacity of the fluid, V_{fi} is the velocity vector, l_{Tm} is the thermal conductivity of the material consisting of liquid and solid particles, H is the latent heat of vaporization and $D_{\theta v}$ is the isothermal vapor diffusivity, θ is the volumetric water content, y is the potential, $\partial\theta/\partial\psi$ is the differential or specific water capacity, h, z are the total head and the position head respectively, β_T is the thermal expansion coefficient of water, β_p is the compressibility of water, u_i is the displacement vector, T is temperature and t is time.

By applying the model suggested by Philips & de Vries [9] to both the saturated and unsaturated regions, the fluid flow is described by the continuity equation :

$$
\left\{ D_{\theta ij} \frac{\partial \theta}{\partial \psi}\left(h_{,j} - z_{,j}\right) + \frac{k_{satij} K_r}{\mu} h_{,j} \right\} + \left\{ D_T T_{,j} \right\}
$$
$$
- \rho_{f0} nSr\, \rho_f\, g\, \beta_p \frac{\partial h}{\partial t} - \rho_f C(\psi)\frac{\partial h}{\partial t} - \rho_f Sr \frac{\partial u_{ij}}{\partial t} + \rho_{f0} nSr\, \rho_f\, \beta_T \frac{\partial T}{\partial t} = 0
\tag{12}
$$

where k_{sat} is the saturated permeability, K_r is the relative permeability of unsaturated zone, μ is the kinematic viscosity, D_T is the thermal water diffusivity, g is the gravity, and $C(\psi)$ is the reciprocal of the specific water capacity, $\partial\theta/\partial\psi$. The subscript, 0, stands for the reference state.

The stress equation considering a swelling behavior of the buffer material is given as

$$
\sigma'_{ij} = C_{ijkl}\, \varepsilon_{kl} - \pi\delta_{ij}
\tag{13}
$$

where C_{ijkl} is the elastic matrix tensor, ε_{ij} is the strain tensor, p is the swelling stress, and δ_{ij} is the Kronecker's delta.

5. AUXILIARY VARIABLES

A number of auxiliary variable are requested by each model. Most of the input data used in each model are given in Chapter 3. However, there are unspecified ones in each model. These data are evaluated from additional laboratory tests and numerical calibration made in each team.

The thermal conductivity, the specific heat and the density of the buffer material are used in all models. However, CNWRA team treated these variable as parameters, i.e. they are kept constant throughout the run if the model.

CLAY TECH. team calculated the thermal conductivity as a function of the porosity, temperature and the degree of saturation based on De Vries model [10].

$$\lambda_{cv} = \lambda^{T_{20}} \left(\frac{\lambda_v^T}{\lambda_v^{T_{20}}} \right)^{\frac{e}{1+e}(1-Sr)} \tag{14}$$

where λcv is the apparent thermal conductivity of the buffer material corrected for vapor flow , l^{T20} is the thermal conductivity of the buffer material at T=20°C, λ^{T20}_v is the thermal conductivity of pore air at T=20°C, and λ^T_v is the apparent thermal conductivity of pore air at temperature T. The influence of the void ratio is small in the limited range of test and the initial value $e = 0.7$ is used.

The dependence of thermal conductivity and heat capacity on temperature is small below 100°C. Thermal conductivity and heat capacity are given by the function of volumetric water content θ in the KPH model as follows;

$$\lambda = 0.33 + 3.1\,\theta \tag{15}$$

$$\rho C = \frac{0.2\,\rho_d + \theta}{\rho_d + \theta} \tag{16}$$

where ρ_d is dry density. All other variable have been previously defined

The values of the above parameters / auxiliary variables are summarized in Table 4.

Saturated hydraulic conductivity k_{sat} is used in all models. The value used by CNWRA and KPH is 4.0×10^{-13}(m/s). CLAY TECH. carried out the hydraulic test in their own laboratory and got 8.0×10^{-13}(m/s) at T=20°C and 1.6×10^{-12}(m/s) at T=50°C.

The unsaturated hydraulic conductivity k_{unsat} is calculated in the CNWRA model using the following equation :

$$k_{unsat} = K_r k_{sat} \tag{17}$$

Table 4 Summary of thermal parameters / auxiliary variables used by different team

Parameters	CLAY TECH.	CNWRA	KPH
Density [kg/m3]	186	186	$\rho_d + \theta$
Specific heat [kJ/kg °C]	1.53	1.45	$\dfrac{0.2\,\rho_d + \theta}{(\rho_d + \theta)^2}$
Thermal conductivity [W/m °C]	$\lambda^{T_{20}} \left(\dfrac{\lambda_v^T}{\lambda_v^{T_{20}}} \right)^{\frac{e}{1+e}(1-Sr)}$	1.15	$0.33 + 3.1\,\theta$

where Kr is the relative conductivity.

The Mualem-van Genuchten model is used to obtain the relative conductivity as a function of the degree of saturation

$$K_r = \sqrt{Se}\left\{1-\left(1-Se^{1/\xi}\right)^{\xi}\right\}^2 \tag{18}$$

where Se is the effective saturation and ξ is a constant.

$$Se = \frac{Sr - Sr_{min}}{Sr_{max} - Sr_{min}} \tag{19}$$

$$\xi = 1 - \frac{1}{\chi} \tag{20}$$

where Sr_{max} is the maximum saturation, Sr_{min} is the residual saturation and χ is a constant. The degree of saturation is calculated using van Genuchten relationship defined as follows :

$$Sr = (Sr_{max} - Sr_{min})\left\{\frac{1}{1+(\zeta\psi)^{\beta}}\right\}^{\left(1-\frac{1}{\chi}\right)} + Sr_{min} \tag{21}$$

where ψ is the suction and ζ is a constant.

The values of these parameters are shown in Table 5.

The isothermal water diffusivity, D_{θ}, presented in Figure 3, and the suction, ψ, presented in Figure 4, are used to unsaturated flow in the KPH model.

The thermal water diffusivity, D_T, is used in both CLAY TECH. and KPH models. D_T were evaluated from additional laboratory test made by each team. The moisture redistribution tests were made under different temperature gradients and at different temperature levels and were terminated at different time by each team. These laboratory tests were simulated with each model to give D_T. The variables are summarized in Table 6.

The buffer material is considered to be an elastic body in the CNWRA and KPH models and as an elasto-plastic body in the CLAY TECH. model. The mathematical formulation of mechanical behavior of the buffer material considers the characteristics of the structure of the solid, soil particles and water in the CLAY TECH. model. Table 7 indicate the mechanical properties of buffer material used by each team.

CLAY TECH team used in their model the relationship between the suction and the moisture swelling to model the swelling process. Table 8 shows the results obtained from the drying test [7].

Table 5 Parameters in Mualem-Van Genuchten model used by CNWRA

Parameter	Value	Parameter	Value
Sr_{max}	1.0	Sr_{min}	0.02
ζ	0.079	χ	2.0785

Table 6 Thermal water diffusivity, D_T

CLAY TECH.		KPH
$D_T = 8.0E-11 \times \sin^6\left(\dfrac{Sr}{0.3} \cdot \dfrac{\pi}{2}\right)$	$Sr \leq 0.3$	
$D_T = 8.0E-11$	$0.3 \leq Sr \leq 0.7$	$D_T = 9.0E-10 \times exp\left(1.8 \times \dfrac{T-25}{25}\right)$
$D_T = 8.0E-11 \times \cdot \cos^6\left(\dfrac{Sr-0.7}{0.3} \cdot \dfrac{\pi}{2}\right)$	$0.7 \leq Sr$	

Table 7 Mechanical properties of buffer material used by the research teams in their models

Parameters	CLAY TECH.	CNWRA	KPH
Young modulus [MPa]	-	27.0	27.0
Poisson's ratio [-]	0.4	0.4	0.4
Porous bulk modulus [-]	0.1	-	-
Friction angle [°]	50	-	-
Dilation angle [°]	5	-	-
Density [kg/m³]	1860	1860	1860
Bulk modulus (particles) [MPa]	$2.0\text{x}10^5$	-	-
Bulk modulus (water) [MPa]	$2.0\text{x}10^3$	-	-
Thermal expansion (water) [/°C]	$3.0\text{x}10^{-4}$		

Table 8 The suction and the moisture swelling used in CLAY TECH model.

Sr	ψ [MPa]	Moisture swelling	Sr	ψ [MPa]	Moisture swelling
0.01	500	-	0.55	3.102	-
0.05	314.995	-	0.60	1.954	0.025
0.10	198.444	0.111	0.65	1.231	-
0.15	125.018		0.70	o.7756	0.000
0.20	78.760	0.111	0.75	0.4486	-
0.25	49.618	-	0.80	0.3078	-0.017
0.30	31.259	0.105	0.85	0.1939	-
0.35	19.693	-	0.90	0.1222	-0.035
0.40	12.406	0.085	0.95	0.077	-
0.45	7.816	-	1.00	0.0485	-0.052
0.50	4.912	0.059			

A liner relationship between the volumetric swelling strain and the degree of saturation is assumed in the CNWRA model.

$$\varepsilon^{vs}(Sr) = \gamma(Sr - Sr_0) \tag{22}$$

where Sr_0 is the initial saturation.

The coefficient, g, was chosen to be 0.024, referring to the approach taken by Fujita et al[11]. In the KPH model the swelling stress, p in the equation (13), is calculated using the follow equation :

$$\pi = \upsilon \Delta \theta \tag{23}$$

where D_θ is the increment of volumetric water content and υ is the coefficient relating to the swelling pressure process.

The value of υ was estimated to be 17000.0 based on back calculation using data from laboratory experiments.

The parameters on man-made rock, overpack and glass beads specified in the problem are shown in Table 3. Bulk modulus of fluid and that of solid skeleton of the man-made rock used by CNWRA are assumed to be 5.0E-4 (/MPa) and 2.5E-5 (/MPa), respectively.

The annular space between the overpack and the buffer material (gap of 2 cm) was filled with quartz sand. The parameters for the annular spaces are set with the same values as those of the buffer material in CNWRA and KPH. The parameters for both gaps used by CLAY TECH. is shown in Table 9, in which the parameters of the heater is also given.

Table 9 Parameters of the quarts sand in the gaps and the heater by CLAY TECH.

Parameters	Inner gap	Outer gap	Heater
Young's modulus [MPa]	5.00	5.00	-
Poisson's ratio [-]	0.45	0.30	-
Density [kg/m^3]	1.00	2000.00	5000.00
Specific heat [kJ/kg °C]	1.00	1.60	1.00
Thermal conductivity [W/m °C]	0.30	1.50	10.0

6. COMPARISON BETWEEN OBSERVED AND SIMULATED VALUES.

All teams use a two dimensional axisymmetric model (Figure 11). Summary of mesh elements is presented in Table 10. The initial conditions and the boundary conditions are summarized in Tables 11 and 13, respectively. Only the heat calculation were made for evaluating the heat transfer coefficient in CLAY TECH. and KPH models. The heat transfer coefficient were given a fairly low value in CNWRA model. This was meant to take into account some resistance to heat flow out of the concrete due to the thin layer of insulation surrounding the apparatus.

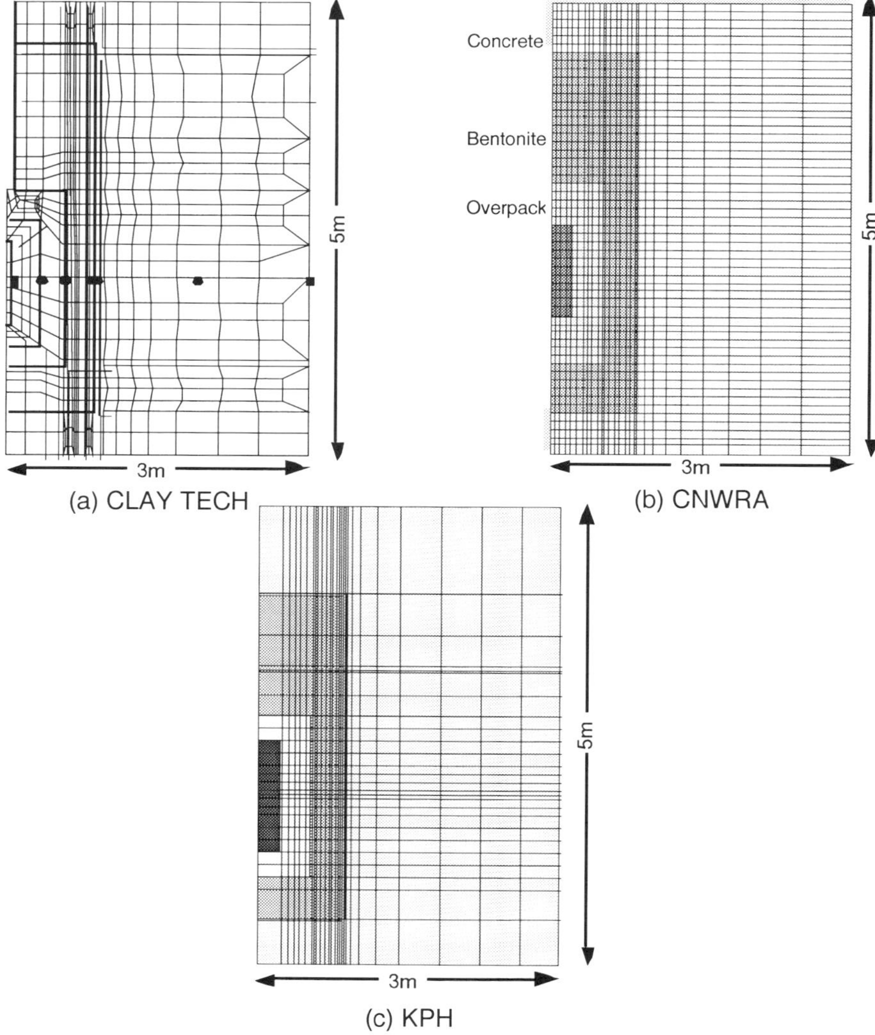

Figure 11 Meshes used for TC-3 ; (a) CLAY TECH., (b) CNWRA, (c) KPH

Table 11 Summary of mesh elements ; element type, the number of elements and nodes

	CLAY TECH.	CNWRA	KPH
Element type	8-nodes (Temperature, Hydromechanical) 4-nodes (Moisture)	8-nodes	8-nodes (Deformation) 4-nodes (Temperature, Pressure)
Number of elements	750 (Temperature) 364 (Hydromechanical, Moisture)	1485	806
Number of nodes	6000 (Temperature) 2912 (Hydromechanical) 1465 (Moisture)	4620	2533

Table 12 Initial conditions

Variable	CLAY TECH.	CNWRA	KPH
Temperature [°C]	15	15	15
Water content [%]	16.6	16.5	16.5
Void ratio	0.7	-	0.69
Effective mean stress [MPa]	0.8	-	0.02 (radial) 0.05 (vertical)
Pore water pressure [MPa]	-1.231	-	-14.31

Table 13 Boundary conditions

Parameters	CLAY TECH.	CNWRA	KPH
Boundary condition for heat calculation	Heat transfer coefficient [W/m^{2}K]; 0.1 (upper) 0.5 (middle) 0.1 (lower)	Heat transfer coefficient [W/m^{2}K]; 0.5	Heat transfer coefficient [W/m^{2}K]; 1.16
Boundary condition for hydraulic calculation	Hydrostatic pressure	0.05 [MPa]	0.05 [MPa]
Boundary condition for mechanical calculation	stiff	stiff	slider

6.1 TEMPERATURE

Figures 12 and 13 show the temperature distribution both measured and calculated values in the buffer material after 1 and 5 months, respectively. The solid lines represent the temperatures calculated by each team, while the dotted line indicates the measured one.

The calculated temperature values by all teams at G.L.-1.5m are a little lower than the measured values. This may be caused by different boundary conditions at the top of the pit then the boundary conditions assumed in the models. For the other levels, the temperature values calculated by CNWRA is a little higher than the temperature observed as well as the values calculated by other teams. Calculated values by the KPH team are slightly lower that the calculated by other teams. Calculated values by the KPH team are slightly lower that the observed temperature values. The maximum temperature difference between observed value and calculated value was 5 °C. The results of this study suggested that the differences between observed and calculated temperature, values were mainly due to the differences between the exited boundary conditions and the assumed boundary conditions in the simulations. However, the overall changes of temperature, both in space and time domains, within the buffer over the duration of the experiment were properly estimated by all three models and simulated thermal gradients were in a good agreement with the observed values.

6.2 WATER CONTENT

Figure 14 shows the calculated gravimetric water contents in the buffer after one month. For the region above the heater (G.L. = -1.5 m) all three models predicted very little changes in the moisture content. For the central region (G.L. = - 3.0 m) and the lower region (G.L. = -4.0 m) the CNWRA model predicted also only small changes in the moisture content whereas the CLAY TECH. and KPH models indicted an increase in moisture content with the higher values at the buffer - man-made rock interface (i.e. water uptake).

Figure 15 shows the calculated and measured values of moisture content after five months, at the end of experiment. For the region above the heater, all three models predicted an increase in moisture contents (water uptake). The simulated values were in a good agreement with the observed values. For the central region, again all three models predicted change in moisture contents that were very close to the observed values. However, for the area adjacent to the overpack, the observe data indicated a decrease in moisture contents due to the thermal gradient. This phenomenon was properly described by the CLAY TECH. and KPH models although the KPH model predicted a larger effect of thermally induced water flux than observed. The CNWRA model did not described this phenomenon since the model disregarded the thermally induced moisture flux. For the region above the heater, all the models predicted similar change in the moisture contents and were in good agreement with the observed values in the area close to the overpack but much lower in the area close to the buffer - man-made rock interface. Overall the models simulated reasonable well the changes in the moisture content in the buffer, although problems associated with accurate in situ measurement of moisture contents as well as the relatively short duration of the experiment do not permitted to draw a definitive conclusion regarding the ability of the models tested to describe moisture redistribution in the buffer. The effect of boundary condition is small. This is likely because the suction values in the buffer is very large.

6.3 STRESS

Measured radial and vertical stress values were not available. Figures 16 - 19 show some of the simulated values; positive values are assigned to the compressive stress. Figures 16 and 17 show the horizontal and vertical stress values after one month. The radial and vertical predicted stress values by the CLAY TECH. and KPH models were similar. Both models predicted very small changes in stresses after one month in the range of 0.02 to 0.15 MPa. The CNWRA model predicted changes in stress in the range of -8.0 to 10.2 MPa that are one order of magnitude larger than the predicted values by the CLAY TECH. and KPH models. Figures 18 and 19 show the simulated stress values after five months by CLAY TECH. and KPH models. The predicted values by the KPH model suggested that only a compression mechanism (i.e. positive stress values) whereas the CLAY TECH. model suggest some vertical expansion in the central and below the heater regions.

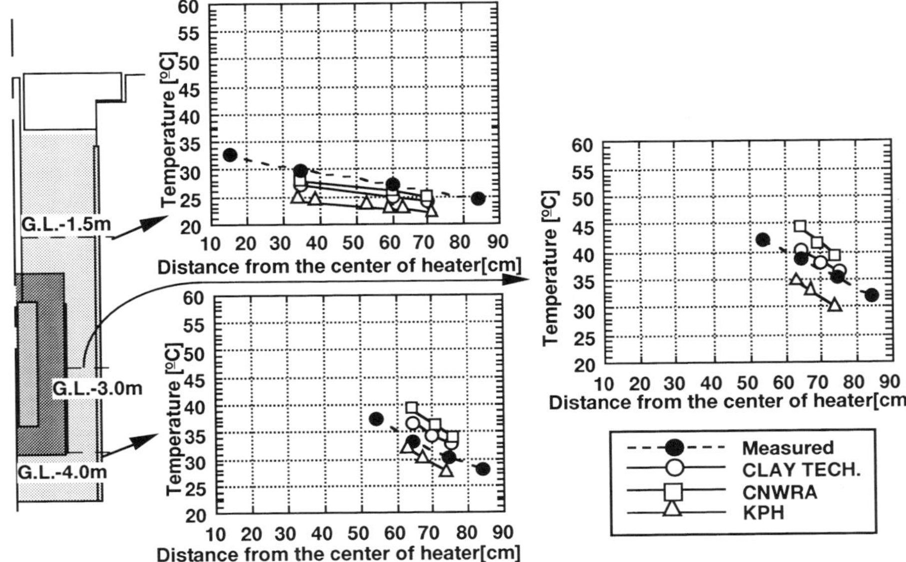

Figure 12. Observed and calculated temperature distribution in radial direction after one month

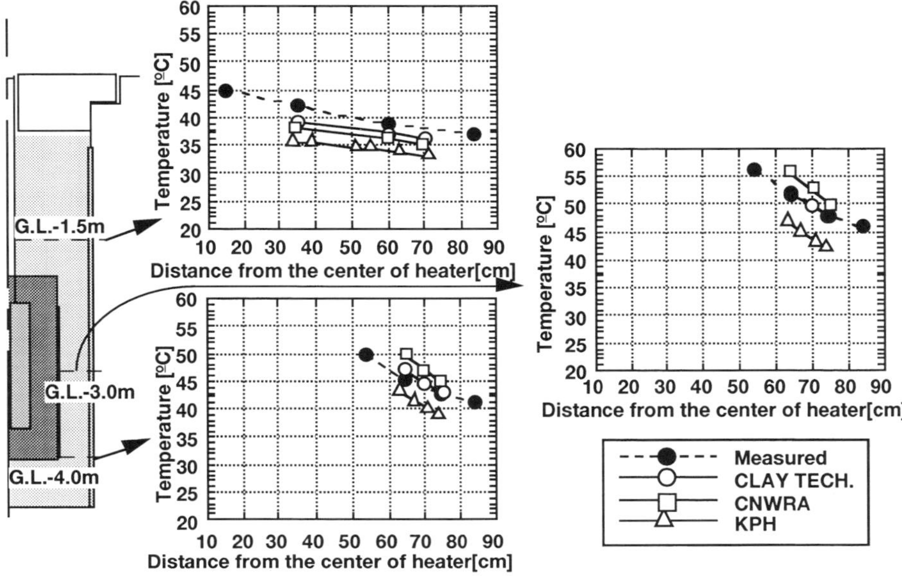

Figure 13 Observed and calculated temperature distribution in radial direction after five months

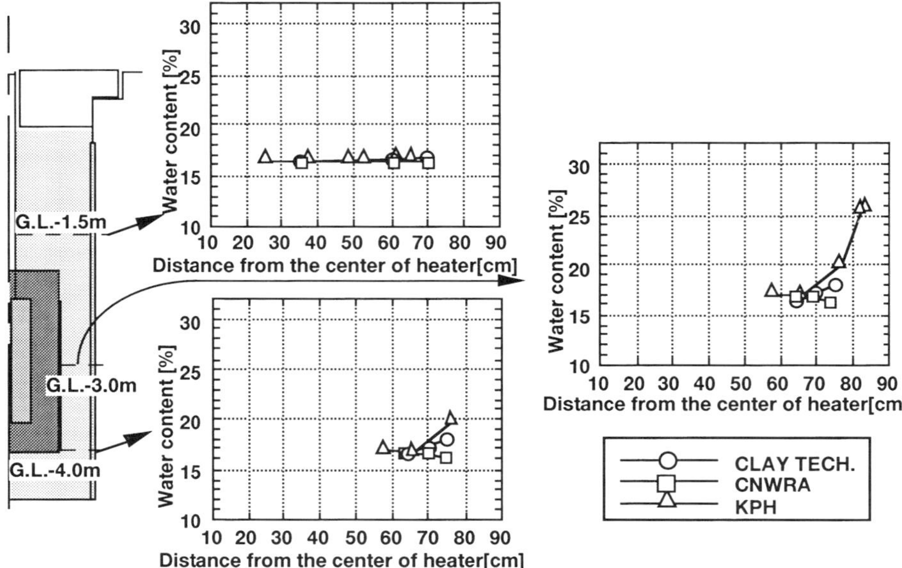

Figure 14 Calculated gravimetric water content distribution in radial direction after one month

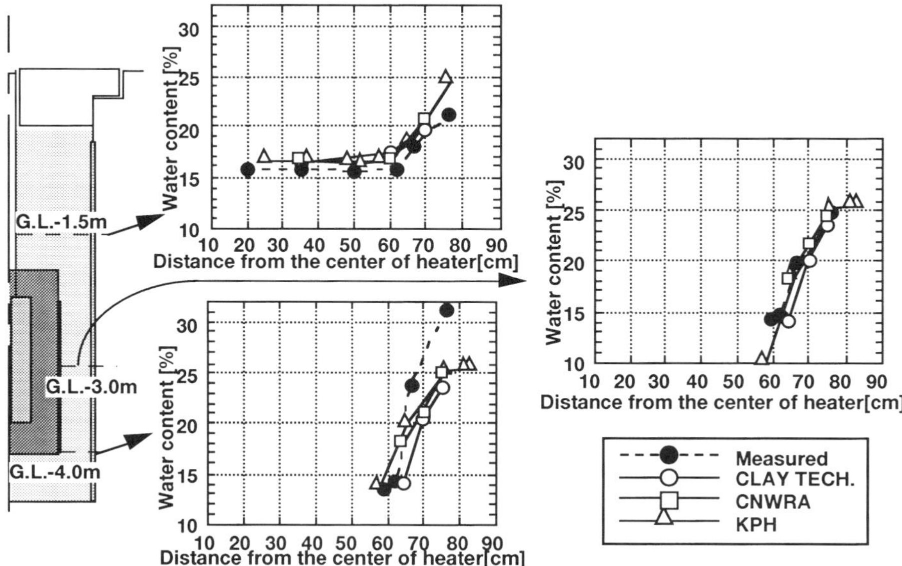

Figure 15 Calculated and measured gravimetric water content distribution in radial direction after one months

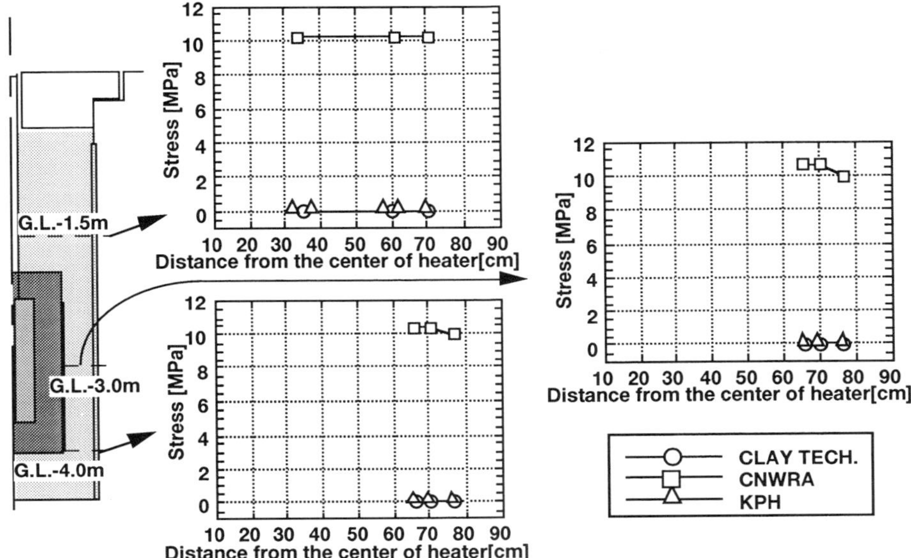

Figure 16 Simulated horizontal stress distribution in radial direction after one month

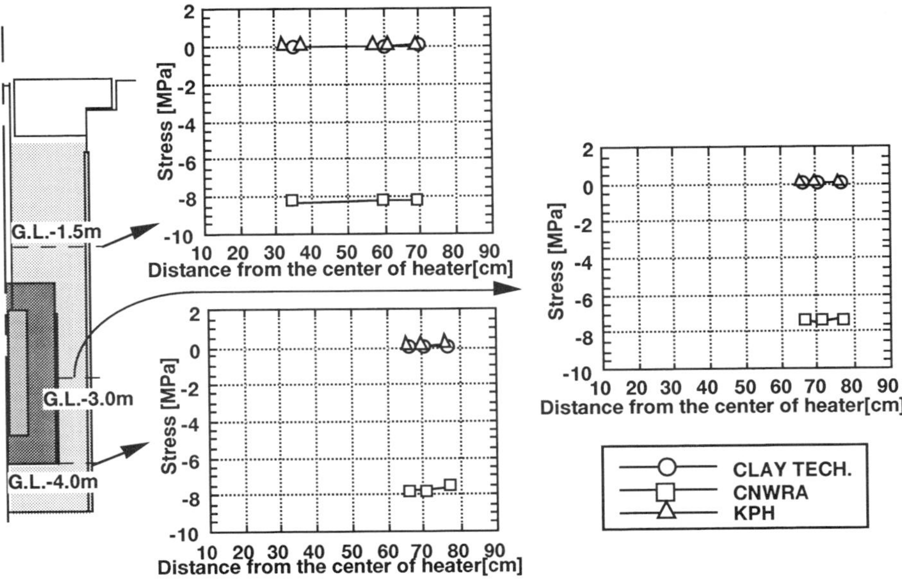

Figure 17 Simulated vertical stress distribution in radial distance after one month

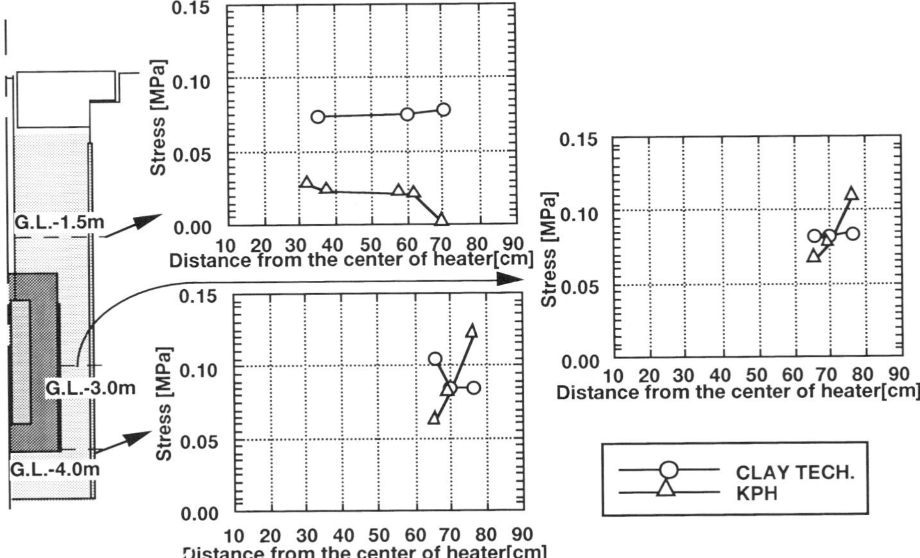

Figure 18 Simulated horizontal stress distribution in radial direction after five months

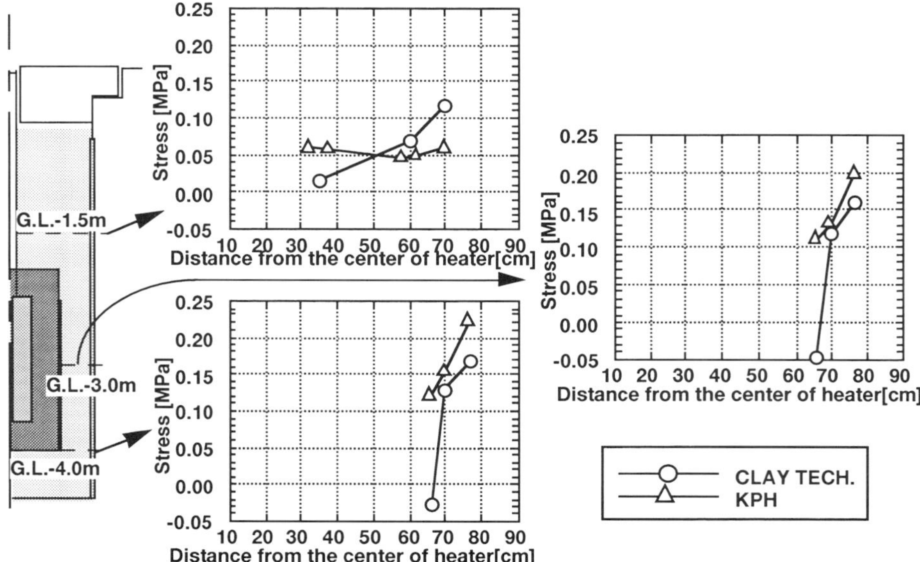

Figure 19 Simulated vertical stress distribution in radial direction after five months

7. CLOSING REMARKS

TC-3 is the problem to simulate the coupled T-H-M process of engineered barrier system and focuses on the behavior of the bentonite. The research teams used different approaches and assumptions regarding some of the material properties and/or boundary conditions. The coupled T-H-M process is simulated with combined applications of thermal and coupled hydraulic-mechanical calculation by the CLAY TECH. and CNWRA teams and with fully coupled calculation by the KPH team. From the results of this test case, the following conclusions are obtained.

(1) Although the maximum difference between the measured and calculated temperature is about 5°, the overall changes of temperature, both in space and time domains, within the buffer over the duration of the experiment were properly estimated by all three models and simulated thermal gradients were in a good agreement with the observed values. The results of this study suggested that the differences between observed and calculated temperature, values were mainly due to the differences between the exited boundary conditions and the assumed boundary conditions in the simulations.

(2) Overall the models simulated reasonable well the changes in the moisture content in the buffer, although problems associated with accurate in situ measurement of moisture contents as well as the relatively short duration of the experiment do not permitted to draw a definitive conclusion regarding the ability of the models tested to describe moisture redistribution in the buffer.

(3) The comparison of the stresses among models show little or no agreement.

It appears that future study is necessary to model the swelling behavior and coupling two phase flow of buffer material. In the coupled T-H-M experiment at Kamaishi, the experiences in TC-3 will be made good use of.

REFERENCE

1. A.Oyama, Nuclear Fuel Cycle Policy in Japan, The Third International Conference on Nuclear Fuel Reprocessing and Waste Management, RECOD'91, Sendai, Japan, April 14-18, 1991.

2. Sato,S. Kobayashi,A. Hara,K. Ishikawa,H. Sasaki,N., Full Scale Test on Coupled Thermo-Hydro-Mechanical Process in Engineered Barrier System, Proc. of '91 Joint International Waste Management Conference, ASME. Seoul, Korea, Oct., 1991.

3. Isikawa,H., Amemiya,K., Yusa,Y., and Sasaki,N.; Comparison of Fundamental Properties of Japanese Bentonite as Buffer Material for Waste Disposal, Proceedings of the 9th International Clay Conference, Strasbourg, 1989 Sci. Grol., Mem. 87, pp.107~115.

4. Richards,L.R., Capillary conduction of liquids through porous mediums, Physics, 1, 318-333, 1931

5. Ohnisi,Y. Shibata,H. Kobayasi,A., Coupled Processes Associated with Nuclear Waste Repositories, edited by Tsang, C.F., pp. 679-697, 1989.

6. Börgesson,L. and Hernelind,J., DECOVALEX TEST CASE 3 -CALCULATION OF THE BIG-BEN EXPERIMENT-, Coupled modelling of the thermal, mechanical and hydraulic behavior of water-unsaturated buffer material in a simulated deposition hole, SKB ARBETSRAPPORT 94-49, 1994

7. Ahola,M., Ofoegbu,G., Chowdhury,A. and Hsiung,S., THERMO-HYDRO-MECHANICAL COUPLED MODELING ; BIG-BEN EXPERIMENT, TC3, DECOVALEX - PHASE III, Nuclear Regulatory Commission Contract NRC-02-93-005, 1994

8. Fujita,T., Moro,Y, Kobayashi,A., Hara,K. and Ohnishi,Y., Current Status and Future Plan of Thermal-Hydro-Mechanical Process Modeling and Validation Experiments, INTERNATIONAL WORKSHOP on RESEARCH & DEVELOPMENT of GEOLOGICAL DISPOSAL, Proceedings of Technical Session, pp.II-1-11, 1993

9. Philip,J.R. and de Vries,D.A, Am. Geophys. Union Trans, 38(2),pp.222-232.,1957

10. De Vries, D.A., Heat transfer in soils. In Heat and Mass Transfer in the Biosphere. 1. Transfer Processes in Plant Environment (D.A. De Vries and N.H.Afghan, Eds.) New York; John Wiley & Sons Inc.,1974

11. Fujita,T., Hara,K., Yusa,Y. and Sasaki,N., Application of Elasto-Plastic Model to mechanical and Hydraulic behavior of Buffer Material under Water Uptake in a Repository, Materials Research Society Symposium, Vol.212, 1991

O. Stephansson, L. Jing and C.-F. Tsang (Editors)
Coupled Thermo-Hydro-Mechanical Processes of Fractured Media
Developments in Geotechnical Engineering, vol. 79
© 1996 Elsevier Science B.V. All rights reserved.

Coupled mechanical shear and hydraulic flow behavior of natural rock joints

Mikko P. Ahola, Sitakanta Mohanty[a], and Axel Makurat[b]

[a]Center for Nuclear Waste Regulatory Analyses, San Antonio, Texas, 78238
[b]Norwegian Geotechnical Institute, P.O. Box 3930, Ullevaal Hageby, N-0806, Oslo, Norway

Abstract

A comprehensive summary of the coupled mechanical hydraulic (MH) behavior of natural rock joints as investigated in two different laboratory experiments (test cases) under the international cooperative project DECOVALEX is presented. The first test case experiment (TC1) was conducted by the Norwegian Geotechnical Institute (NGI) using their coupled shear-flow test apparatus. With this apparatus, joints can be closed and sheared under stress controlled conditions while fluid is injected into the fracture. Two different tests were conducted on joints of varying roughness denoted as TC1:1 and TC1:2. Numerical simulation of both TC1:1 and TC1:2 was also conducted using several computer codes with different constitutive relations for the joint. The second test case experiment (TC5) was conducted at the Center for Nuclear Waste Regulatory Analyses (CNWRA) using a direct-shear apparatus modified to include fluid flow within the joint. In this apparatus, the fluid is injected along one edge (inlet) of the rectangular joint and simultaneously collected form the opposite edge (outlet), while the remaining two joint edges are sealed to prevent leakage. Linear flow measurements were conducted under normal load as well as under combined normal and shear loading on a rock comprised of two blocks of rock bounding a naturally fractured, welded tuff joint. Numerical simulation of both test cases required assumptions to be made with respect to the material properties and specifically the boundary conditions. For TC1:1 and TC1:2, numerical modeling showed some deviation between the Barton-Bandis joint model and the experimental results. The deviation might be caused by the joint itself, or the way the joint behaves in the apparatus. With regard to the shear deformation behavior, the modeling of both TC1:1 and TC1:2 shows that the dilation and changes in hydraulic aperture are not well modeled. However, some of the numerical results did tend to agree with the upper bound of the measured hydraulic aperture as well as the residual hydraulic aperture. The experimental results for TC5 show that the hydraulic conductivity of the joint can change by up to a factor of 3 under shear deformation. However, it was determined that additional tests and improvements to the experimental apparatus are necessary in order to make more quantitative assessments of changes in such properties during shear and production of gouge within the joint.

1. INTRODUCTION

The coupled mechanical-hydraulic (MH) behavior of rock joints is an important issue in many applications, including the analysis of oil and gas reservoirs as well as deep geologic disposal of nuclear waste. In the case of underground disposal of nuclear waste, accurate prediction of the fluid flux within the near-field waste emplacement area is necessary for assessment of waste package corrosion rates as well as calculations on radionuclide migration to the accessible environment. The host rock environment for siting a repository for nuclear waste has predominantly focused on competent hard rock formations such as granite (e.g., Canadian and Finnish Programs) or welded volcanic tuff (e.g., U.S. Program). Such rock units have predominantly low porosities and matrix permeabilities, such that the majority of the fluid flow through the waste emplacement horizon is likely to be along natural and induced joints. Joints can be induced by excavation of the tunnels, thermal expansion due to heating, seismic motion, and overall long-term deterioration of the excavation disturbed zone. The primary or natural joints can be formed due to compressional or extensional tectonics, uplift, etc.

To better understand the coupled MH behavior of natural rock joints as well as to establish a basis for comparison with numerical codes, two experiments have been conducted under DECOVALEX. The focus of this chapter is to provide a comprehensive review of the experimental results, as well as the theories and numerical analyses used by various computer codes to simulate the response of a jointed rock mass to excavation and nuclear waste storage induced temperature changes. The first test case experiment (TC1) was conducted by the Norwegian Geotechnical Institute (NGI) using their coupled shear-flow test apparatus. With this apparatus, joints can be closed and sheared under controlled stress conditions while fluid is injected into the joint. Numerical simulation of this TC1 experiment was conducted using several computer programs with different joint constitutive relations. The second test case experiment (TC5) was conducted at the Center for Nuclear Waste Regulatory Analyses (CNWRA) using a direct-shear apparatus, modified to include fluid flow in the joint. In this apparatus the fluid is injected along one edge (inlet) of the rectangular joint and simultaneously collected from the opposite edge (outlet), while the remaining two joint edges are sealed to prevent leakage. Thus, it is expected that linear flow takes place between these two inlet and outlet edges. Linear flow experiments were conducted under normal load as well as under combined normal and shear loading on a rock specimen comprised of two blocks of rock bounding a naturally fractured, welded tuff joint.

2. BACKGROUND

Changes in stress conditions in the near-field rock mass surrounding underground tunnels due to excavation, thermal expansion (in the case of nuclear waste storage), and possibly seismic loadings result in incremental deformation along the joints both in the normal and shear directions [1-3]. Laboratory studies on single rock joints have shown that mechanical deformation can have a significant influence on the hydraulic properties of rock joints [4,5]. The MH behavior of a deformable rock joint under pure normal stress has been investigated by Tsang and Witherspoon [5] and Cook [6] among others, and appears to be fairly well understood. In most cases under laminar flow conditions in the joint, the cubic law which relates the flow to the cube of the aperture can be applied with fairly reasonable accuracy. However, for joints with high surface roughness or small apertures, modification of the cubic law becomes necessary [5,7], perhaps because of differences between the mechanical aperture and theoretical smooth wall aperture, turbulent flow, increasing tortuosity, and boundary layer effects. Both the mechanical and hydraulic response of rock joints to variation in normal stress are fairly repeatable after the first 3-4 normal loading cycles, namely behaving in a nonlinear elastic manner with little hysteresis between loading and unloading of the joint. Also, in most cases, normal loading of the joint creates little gouge material due to crushing of asperities as compared to the more destructive shear loading, all of which make it easier to understand the coupled MH behavior of rock joints under pure normal loading.

It is known that during joint shear displacement, dilatancy and asperity degradation can substantially modify joint flow characteristics [4,8]. In most cases, shear displacement has a much greater impact on the hydraulic properties of joints than normal deformation. This is specifically valid for shear displacement and high normal stresses compared to the strength of the joint surface. There has been an increasing effort to develop a better fundamental understanding of the role of shearing of a fracture on its effective hydraulic conductivity aperture. However, asperity degradation due to shearing and the resulting changes in hydraulic aperture seem to be much more difficult to predict and model than those primarily caused by joint normal displacements. Gouge formation may in some cases restrict the flow, thus reducing the hydraulic conductivity even though the joint is undergoing dilation. The amount of asperity degradation and subsequent gouge production is dependent of the mechanical properties of the rock joint as well as the applied loading state.

Experimental studies have been conducted over the past several years to investigate permeability changes during shear displacement of rock joints. Teufel [9] conducted shear-flow coupling tests on pre-fractured samples of Coconino sandstone using a triaxial experimental apparatus and showed that the hydraulic conductivity across (i.e., perpendicular to) a joint decreases with increasing shear displacement because of localized deformation along the joints and the evolution of a gouge zone. These studies showed that the reduction in permeability across the joint during sliding on these artificially fractured specimens increased markedly with an increase in effective confining pressure and normal stress across the joint. Teufel found that for a test conducted at 60 MPa confining pressure and with a normal stress across the joint of 118 MPa to 132 MPa, the permeability decreased nearly 3 orders of magnitude after a shear displacement of 7.1 mm. This was attributed to the development of a gouge zone between the fractured surfaces, and the progressive decrease in grain size and porosity of the gouge during shear and increasing normal stress. Teufel also observed the creation of microfractures adjacent and subparallel to the sliding surface, the density of which also increased with normal stress. Based on this observation, he suggested that, in sharp contrast to a gouge zone which decreases permeability perpendicular to the fracture, localized microfracturing may create a narrow channel of high permeability parallel to the fracture. These results, however, may not be directly applicable to similar studies for nuclear waste disposal due to the different rock types involved.

Makurat [4] conducted coupled shear displacement and conductivity tests on natural joints for a number of rock types using a biaxial cell. In contrast to Teufel [9], they measured the hydraulic conductivity along the joint, and concluded that whether the joint conductivity increases or decreases with ongoing shear is dependent on both the joint and rock properties, as well as the stresses applied. The main parameters of importance were suggested to be the uniaxial compressive strength of the intact rock, joint compressive strength, joint roughness coefficient (JRC), normal stress, and shear displacement. It was found that relatively small shear displacements are sufficient to dilate hard rocks and cause the joint hydraulic conductivity to increase up to two orders of magnitude. They determined that this type of behavior occurred when there was a high ratio of joint compressive strength to applied normal stress ratio as well as a distinct joint roughness morphology. However, it was found that strong mineralization and repeated shearing tended to reduce joint conductivities even in high strength rocks. For soft sedimentary rocks, it was found that even though dilation occurred during shearing on all joints, not all joints experienced a resulting increase in hydraulic

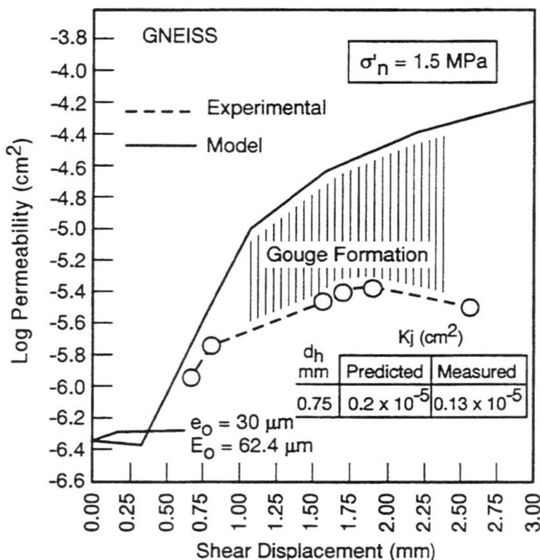

d_h mm	Predicted	Measured
0.75	0.2×10^{-5}	0.13×10^{-5}

Figure 1. Comparison between predicted and measured joint conductivity during joint shearing [4].

conductivity. Only those joints with a high value of JRC experienced an increase in permeability. Makurat [4] determined decreases in hydraulic conductivity during shearing to be a result of gouge production, which tended to block flow paths and disturb the parallel plate analogy. This is apparently the main reason why existing models developed to predict the coupled flow response during shearing have not agreed well with experimental measurements, as depicted in Figure 11, since most do not account for cumulative damage along the joint surface during shear. As shown in Figure 1, even if the numerical model takes into account the reduction in asperity height during shear by reducing the joint roughness coefficient, the joint hydraulic conductivity is still overestimated due to the inability to take into account the influence of gouge production.

Experiments were conducted by Esaki [10] to investigate the shear-dilation-flow characteristics of artificially created granite joints extended over displacements well past those corresponding to peak shear stress. Their goal was to expand the experimental database developed from previous studies to include larger shear displacements and higher normal loads, which would more adequately represent the conditions of actual joints. In this study, rectangular joint specimens (4.72 cm long by 3.94 cm wide) were tested under normal stresses ranging from 0.2 to 20 MPa with a maximum shear displacement of 20 mm. Results of the coupled shear displacement-flow experiments show that the hydraulic conductivities through the joint increase by about 1 order of magnitude for the first 5 mm of shear displacement (Figure 2a). In the case of high normal stress (i.e., 20 MPa), some joint surfaces were broken by shearing without riding over each other and the hydraulic conductivity increased significantly. During reverse shearing, the hydraulic conductivity was found to be slightly lower than that during forward shearing, essentially following the same response as the dilation behavior. It was found that the hydraulic conductivity was somewhat higher at the initial shearing position after one complete shear cycle was completed. It should be noted that, in these experiments, the flow was radially injected from a hole in the bottom block, and the joint area was not conserved during shearing, the latter of which may be partially responsible for the reported changes in hydraulic conductivity.

Esaki et al. [11] conducted further studies which concluded that the shear-dilation-flow characteristics of artificially fractured granite are very different from those of sandstone because of the difference in the uniaxial compressive strength of the two rock types (162 MPa for granite and 37 MPa for sandstone). In the case of sandstone, the hydraulic conductivity increased rapidly and over a much broader range than in a similar test on granite (Figure 2b). This is in spite of the fact that the dilation in the sandstone was much more restrained and consequently lower than that in the granite due to its higher ratio of normal stress to joint compressive strength. In addition, because of the larger amount of gouge created by shear deformation in the case of sandstone, the decrease of hydraulic conductivity during reverse shearing is more dramatic, as shown in Figure 2b.

Boulon et al. [12] also studied the influence of rock joint degradation during shearing on the hydraulic conductivity of granitic rock joints. They proposed a model describing the flow changes which is locally based on the cubic law and taking into account the asperity degradation. They determined that their model generally exhibited a smaller deviation from experimental measurements than the cubic law.

Many investigators [13,14] conducted MH numerical modeling studies on fractures. Most, if not all, numerical models of rock joints do not account for the production of gouge within the joints and its effect on the subsequent flow or pressure drop along the joint. Consequently, it is not surprising that the numerical predictions of the hydromechanical response of rock joints under shear have not tended to agree well with experimental measurements (Jing et al., 1994).

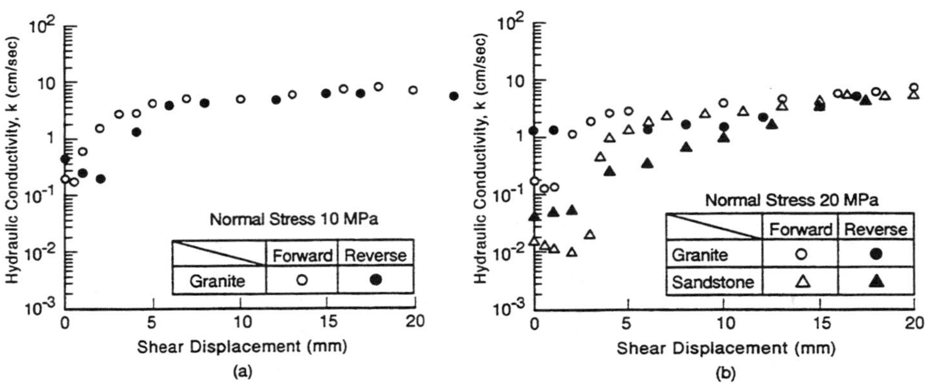

Figure 2. Hydraulic conductivity-shear displacement diagram for (a) artificially fractured granite under 10 MPa normal load, and (b) sandstone and granite under 20 MPa normal load [10,11].

3. MECHANICAL-HYDROLOGICAL STUDIES ON NATURAL ROCK JOINTS UNDER DECOVALEX

The first TC was formulated by NGI making use of their Coupled Shear Flow Temperature (CSFT) test apparatus. Two individual experiments were conducted under this TC. The first, referred to as Test Case 1, Phase 1 (TC1:1), consisted of both normal and shear deformation coupled with fluid flow through a natural rock joint with fairly low roughness (JRC=1.9) as determined using the standard approach proposed by Barton and Choubey [15]. For TC1:1, a very simplified representation of the experimental apparatus was supplied to the modeling teams. For the second experiment, referred to as Test Case 1, Phase 2 (TC1:2), a natural joint with much higher roughness (JRC=6.3) was tested to provide higher peak shear response of the joint under coupled mechanical-hydrological shear deformation. Based on lessons learned in the modeling of TC1:1, it was recommended for this second experiment that a much more detailed representation of the experimental apparatus should be provided for the modeling of TC1:2.

The second TC, referred to as TC5, was formulated by the CNWRA making use of their direct shear testing apparatus modified to include fluid flow through a joint. Similar to TC1, this experiment intended to study the hydro-mechanical response of natural rock joints under combined normal and shear loading. The experimental apparatus and type of rock tested were much different from those of TC1, and the magnitude of shear displacement was significantly higher. Sections 3.1 and 3.2 discuss in detail the two test cases.

3.1. Coupled Shear-Flow Test (DECOVALEX Test Case 1)

To better understand the relationship between joint displacement and joint conductivity, a CSFT testing facility was designed and built by NGI [16]. TC1 was conducted using this apparatus, which involved a single rock joint. The test involved several normal stress cycles followed by shear cycling; it involved modeling of specific stress and fluid pressure boundary conditions, and materials (steel, epoxy cement, rock) with different mechanical properties and interfaces. The two experiments conducted by NGI within DECOVALEX serve as calibration exercises for the different joint behavior models used in the codes (continuum and discontinuum) of the DECOVALEX participants.

3.1.1. Experimental Apparatus

The CSFT testing equipment is able to close and shear (maximum 5 mm) rough joints under controlled normal stress conditions while simultaneously injecting fluid into the joint. The CSFT test is designed to simulate as closely as possible the in situ (stressed, "closed") state of single joints and their alteration by increasing or decreasing normal and shear stresses. Normal-load induced closure and shear-induced dilation of the joint can be caused by these stress changes.

A horizontal cross section through NGI's CSFT apparatus for testing single joints is shown in Figure 3. The maximum normal stress acting across the joint depends on the joint surface area, which for the sample in TC1:2 was approximately 135 cm^2. The maximum allowable flatjack pressure is 25 MPa operating on an area of approximately 300 cm^2. Each part of the sample is cast into a reinforced concrete block, such that the joint is oriented 45 degrees with respect to the sides of the concrete block and the principal stress direction. The two blocks containing the sample are mounted into the apparatus with flatjacks acting on each

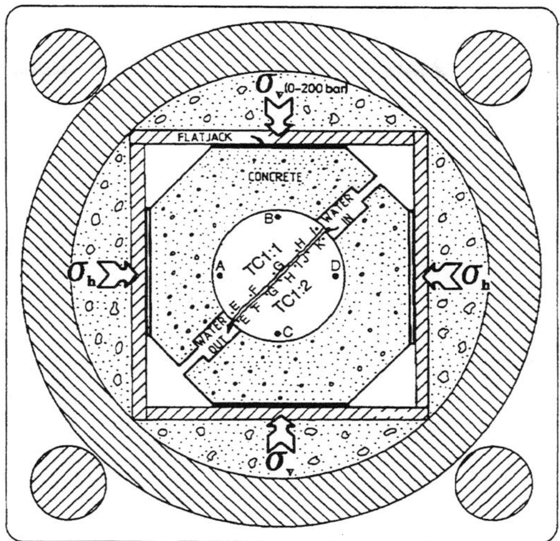

Figure 3. NGI's biaxial cell for coupled shear flow testing (CSFT) of natural rough joints [17].

of the four sides as shown in Figure 3. Flatjack pressures are controlled by two hydraulic pumps.

Displacements normal to the joint (normal displacement) and displacements along the joint (shear displacement) are measured during all stages of the test by four normal displacement and two shear displacement Linear Variable Differential Transducers (LVDT's). LVDT mounting points are referred to as points A, B, C, and D as illustrated in Figure 3.

The effective hydraulic aperture of the joint is calculated from the fluid flow rate that passes through the joint (in the horizontal direction) under a constant fluid pressure according to the following relation (e.g., cubic law)

$$e = \left[\frac{12Qv}{gwi}\right]^{\frac{1}{3}} \tag{1}$$

where

e = effective hydraulic joint aperture (m)

Q = flow rate (m^3/s)

v = kinematic viscosity (m^2/s)

g = gravitational acceleration (m/s^2)

w = width of flow path (m)

i = hydraulic gradient between joint ends (-)

The joint is subjected to three normal stress cycles with the maximum normal stress being equal to 60 percent of the joint compressive strength (JCS) or the normal stress generated at maximum obtainable flatjack pressure. For the TC1:2 experiment, the maximum normal stress applied to the rock joint was 25 MPa. During the fourth cycle, the joint is loaded to the normal stress level under which the shear part of the test was conducted. As long as the same oil pressure is applied to the four flatjacks, only normal stress acts across the joint. This is followed by a shearing stage conducted under constant normal stress. Shear displacement along the joint is created by reducing the oil pressure in two opposite flatjacks by the same increment as it is increased in the other two, so that the normal stress acting on the joint remains approximately constant.

3.1.2. Numerical Simulation of Joint Behavior under Shearing

The two CSFT tests simulated during the DECOVALEX are referred to as TC1:1 and TC1:2. Both test cases consist of a sequence A (normal loading stage) and a sequence B (combined normal and shear loading stage). The focus of this chapter is only on the coupled mechanical and hydraulic behavior of joints under shear, and thus only Sequence B is discussed. A complete discussion of the normal and shear loading results for TC1:1 and TC1:2 as well as their numerical simulations is given by Makurat et al. [17] and Jing et al. [13,18]. TC1:1 and TC1:2 were modeled by four and three research teams, respectively (see Table 1). In the case of the two experiments, pore pressures as predicted by various research teams were small compared to the boundary stresses applied. Several teams thus chose to simulate the tests as uncoupled processes and switch on fluid flow after every normal load variation or shear increment.

Table 1.
The TC1:1 and TC1:2 research teams and computer codes

Research Team	Code	Hardware	Comments
AECL1	MOTIF	VAX Station 3100M38	TC1:2: Only Sequence A
CNWRA2	UDEC	Sun IPX Sparcstation	TC1:1
ITASCA3	UDEC	Gateway 486, 33 MHz	TC1:1
LBL4	ROCMAS	IBM Risk 6000	TC1:1 Only Sequence B TC1:2
NGI5	UDEC	DEC 5000/125 Workstation	TC1:1 TC1:2

1 Atomic Energy of Canada Limited, Pinawa, Manitoba
2 Center for Nuclear Waste Regulatory Analyses, San Antonio, Texas
3 Itasca Consulting Group, Inc., Minneapolis, Minnesota
4 Lawrence Berkeley Laboratory, Earth Science Division, Berkeley, California
5 Norwegian Geotechnical Institute, Oslo, Norway

UDEC was used by three teams (Itasca, CNWRA, and NGI). UDEC is a two-dimensional code for coupled thermal-mechanical (TM) analysis of discrete block systems and coupled MH analysis through discontinuities [19]. The simulated rock mass is assumed to consist of an assemblage of discrete blocks interfaced by discontinuities. The parallel plate analogy is assumed for fluid flow through the joints, while the matrix is impermeable. All three teams used the Barton-Bandis (BB) joint model [20] for their simulations.

In the BB-model, the two components of joint deformation, namely, normal and shear displacements are both based on the scale dependent index properties JRC and JCS, [17]. The principal shear strength-displacement behavior is described by the following two generalized equations:

$$\sigma'_s = \sigma'_n \tan \left[JRC_{mob} \log \left(\frac{JCS}{\sigma_n} \right) + \phi_r \right] \tag{2}$$

$$d_{n\,(mob)} = \frac{1}{2} JRC_{mob} \log \left(\frac{\sigma_1 - \sigma_3}{\sigma'_n} \right) \tag{3}$$

where

σ'_s = effective shear stress (MPa)

JRC_{mob} = the full-scale mobilized JRC at a given displacement (-)

JCS = joint compressive strength (MPa)

ϕr = residual friction angle (degrees)

σ_n = effective normal stress (MPa)

$d_{n\,(mob)}$ = full-scale mobilized dilation angle at any given displacement (degrees)

Lawrence Berkeley Laboratory (LBL) used the code ROCMAS which is a three-dimensional (3D) finite element code for solution of coupled thermal-mechanical-hydrologic (TMH) processes in geological systems [21]. The code considers the stress-strain equation and the law of static equilibrium for both the intact rock elements and the joint elements. The discontinuities are represented explicitly as four-noded joint elements with strain-softening behavior for stress analysis and as one-dimensional line elements for fluid flow in discontinuities. The peak shear stress of the joints is based on the Landanyi and Archambault [22] criterion. The peak shear strength τ_p is given by:

$$\tau_p = \frac{\sigma \,(1 - a_s) \,(\dot{v} + \tan\phi_\mu) + a_s S_R}{1 - (1 - a_s) \,\dot{v} \tan\phi_\mu} \tag{4}$$

where

σ = joint normal stress

a_s = proportion of joint area sheared through the asperities (-)

$\dot{v}$ = dilation rate at peak shear stress (-)

ϕ_μ = friction angle of the sliding surface (degrees)

S_R = shear strength of the rock composing the asperities (MPa)

The asperity shear strength S_R is calculated by Ladanyi's equation as:

$$S_R = q_u \frac{\sqrt{1+n}-1}{n}\left[1 + (n\sigma) \S q_u\right]^{\frac{1}{2}} \tag{5}$$

where

$\quad q_u \quad = \quad$ unconfined compressive strength (MPa)

$\quad n \quad = \quad$ ratio of compressive strength to tensile strength of the rock composing the asperities (-)

For $\sigma < \sigma_T$ Landanyi and Archambault [22] suggest the following power laws for $\dot{v}$ and a_s:

$$a_s = 1 - \left(1 - \frac{\sigma}{\sigma_T}\right)^{K_1} \tag{6}$$

$$\dot{v} = \left(1 - \frac{\sigma}{\sigma_T}\right)^{K_2} \tan i_o \tag{7}$$

where

$\quad \sigma_T = \quad$ transition stress at which the joint ceases to be weaker than the rock (MPa)

$\quad K_1 = \quad$ 1.5 (-)

$\quad K_2 = \quad$ 4.0 (-)

$\quad i_o = \quad$ effective roughness at $\sigma = 0$ (degrees)

The parallel plate model is used for the joint permeability calculations.

Atomic Energy of Canada, Ltd. (AECL) used the 3D finite element code MOTIF, which solves for the coupled fluid flow, heat- and solute-transport processes, and the mechanical deformation of the rock [23]. The solid matrix is assumed to be linearly poroelastic, transversely isotropic, and thermoelastic. The joints are modeled using the BB joint model and both conduction and convection are considered for heat transport. The solid matrix is represented by 8-noded hexahedral elements. Joints are represented by 4-noded quadrilateral elements for the flow and heat analysis, and by 8-noded quadrilateral joint elements for stress analysis.

3.1.3. Material Properties and Boundary Conditions

For TC1:1 the modeling teams were supplied with a simplified geometry of the experiment, whereas for TC1:2 a detailed description of the biaxial experiment was provided. Table 2 summarizes the rock and joint material properties of TC2:1 and TC1:2 as specified to the modeling teams. Table 3 gives the loading sequences for TC1:1 and TC1:2 as specified to the modeling teams.

3.1.4. Experimental and Numerical Simulation Results for TC1:1

Results are presented for selected shear displacement values for sequence B (i.e., combined normal and shearing portion of the experiment). The monitoring points A to D in Figure 3 correspond to the LVDT positions in the CSFT setup, whereas points E to I (TC1:1)

Table 2.
Material properties for TC1:1 and TC1:2

Property		Steel	Epoxy & Epoxy-Rock Interace	Rock & Rock Joint	Fluid	Unit
E						
Young's	TC1:1	200000	10000	55000		
modulus	TC1:2	200000	25000	55000		MPa
ν						
Poisson's ratio		0.27	0.30	0.25		
ρ	TC1:1	7.00	2.25	2.60		
density	TC1:2	7.00	2.40	2.60	1.00	10^3kg/m^3
JCS_n						
Joint wall						
compressive	TC1:1			150		
strength	TC1:2			87.1		MPa
JRC_n						
Joint surface	TC1:1			1.9		
roughness	TC1:2			6.3		
L_n						
Sample joint	TC1:1			0.19		
length	TC1:2			0.09		m
ϕ_r						
Residual friction	TC1:1			26.5		
angle	TC1:2			28.7		degree
Dynamic fluid						
viscosity					0.001	Ns/m^2

and E to K (TC1:2) are fictitious monitoring points in the joint plane which were used to compare results from different codes. Results referring to points A to D are not corrected for intact rock deformation. Figure 4 illustrates the discretizations of the model adopted by the different teams. Itasca, CNWRA, and NGI included the steel plates into their model, whereas LBL modeled only the epoxy block. All teams used stress boundary conditions during the normal joint loading prior to shear. However, the modeling teams used different approaches and assumptions to simulate the shear stage of TC1:1. Itasca and the CNWRA used displacement boundaries and assumed a dilation angle of 0.5 degrees in order to maintain the constant normal stress condition. NGI used stress boundary conditions, accepting that the

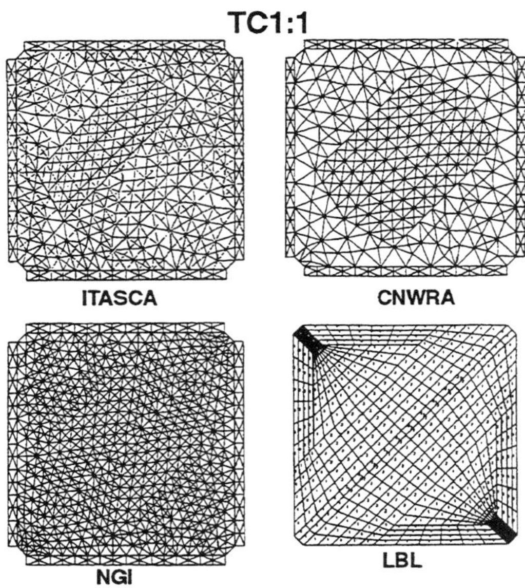

Figure 4. Discretization of the TC1:1 model geometry.

Table 3.
Loading sequence for TC1:1 and TC1:2

Sequence	TC1:1		TC1:2	
A	one normal load-ing cycle (4th cycle in CSFT experiment)	nominal joint normal stress $0\rightarrow25\rightarrow0$ MPa	one normal load-ing cycle (3rd cycle in CSFT experiment)	nominal joint normal stress $0\rightarrow26\rightarrow0$ MPa
B	4 mm forward shear, followed by 4 mm reverse shear	nominal joint normal stress 25 MPa	2.8 mm shear	nominal joint normal stress 16.5 MPa

equilibrium condition might not be satisfied after the peak shear stress is reached (after 1.68 mm of shear displacement), UDEC tries to satisfy the tabulated U_s - σ_s relationship in the BB-model and thus gives the correct joint apertures.

The LBL team did not include the steel plates in their model geometry. This resulted in large stress concentrations which prevented shear failure. Hence only the three UDEC simulations by CNWRA, Itasca, and NGI are compared. Due to the same modeling approach of CNWRA and Itasca with respect to the boundary conditions and an assumed constant dilation angle of 0.5 degrees over the total joint length, their shear stress-shear displacement curves are rather similar (Figure 5). Compared to these, the NGI shear stresses are 5 to 10

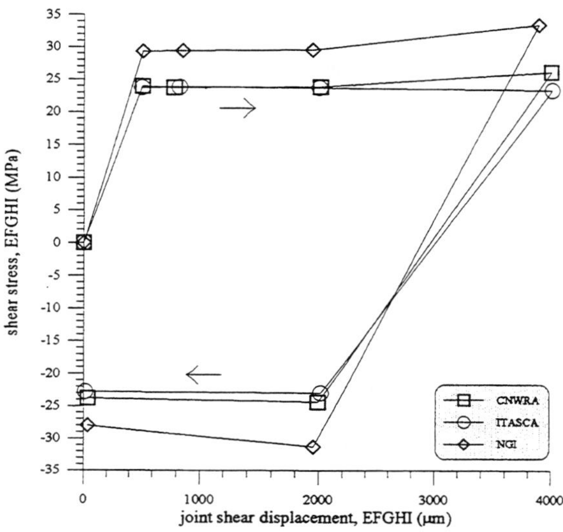

Figure 5. Forward and reverse shear stress shear displacement curve for TC1:1.

MPa higher. This can be explained by the stress boundary-induced higher joint end normal stresses. Due to the low JRC, none of the teams predicted substantial joint dilation (< 1 mm), measured at points E to I. This is in contrast to the experimental results, which indicate about 90 mm dilation over the first 1000 mm of shear displacement (Figure 6a). Both Itasca and CNWRA predict an 8-10 mm increase in hydraulic aperture from the initial value during forward shear (Figure 6b). The experimental results also show a maximum change in hydraulic aperture of around 8 mm during shear. The NGI calculated hydraulic apertures are close to the upper part of the experimental range. The results of the other teams are 10-20 µm higher. No experimental data exist for the reverse shear part of the cycle. All teams predicted a substantial increase of hydraulic aperture during reverse shear, which is in contrast to experimental data demonstrating that joint conductivity decreased due to continuous joint surface degradation and associated gouge production during shear cycling [4,8].

3.1.5. Experimental and Numerical Simulation Results for TC1:2

The combination of low joint surface roughness ($JRC_o=1.9$) and high normal stresses during shear resulted in non-peak dominated stress strain curves and little dilation in TC1:1. A joint with higher roughness was therefore chosen for TC1:2 (see Table 2). The sample comes from the Borrowdale Volcanic Group and challenges the teams to model dilation and strain softening behavior after the peak shear stress.

TC1:2 was modeled by AECL, LBL, and NGI. However, AECL simulated only the normal loading sequence of TC1:2 (sequence A). Again, only the results for the shearing portion of the experiment are presented herein. Figure 3 shows the location of the measurement points. A to D (now inside the epoxy block), and E to K along the joint plane. In order to avoid over-representation of the over-stressed joint edge points, average values along the joint plane included only points F to J.

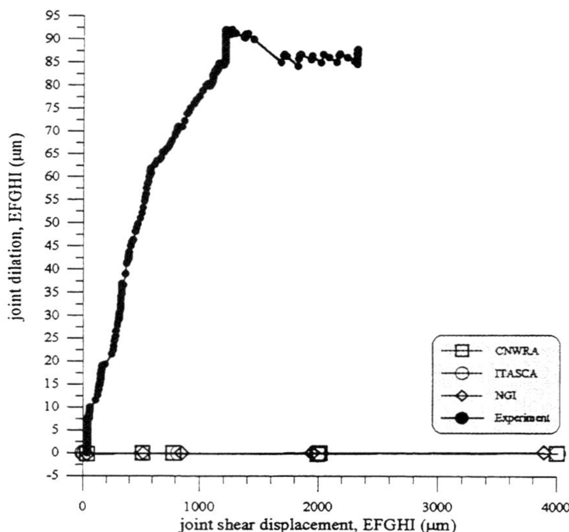

Figure 6a. Comparison between experimental data and the simulated average joint dilation at points E-I.

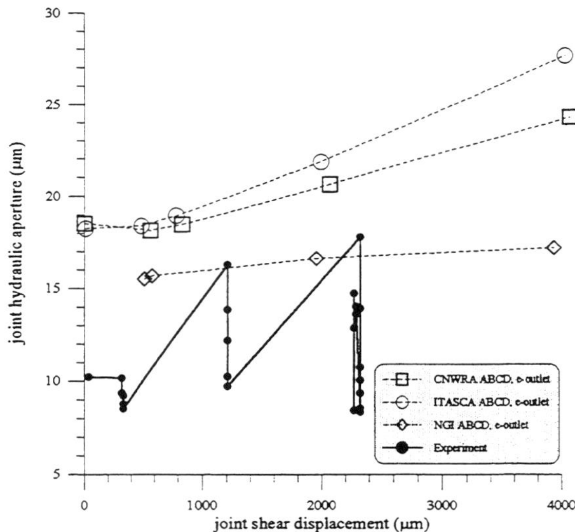

Figure 6b. Comparison between experimental data and the simulated variation of the joint hydraulic aperture at outlet vs. shear displacement between points A-D.

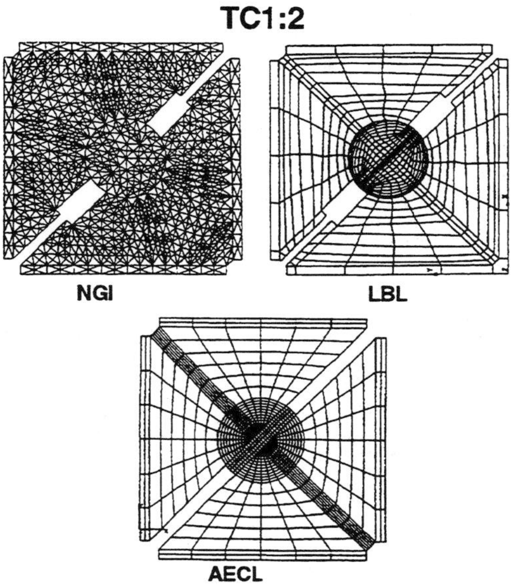

Figure 7. Discretization of the TC1:2 model geometry.

All teams included the steel plates into their models. LBL used four-node finite elements with linear displacement interpolation for the intact material, and four-node joint elements with linear interpolation of the displacements and fluid pressure fields. NGI split the model into two separate blocks with the joint as the interface (Figure 7). LBL modeled shearing by using a "stabilized boundary load drive" system. The system consists of a combination of stress boundaries and a small high shear stiffness element in the joint plane. In the pre-failure range, every joint element acts as an individual spring, and the stabilizer, which deforms much less, will have no bearing on the response of the other joint elements. After failure of an element, all released load of the failed element is absorbed by the stiff linear stabilizer element. NGI modeled sequence B by applying a mixed boundary system, similar to a direct shear box loading principle. This is achieved by combining the direct application of the required normal stress to the outer boundary of one block with a velocity boundary condition on the other block. In order to avoid block tilting, the system is completed by zero displacement boundaries perpendicular to the applied normal stress and perpendicular to the applied velocity boundary.

As stated earlier, sequence B was modeled only by LBL and NGI for TC1:2. The experimental results indicate 500 mm mechanical dilation during 2800 mm shear displacement (Figure 8a). The BB model predicts about 80 mm dilation at 2800 mm shear displacement, and 10.8 MPa peak shear strength at 730 mm peak shear displacement. Consequently, the constitutive models in UDEC and ROCMAS are not capable of modeling this obviously non-BB joint behavior. The LBL simulation (65 mm dilation) is rather close to the 1D-predicted 80 mm dilation obtained by NGI. NGI's results are strongly affected by tilting of the upper block, resulting in 30 mm dilation at one joint end and 100 mm closure at the other end. Experimental results also indicate a dramatic increase in hydraulic joint aperture from 52 to 222 mm during 0 to 100 mm shear, and a residual hydraulic aperture of 72 mm after 2100 mm shear (Figure 8b). The NGI results are again influenced by the tilting of

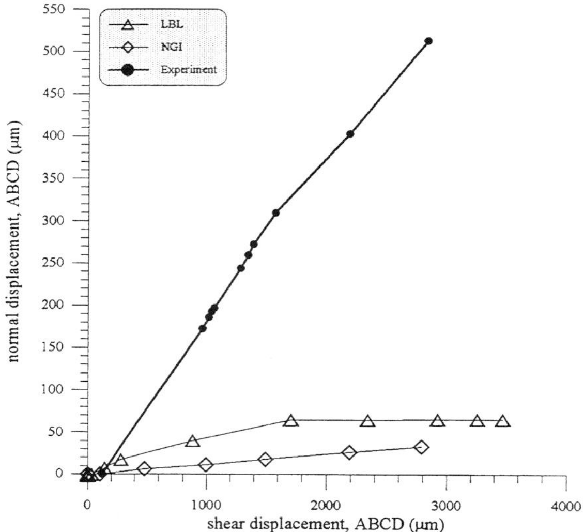

Figure 8a. Comparison between experimental data and the simulated joint dilation between points A-D.

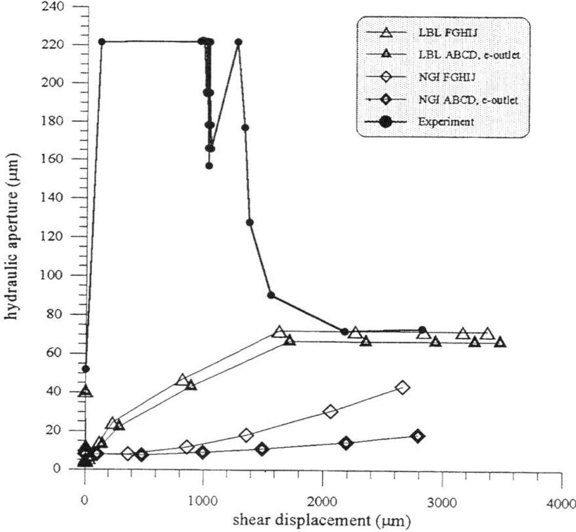

Figure 8b. Comparison between experimental data and the simulated variation of the joint hydraulic aperture at outlet vs. shear displacement between points A-D and F-J.

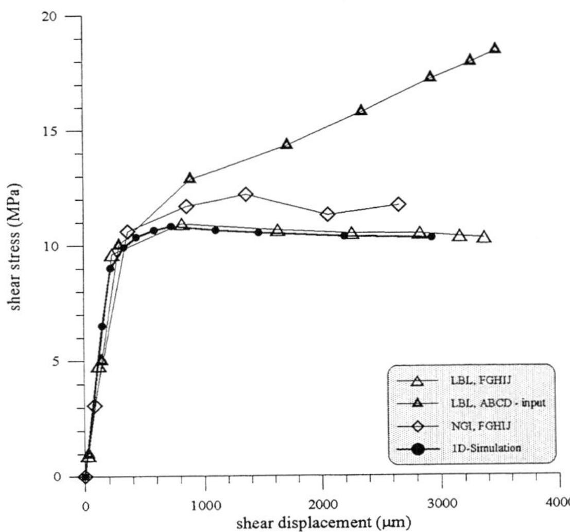

Figure 9. Comparison between the 2D simulated stress strain behavior and the 1D joint stress strain curve.

the top block, and show a continuous increase in joint aperture with shear. The LBL simulation seems to settle at a residual aperture similar to the experimental residual aperture. The stress strain curves are given in Figure 9. Since the LBL model did not experience any type of block rotation during shear, the results for the joint elements match perfectly the one-dimensional simulated BB stress strain curve. NGI overpredicts the peak shear stress by about 2 MPa and peak shear displacement by about 600 mm.

3.2. Direct Shear-Flow Test (DECOVALEX Test Case 5)

As a part of Phase III of DECOVALEX, a Test Case was also proposed by the CNWRA to investigate the coupled MH fluid flow behavior through a natural rough joint as a result of both normal and shear displacement along the joint. The experiment was conducted using the CNWRA direct shear testing apparatus [24] which was modified to allow linear fluid flow experiments to be conducted within the joint under combined normal and shear loads [8]. Linear fluid flow experiments were conducted under normal stresses up to 8.0 MPa, as well as shear displacements up to 2.54 cm under constant normal stresses of 2.0, 4.0, and 5.0 MPa.

3.2.1. General Description of Apparatus and Experimental Setup

To conduct coupled MH experiments on single jointed rock specimens, the CNWRA basic servocontrolled direct shear test apparatus with combined normal and shear loading capability was modified to include the necessary hydraulic system. A brief description of this basic apparatus is given in the following sections. A more detailed discussion is given by Kana [25] and Hsiung [24]. Figure 10 shows the overall mechanical loading apparatus and associated data channels. Modification of the basic apparatus to include the necessary hydraulic system is discussed in Section 3.2.4. A more in-depth discussion is given by Mohanty [8].

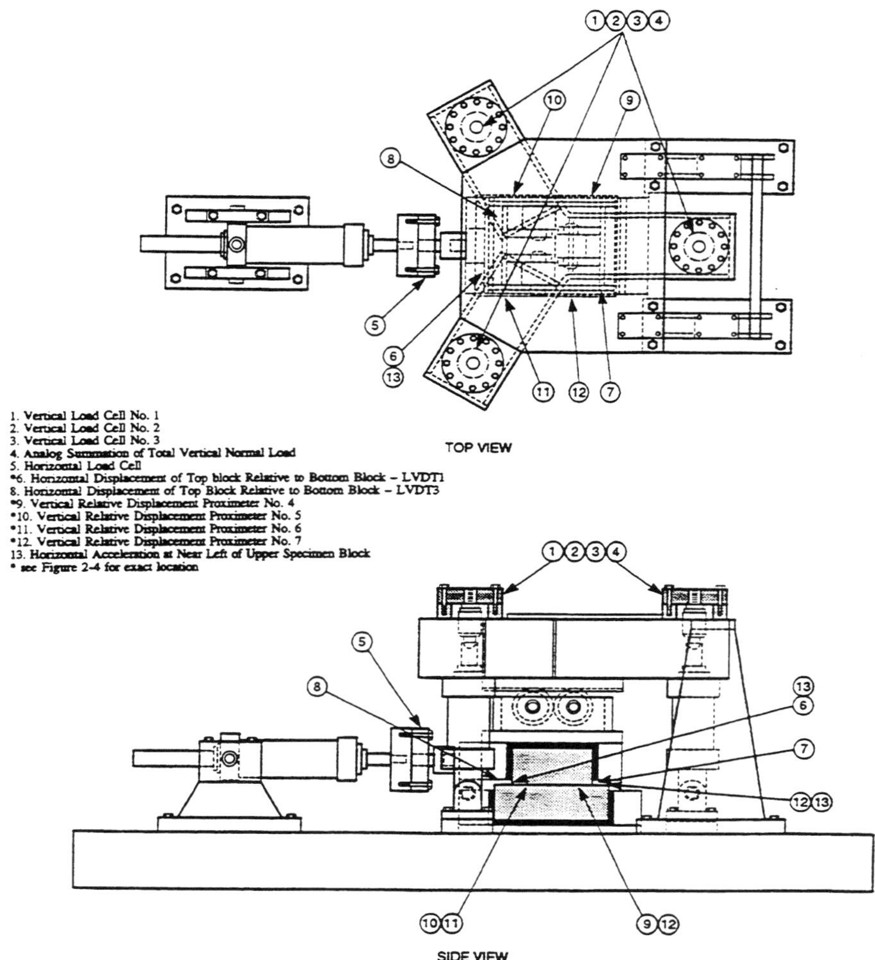

Figure 10. Loading apparatus for normal and direct shear testing of rock joints (basic mechanical apparatus)

The apparatus consists of vertical and horizontal servocontrolled loading actuators, reaction frames, shear box fixtures, and an instrumented jointed tuff specimen. The loading capacity for each of the three vertical actuators is 0.133 MN. The loading capacity of the horizontal actuator is 0.222 MN. For the MH experiments, the horizontal actuator is operated in a displacement control mode, with the displacement ramped pseudostatically. Each vertical actuator is equipped with a 0.111-MN capacity load cell for monitoring the applied forces. The instrumentation for monitoring the applied normal load is arranged to provide an analog output for the sum of the three load cells, as well as for the individual signals. The bottom shear box was designed to house a specimen with maximum dimensions of 0.305×0.203×0.102 m. The top shear box houses a specimen with maximum dimensions of 0.203×0.203×0.102 m. Both are grouted in their respective specimen boxes. The bottom shear box and other fixed devices are bolted to a 1.22×2.13×0.15 m thick steel base plate for rigidity. The horizontal translation of the top shear box along the direction of shearing is guided through three rollers between the top shear box and normal load frame. It is also guided through side rollers. Thus, the normal load frame and the side rollers prevent rotation of the vertical actuators (and therefore also the top specimen block) about a vertical axis perpendicular to the direction of shearing.

3.2.2. Normal Mechanical Load System

Normal compression is applied to the specimen by three vertical actuators set at $120°$ about the specimen's vertical centerline. These actuators act through individual load cells whose output is summed and used as the control signal. Thus, the total normal load is controlled at a preselected static or slowly ramped value. This total resultant load is ultimately applied to the specimen via the normal load frame which acts on the three normal load rollers (see top view of Figure 10) and thereby on the upper specimen box. The line of action for this normal load is through the null position of the upper specimen box. Thus the normal load frame is constrained to three degrees of freedom:

    (i)   Vertical translation
    (ii)   Rotation about the horizontal axis in line with the shear
    (iii) Rotation about the horizontal axis transverse to the shear

These constraints are assured by two double flexures which connect the normal load frame to a fixed reaction brace, and by the two side roller assemblies, which act on the upper specimen box. Thus, the upper specimen block is constrained to these same degrees of freedom, plus a fourth, which is translation in the direction of shear.

As indicated in Figure 10, each of the three vertical actuators is pinned at the bottom to a clevis which is bolted to the base plate. At the top, each is connected to its associated load cell through a spherical coupling. This arrangement is consistent with the three degrees of freedom identified above. Furthermore, for quick disassembly, the three actuator pins are removed, the two double flexures are detached, and the entire normal load frame with actuators attached can be hoisted away from the specimen/roller box assembly. The apparatus is capable of applying mechanical normal loads of up to approximately 0.333 MN.

3.2.3. Horizontal Mechanical Load System

The horizontal actuator produces direct shear to the upper specimen box via the horizontal load cell, which acts through a spherical coupling. This coupling allows for slight misalignment in the horizontal shearing motion. It also allows for elevation changes of the upper specimen due to shear induced dilation. Control of the horizontal actuator load for all tests described herein is based on the horizontal shear displacement.

3.2.4. Hydraulic System

The hydraulic portion of the apparatus, which represents the modification to the basic direct shear apparatus, is designed to allow linear flow experiments to be conducted while the rock joint is undergoing normal or shear loading. As shown in Figure 11, The fluid is injected

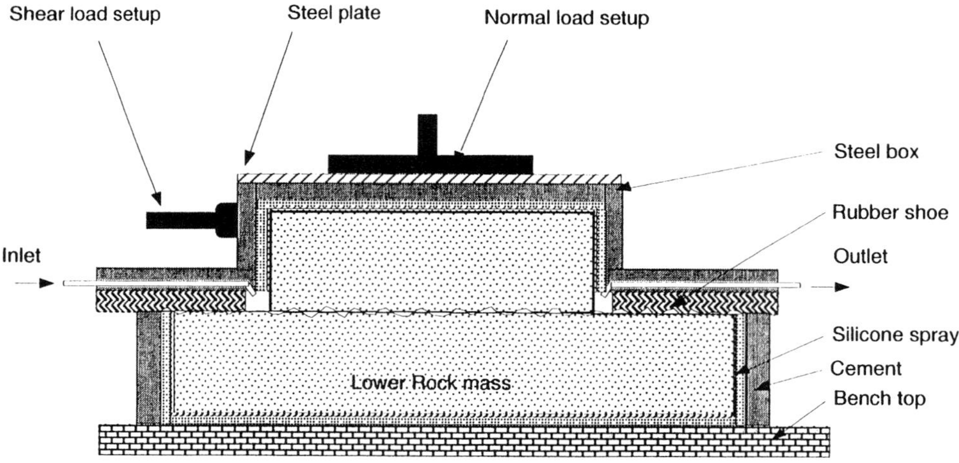

Figure 11. Schematic diagram of the linear flow apparatus with normal and shear loading arrangements.

at a constant flowrate over the entire width of the joint on the left edge of the top specimen, and collected over the entire width of the joint from the right edge of the top specimen. The flowrate is chosen such that laminar flow is maintained within the joint. For these MH experiments, a flowrate of 4.0 cm^3/min was determined to meet this criteria for all normal mechanical loadings. In other words, even at the highest load, the expected aperture is such, that laminar flow is maintained. Absolute and differential pressure transducers are used to measure the inlet and outlet fluid pressures.

A rectangular rubber gasket is placed around the specimen to prevent water leakage. The reason for this gasket is that the lower block extends beyond the upper block, thus making it difficult to prevent water leakage especially during shearing of the upper block. Silicon grease and rubber cement are used to ensure that water does not leak out from above or below the rubber gasket. The apparatus is designed such that a finite normal load must be applied to compress the rubber gasket and allow the upper and lower rock joint surfaces to come into contact. This ensures adequate sealing of the system against leakage. The portion of the rubber gasket covering the joint along the front and back surfaces of the block is not shown in Figure 11 for clarity. In addition, the rubber along these two front and back surfaces is compressed with thin metal plates to prevent leakage of fluid from the joint along these surfaces.

Prior to grouting the specimens into the steel boxes, all five sides of the top and bottom rock blocks, except the joint surfaces themselves, are coated with a silicon rubber cement. The specimen (rock matrix and joint) is then completely saturated with water prior to the MH experiment. This is done to concentrate fluid flow into the joint, and prevent water leakage from the matrix.

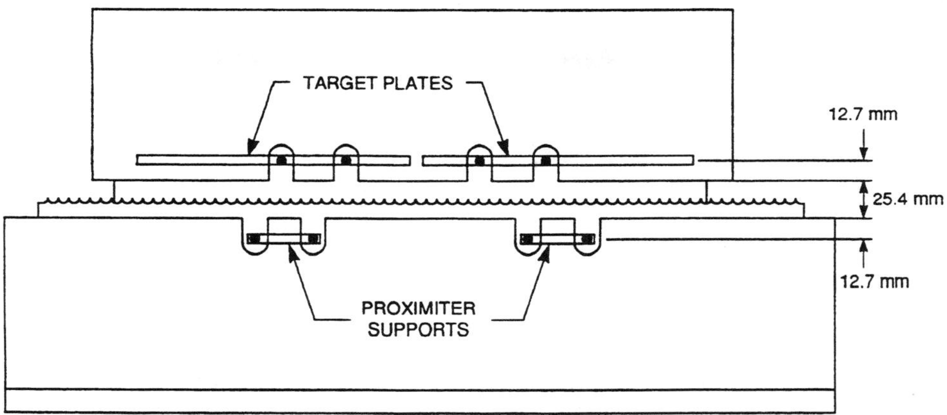

Figure 12. Side view of vertical displacement instrumentation supports and targets.

3.2.5. Instrumentation and Control

Instrumentation channels for the mechanical loading portion of the experiment are identified in Figure 10. The locations of various relative displacement sensors on the specimen are shown in Figures 12 and 13.

All load cells are typical commercial strain gage units with dominant sensitivity to tension/compression along one axis. Reaction to the applied static normal load is measured in terms of relative vertical displacements of the two blocks at four locations near the interface. Measurements near the interface are desirable to reduce the effect of the deformation of the intact rock, such that only the joint deformation is measured. For this, the transducers are of proximity (noncontacting) eddy-current sensing type, since horizontal movement of the two surfaces must be allowed, but only vertical displacement changes must be sensed. Hence, the four vertical measurement points can be used to resolve the rigid body displacement of the upper block relative to the lower specimen block, according to the first three degrees of freedom identified earlier.

Vertical proximity transducers were mounted on each side near the joint interface as shown in Figures 12 and 13 As indicated in Figure 12, the specimen is grouted into the upper and lower boxes so that a 2.54 cm gap is left between the box faces. The interface, which varies from one specimen to another, is nominally enclosed within this gap. The sideplates of each half of the specimen box are slotted, so that vertical proximeter supports and target plates can be mounted directly onto the sides of respective halves of the specimen near the interface. Two prongs which support each plate component are cemented into lateral holes that are drilled into the specimen sides. Although some movement of the specimen within the grout occurs during loading, the side slots are large enough so that no interference occurs between the support prongs and the box side plates. Thus, as the upper box and associated target plates move horizontally relative to the lower specimen, change in vertical relative position is also sensed continuously. Furthermore, the heavy mounting frame for the upper box side rollers is slotted so that there is no interference between the frame and the target plates as the upper box displaces both horizontally due to shear and vertically due to unevenness and wear of the interface.

Relative shear displacements of the specimen blocks are measured by two LVDT's as indicated in Figure 13. LVDT1 is located at the near end of the specimen and measures displacement of the upper block relative to the lower block. Each half of the transducer is cemented directly into a hole drilled into the respective specimen block. LVDT2 is similarly mounted on the far side of the specimen pair, as shown in Figure 13, The output of these two transducers provides the direct shear movement of the joint interface. Control of the MH joint experiments was imposed on a prescribed horizontal displacement signal by applying a slow ramp to the relative displacement, while the force required was simultaneously recorded.

3.2.6. Mechanical Properties of the Intact Rock

The rock specimens for the MH experiments were prepared from a 45.7-cm-diameter core that was drilled from a road cut near Superior, Arizona (USA). The rock is a welded tuff of the Apache Leap formation. Statistical analysis on 113 uniaxial Apache Leap welded tuff samples and 72 triaxial specimens yielded the following average values for the mechanical rock properties, as shown in Table 4.

Table 4.
Intact rock properties obtained from testing of welded tuff samples

Property	Value	Units
Young's Modulus E [Dry]	38.6±3.4	Gpa
Poisson's Ratio	0.20±0.029	
Uniaxial Compressive Strength C_o [Dry]	161.0±26.0	MPa
Triaxial Compressive Strength [Dry]		
- Confining Pressure 3.4 MPa	202.0±27.0	MPa
- Confining Pressure 6.9 MPa	248.0±22.0	MPa
- Confining Pressure 10.3 MPa	271.0±18.0	MPa
Uniaxial Tensile Strength To [Dry]	10.3±2.2	MPa
Density [Dry]	2420.0	Kg/m^3
Saturated Hydraulic Conductivity k_s	2.131 E-08	m/s
Effective Porosity	0.175	

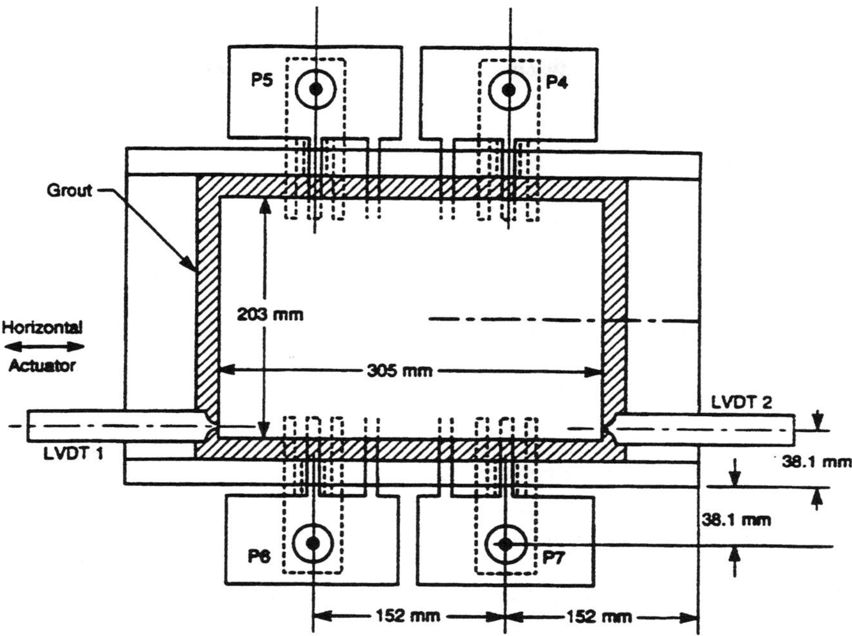

Figure 13. Location of relative displacement sensors on specimen lower block.

3.2.7. Mechanical Properties of the Rock Joint

The mechanical displacement of a typical rock joint is nonlinear in that the normal joint stiffness (K_n) increases with increasing normal stress in a nonlinear fashion. The normal stiffness of the joint surface has been calculated from the experimental data obtained in normal displacement tests in the laboratory. Values of the normal stiffness have been calculated at the expected normal loads during each shear test based on the global slope of sn versus un curve (equivalent to the secant stiffness). These values are provided in Table 5 and are based on mean values from the fifth loading cycle for 19 jointed specimens tested. The shear stiffness, K_s, is the slope of the shear stress, σ_s, versus shear displacement, u_s, curve in the elastic (pre-peak) region. The shear stiffness is estimated from the experimental shear stress versus shear displacement curve obtained in the laboratory for each specimen with each normal stress applied on the specimen. The estimated mean values from the 19 specimens tested at the expected normal loads to be conducted in the MH test are also given in Table 5. The estimated values are the global slope of the curve in the elastic region, that is, they are the secant stiffnesses of the joint at the given normal stress.

During shear displacement, the amount of joint dilation depends on many factors including the joint compressive strength, joint roughness, and applied normal load. For the mechanical experiments conducted on 19 naturally jointed, welded tuff samples, Table 6 provides the mean dilation angle as well as the maximum dilation and shear displacement corresponding to the maximum dilation at each of the expected normal loads for the MH experiment. Typically, the joints experience a small amount of closure at the onset of shear followed by joint dilation. The amount of initial closure increases with normal load. The dilation angle, ψ, has been calculated as the global slope (i.e., secant slope) of the dilation curve between zero dilation and maximum dilation.

Table 5.
Estimated joint normal (K_n) and shear stiffnesses (K_s) at expected normal loads.

Expected Normal Stress (MPa)	Normal Stiffness (GPa/m) (Secant)	Shear Stiffness (GPa/m) (Secant)
2.0	37.7±19.9	5.7±4.6
4.0	46.8±26.1	10.6±7.0
5.0	49.0±25.4	12.4±4.9

Table 6.
Estimated values of dilation angle, ψ, with other parameters to describe the u_n versus u_s curves.

Normal Stress (MPa)	Peak Dilation unmax (mm)	Shear Disp. at Peak Dilation u_{sdmax} (mm)	Dilation Angle, ψ (degrees)
2.0	1.58±0.91	35.07±8.51	2.74±1.27
4.0	1.29±0.85	38.00±6.52	2.18±1.11
5.0	1.28±0.91	38.02±7.75	2.27±1.12

The cohesion, c, and friction angle, ϕ, represent the strength of the rock joint. The mean values of the residual joint cohesion, c_r, and residual joint friction angle, ϕ_r, were experimentally determined to be 0.10 MPa and 39.6±4.5°, respectively. Surface profile measurements were taken of each joint surface prior to conducting the direct shear tests. Based on the profile measurements, a numerical procedure was utilized to determine the approximate hydraulic joint aperture under zero applied normal stress. This was determined to be 0.81 mm.

Parameters for the BB joint model were also determined for both joint closure behavior and joint shear behavior. The joint closure under normal loads can be modeled empirically using hyperbolic loading and unloading curves relating the effective normal stress, σ'_n, and joint closure, u_n as:

$$\sigma_n = \frac{u_n}{a - bu_n} \tag{14}$$

where *a* and *b* are empirical constants. The initial normal stiffness (K_{ni}) and maximum allowable closure (u_m) for each loading/unloading were determined through laboratory cyclic normal loading/unloading tests. Bandis [20] showed that K_{ni} of the joint is equal to the inverse of the constant a in Equation (8) and u_m is defined by a/b. The mean values of K_{ni} and u_m were determined from experimental tests of 19 welded tuff joints for the 5th normal load cycle to be 24.7±12.7 GPa/m and 0.19±.08 mm, respectively. The joint shear behavior for the BB joint model is given by equation (2).

Barton and Choubey [15] have proposed relations for JCS and ϕ_r in this equation based on output from Schmidt-Hammer rebound tests. Using these relations, the mean values for JCS and ϕ_r from 26 joint tuff samples were determined to be 123.6±18.3 MPa and 26.6±1.2°, respectively. The value of ϕ_r using Barton and Choubey's relation is somewhat lower than previously derived applying the Coulomb relation. The initial JRC_0 can be estimated using the following relation proposed by Barton and Choubey [15]:

$$JRC_o = \frac{\alpha - \phi_r}{\log\left(\dfrac{JCS}{\sigma'_{no}}\right)} \tag{15}$$

where α is the tilt angle, and σ'_{no} is the corresponding effective stress calculated from the weight of the top block of the joint specimen when sliding occurs. The mean value for JRC_0 from 12 specimens tested was calculated using Equation 9 to be 5.93±1.18.

3.3. Experimental Results for Shear Loading

The joint shear displacement experiment initially encountered some problems with fluid leakage from the apparatus. The experiment was originally designed to conduct coupled shear-flow tests under normal stresses of 2.0, 4.0, and 5.0 MPa. However, it was not until the fourth direct shear cycle under a constant normal stress of 2.0 MPa that confidence was reached regarding the results of the flow measurements. In other words, the joint surface was already damaged before accurate flow measurements were obtained. During each shear cycle, the top block was sheared approximately 2.54 cm (1 in.) in the forward direction followed by shearing in the reverse direction back to the initial zero point. It should be pointed out that the top block is centered over the bottom block such that the bottom block extends 5.08 cm (2 in.) on either side in the direction of shear. Thus, during the shearing, the area of the sheared surface remains constant.

Figure 14 represents the first of the four mentioned shear stress versus shear displacement cycles conducted under a constant normal stress of 2.0 MPa. For this particular specimen, a peak shear stress response does not occur. This is somewhat peculiar, since in almost all previous mechanical tests done on very similar jointed specimens of the same rock type, a peak shear stress response was observed during the first cycle. Thus, this particular test may not be representative. However, one possible reason for the absence of a peak shear stress is that there may have been some slight mismatch of the two joint surfaces at the start of shearing. It is also observed in Figure 14 that the residual shear stress upon reverse shearing is lower than that during forward shearing. Figure 15 shows the normal displacement (i.e., joint dilation) as a function of shear displacement for the four cycles conducted under a normal stress of 2.0 MPa. Some initial closure (i.e., negative displacement) occurs upon initiation of

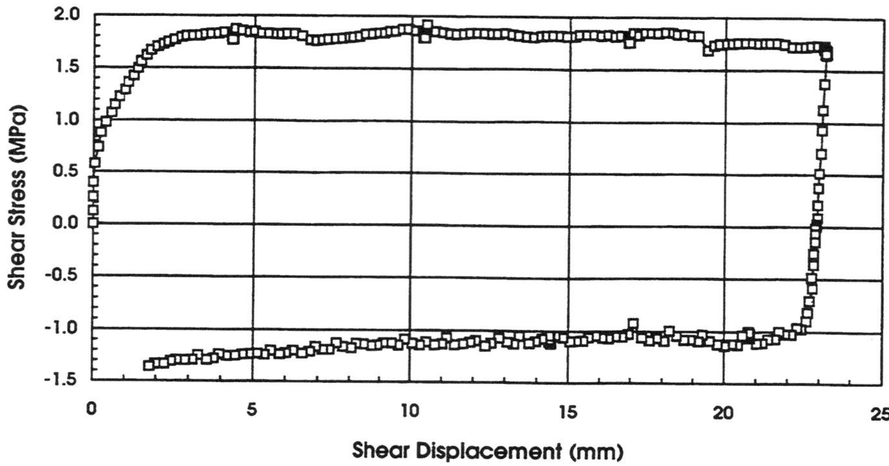

Figure 14. Mechanical response of rock joint for first shear loading cycle.

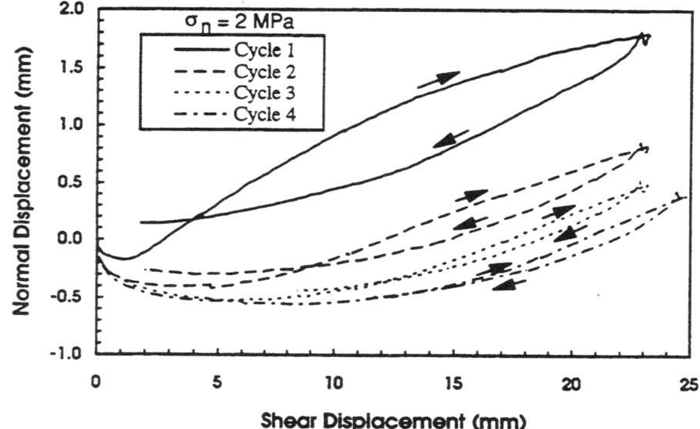

Figure 15. Dilation response of rock joint for first four shear cycles.

shearing, followed by opening of the joint. While the shear stress versus shear displacement curves are very similar for all four cycles under 2.0 MPa constant normal stress, the normal displacement versus shear displacement curves are remarkably different. As the surface is worn down under repeated shear cycling, the amount of joint dilation also decreases, as shown in Figure 15. Presumably, at the same time, there is more and more gouge buildup within the joint. At the completion of four cycles of shearing, the apparatus was disassembled and it was verified that a significant amount of gouge material had collected within the joint. The maximum joint dilation is measured to be 1.8, 0.8, 0.45, and 0.35 mm for the first four shear cycles, respectively. Again, this maximum dilation represents an average of the four vertical displacement proximeters.

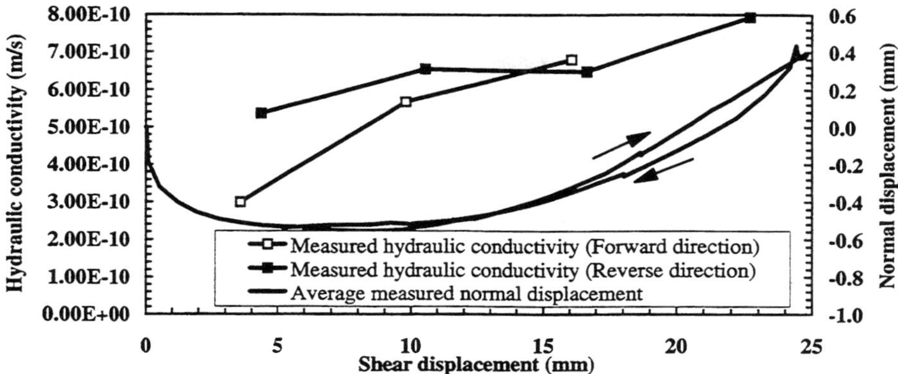

Figure 16. Hydraulic response and mechanical dilation of rock joint during fourth shear cycle under constant applied normal stress of 2.0 MPa.

Only for the fourth cycle, hydraulic data were obtained under the 2.0 MPa applied normal stress. Figure 16 shows the change in joint hydraulic conductivity with shear displacement in both the forward and reverse directions for this fourth cycle. The dilation curve has also been superimposed on the figure to aid in the interpretation. To obtain the hydraulic data, steady state pressure measurements across the length of the joint were taken after each 6.35-mm (0.25-in.) increment in shear displacement. During the forward shear movement of 25 mm, the available data show an increase in hydraulic conductivity of approximately 167 percent under this particular normal load. Again, one would expect a much larger increase in hydraulic conductivity if data were available for the first shear cycle, in which the maximum dilation was over five times greater than that for the fourth cycle. During reverse shearing, the hydraulic conductivity decreases to a lesser extent, ending up with a value significantly higher than at the start of the test. The change in hydraulic conductivity is attributed to both dilation as well as gouge formation due to wearing down of asperities. However, the experiment was not able to discriminate between these two effects.

Additional shearing was conducted under normal stresses of 4.0 and 5.0 MPa. Figure 17 shows the shear stress versus shear displacement response under these two applied normal stresses (cycles 5 and 6) in addition to that just described for the fourth shear cycle. During these latter shear cycles, the reverse shear displacement was limited to only 19 mm (0.75 in.). This was done because large rises in fluid pressure were experienced at the end of the shear cycles (back near the initial starting point), which began to exceed the range of the pressure gauges. This was found to be due to the accumulation of gouge material within the joint. Also, at the higher normal stresses, even though the maximum displacement of 25 mm was measured from the horizontal actuator, the measured relative shear displacement along the joint fell short of this value at the higher normal stress loadings of 4.0 and 5.0 MPa as shown in Figure 17. The reason for this is that there is some amount of compliance in the apparatus, especially in the steel boxes and grout surrounding welded tuff specimens. The 4th shear cycle

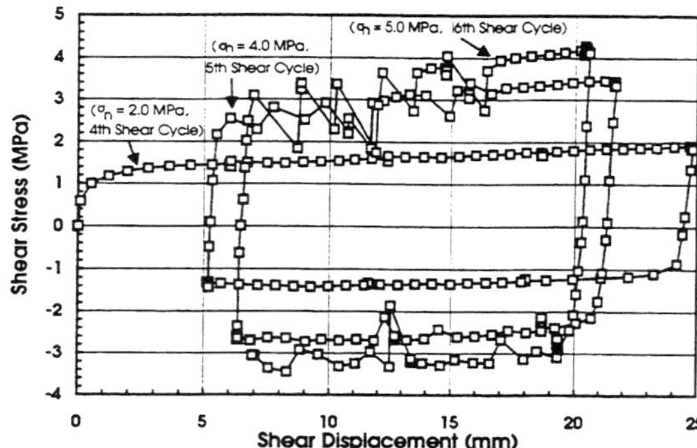

Figure 17. Mechanical response for shear loading under different applied normal stresses.

is very smooth since repeated cycling was done at this normal stress loading of 2.0 MPa and one would not expect much additional asperity breakage. However, the 5th and 6th shear cycles under higher normal loadings show fluctuations in the shear load due to stick-slip behavior as well as shearing (breakage) of asperities causing a temporary reduction in shear load.

Figures 18 and 19 show both the normal displacement and hydraulic conductivity plotted against shear displacement for shear cycles 5 and 6, which correspond to normal applied stresses of 4.0 and 5.0 MPa. In these plots, the normal displacement is the average reading from the four vertical displacement proximeters. Again, each of these two shear cycles began at approximately 6.35 mm to the right of the initial mated position for reasons explained earlier. At these higher normal loads, more problems were encountered with water leakage such that fluid pressure drop measurements could be made at only a few discrete shear displacement locations as indicated in Figures 18 and 19. From the available data, there appears to be consistency with hydraulic conductivity measurements obtained under the normal stress of 2.0 MPa (i.e., Figure 16), namely that during the reverse shearing there is a noticeable decrease in the joint hydraulic conductivity.

3.4. CONCLUSIONS

The results of two separate experimental studies on the MH behavior of natural rock joints under shear deformation have been presented. The joints tested had varying degrees of surface roughness as well as different material properties. In addition, to the extent possible, attempts were made to compare various computer codes with different joint constitutive models against the experimental results to assess whether these constitutive models correctly model the actual joint behavior.

For TC1, despite the fact that the NGI CSFT test is a well defined laboratory experiment, several assumptions with respect to the material properties and specifically the boundary conditions had to be made by the modeling teams. Both modeled joints show some deviation from the BB joint model. The deviation might be caused by the joint itself, or the way the joint

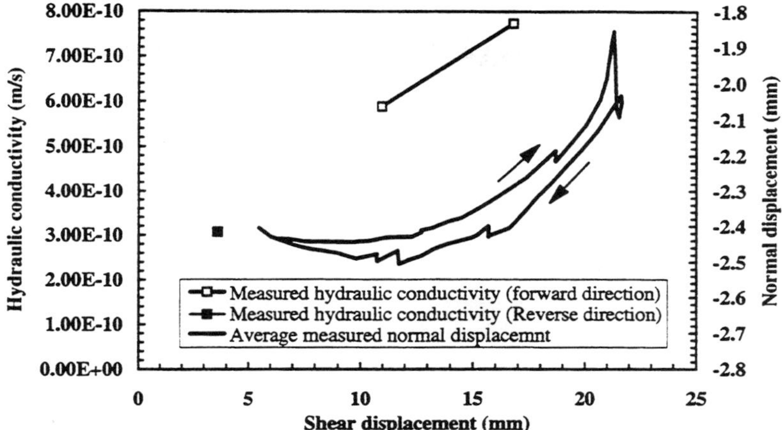

Figure 18. Hydraulic response and mechanical dilation of rock joint for the sixth shear loading cycle under 5.0 MPa normal stress.

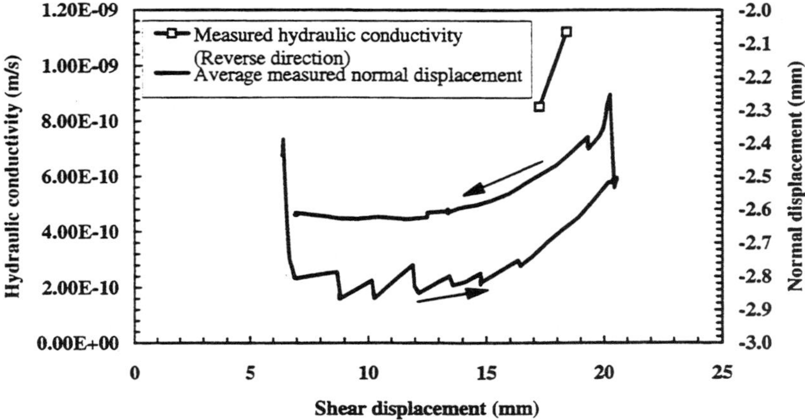

Figure 19. Hydraulic response and mechanical dilation of rock joint for the fifth shear loading cycle under 4.0 MPa normal stress.

behaves in the CSFT apparatus. A computer program will only be able to reproduce the measured joint behavior if the assigned constitutive joint behavior model is appropriate. With regard to shear deformation, both TC1:1 and TC1:2 results show that dilation and changes in hydraulic aperture are not well modeled. However, NGI is close to the upper bound of the experimentally measured hydraulic apertures in TC1:1, whereas LBL is in close agreement with experimentally measured residual hydraulic apertures in TC1:2. All teams had to overcome difficulties related to the application of the boundary conditions. The strong influence of the various approaches on the final results has been demonstrated.

The TC5 experimental results show that the hydraulic properties of the joint can change by up to a factor of 3 under shear deformation. However, additional tests and improvements to the experimental apparatus are necessary in order to make more quantitative assessments of changes in such properties during shear and production of gouge.

3.5. REFERENCES

1   D.D. Kana, B.H.G. Brady, B.W. Vanzant, and P.K. Nair, Critical Assessment of Seismic and Geomechanics Literature Related to a High-Level Nuclear Waste Underground Repository, NUREG/CR-5440, Washington, DC: Nuclear Regulatory Commission (1991).

2   S.M. Hsiung, W. Blake, A.H. Chowdhury, and T.J. Williams, Effects of Mining-Induced Seismic Events on a Deep Underground Mine, Pure and Applied Geophysics, Vol. 139, (1992) pp. 741-762.

3   S.M. Hsiung, A.H. Chowdhury, W. Blake, M.P. Ahola, and A. Ghosh, Field Site Investigation: Effect of Mine Seismicity on a Jointed Rock Mass, CNWRA 92-012, San Antonio, TX: Center for Nuclear Waste Regulatory Analyses (1992).

4   A. Makurat, N. Barton, N.S. Rad, and S. Bandis, Joint conductivity variation due to normal and shear deformation, Proceedings of the International Symposium on Rock Joints, Leon, Norway, N. Barton and O. Stephansson (eds), Netherlands, Rotterdam: A.A. Balkema (1990), pp. 535-540.

5   Y.W. Tsang and P.A. Witherspoon, Hydromechanical behavior of a deformable rock fracture subject to normal stress, Journal of Geophysical Research, Vol. 86, No. B10, American Geophysical Union (1981) pp. 9287-9298.

6   N.G.W. Cook, Natural Joints in Rock: Mechanical, Hydraulic, and Seismic Behavior and Properties under Normal Stress, Int. J. Rock Mech. Min. Sci. & Geomech. Abstr. Vol. 29 No. 3 (1992) pp. 198-223.

7   N. Barton, S. Bandis, and K. Bakhtar, Strength Deformation and Conductivity Coupling of Rock Joints, Int. J. Rock Mech. Min. Sci. & Geomech. Abstr. Vol. 22 (1985) pp. 121-140.

8   S. Mohanty, A.H. Chowdhury, S.M. Hsiung, and M.P. Ahola, Single Fracture Flow Behavior of Apache Leap Tuff Under Normal and Shear Loads, CNWRA 94-024, San Antonio, TX: Center for Nuclear Waste Regulatory Analyses (1994).

9   W.T. Teufel, Permeability changes during shear deformation of fractured rock, 28th US Symposium on Rock Mechanics, Rotterdam, Netherlands: A.A. Balkema (1997) pp. 473-480.

10 T. Esaki, H. Hojo, T. Kimura, and N. Kamedam, Shear-flow coupling test on rock joints. Seventh International Congress on Rock Mechanics, Aachen, Germany, Vol. 1, Ed. W. Wittke, Rotterdam, Netherlands: A.A. Balkema, (1991) pp. 389-392.

11 T. Esaki, K. Ikusada, and A. Aikaw, Surface roughness and hydraulic properties of sheared rock. Fractured and Jointed Rock Masses, Lake Tahoe, California, Vol. 2 (1992) pp. 366-372.

12 M.J. Boulon, A.P.S. Selvadurai, H. Benjelloun, and B. Feuga, Influence of Rock Joint Degradation on Hydraulic Conductivity, Int. J. Rock Mech. Min. Sci. & Geomech. Abstr. Vol. 30, No. 7 (1993) pp 1311-1317.

13 L. Jing, J. Rutqvist, O. Stephansson, C.F. Tsang, and F. Kautsky, DECOVALEX - Mathematical Models of Coupled T-H-M Processes for Nuclear Waste Repositories - Report of Phase II, (in publication), Stockholm, Sweden: Swedish Nuclear Power Inspectorate (SKI) (1994).

14 A. Makurat, N. Barton, G. Vik, P. Chryssanthakis, and K. Monsen, Jointed rock mass modeling, Proceedings of the International Conference on Rock Joint, N. Barton and O. Stephansson, eds, Leon, Norway. (1990) pp. 647-656.

15 N.R. Barton and V. Choubey, The shear strength of rock joints in theory and practice, Rock Mechanics, Vol. 10 (1977) pp. 1-54.

16 A. Makurat, The effect of shear displacement on permeability of natural rough joints. Hydrogeology of rocks of low permeability, Proc. 17th Int. Congr. Hydrogeol., Tucson, Ariz. (1985) pp. 99-106.

17 A. Makurat, M. Ahola, K. Khair, J. Noorishad, L. Rosengren, and J. Rutqvist, The DECOVALEX Test-Case One, Int. J. Rock Mech. Min. Sci. & Geomech. Abstr. (in publication) (1995).

18 L. Jing, J. Rutqvist, O. Stephansson, C.F. Tsang, and F. Kautsky, DECOVALEX - Mathematical Models of Coupled T-H-M Processes for Nuclear Waste Repositories - Report of Phase I. SKI Technical Report 93:31, Stockholm, Sweden: Swedish Nuclear Power Inspectorate (SKI) (1993).

19 P. Cundall and R.D. Hart, Development of Generalized 2-D and 3-D Distinct Element Programs for Modeling Jointed Rock. Itasca Consulting Group. Misc. paper SL-85-1. Vicksburg, MS: U.S. Army Corps of Engineers (1985).

20 S. Bandis, A.C. Lumsden, and N. Barton, Fundamentals of Rock Joint Deformation, Int. J. Rock Mech. Min. Sci. & Geomech. Abstr. Vol. 20 (1983) pp. 244-268.

21 J. Noorishad, C.F. Tsang, and P.A. Witherspoon, A coupled thermal-hydraulic-mechanical finite element model for saturated fractured rocks, J. Geoph. Res. Vol. 89, No. B12, (1984) pp. 10365-10373.

22 B. Landanyi and G. Archambault, Simulation of shear behavior of a jointed rock mass, Proc. 11th Symp. Rock Mech, Rotterdam: Netherlands, A.A. Balkama, (1970) p. 105.

23 V. Guvanasen and T. Chan, Three-dimensional Finite Element Solution for Heat and Fluid Transport in Deformable Rock Masses with Discrete Fractures, Proc. Int. Conf. Computer Methods and Advances in Geomechanics (1990) pp. 1547-1552.

24 S.M. Hsiung, D.D. Kana, M.P. Ahola, A.H. Chowdhury, and A. Ghosh, Laboratory Characterization of Rock Joints, CNWRA 93-013, San Antonio, TX: Center for Nuclear Waste Regulatory Analyses (1993).

25 D.D. Kana, D.C. Scheidt, B.H.G. Brady, A.H. Chowdhury, S.M. Hsiung, and B.W. Vanzant, Development of a Rock Joint Shear Test Apparatus CNWRA 90-005 San Antonio, TX: Center for Nuclear Waste Regulatory Analyses (1990).

O. Stephansson, L. Jing and C.-F. Tsang (Editors)
Coupled Thermo-Hydro-Mechanical Processes of Fractured Media
Developments in Geotechnical Engineering, vol. 79
© 1996 Elsevier Science B.V. All rights reserved.

425

Experimental investigation and mathematical simulation of a borehole injection test in deformable rocks

J. Rutqvist[a], S. Follin[b], K. Khair[c], S. Nguyen[d] and P. Wilcock[e]

[a]Division of Engineering Geology, Royal Institute of Technology, S-100 44 Stockholm, Sweden

[b] Department of Engineering Geology, Lund University, S-211 00 Lund, Sweden

[c]Applied Geoscience Branch, Whiteshell Laboratories, Pinawa, Manitoba R0E 1L0, Canada

[d]Atomic Energy Control Board, 280 Slater, Ottawa, Canada, K1P 5S9

[e]AEA Technology, 424.4 Harwell , Didcot, Oxfordshire, United Kingdom, OX11 ORA

Abstract

Coupled hydromechanical computer codes are evaluated by modelling of hydraulic injection field tests. The hydraulic injection tests were conducted on a joint intersecting a vertical borehole at depth of 350 metres in crystalline rocks. Three types of injection tests were conducted with pressures ranging from below to above the overburden pressure. Important processes, parameters and boundary conditions dictating the pressure flow response are identified. Modelling results shows that for well pressures lower than the overeburden stress, the pressure versus flow response is dominated by the stress-transmissivity relation of the joint plane intersecting the borehole. The dominating parameters dictating the stress-transmissivity relation of a joint is the hydraulic aperture and the normal stiffness. The pressure versus flow response could be simulated fairly well by all computer codes in when the well pressure is lower than the virgin stress normal to the joint. The initial hydraulic aperture and normal stiffness back-calculated by different computer codes is fairly consistent. This indicate that the conceptual models of the test site and joint constitutive models are sound for the hydromechanical processes involved in this test.

1. INTRODUCTION

A porous medium containing fluids generally undergoes deformation as fluid pressure changes [1]. This phenomenon is a coupled process between mechanical deformation and fluid flow and is known as coupled hydromechanical effects [2]. The coupled hydromechanical behaviour of rock joints has experienced an increasing interest as an issue of concern in the performance assessment of nuclear waste repositories [3]. The safety assessment of the repository requires an understanding of the hydromechanical processes in the rock barrier. It must be assured that changes in fracture apertures in the rock mass around the nuclear waste repository do not impair the protecting function of the rock barrier more than tolerable limits to jeopardise the safety.

Experimental and field studies [4 , 5 , 6 , 7] have shown that the fluid through natural joints is strongly affected by the stress normal it. This implies that deformation of fractures, due to some alteration of the stress field, have significant effect on water flow through the rock mass (Figure 1). The phenomenon of opening and propagation of fractures during fluid injection in boreholes is well known and utilised in geothermal reservoir stimulation [8] and in hydraulic fracturing stress measurements [9]. In those tests, the fluid pressure is frequently increased to above the *in situ* stresses to open and propagate the fractures with high injection flow rate. On the other hand, geohydrological well testing are conducted at low fluid pressures to avoid joint opening as a result of reduced effective stress. Such joint opening can introduce errors in the determination of hydraulic conductivity.

Numerical experiments of coupled hydromechanical fluid flow behaviour during well testing of single joints was carried out by Noorishad and Doe [10]. These studies indicated an increasing flow rate with time for a constant pressure injection test. This is a complete deviation from what is predicted by traditional fluid flow analysis where the flow rate is expected to decrease with time. However, an increased flow rate with time was observed in a series of hydraulic field tests by Rutqvist *et al.* [11]. The field test results were analysed with coupled hydromechanical modelling and it was concluded that joint opening may occur during injection even if the fluid pressure is lower than the virgin stress. It was also concluded that the joint stiffness is the most important factor controlling the effect of stress on the fluid flow though joints. The field tests were later defined as a test case (Test Case 6) in the DECOVALEX project [12]. The purpose of Test Case 6 (TC6) is to investigate the important coupled hydromechanical processes and parameters in high pressure hydraulic testing, and to validate the soundness of constitutive hydromechanical relations used in various models.

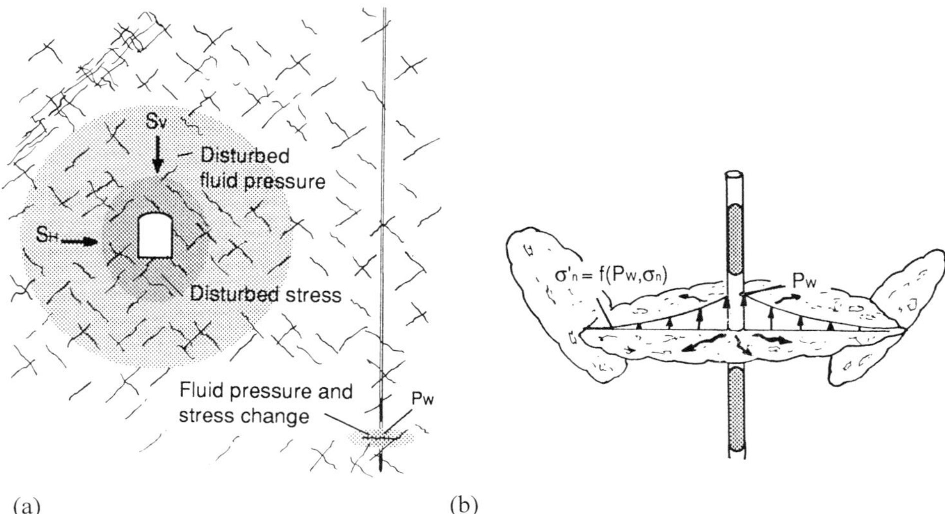

(a)   (b)

Figure 1. Coupled hydromechanical processes in the rock mass: (a) Changes of stress and fluid pressure due to excavation and fluid injection. (b) Hydraulic injection test of a single joint.

This article presents the results and experience of the modelling of TC6. Section 2 presents the field tests according to the Test Case definition. Section 3 presents the numerical simulation of the injection test conducted by Research Teams within the DECOVALEX project.

2. HYDRAULIC FIELD TESTING OF A JOINT IN GRANITE

Hydraulic testing was conducted at 356 metres depth on a joint intersecting a 500 metres deep vertical borehole at the Rock Mechanics Laboratory of Luleå University of Technology, Sweden. Three types of the hydraulic test are included in this study; i) constant pressure injection test, ii) pulse test and iii) step-pressure test.

2.1 Test site and equipment

The rock mass in the vicinity of the borehole consists mainly of granite and gneiss with a very high Quality Designation Index (RQD). The fractures in the drill core samples from the borehole are clean with slight coating and filling of calcite, epidot and chlorite. Laboratory experiments were conducted on the drill core at 356 metres depth and the following properties of the intact rock were obtained:

- Young's modulus, $E = 80$ GPa
- Poisson's ratio, $v = 0.25$
- Uniaxial compressive strength, $\sigma_c = 140$ MPa

In addition, a compression test was conducted on the drill core containing the fracture at 356 metres depth to obtain a curve that relates the fracture normal stress to the normal displacement (Figure 2). The in-situ normal stress across the fracture has been estimated by different methods of hydraulic fracturing stress measurement to be between 9.5 - 10.2 MPa.

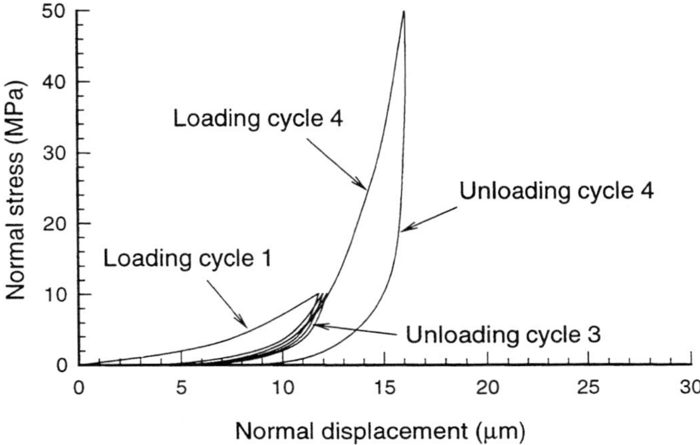

Figure 2. Normal stress as a function of normal displacement for the compression test of the core sample containing the fracture.

The test equipment is designed for hydro-fracturing stress measurements [13]. It is permanently installed on a truck and equipped with a multi-hose system which permits measurements down to a depth of 1000 metres in boreholes greater than 56 mm in diameter. The down hole equipment consists of a straddle packer with 0.65 m packer separation and a pressure transducer (0-50 MPa) (Figure 3). A downhole valve is located above the packer section to allow for fast pressure pulse testing. The valve, which is triggered with compressed air, can create a pressure pulse of at least 20 MPa within a fraction of a second. The flow rate is monitored by a flow meter (0.02 - 5 l/min) at the ground surface.

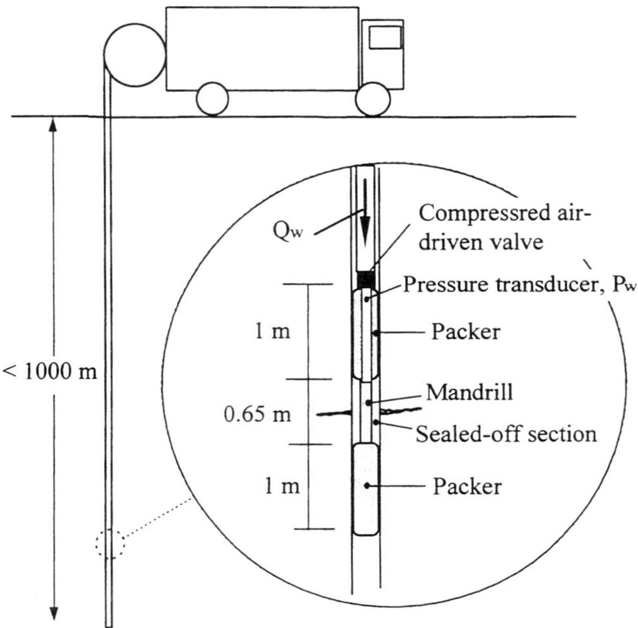

Figure 3. Equipment for conducting hydraulic injection tests of single joints.

2.2 Pulse test

This test is conducted by first instantaneously increasing the pressure in the isolated borehole section and thereafter closing the valve immediate above it. The well pressure is then declining to the pre-pulse pressure as water is flowing into the joint, Figure 4. The rate of pressure decay is dependent on the fracture properties, and the volume and compliance of the sealed-off wellbore section. The volume of the sealed-off section is $6.2 \cdot 10^{-4}$ m^3. An effective compressibility of the sealed off volume has been determined in a separate experiment to be $C_{eff} = 2.2 \cdot 10^{-9}$ Pa^{-1}. This number takes into account the compliance of water and equipment.

The transmissivity and storativity of the joint can be determined according to Bredehoeft and Papadopulos [14] or Wang [15]. These solutions are valid for an ideal fracture with homogenous hydraulic properties in an impermeable and incompressible rock matrix. The interpretation of field test results with these solutions may therefore give rise to errors in the derived hydraulic properties of a natural joint [16]. One such error may arise from the coupling of stress and transmissivity. The pulse pressure in this case has a peak value of 9 MPa which is close to the in situ stress magnitude at the test site. However, the pulse has a very short duration and radius of influence. Therefore, the forcing effect of the fluid pressure causing joint opening may not be significant in this case.

Table 1
Data points of hydraulic pulse test in Figure 4.

Time (Sec)	Pressure (MPa)	Time (Sec)	Pressure (MPa)
0.1	7.8	2.0	4.8
0.2	7.0	2.5	4.6
0.3	6.6	3.0	4.5
0.4	6.3	4.0	4.3
0.5	6.0	5.0	4.2
0.6	5.8	7.0	4.0
0.8	5.5	10.0	3.9
1.0	5.4	15.0	3.8
1.5	5.0	20.0	3.7

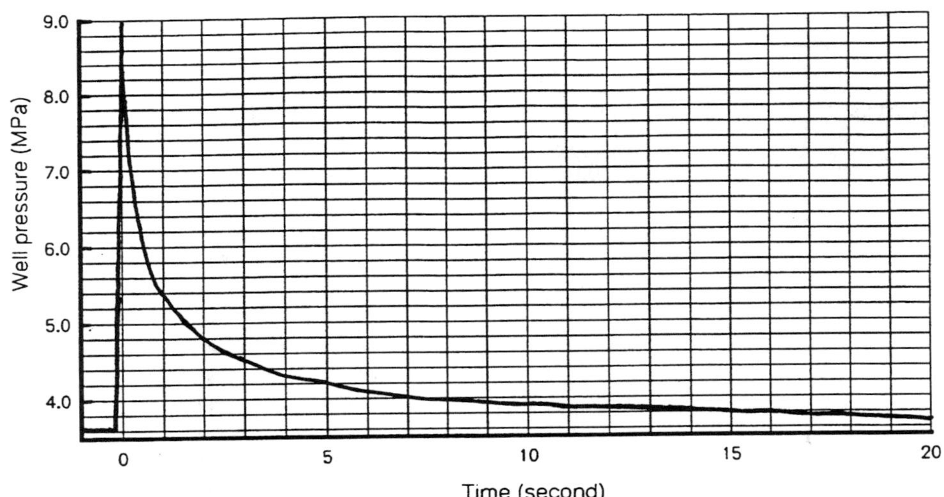

Figure 4. Fluid pressure as a function of time for the pulse test. Data points are given Table 1.

2.3 Hydraulic jacking test

This test is carried out by increasing the well pressure stepwise and monitoring the rate of water injection into the joint. At each pressure step an apparent stationary flow is obtained within 3 minutes before the pressure is increased for the next step, Figure 5a. The test result is presented as pressure versus flow rate in Figure 5b and Table 2.

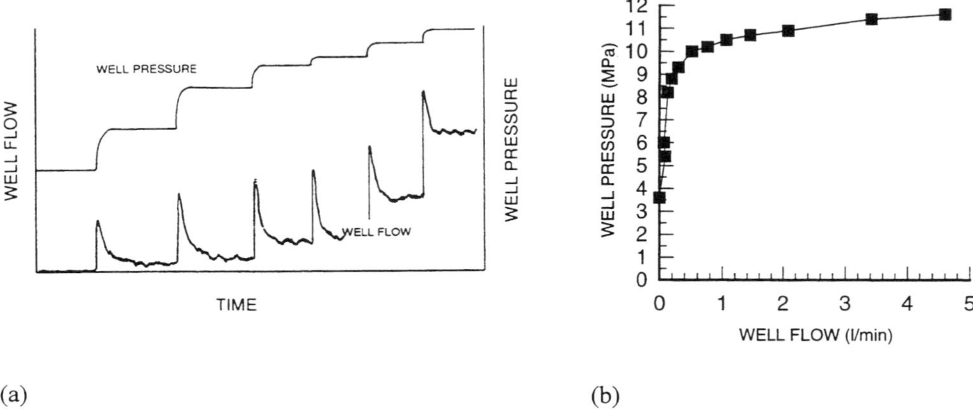

Figure 5. Hydraulic jacking test: (a) a typical strip chart record of the pressure and flow rate in a hydraulic jacking test and (b) Fluid pressure versus flow rate for TC6. Data point are give in Table 2.

Table 2
Values of data points in Figure 5 and total volume of injected water.

Fluid pressure (MPa)	Flow rate (l/min)	Injected volume (litres)
0.0	0.0	0.0
5.4	0.09	0.4
6.0	0.07	0.9
8.2	0.13	1.3
8.8	0.19	1.6
9.3	0.30	2.2
10.0	0.51	3.2
10.2	0.76	4.3
10.5	1.06	5.4
10.7	1.45	6.8
10.9	2.08	8.9
11.4	3.42	12.3
11.6	4.6	16.9

For a non-deformable fracture the well flow would be a linear function of well pressure. However, in a field test, the well flow increases more than would be expected for a non-deformable fracture. The curve is typically characterised by two slightly non-linear slopes connected with an inflexion part (Figure 5b). This inflexion part is often used for determination of in situ stresses normal to the fracture plane [17].

The shape of the curve reflects the hyperbolic stress-deformation behaviour of rock joints. At each pressure level, the effective stresses inside the joint are reduced. This implies that the fracture is mechanically unloaded as the fluid pressure increases. Opening of the joint will also occur for fluid pressures below the *in situ* stresses since water is always able to diffuse into the fracture and change the effective stress.

2.4 Constant pressure injection test

This test was conducted by increasing the downhole pressure to 11.2 MPa and then maintaining it constant during the entire experiment, Figure 6a. The resulting flow rate response in Figure b shows an increased flow rate with time. This. is a complete deviation from what is expected using the classical fluid flow equation where the fluid flow is decreasing with time [18]. However, in the classical equations, the transmissivity is assumed to be constant and independent of the fluid pressure. This assumption is valid if the fluid pressure is low and the joint aperture remains constant during the test. In this case the well pressure is higher than the virgin normal stress and the joint may open considerable as a result of the normal pressure acting on the fracture plane.

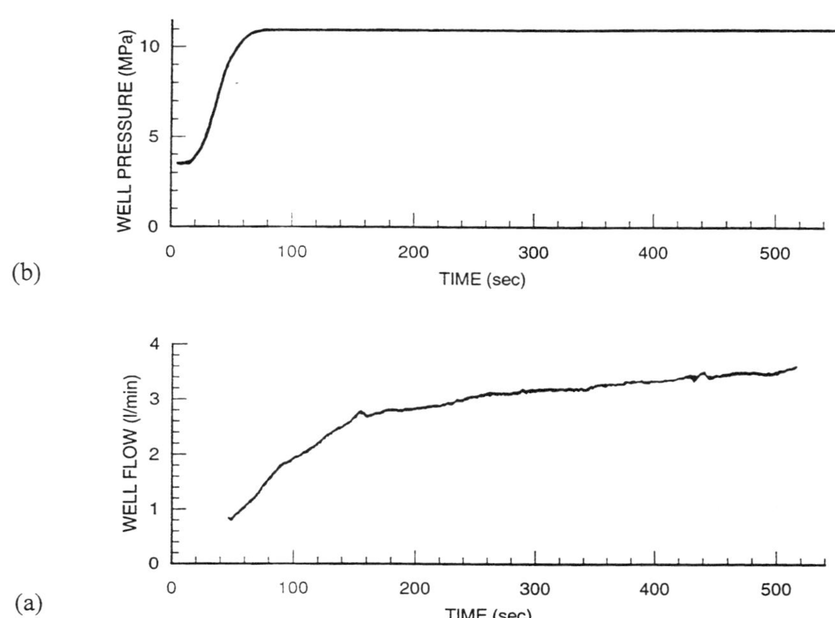

Figure 6. Constant pressure injection test: a) Well pressure and b) well flow.

3. NUMERICAL SIMULATION OF FIELD TESTS

3.1 Modelling methodologies

Test Case 6 were simulated by five Research Teams with the models presented in

Table 3. ROCMAS [19], FRACON [20] and MOTIF [21] are all finite element programs based on Biot's theory of consolidation [22]. In those computer codes, the intact rock material between individual fractures in the rock mass is modelled as porous linear elastic material and the fractures are modelled as hydroelastic material with non-linear stress-strain behaviour. This implies that the hydraulic conductivity is a function of effective stress in both the joint and the surrounding rock. The mechanical joint opening, Δv, is dependent on the effective normal stress, σ'_n across the fracture according to empirical relationships by Goodman [23] for ROCMAS and Bandis [24] for FRACON and MOTIF. The relation of Δv and σ'_n of the two models is similar and is defined by the constants initial joint stiffness k_{ni}, initial maximum closure V_{mi}, and initial effective stress σ'_{ni}. That is:

$$\Delta v = F\left(k_{ni}, V_{mi}, \sigma'_{ni}, \sigma'_n\right) \tag{1}$$

Table 3
Research Teams and models for modelling of TC6.

Research Team	Computer code	Type of model
KTH[1]	ROCMAS	FEM of single joint and surrounding rock
AECB[2]	FRACON	FEM of single joint and surrounding rock
AECL[3]	MOTIF	FEM of single joint and surrounding rock
AEA[4]	NAPSAC	Fracture network model
LTH[5]		FEM of single joint

[1] Royal Institute of Technology, Division of Engineering Geology, Stockholm, Sweden
[2] Atomic Energy Control Board, Ottawa, Canada
[3] Atomic Energy of Canada Limited, Manitoba, Canada
[4] AEA Engineering Services Division, Harwell Laboratory, Oxfordshire, United Kingdom
[5] Lund University, Department of Engineering Geology, Lund, Sweden

Fluid flow takes place dominantly along fractures and the hydraulic permeability, k, is a function of the square of the hydraulic fracture aperture, e, according to the "cubic law":

$$k = \frac{e^2}{12} \tag{2}$$

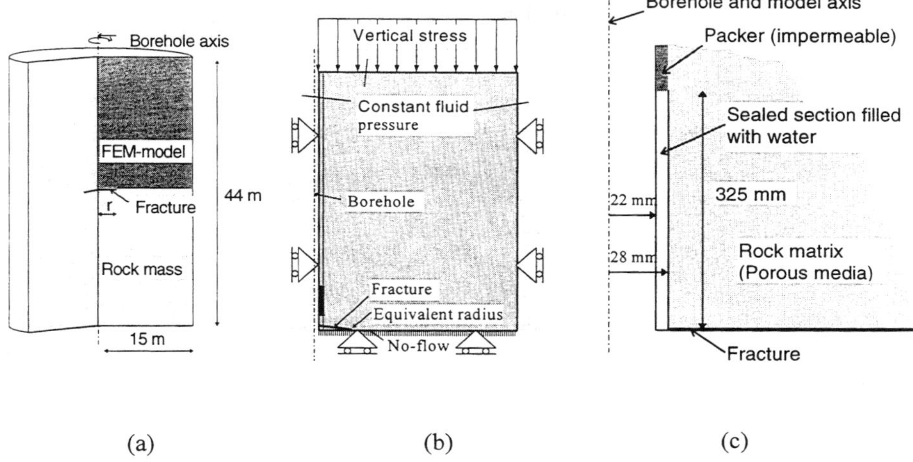

Figure 7. KTH's finite element model of the test site: (a) Axisymmetric model, (b) mechanical and hydraulic boundary conditions and (c) detail of the test interval.

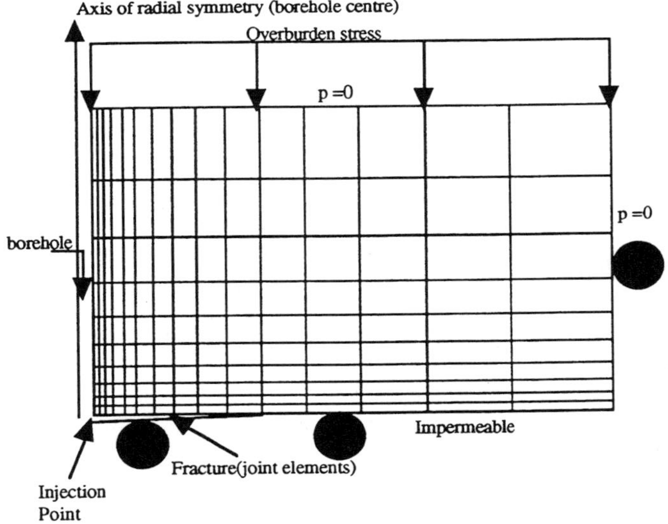

Figure 8. AECB's finite element model of the test site.

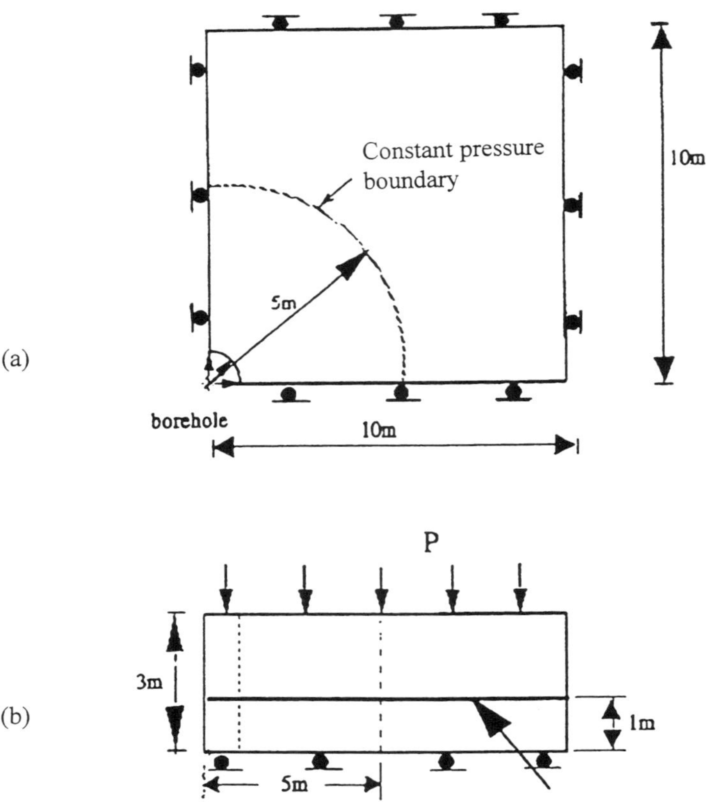

Figure 9. AECL's model of the test site: (a) Vertical cross section and (b) horizontal cross section.

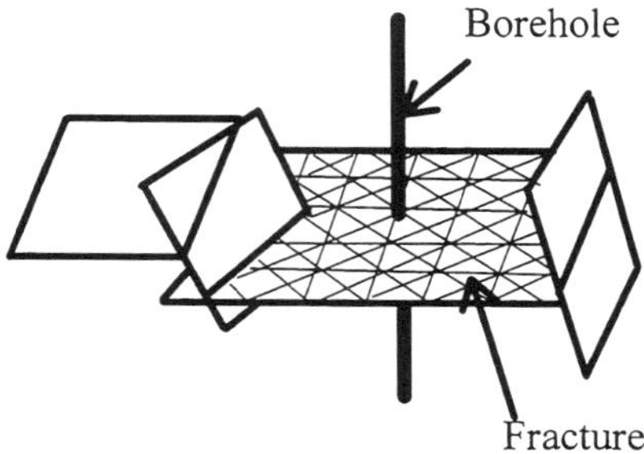

Figure 10. Sketch of AEA's discrete fracture network model of the test site.

The hydraulic aperture, e, at a certain effective stress, σ'_n is:

$$e = e_i + \Delta v = e_i + F\left(k_{ni}, V_{mi}, \sigma'_{ni}, \sigma'_n\right) \qquad (3)$$

where e_i is the initial hydraulic aperture at the initial reference stress, σ'_{ni}.

KTH and AECB discretized TC6 to an axisymmetric model with the vertical borehole in its centre (Figure 7 and 8). The sub-horizontal fracture is assumed to be penny-shaped and centred at the borehole intersection. AECL used a three-dimensional square model with the borehole in one of the corners (Figure 9). ROCMAS, FRACON and MOTIF can model the interaction between joint opening and deformation of the intact rock surrounding the fracture. However, the modelling is limited to the coupled hydromechanical behaviour in one single fracture near the borehole.

AEA Technology simulated TC6 with the fracture network model NAPSAC [25] (Figure 10). A single fracture was made to intersect the borehole and this fracture was surrounded by a realistic fracture network. Hydro-mechanical coupling is provide by a relation between transmissivity, T, and the effective normal stress, σ'_n, acting on the fracture. The effective normal stress is calculated from the normal stress acting on the fracture, σ_n, and the fluid pressure, P, of the fluid flowing through it by $\sigma'_n = \sigma_n - P$. The resulting change in effective stress is converted to a change in aperture via one of the following couplings

1. $T = \mathrm{cons\,tan\,t}$ $\qquad (4)$

2. $\dfrac{T}{T_i} = \left(\dfrac{\sigma'_n}{\sigma'_{ni}}\right)^{-\alpha}$ $\qquad (5)$

3. $e - e_i = \max\left(-\dfrac{\sigma'_n - \sigma'_{ni}}{k_n}, e_{min}\right)$ $\qquad (6)$

where T_i is initial transmissivity , k_n is joint normal stiffness and e_{min} is a minimum aperture set to prevent a negative aperture.

There are two levels of discretisation available. The fracture plane intersecting the borehole is discretized into a number of finite elements. At the centre of each element, an effective stress and hence an aperture is calculated. For fractures away from the borehole, an average aperture of the whole fracture plane is calculated from the effective stress in the centre of the fracture plane.

The NAPSAC model can thus simulate the interaction between the fluid flow in the single fracture and the rest of the fracture network around the borehole. However, it cannot model the mechanical interaction of joint deformation with the surrounding rock mass. The total stress normal to the fracture is assumed to be constant and not affected by the joint deformation. This implies that the effective compressive stress becomes negative when the fluid pressure exceeds the initial total stress. NAPSAC can therefore not model fluid injection with pressures greater than the virgin normal stress.

LTH used the diffusion equation together with the "cubic law" and an empirical relationship between rock porosity and rock hydraulic conductivity [26]. In the numerical simulation the hydraulic aperture and transmissivity were increased at each time step in proportion to the increase in fluid pressure as:

$$e = e_i + P(r,t)\frac{\partial e}{\partial P} \tag{7}$$

where e_i is the initial uniform fracture aperture at time $t < 0$. $P(r,t)$ is the local fluid pressure at time, t, and at a distance, r , from the wellbore. Equation (7) implies that the hydraulic aperture is dependent on the local fluid pressure and a constant value on $\partial e/\partial P$. The sub-horizontal fracture is discretized into an axisymmetric finite element model (Figure 11). The model does not inlcude the rock mass surrounding the fracture.

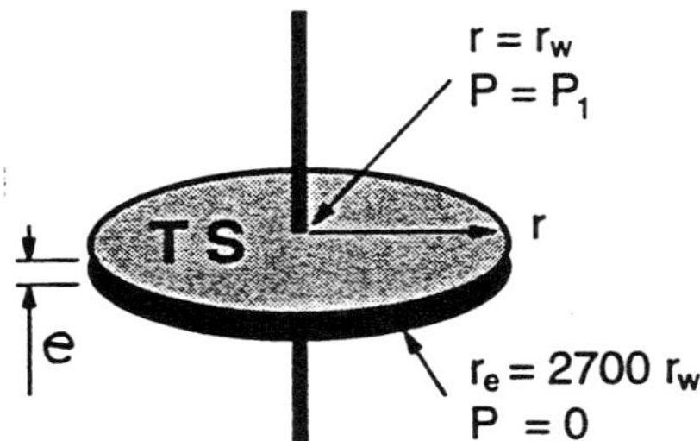

Figure 11. LTH's finite element model of the test site.

3.2 Comparison of field tests and numerical modelling

The purpose of TC6 is to investigate important coupled hydromechanical processes and parameters during high pressure hydraulic testing, and to validate the soundness of constitutive hydromechanical relations used in various models. There are a large number of material parameters and boundary condition which can affect the pressure flow response. The exact geometry of the fracture and the boundary conditions at the test site is not known. Therefore, a complete validation of the computer codes is not possible. However, variation of input parameters of the different models can be used to identify what parameters are of importance for the pressure flow response. The influence of different boundary condition can also be investigated. If certain processes or constitutive relations are dominating, it may be possible to validate the soundness of those relations by comparing the field test response with the numerical modelling. During high pressure water injection into a joint, the hydromechanical relation of the fracture plane has a dominating affect on the pressure flow response. It is therefore possible to investigate the soundness of the constitutive hydromechanical relations of rock joints used in the computer codes.

3.3 Modelling of pressure pulse test

Figure 12 presents the results of the field experiment and the modelling of three research teams. Modelling results of AECB exhibit a delayed and comparatively steeper pressure versus time response than KTH. The reason is that AECB used a smaller much smaller joint normal stiffness (Table 5) than KTH. The storativity of the test, is determined by the deformability of the fracture which in turn is dependent on the normal stiffness. A more deformable fracture provides greater storativity which makes the pressure drop smoother. The storativities in Table 5 are estimated according to [27]:

$$S_f = \rho g(1/k_n + eC_w)$$ (8)

where C_w is the compressibility of water.

A parametric variation conducted by KTH showed that the steepness of the pressure time curve is determined by the stiffness of the joint and adjacent rock (Figure 13). This explains the difference in normal stiffness obtained by KTH and AECB.

Table 4

Input parameters for modelling of pulse test.

Parameter	Research Team		
	KTH	AECB	AEA
Initial hydraulic aperture, e (μm)	16	17	13.5
Initial normal stiffness, k_n (GPa/m)	370	5400	-
Young's modulus of rock, E_{rm} (GPa)	60	70	-
Fracture transmissivity, T_f (m²/s)	$3.4 \cdot 10^{-9}$	$4.1 \cdot 10^{-9}$	$2.0 \cdot 10^{-9}$
Fracture storativity, S_f	$2.7 \cdot 10^{-8}$	$1.9 \cdot 10^{-9}$	$2.4 \cdot 10^{-6}$

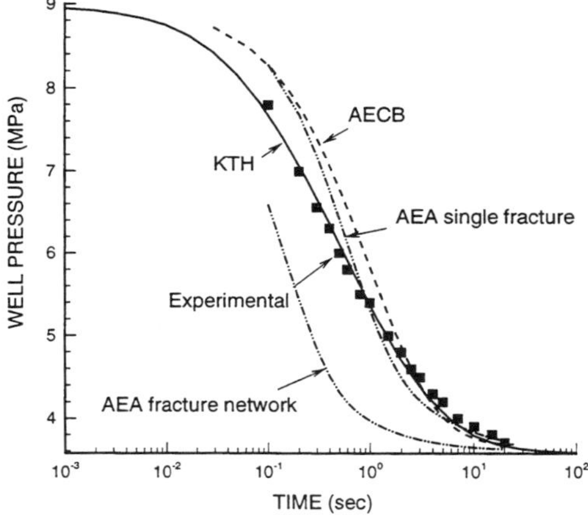

Figure 12. Results of modelling and field test of pulse injection.

3.4 Hydraulic jacking test

The results of modelling and field experiments of the hydraulic jacking test are compared in Figure 14 and Figure 15, and the input parameters are presented in Table 5. KTH obtained best match using the initial aperture of 16 μm and the initial normal stiffness of 1000 GPa/m. For well pressures below the virgin normal stress, the boundary condition at the outer edge of the fracture was assumed to be equal to the ambient pressure at the given depth. However, at maximum well flow, the best fit was obtained with the pressure at the outer edge of the fracture equal to the stress normal to the fracture. The pressure equal to the virgin normal stress is an upper limit for the fluid pressure. If the fluid pressure is higher the fracture will propagate outwards.

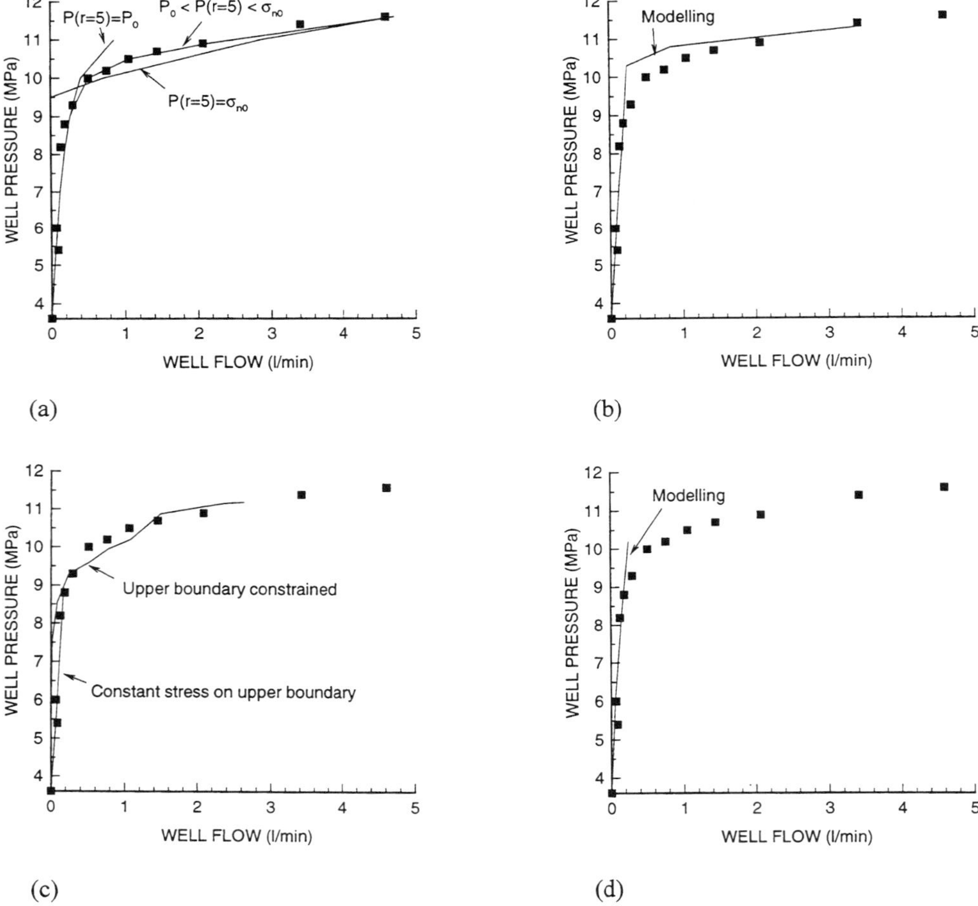

Figure 14. Results of modelling and field test of hydraulic jacking: (a) KTH, (b) AECB, (c) AECL and (d) AEA.

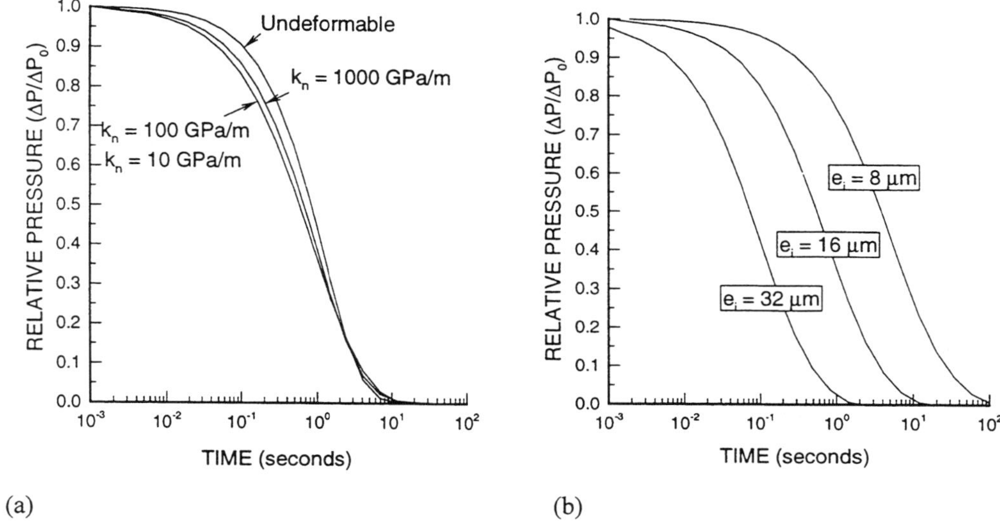

(a)   (b)

Figure 13. Sensitivity analysis of a pulse test by KTH using ROCMAS with variation of: (a) Initial joint normal stiffness and (b) Initial hydraulic aperture.

AEA Technology assumed that although the pulse pressure of 9 MPa was high, the short duration of the pulse test (20s) meant that the effect of stress could be neglected. This test was interpreted in the conventional way described above and the values of transmissivity and storativity, shown in Table 4 give a hydraulic diffusivity, D of $8.5 \cdot 10^{-4}$ m^2s and an initial hydraulic aperture of 13.5 µm.

For the NAPSAC simulations, the model of storativity adopted was S = T/D, which means that D was kept constant. The values obtained from interpreting the pulse test were then put into NAPSAC and the pulse test was simulated on the single fracture plane, as shown in Figure 12. The test was simulated for different values of D but the best fit with the experimental results was for D = $8.5 \cdot 10^{-4}$ m^2s. A fracture network, loosely based on data obtained at Stripa mine was then built around the single fracture an the pulse test was again simulated and the results are shown in Figure 12. These show that the pressure decay is faster because the effect of fractures intersecting the single plane would be to open up more paths along which the fluid can travel, hence speeding up the flow from the borehole.

In conclusion the modelling of the pulse test shows that the hydraulic aperture is the most important factor controlling the pressure versus time response. The joint normal stiffness is also important since it affects the storativity of the fracture which affects the steepness of the pressure versus time curve. Outer boundary conditions of fluid pressure and stress have little effect on the test results. The reason for this is that the pulse test affects only a small part of the fracture plane near the well bore with a small radius of influence.

Table 5
Input parameters used for modelling of the hydraulic jacking test.

Parameter	Research Team			
	KTH	AECB	AECL	AEA
Initial hydraulic aperture, e_i (μm)	16	17	14	13.5
Initial normal stiffness, k_{ni} (GPa/m)	1000[*]	5400[**]	2000[**]	400[***]
Initial normal stress, σ_{ni} (MPa)	9.5	8.6	9.5	10.2
Fracture radius, r (m)	5.0	4.9	5.0	-
Young's modulus of rock, E_{rm} (GPa)	60	70	80	-
Permeability of rock matrix, k_r (m^2)	$1.0 \cdot 10^{-19}$	$1.0 \cdot 10^{-19}$	$1.0 \cdot 10^{-14}$	-

[*]Initial normal stiffness at the initial normal stress for Goodmans joint model
[**]Initial normal stiffness at the initial normal stress for Bandis joint model
[***]Constant normal stiffness model

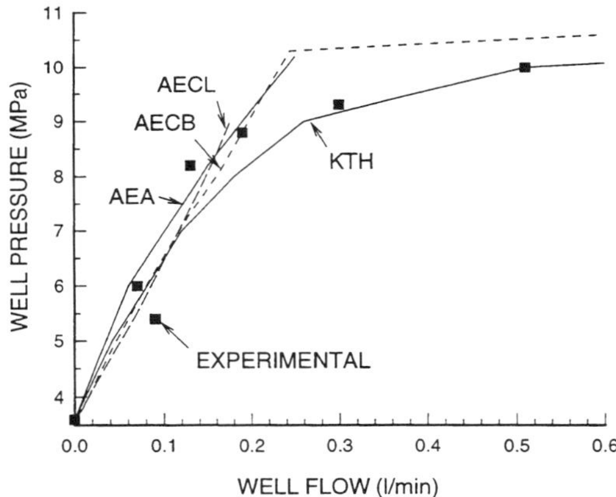

Figure 15. Comparison of modelling and field tests for the low flow rate regime of the hydraulic jacking test.

AECB used an initial hydraulic aperture of 17 μm and an initial normal stiffness of 5400 GPa/m. In order to fit the high pressure part of the test, the virgin normal stress had to be set to 8.6 MPa. The boundary condition at the outer edge of the fracture was limited to be lower than the initial normal stress. A higher pressure is not admissible since the rock mass is assumed to have a negligible tensile strength. The agreement with the field test result is satisfactory. However, the transition from low to high flow from the FRACON simulation occurs more abruptly and at a higher injection pressure as compared to the experimental data.

AECL divided the simulation of the hydraulic jacking into two parts: (1) simulation at low pressure; (2) simulation at high pressure. The first part was simulated with a constant vertical normal stress of 9.5 MPa at the top boundary of the model. This resulted in a good agreement with the field data for well pressures up to 9 MPa. The second part was simulated by constraining the top boundary from movement in the vertical direction. The different results for the different mechanical boundary conditions implies that the results is influenced by the mechanical boundary condition on the top of the model. If the dimension of the model is increased the influence of the upper mechanical boundary condition would diminish.

AEA Technology made a sensitivity study using the three stress transmissivity models in equation (4) to (6). To get initial guesses for α and k_n, the models were fitted to the results from laboratory compression tests giving the values of k_n = 449.9 GPa/m and α = -0.25. The effect of the three models of flow field was investigated by simulating the steady-state flow for the hydraulic jacking tests. They concluded that the third model gave the best match with a normal stiffness k_n of 449.9 GPa/m and this was the model of stress adopted. These fits were calculated using the single plane model. The addition of a fracture network around the plane slightly reduced flow but showed little difference. Figure 14d shows that the modelling result is in close agreement with the experimental ones when the pressure was below overburden.

The simulation of the hydraulic jacking test for single plane model attempted to recreate the experiment numerically in that each stage continued for a period of fifteen minutes in order to reach a steady state. In the actual experiment, each stage lasted only about three minutes and a true steady-state was not reached. Figure 16a shows the results of the simulation described above, with and without the effect of stress. It can be seen that for low pressures there is not much difference between the two simulations. A large jump in the injection pressure (from 6.6 MPa to 10.2 MPa) produces a large jump in the flow in the case of the stress model. The qualitative behaviour of the simulation is the same as the experimental that there is a large initial flux which then decays. Figure 16b shows the simulation of this experiment for the fracture network model, with stress effects considered. Each stage was simulated for 200s which is of the same order as the experiment. The results show the same qualitative behaviour as the single plane model and the experiment and the flows have been slightly reduced.

NAPSAC has not been designed to model crack propagation and uplift and as the stress acting on the individual blocks is not modelled exactly, there is, in theory, nothing to stop the fracture opening up indefinitely, for negative effective stress. Therefore the models of stress described above are not appropriate when the pressure rises above overburden.

Parametric variations performed by AECB showed that at well pressure below the virgin total stress, the resistance to opening comes mainly from fracture normal stiffness without much contribution from the surrounding rock mass. KTH came to the same conclusion and the results of variation of stiffness and initial aperture is presented in Figure 17. There are therefore two parameters determining the pressure versus flow response at low injection pressures: hydraulic aperture and joint normal stiffness. For fluid pressures above the virgin normal stress, the fracture normal stiffness becomes very low and its aperture increases drastically, resulting in a rapid increase in flow rate. At this stage, the hydromechanical properties of the joint plane near the well bore is no longer dominating. There are a large number of parameters and boundary conditions affecting the response and it is not possible to back-calculate a unique set of input parameters.

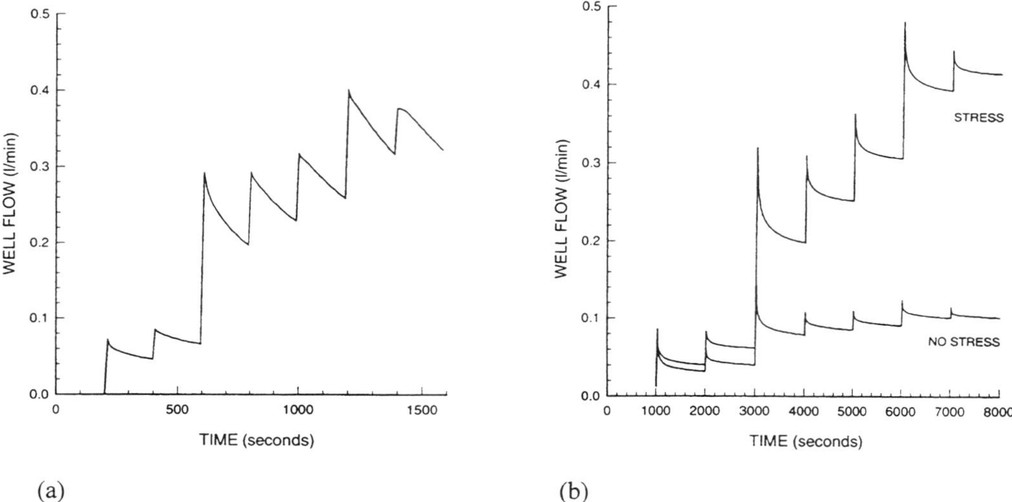

Figure 16. Results of the transient hydraulic jacking simulations by AEA Technology for (a) the single plane model with and without stress and (b) the fracture network model with stress effects.

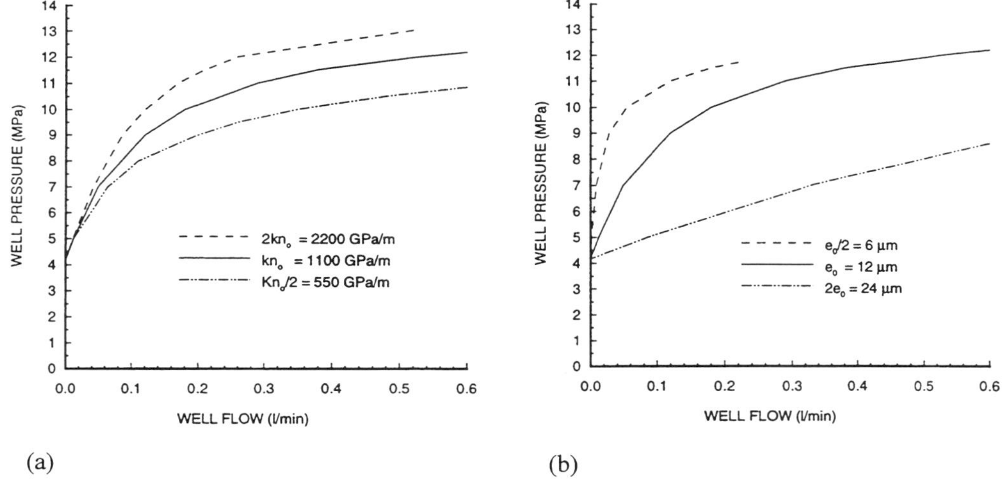

Figure 17. Sensitivity analysis of a hydraulic jacking test by KTH using ROCMAS with variation of: (a) Initial joint normal stiffness and (b) Initial hydraulic aperture.

The initial hydraulic aperture and initial normal stiffness in Table 5 may be meaningful to compare because these parameters are dominating the hydromechanical response of the fracture plane when the well pressure is lower than the virgin normal stress. The values initial hydraulic aperture ranges between 13.5 and 17 microns. These differences lasts from the interpretation of the pulse injection test. The initial normal stiffness for AEA is lower than other teams. However, AEA is using a constant stiffness model which implies that is will not change when the effective stress is reduced. For KTH, AECB and AECL, on the other hand, the stiffness decreases when the effective stresses are reduced. Therefore it may not be fair to compare the initial stiffness of AEA with other teams. The difference in modelling results of different Research Teams for the low flow rate regime in Figure 15 can be explained by the stress transmissivity relation of their joint models. Figure 18 presents the relation between aperture and effective stress of each Research Team. AECB used a much higher stiffness than other teams and obtained a very small increase of aperture for fluid pressures below the virgin stress. However, an abrupt increase of the flow rate is obtained at higher fluid pressure. AEA has a constant normal stiffness leading to good fit on average for pressures lower than the in situ normal stress. When the fluid pressure is higher than 9 MPa the constant stiffness model results in a lower flow rate than the experimental results. In reality the stiffness is not constant but decreases when the effective stress becomes lower due to the reduced contact area between the fracture surfaces. However, it appears that the simpler linear stiffness model is satisfactory for low fluid pressures.

The remaining parameters such as Young's modulus and fracture radius affects the high flow rate regime of the hydraulic jacking test. They can not be back-calculated uniquely from the pressure versus flow response and it may therefore be meaningless to compare the results of different research teams. However, it can be concluded that the values of Young's modulus in Table 5 are reasonable for the rock at the test site and that the numerical simulations with those values are in agreement with the field test data.

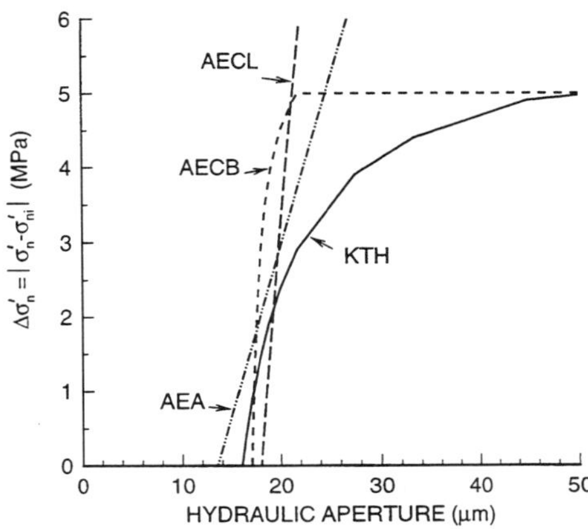

Figure 18. Aperture versus change in effective normal stress for various hydromechanical models.

3.5 Constant pressure injection test

The constant pressure injection test is the most difficult to model of the three types of test in this study. It involves simulation of fluid flow as a response to fluid pressures ranging from below to above the in situ stress. The simulation has also to match the transient rate of joint opening and increase of flow rate. Such analysis cannot be done with a conventional fluid flow analysis but requires transient coupled hydromechanical modelling.

The field test results shows an increasing flow rate with time. However, LTH Research Team proved that an equation of diffusion type for the water flow in fracture cannot produce an increasing water flow versus time in a constant pressure test [26]. This holds true for any space- and pressure-dependent properties of the fracture. That is, even if the joint aperture and transmissivity is assumed to increase as a function of the local pressure. This observation was verified by numerical modelling assuming that the aperture increased in proportion to the increased fluid pressure according to equation (7). The modelling results clearly showed a decreased flow rate as a function of time (Figure 19).

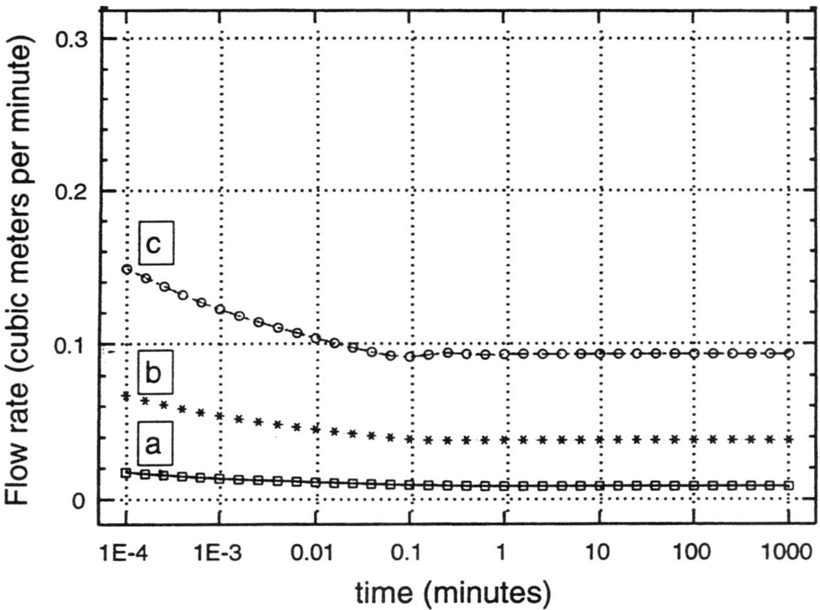

Figure 19. LTH results of flow rate as a function of time for constant pressure at the borehole assuming diffusional flow. Curve a assumes constant hydraulic aperture ($e=e_i$). Curve b and c assumes that the hydraulic aperture is dependent on the fluid pressure according to equation (7). In curve c, the constant value of $\partial e/\partial P$ is greater than in curve b.

However, in the field the transmissivity is not only dependent on the local fluid pressure. An increase of the fluid pressure in a region of the fracture may through the surrounding rock influence other regions. This may cause closing or opening of the fracture as a response to pressure changes at other parts of the fracture area. Therefore, this high pressure injection test can only be properly simulated by including the deformation of both the fracture and surrounding rock mass into the model. ROCMAS, FRACON and MOTIF have such capabilities and the modelling results from those simulations clearly showed an increased flow rate with time (Figure 20).

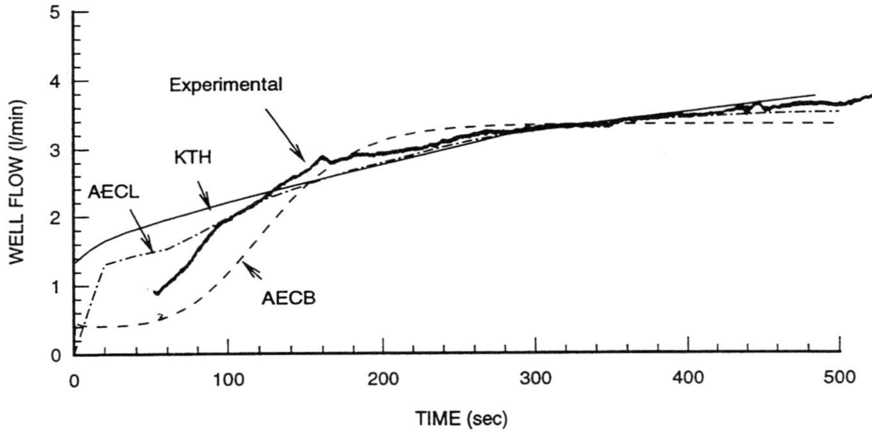

Figure 20. Results of modelling and field test of constant pressure injection.

KTH obtained a faster increase during the first 50 seconds. This can partly be explained by that the KTH modelling started with 11.2 MPa fluid pressure at time zero while it took 50 seconds to attain the maximum pressure in the field test. AECB obtained a delayed but faster rate of flow increase than the field data. The reason for this may be the relatively high normal stiffness which gives a lower storativity and also a higher resistance to joint opening. A reasonable fit to the field test results for constant pressure injection test could be obtained by a number of different combinations of the values of the material properties. The initially low flow rate into a relatively closed fracture with the flow response controlled by the hydraulic aperture and joint stiffness. However, after later time in the test, the fracture opens completely and with loss of contacts between the fracture surfaces leading to a loss of joint stiffness. The flow response becomes sensitive to the rock stiffness and the radius of influence increases dramatically. The pressure increases at the outer edge of the fracture and may be high enough to cause fracture propagation. A transient analysis of a high pressure injection test is complicated an involves a mix of coupled hydromechanical processes which may not be fully captured in the above simplified conceptual models. In a transient constant high pressure injection test there are too many parameters involved at the same simultaneously making it difficult to back-calculate at unique set of hydromechanical parameters.

4. CONCLUDING REMARKS

A high pressure injection test involves mechanical joint opening as a result of reduced effective stress. The pressure versus flow response depends on mechanical and hydraulic properties of both the joint and the surrounding rock mass. A calibration of the models against the field test involves variation of more than 5 parameters and it may therefore be difficult to obtain a unique set of input parameters. However, the numerical modelling of TC6 shows that for injection pressures below the virgin stress normal to the fracture, the hydromechanical properties of the joint are dictating the pressure versus flow response. The key parameters are:

- hydraulic aperture
- joint normal stiffness
- effective normal stress

The modelling of the TC6 showed a fairly good agreement to the field test data. This indicates that the stress-transmissivity relations in the joint models are sound. The discrepancies between different teams are due to different choice of initial hydraulic aperture and joint normal stiffness.

For well pressure higher than the virgin normal stress, the pressure flow response is affected by a number of parameters and also by the outer hydraulic and mechanical boundary conditions. It is not possible to back-calculate uniquely the parameters and a validation based on the high flow rate data is therefore difficult to conduct.

Hydromechanical modelling of injection test can be a used for in situ determination of coupled hydromechanical properties of single joints. The hydraulic aperture can be determined from pulse test and the normal stiffness can be back-calculated from the low flow rate regime of a hydraulic jacking test. Such in situ determination has the advantages over laboratory test that it involves a large area of the joint plane which is undisturbed from core sampling. Furthermore, it has the advantages to measure the hydraulic aperture directly from the flow rate. The mechanical aperture cannot be determined because the exact relation between hydraulic and mechanical aperture is not known. However, the hydraulic testing and coupled modelling provides a stress-transmissivity relation of the rock joint which is a property that can be included into the analysis of fluid flow through large scale rock masses.

5. REFERENCES

1   Detournay E. and H.-D. Cheng. Fundamentals of poroelasticity. In *Comprehensive Rock Engineering,* Vol. 1 (J.A. Hudson, Ed), 395-412. Pergamon Press, Oxford (1993).

2   Tsang C.-F. Coupled hydromechanical-thermochemical processes in rock fractures. Reviews of Geophysics. **29**, 537-551 (1991).

3   Tsang C -F. *Coupled Processes Associated with Nuclear Waste Repositories* (C. -F. Tsang, Ed), p. 801. Academic Press (1987).

4   Raven K. G. and Gale J. E. Water flow in a natural rock fracture as a function of stress and sample size. *Int. J. Rock Mech. Min. Sci. & Geomech. Abstr.* **22**, 251-261 (1985).

5   Makurat A., Barton N. and Rad N. S. Joint conductivity variation due to normal and shear deformation. In Rock Joints. Proc. Int. Symp. on. Rock Joints (Barton and Stephansson, Eds), Loen, 535-540. Balkema (1990).

6   Carlsson A. and Olsson T. Large scale in-situ test on stress and water flow relationships in fractured rock. Swedish State Power Board, Technical Report, P 152. (1986).

7   Jung R. Propagation and hydraulic testing of a large unpropped hydraulic fracture in granite, Institute of Geophysics, Ruhr University, Germany, A, p. 20. (1986).

8   Nemat-Nasser S., Abé H. and Hirakawa S. *Hydraulic Fracturing and Geothermal Energy.* p. 501. Martinus Nijhoff Publishers, The Hauge (1983).

9   Haimson B. C. The hydraulic fracturing method of stress measurement: Theory and practice. In *Comprehensive Rock Engineering* (J. Hudson, Ed), 395-412. Pergamon Press, Oxford (1993).

10   Noorishad, J. and Doe, T. Numerical simulation of fluid injection into deformable fractures. *Proc. 23rd U.S. Rock Mechanics Symp., Berkeley, California.* 664-654 (1982).

11   Rutqvist J., Noorishad J., Stephansson O. and Tsang C. -F. Theoretical and field studies of coupled hydromechanical behaviour of fractured rocks - 2. Field experiment and modelling. *Int. J. Rock mech. Min. Sci. & Geomech. Abstr.* **29**, 411-419 (1992).

12   Rutqvist J. DECOVALEX, Test Case 6, Borehole Injection Test, Doc 93/139, DECOVALEX Secretariat, Engineering Geology, Royal Institute of Technology, Stockholm.

13   Bjarnasson, B., Ljunggren C. and Stephansson O. New development in hydrofracturing stress measurements at Luleå University of Technology. *Int. J. Rock mech. Min. Sci. & Geomech. Abstr.* **26**, 579-586 (1989).

14   Bredehoeft, J. D. & Papadopulos, S. 1980. A method to determine the hydraulic properties of tight formations. Water Resources Research. **16**, 233-238.

15   Wang J. S. Y., Narasimhan T. N., Tsang C.-F. and Witherspoon P. A. Transient flow in tight fractures. *Proc. Invitational Well-testing Symposium,* Berkeley, 103-116. Lawrence Berkeley Laboratory (1977).

16   Black, J. H. & Barker, J. A. *Design of a single-borehole hydraulic test programme allowing for interpretation based errors.* Nationale Genosseschaft für die Lageruing radioaktiver Abfälle, Technischer bersicht 87-03, NAGRA, Bern, Switzerland (1987).

17   Doe T. W. and Korbin G. E. A comparison of hydraulic fracturing and hydraulic jacking stress measurements. *Proc. 28th U.S. Rock Mechanics Symp.,* Tucson, 283-290 (1987).

18   Jacob C. E. and Lohman S. W. Nonsteady flow to a well or constant drawdown in an extensive aquifer. Trans. Am. Geophys. Union, 33, 559-569 (1967).

19   Noorishad J., Tsang C. -F. and Witherspoon P. A. Theoretical and field studies of coupled hydromechanical behaviour of fractured rocks - 1. Development and verification of a numerical simulator. Int. J. Rock mech. Min. Sci. & Geomech. Abstr. 29, 401-409 (1992).

20   Selvadurai A.P.S. and Nguyen T.S. Finite element modelling of consolidation of fractured porous media. Proc. of the 1993 Canadian Geotechnical Conference, 79-88. (1993).

21   Guvanasen, V. and Chan T. Three-dimensional finite element solutions for heat and fluid transport in deformable rock mass eith discrete fractures. Proceedings of the international conference on computer methods and advances in geomechanics, Cairns,6-10 May, 1547-1552, Balkema Press.

22   Biot M. A. General theory of three dimensional consolidation. J. Applied Physics, 12, 155-164 (1941).

23 Goodman R. E. Methods of geological engineering in discontinuous rock, p. 472. West Publishing, New York (1976).

24 Bandis S. C., Lumsden A. C. and Barton N. R. Fundamentals of rock joint deformation. Int. J. Rock mech. Min. Sci. & Geomech. Abstr. 20, 249-268 (1983).

25 Grindrod P., Herbert A., Roberts D. and Robinson P. NAPSAC technical document. Stripa project, TR 91-03, SKB, Stockholm (1991).

26 Claesson J., Follin S. and Hellström G. High pressure testing on fractured crystalline rock-A note on the diffusion equation regarding Test Case 6 of DECOVALEX. Int. J. Rock mech. Min. Sci. & Geomech. Abstr. Special Issue on Thermoe-hydro-mechanical Coupling in Rock Mechanics. July, (1995).

27 Doe T. W. and Osnes J. D. Interpretation of fracture geometry from well test. In *Rock Joints. Proc. Int. Symp. on Fundamentals of Rock Joints* (O. Stephansson, Ed), Björkliden, 281-292. Balkema (1985).

O. Stephansson, L. Jing and C.-F. Tsang (Editors)
Coupled Thermo-Hydro-Mechanical Processes of Fractured Media
Developments in Geotechnical Engineering, vol. 79
© 1996 Elsevier Science B.V. All rights reserved.

449

Experimental study on the coupled T-H-M-processes of single rock joint with a triaxial test chamber

J. Pöllä[a], A. Kuusela-Lahtinen[a] and J. Kajanen[b]

[a]Technical Research Centre of Finland, Communities and Infrastructure, P.O. Box 19041, FIN-02044 VTT, Espoo, Finland

[b]Helsinki University of Technology, Laboratory of Rock Engineering, Vuorimiehentie 2, FIN-02150 Espoo, Finland

Abstract
Laboratory tests have been carried out to determine the effect of normal stresses to the hydraulic conductivity of a single rock fracture. Tests were carried out in triaxial cell and both artificial and natural fractures have been used. Normal stresses were 2 - 16 Mpa and the water flow rate was kept constant during test. Hydraulic conductivities of fractures were calculated using cubic law. These results were compared to different theoretical models and there were no significant deviation between measured values and theoretical results.

1. INTRODUCTION

Underground repository is one concept in solving the final disposal of spent nuclear fuel. The canisters containing spent fuel are placed into the holes at the bottom of repository tunnel. The rock surrounding repository tunnels is subject to changes in mechanical stresses due to excavation work and thermal induced stresses due to heating effect of spent fuel. Changes in stress fields cause both normal and shear displacements in rock fractures. These displacements change hydraulic properties of the fractures, because the fractures may open or close and therefore basically determine the flow field in the vicinity of the disposal holes.

In order to understand the effects of spent fuel on the hydraulic behaviour of the rock mass it is necessary to have knowledge about the relation between stresses and hydraulical properties of the fractures. DECOVALEX-project is concentrating on coupled thermo-hydro-mechanical processes, and during the project it has been obvious that more knowledge about the stress - water flow relationship of the fractures is required.

Because of the variation of fracture surface geometry and channelling effects experimental data is needed as a basis for theories. This study is a first part of tests on relationship between normal stress-hydraulic conductivity in a single fracture without shear displacements. In this first attempt temperature effect has not been studied. The basic idea is to find out the changes in hydraulic conductivity when normal stress across the fracture is changing. When the hydraulic conductivity versus normal stresses is known, it is possible to relate hydraulical and mechanical apertures and thus increase the basic understanding of the phenomena.

2. BACKGROUND

Theories for the fluid flow through fractures in rock are normally based on one-dimensional laminar flow between two parallel plates separated by a distance e. This is called the cubic law and it is developed from Darcy's law and can be written

$$\frac{Q}{\Delta h} = Ce^3 \tag{1}$$

In the case of laminar flow the expression for C is

$$C = \frac{W\rho g}{L\ 12\mu} \tag{2}$$

where Q is the flow rate, Δh is the difference in hydraulic head, e is fracture aperture, w is width of fracture, L is length of fracture, μ is viscosity of the fluid, ρ is density of fluid and g is acceleration of gravity.

The hydraulic conductivity of a fracture with an aperture e can be determined from equation

$$K_f = e^2\frac{\rho g}{12\mu} \tag{3}$$

Witherspoon et al. [1] have suggested that, when the cubic law is applied to natural fractures with rough surfaces, constant C can be replaced by C/F. F is a factor which describes the effect of deviations from the ideal parallel plate concept.

Transition from laminar to turbulent flow can be determined by the Reynolds number (Re). Re is a dimensionless number and it is defined by equation

$$\text{Re} = \frac{qd}{\mu} \tag{4}$$

where q is the specific discharge, d is some length dimension of fracture.

At low Re-numbers, there is a region where the flow is laminar, viscous forces are predominant and the linear Darcy law is valid [2]. When the Re-number is lower than 1 - 10, then the flow is laminar.

The fracture is closing i.e. aperture e is decreasing when normal stress is applied to the fracture. Normal stress - closure relation is non-linear and Goodman [3] has given hyperbolic function

$$\frac{\sigma_n' - \zeta}{\zeta} = S\left(\frac{\Delta V}{V_m - \Delta V}\right)^t \tag{5}$$

where
σ_n' = effective normal stress
ζ = seating pressure
V_m = maximum closure of fracture
S, t = constants.

and Zhao and Brown [4] has presented an alternative model of logarithmic type [4]

$$\frac{V_m - \Delta V}{V_m - \Delta V_r} = 1 - B\ln\left(\frac{\sigma_n'}{\sigma_r'}\right) \tag{6}$$

where
σ_n' = effective normal stress
σ_r' = reference value of effective stress
ΔV_r = reference value of fracture closure
B = constant.

Effective normal stress is calculated subtracting the water pressure from total normal stress.

Relation between hydraulic properties of a fracture and effective normal stress has been presented, for example, by Gangi [5], Swan [6] and Gale [7]. Their models are as follows:

Gangi [5]

$$k_f = k_0\left[1 - \left(\frac{\sigma_n'}{D}\right)^n\right]^3 \tag{7}$$

Swan [6]

$$k_f = k_{0,c} - m\ln\sigma_n'^{,2} \tag{8}$$

Gale [7]

$$k_f = s\sigma_{n'}{}^{a} \tag{9}$$

where
k_f = conductivity
k_o = initial conductivity
$\sigma_{n'}$ = effective normal stress
D, c, m, n , s and a = constants

These expressions are based on at least one experimental flow test [4]. However, large variation in model parameters causes difficulties when applied for predictioms of water flow. For example, Zhao and Brown [4] found parameters D and n in Gangi's model to vary from D = 52.46; n = 6.00   to D = 4.52; n = 0.88 for natural joints in granite. The use of these relationships obviously requires several tests in order to determine the constants.

Hakami and Larsson [8] carried out flow experiments and aperture measurements on the same specimen of single natural fracture in order to compare measured flow with predicted flow based on geometrical description of the fracture void space. The predicted and measured flow through the fracture specimen were in good agreement.

3. DESCRIPTION OF TEST

3.1 Test system
Laboratory tests were conducted using MTS 815 Rock Mechanics Test System of the Laboratory of Rock Engineering at Helsinki University of Technology. The testing system is servo-controlled, and it consists of 2600 kN load frame, triaxial cell, hydraulic power supply, confining and pore pressure intensifiers, and TestStar digital controller. Data collection and recording is done using TestWare-SX application programme. It is possible to perform both uniaxial and triaxial tests with the system. Temperature control chamber (from - 60 °C to + 200 °C) can be used in uniaxial testing. The triaxial cell can be heated up to + 200 °C in triaxial testing.

3.2 Test samples and their preparation
The samples for the laboratory tests were cored from the construction stone quarry at Kuru in the central Finland. The samples are fine-grained grey granite (Kuru Grey). This granite is well known and vast numbers rock mechanical test results are available. Some properties are shown in Table 1. Weathering and wear resistance of Kuru Grey is good as for most granites. Kuru Grey granite is isotropic and homogeneous, and contains about 36% alkali feldspar, 21% plagioclase, 35% quartz, and 8% amphibole and micas [9].

Samples were cored from six boreholes marked KR1 - KR6 at the depth between 0.5 - 1.0 m below surface. Sample diameter was 102 mm and length of the prepared samples was 100 mm. Samples from boreholes KR1 - KR3 contained natural fracture. The fracture is parallel to the axis of the cored sample and divides the samples into the two equal parts along the centreline.

Table 1.   Rock Mechanical Properties of Kuru Grey granite [9].

Density kg/m^3	Water Absorption %	Uniaxial Compressive Strength, MPa	Tensile strength, MPa	Young's Modulus, MPa	Hardness, Mohr's
2660	0.11	300	25	52,000	6.1

An artificial fracture parallel to the sample axis was prepared at the laboratory for the samples from boreholes KR4 - KR6. The artificial fracture was made using indirect tension test (Brazilian test). The splitting tool was a sharp steel edge and the splitting was done by fast axial loading by MTS 815 actuator.

On the preparation of the samples sample ends were first cut with a diamond saw and then ground to flatness of 0.02 mm or better. The diameter and the length of the specimen were measured with a sliding caliper. The types and dimensions of samples are shown in table 2.

Table 2.   The types and dimensions of samples.

Test	Sample	Fracture type	Diameter (mm)	Length (mm)
1711	KR5-1[*]	artificial	102.4	100.7
1811	KR2-1	natural	101.1	98.9
2311	KR2-2	natural	101.1	101.1
2411	KR3-3[*]	natural	101.4	96.7

[*] roughness profiles have been measured

Roughness of the fracture surfaces were measured using optical laser-profilometer. The measuring accuracy is ±0,03 mm and readings were taken at 0.25 mm intervals along the surface. Roughness was measured along 10 profiles parallel to the longitudinal axis of the sample. Measuring profiles were 1 cm apart. The profilometer is originally developed for measurements of pavements and other road surfaces. It was first time used for rock fractures in Finland. Due to difficulties in determining the actual position of the measuring point, it not possible to localise any channels or flow paths inside the fractures. Examples of measured profiles are shown in Figs. 1 and 2.

Sample KR5, surface A, test 1711

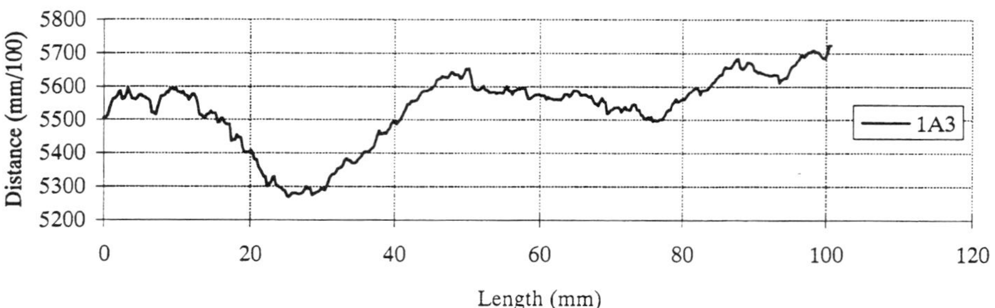

Figure 1.   Profile along sample KR5, with artificial fracture, test 1711. Distance is measured from zerolevel.

Sample KR3, surface A, test 2411

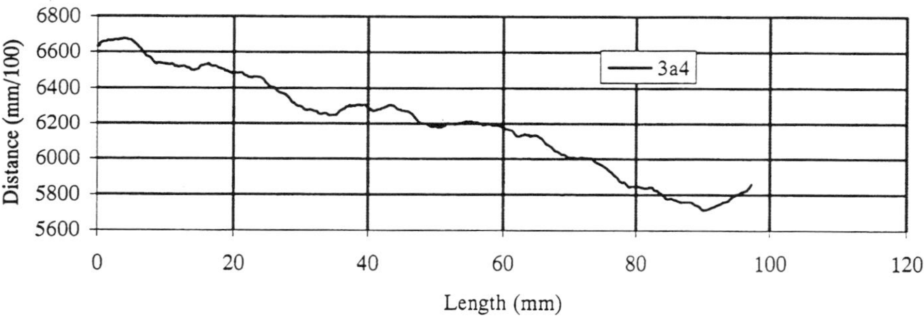

Figure 2.   Profile along sample KR3, with natural fracture, test 2411. Distance is measured from zerolevel.

3.3 Test procedure

The water conductivity test in this study is quite similar to tests conducted by Zhao and Brown [4] except that the effects of the temperature on hydraulic conductivity was not studied.

Instrumentation of the specimen consists of one circumferential extensometer, which was used to measure the radial deformations of the test sample. The total radial deformation is composed of three components: i) the fracture opening/closure due to varying water pressure in the fracture, ii) the circumferential deformation of the sample due to the axial and confining stress, and iii) the deformation of the neoprene jacketing sleeve around the sample.

The normally used jacketing material (shrinking plastic tube) was replaced by neoprene rubber because of the unpredictable behaviour of the shrinking plastic tube. The neoprene rubber sleeve behaves almost elastically and the test results can be calibratated. Errors of deformation caused by the different types of sleeves have been studied by Kuula [10]. The instrumentation of the test specimens and the jacketing neoprene tube is shown in Figure 3.

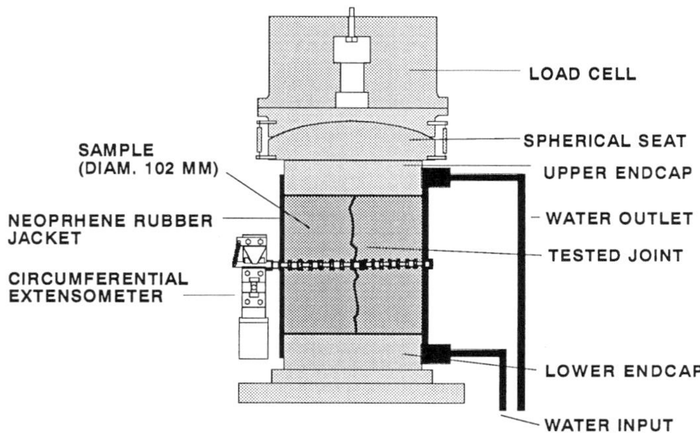

Figure 3.   Sample instrumentation.

During the tests the sample was subjected to 20 MPa axial loading and to 2 - 16 MPa confining pressure. The confining pressure gives a normal stress on the fracture plane during the test. The axial loading was applied to the sample by the actuator of the testing system and the confining pressure by pressure intensifier. Both of these were servo-controlled.

Water flow through the fracture was generated by the servo-controlled pore pressure intensifier. Using the intensifier's displacement control mode, the water flow was kept steady at the rate of 0.05 cm^3/sec. Also, 350 cm^3 of water was pushed trough the fracture before any loading took place in order to get the fracture fully saturated.

The following test procedure was used in the water conductivity test:

<u>Starting phase</u>
> Axial loading was increased up to 20 MPa at a rate of 0.8 MPa/sec. At the same time confining pressure was increased to 2 MPa at a rate of 0.12 Mpa/sec. Continuous water flow at a rate of 0.05 cm^3/sec was pumped through the fracture. The flow was maintained for 600 seconds.

<u>Loading phase</u>

The confining pressure was raised at intervals of 2 MPa at a rate of 0.1 MPa/sec. Continuous water flow at a rate of 0.05 cm^3/sec for 600 seconds during each loading interval was pumped through the fracture

<u>Ending phase</u>

Water pressure, confining pressure and axial loading was decreased to zero level at a rate of 0.25 MPa/sec, 0.25 MPa/sec, and 0.8 MPa/sec, respectively.

The test procedure was repeated using the same sample and procedure. First test is marked by letter A and second test by letter B.

During the tests, eight different parameters were determined, Table 3.

Table 3. Measured parameters in water conductivity tests.

Parameter	Value	Measuring device
Time, t (sec)	time	computer clock
Actuator movement, u_a (mm)	axial strain	actuator LVDT
Axial force, F_a (kN)	axial stress	2500 kN load cell
Circumferential displacement, u_c (mm)	normal strain	circumferential in-vessel extensometer
Confining pressure, P_c (MPa)	normal stress	confining pressure transducer
Confining volume, V_c (cm^3)		confining intensifier piston LVDT
Pore pressure, P_w (MPa)	water pressure	water pressure transducer
Pore volume, V_w (cm^3)	water flow	water intensifier piston LVDT

4. RESULTS

There are two loading parameters, axial force (axial stress) and confining pressure (normal stress) across the fracture. The axial stress is calculated by dividing the axial force by the area of core. To illustrate the steps of test procedure the results of the confining pressure is shown in Figure 4.

The mechanical closure of the fracture is twice the measured normal strain. The measured values of this parameter are shown in Figure 5. According to the increasing normal stress the mechanical closure increases.

The values of water pressure express how water flows through a fracture in steady flow condition. The changes in water pressure during the test is shown in Figure 6. The variation in pressure is due to feedback control system, which keeps the water flow constant regulating the pressure.

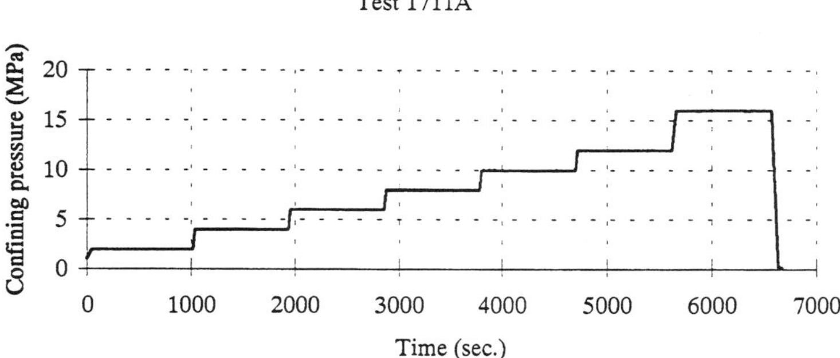

Figure 4.   The values of confining pressure (normal stress across the fracture) versus time, test 1711A.

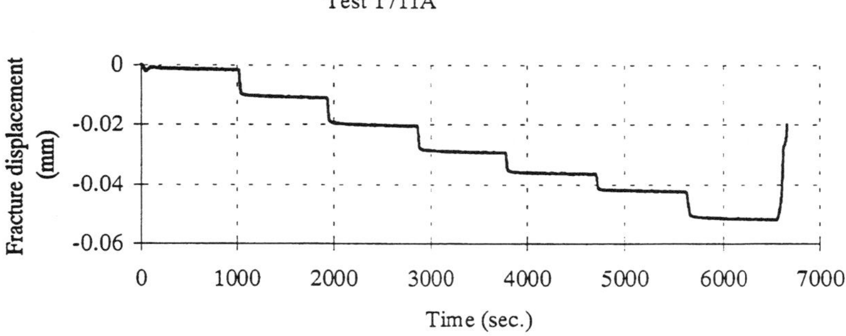

Figure 5.   The values of normal displacement (fracture closure) versus time, test 1711A.

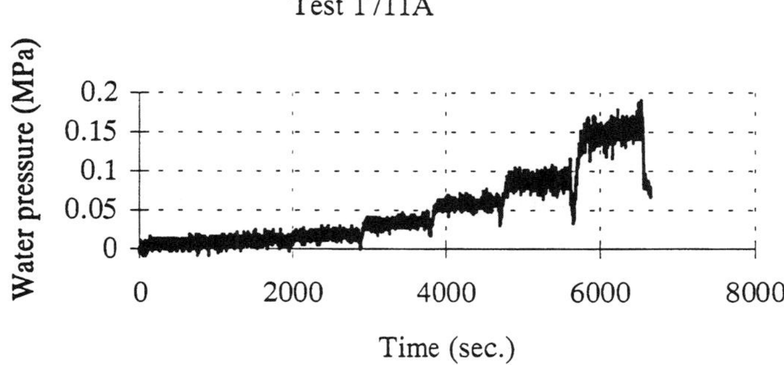

Figure 6.   Values of water pressure (zero level corrected) versus time, test 1711A. Water pressure drops, when confining pressure is increased every 600th second.

Test 2411A

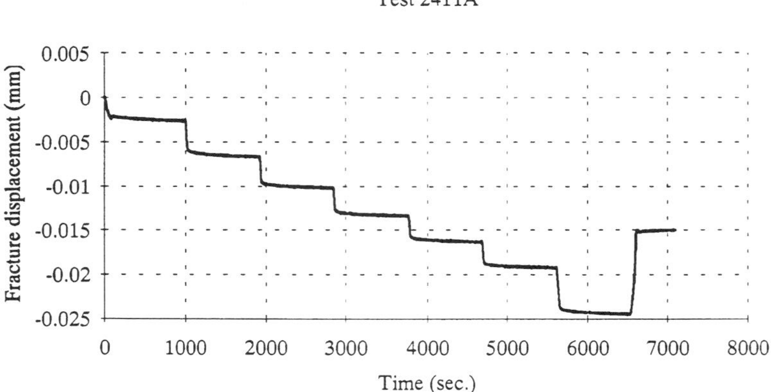

Figure 7.   The values of normal displacement (fracture closure) during the test 2411A.

Test 2411B

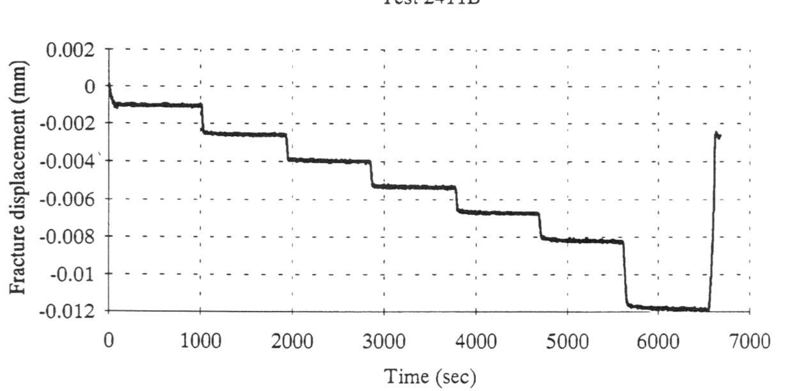

Figure 8.   The values of normal displacement (fracture closure) versus time, test 2411B.

5. DATA ANALYSIS

5.1 Mechanical behaviour of fractures

Preliminary test (sample KR5-1) gave Re-number 0.26 and the flow was laminar. Re-numbers were not calculated for other tests. All the analyses are based on the assumption that the cubic law is valid.

The initial mechanical aperture of fracture is estimated by assuming that it is equivalent in value to the maximum mechanical closure of fracture. It is assumed that the maximum mechanical closure corresponds to complete closure and occurs when the hydraulic conductivity of the fracture is zero.

To define the initial mechanical aperture of fracture, the hydraulic aperture of fracture is calculated from equation (1) for each normal stress. In this investigation all parameters except h (hydraulic head) in the cubic law are constants. Measured parameter P_w is the same as Δh and it varies according to the normal stress. In Figure 9 the calculated hydraulic apertures (e) are plotted against the mechanical closure from test 1711A. The maximum mechanical closure (V_{max}) is determined by fitting the straight line to the data points by regression and its value is obtained from the intersection point of the straight line and mechanical closure-axis.

Calculated values for model parameters are shown in Table 4. Plots of measured values, hyperbolical and logarithmic models are shown in Figures 11 - 14.

Zhao and Brown [4] have collected data concening these parameters. According to that data parameters for natural fractures are S = 1.18 - 27.94, t = 0.60 - 1.22, B = 0.159 - 0.203 at normal temperatures.

Table 4.   Constants of models describing deformation of fracture versus effective normal stress.

Hyperbolic model (Goodman [3])					Logarithmic model (Zhao and Brown, [4])		
Test	S	t	V_m (mm)	ζ (MPa)	B	ΔV_r (mm)	σ_r' (MPa)
1711A	4.894	0.799	0.0735	1.7	0.303	0.0013	1.994
1711B	5.218	0.783	0.0524	2.2	0.238	-0.000586	1.964
1811A	11.607	0.894	0.0816	1.5	0.173	0.00062	1.708
1811B	29.182	1.0002	0.127	1.6	0.0883	0.00026	1.690
2311A	17.094	0.963	0.0717	1.85	0.126	0.00026	1.984
2311B	10.891	0.789	0.0282	1.97	0.145	-7.96E-6	1.971
2411A	61.446	1.020	0.101	0.8	0.09	0.00249	1.96
2411B	81.505	0.932	0.0487	0.6	0.0828	0.00106	1.943

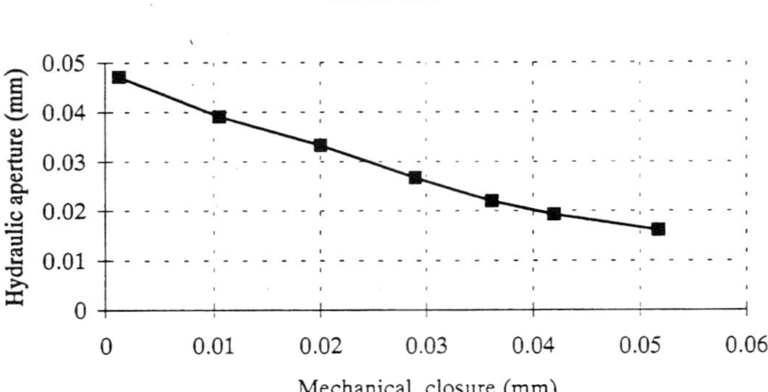

Figure 9.   The relationship between mechanical closure and hydraulic aperture in test 1711A.

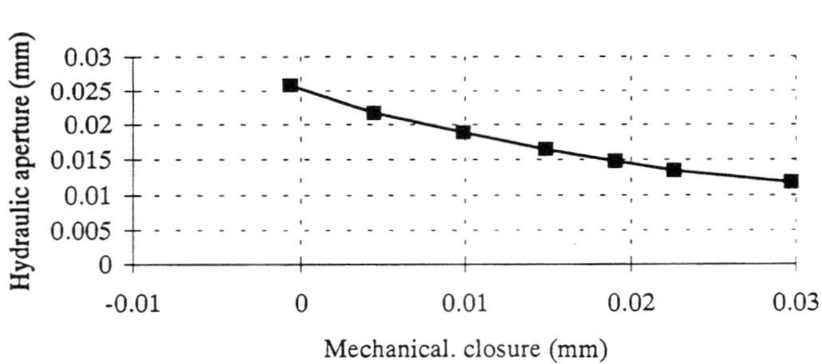

Figure 10.   The relationship between mechanical closure and hydraulic aperture in test 1711B.

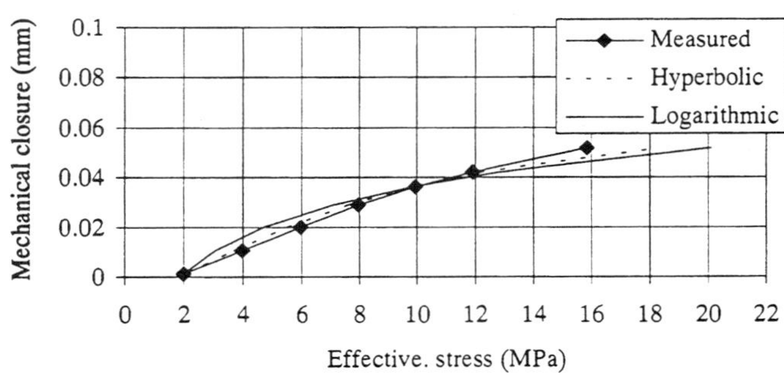

Figure 11.   Measured and calculated mechanical closure, test 1711A.

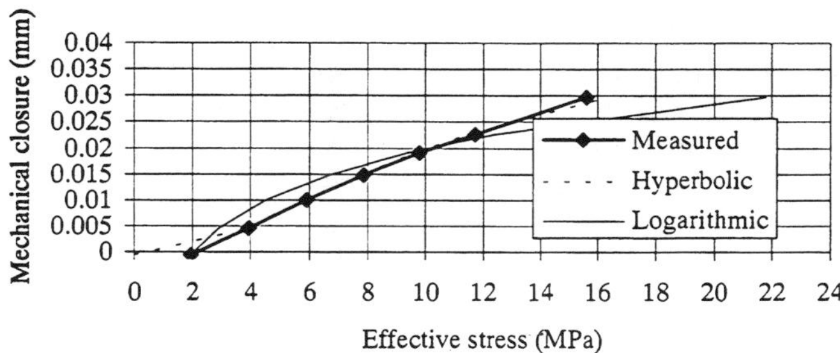

Figure 12.   Measured and calculated mechanical closure, test 1711B.

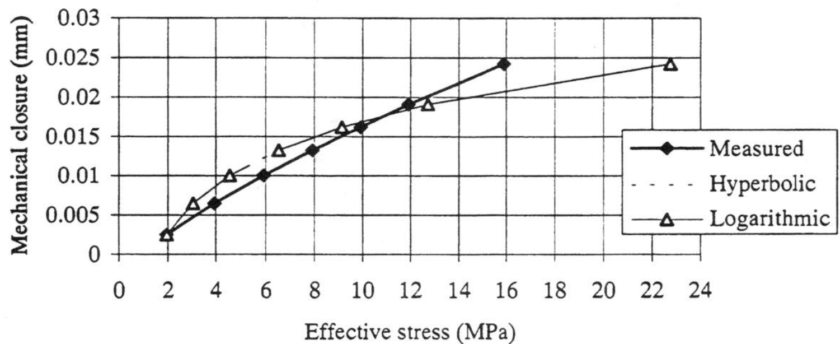

Figure 13.   Measured and calculated mechanical closure, test 2411A.

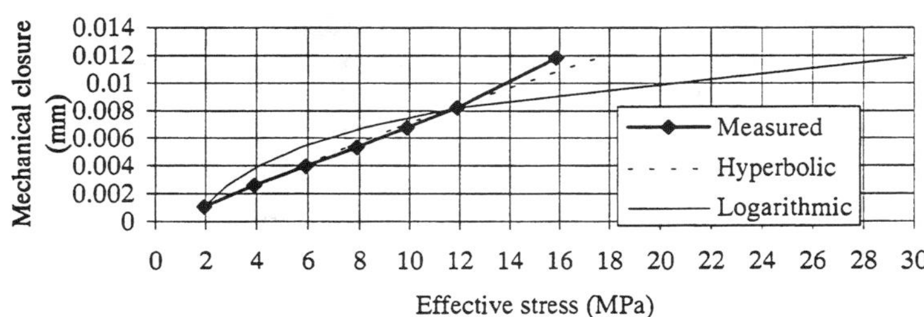

Figure 14.   Measured and calculated mechanical closure, test 2411B.

5.2 Hydraulic behaviour

Plots of conductivities of fractures versus effective stresses are shown in Figures15 - 18. Three different models are fitted to the data obtained from this study.

Table 5.   Parameters of models describing conductivity of fracture against effective normal stress.

	Initial conductivity	Gangi		Swan		Gale	
Test	k_O (m/s)	n	D	m	c	s	a
1711A	0.0023	0.933	24.868	0.293	1.102	0.00464	-1.0782
1711B	0.0008	0.649	41.441	0.219	0.969	0.000991	-0.771
1811A	0.00017	0.681	53.634	0.177	0.975	0.000184	-0.529
1811B	0.00015	0.565	213.253	0.0967	0.968	0.000145	-0.243
2311A	0.0011	0.774	91.055	0.136	1.044	0.0013	-0.349
2311B	0.00075	0.753	60.075	0.169	1.022	0.000883	-0.472
2411A	0.00054	0.688	166.574	0.0964	1.00978	0.000571	-0.233
2411B	0.00044	0.671	144.588	0.0995	0.990	0.000446	-0.244

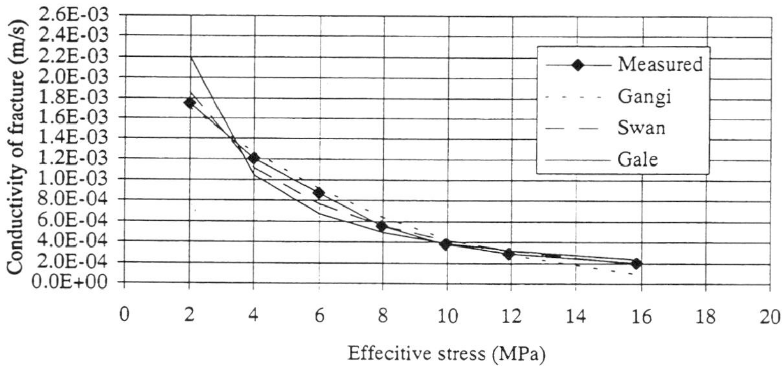

Figure 15.   Measured and calculated hydraulic conductivities, test 1711A.

1711B, Artificially induced fracture

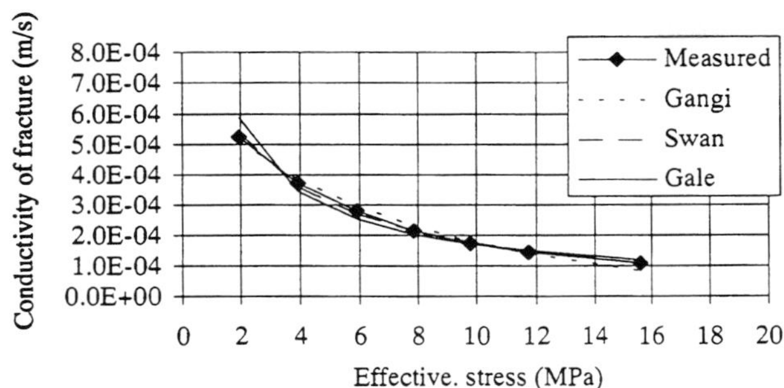

Figure 16.   Measured and calculated hydraulic conductivities, test 1711B.

2411A, Natural fracture

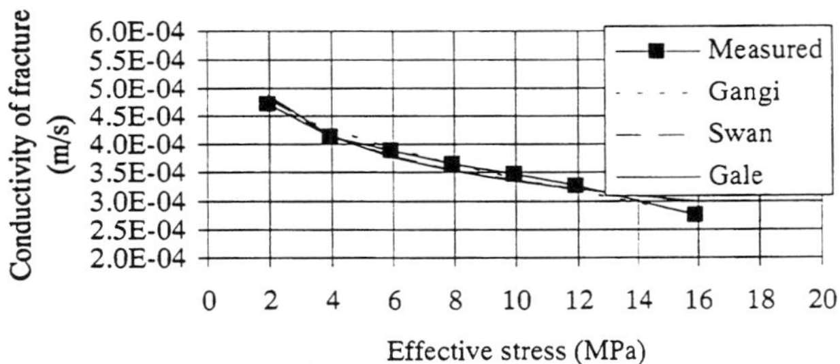

Figure 17.   Measured and calculated hydraulic conductivities, test 2411A.

2411B, Natural fracture

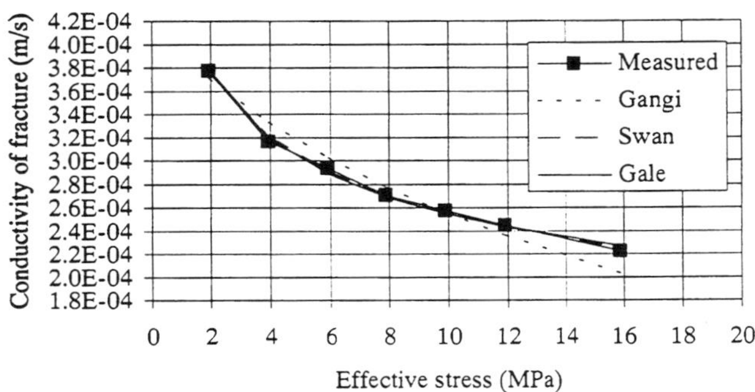

Figure 18.   Measured and calculated hydraulic conductivities, test 2411B.

6. CONCLUSIONS

Laboratory tests have been carried out to determine the effect of normal stresses to the hydraulic conductivity of a single fracture. Tests were carried out in triaxial cell and both artificial and natural fractures have been used. Normal stresses across the fracture were produced by confining pressure in the triaxial cell. Normal stresses were increased from 2 MPa to 16 MPa at the intervals of 2 MPa. The water flow rate was kept constant as 0.05 cm^3/s and the water pressure required to keep the constant flow rate was measured at different normal stresses.

In the preliminary tests, normal stresses larger than 10 MPa were used. Within normal stress range used, the behaviour of mechanical closure of the fracture can be approximated by linear relationship, indicating that normal stiffness of the fracture can be regarded as constant at normal stresses larger than 10 MPa. In the tests carried out by Zhao and Brown [4] clearly non-linear relationship has been observed at low (< 4 MPa) effective stresses.

To find out the behaviour at stresses below 10 MPa, a new test series was carried out. These tests showed, that a hyperbolic function describes quite accurately the actual mechanical behaviour of the fracture subjected to normal stresses in the interval 2 - 16 MPa.

When two tests are carried out using the same sample, the displacement is not fully reversible, and a permanent deformation, about 0.01 - 0.02 mm, has taken place. This means that the closure during the second test is less and thus the normal stiffness of the fracture is higher.

Hydraulic conductivities of the fractures have been calculated based on the assumption that the cubic law is valid. Also, comparison between measured and calculated conductivities given by Gangi [5], Swan [6] and Gale [7] have been made. These calculations show that there are no significant deviations between measured values and theoretical results. However, theoretical models are empirical and are basically best-fit functions of the data set. This causes large variations in modelling parameters and the use of those models is questionable for predictions.

Roughness profiles were measured, but they cannot be used in estimating the initial apertures due to difficulties in determinating the actual co-ordinates of the measured points.

REFERENCES

1   P.A.Witherspoon, J.S Wang, K. Iwai, and J.Gale, Validity of cubic law for fluid flow in a deformable rock fracture. Water Resources Research, Vol 16, No 6, (1988)1016.

2   J. Bear, Dynamics of fluid in porous media. New York, Dovers Publications, Inc., (1972).

3   R. Goodman, The mechanical properties of joints. in Advances in rock mechanics, Proc. of 3rd. Cong. of Int. Soc. for Rock Mech., Denver. National Academy of Sciences (1974).127.

4   J. Zhao, E.T. Brown, Hydro-thermo-mechanical properties of fractures in the Carnmenellis granite. Quarterly Journal of Engineering Geology, 25 (1992), 279.

5   A.F. Gangi, Variation of whole and fractured porous rock permeability with confining pressure. Int. J. of Rock Mech. and Min.Sci. and Geomech. Abstr. Vol. 15, (5) (1978). 249.

6   G. Swan, Determination of stiffness and other joint properties from roughness measurements. Rock Mech. and Rock Eng., 16,(1983) 19.

7   J.E. Gale, The effects of fracture type (induced versus natural) on the stress-fracture closure-fracture permeability relationships. Proc. of 23rd US Symp. on Rock Mech., Berkeley, California, (1982). 290.

8   E.Hakami and E. Larsson, Aperture measurements and flow experiments on a single natural fracture. In: Aperture distribution of rock fractures. Doctoral thesis. Division of Engineering Geology, Royal Institut of Technology, Stockholm, (1995).

9   Luonnonkivikäsikirja, Kiviteollisuusliitto ry, 1994.2.(1994). (in Finnish)

10   H. Kuula, Kolmiaksiaalikokeissa käytettävän kumisukan muodonmuutosominaisuudet. Helsinki, Faculty of Process Engineering and Materials Science, Helsinki University of Technology. (1993). (in Finnish).

O. Stephansson, L. Jing and C.-F. Tsang (Editors)
Coupled Thermo-Hydro-Mechanical Processes of Fractured Media
Developments in Geotechnical Engineering, vol. 79

467

Experimental study on dynamic behavior of rock joints

Mikko P. Ahola[a], Sui-Min Hsiung[a], and Daniel D. Kana[b]

[a]Center for Nuclear Waste Regulatory Analyses, [b]Southwest Research Institute, 6220 Culebra Road, San Antonio, Texas, 78238, USA

Abstract

A comprehensive evaluation of the dynamic behavior of rock joints based on both experimental studies in the laboratory on natural and simulated rock joints as well as an actual underground case study is presented. The laboratory single-jointed dynamic tests made use of natural rock joints in a welded tuff, and were tested under both harmonic and earthquake loading conditions at various frequencies under displacement control. Experimental results showed that the shearing resistance could be markedly different between the forward and reverse shearing directions depending on the joint roughness and interlocking nature of the mated joint surfaces. It was also found that the joint dilation that takes place during foward shearing is fully recovered during shear reversal, with a small offset due to gouge buildup within the joint. A laboratory-scale model experiment was also conducted to study the dynamic behavior of a system of interconnected (artificial) joints around a circular opening in a scaled down rock mass when subjected to earthquake shear wave motion at the base. Results showed that the primary mode of deformation of the rock mass around the tunnel was due to stick-slip behavior along the joints. This type of stick-slip behavior was confirmed during an actual 3 year underground seismic field experimental program designed to study the effect of relatively low-magnitude, repetitive seismic motion (i.e., mining induced) on the behavior of mined excavations. This stick-slip behavior as evidenced in both the field and laboratory seems to explain quite well the phenomenon of the excavations responding to some seismic events but being unresponsive to others. It is believed that the joint stick-slip behavior forms a basis for the progressive accumulation of joint permanent deformation and, consequently, rock mass fatigue. Since materials are normally weaker under fatigue conditions, it is suggested that similar, or even more, damage to an excavation may be realized through a number of seismic events with relatively smaller magnitudes, as opposed to the damage due to a single seismic event with a strong motion.

1. INTRODUCTION

In many cases, the deformation of the rock surrounding an excavation occurs primarily along the natural joints, bedding planes, or blast-induced fractures. The deformation along such interfaces may result from pseudostatic loadings due to excavation of tunnels, as well as thermal loads in the case of underground nuclear waste storage. Deformation along the joints may also be a result of dynamic loadings due to seismic motion from earthquakes. In the case of permanent underground storage of high-level nuclear waste, in which design criteria impose more stringent requirements than ordinary excavations, and long periods of performance are required, the dynamic effects on rock joints need to be considered. This need is especially true in countries such as the United States, where current site investigations for high-level nuclear waste repositories are in seismically active geologic regions. Limited studies on underground excavations have indicated that damage can take place due to fault or joint slip, rock burst, and prolonged or repetitive seismic shaking [1]. Experimental measurements in the field, as well as in laboratory-scale model tests of a jointed rock mass, indicate that successive episodes of dynamic loading on joints result in progressive accumulations of shear displacements (plastic deformation) along the joints [2,3]. Failure of excavations can occur when sufficient accumulated joint shear displacements take place. Permanent cumulative displacements, which occur when the rock joint strength is exceeded, are often estimated based on material properties obtained under static conditions. As a result, if such properties deteriorate during dynamic loading, analyses of excavation behavior under dynamic loading may be nonconservative.

Several studies have been conducted to understand the dynamic behavior of rock joints. Gillette [4] conducted dynamic direct shear experiments on artificial rock joint specimens (Loveland sandstone) under both drained and undrained conditions. Tests were carried out at frequencies ranging from 0 to 10 Hz and normal stresses ranging from 69 to 3,448 kPa. They observed that varying frequencies did not significantly alter the response behavior of the individual specimens. However, the behavior of the different specimens was markedly different due to variation in sample geometry. Gillette [4] also showed that the velocity effects for most rock samples tested were normal stress independent. Although there was some scatter in the experimental data, they did find a general trend indicating increasing shear strength with increasing shear velocity for their tests on Loveland sandstone. This scattering of data was determined to be most likely due to properties of the sample geometry and mating characteristics. Crawford and Curran [5] conducted tests over a wider range of rock types. Their results indicated that, in general, the shear resistance of harder rocks decreased with increasing the velocity over a variable critical velocity. The shear resistance of softer rocks increased with increasing velocity up to a critical velocity. The basic conclusion of these two different studies on dry rock joints is that the rate-dependent strength and shear stress-shear displacement response behavior may have a pronounced influence on rock mass behavior during seismic motion, and it may be important to include such features in dynamic analyses of discontinuous rock masses. Dynamic tests on undrained rock joints by Gillette [4] showed that the interstitial water pressure in a joint subjected to dynamic shear displacements stabilizes early in the process and does not continue to increase with increasing displacement cycles. They did find that the pressure fluctuated in a manner closely related to joint dilation or contraction. The strength of the joint closely followed the effective stress law even during the highly fluctuating water pressures.

Bakhtar and Barton [6] conducted large-scale dynamic friction experiments on artificially fractured blocks of sandstone, tuff, granite, hydrostone, and concrete. The fracture surfaces had surface areas of approximately 1 m^2 and were tested under shear velocities in the range of 400 to 4,000 mm/sec. Using modified stress transformation equations along the angle of inclination of the joint as well as joint property characterization methods developed by Barton and Choubey [7], Bakhtar and Barton were able to predict the measured rock joint strengths to an accuracy of ±15 %. When they partitioned the tests as pseudostatic or dynamic, the average predicted shear strengths were approximately 5% lower than measured under pseudostatic conditions and 10% lower than measured under dynamic conditions. Thus, their joint behavior model is slightly conservative, and the dynamic strength may be approximately 5% higher than the static strength when shear displacement velocities of approximately 0.001

to 0.1 mm/sec (pseudostatic) are compared with the dynamic velocity range of approximately 400 to 4,000 mm/sec.

Direct shear testing of single-rock joints under dynamic loading was also conducted by Barla [8]. They tested the dynamic behavior of saw-cut surfaces of a dry quartzitic sandstone (Monticello sandstone) up to 100 mm in diameter under a single shear load impulse and constant normal load. The dynamic shear strength was observed to be greater than the corresponding static value and also to increase with increasing shear stress rate. The normal stress appeared to decrease the rate of increase of the dynamic shear strength with respect to the static value. In other words, they found for these single shear load impulses that the larger the normal stress, the smaller the dependence of the dynamic shear strength on the shear stress rate. These results tend to contradict those presented by Gillette [4], which showed little velocity dependence on the normal load. However, the type of joint surface was quite different in these studies (i.e., saw cut in one case and artificially fractured in the other). The discrepancies are also likely due to the fact that these dynamic tests were conducted using an impulse load which is different from the cyclic joint loading used by Gillette [4] which assumed the shear velocity built up to the maximum value over a finite time period depending on the input frequency.

Hobbs [9] also studied the dynamic behavior of rock joints and observed changes in the joint frictional response as a result of perturbation in the sliding velocity. They explained the frictional shear response in terms of cohesion and friction angle evolution laws which were observed to be of a softening character.

To develop a comprehensive database for characterizing the dynamic behavior of rock joints and to aid in the validation of existing rock joint models, extensive laboratory experiments were conducted at the Center for Nuclear Waste Regulatory Analyses (CNWRA) on single, natural rock joints. The dynamic direct shearing was conducted in a displacement-control mode under various normal loads, frequencies, and displacement amplitudes. During the test, the top rock block was sheared with respect to the bottom block. In addition to the experiments on single jointed rocks, a laboratory-scale model experiment was conducted to study the dynamic behavior of a system of interconnected (artificial) joints around a circular opening in a scaled down rock mass when subjected to earthquake shear wave motion at the base. Finally, an actual field experimental program was conducted over a period of about 3 yr to investigate the effect of relatively low-magnitude, repetitive seismic motion on the behavior of mined excavations at the Lucky Friday Mine located in the Coeur d'Alene Mining District in the Idaho panhandle region.

2. DYNAMIC BEHAVIOR OF SINGLE NATURAL ROCK JOINTS

Extensive laboratory experiments on the dynamic behavior of single, natural rock joints were recently performed by Hsiung [10, 11]. The natural rock joint specimens for these dynamic tests were collected from the Apache Leap welded tuff near Superior, Arizona, USA, and were tested under two types of dynamic shear-loading conditions with various frequencies. These two loading conditions were harmonic and earthquake loads. This work can be considered as an extension to the dynamic studies discussed earlier, which considered only a single-shear impulse load or at most cyclic harmonic loadings on either saw cut or artificially fractured rock joints. The following sections discuss briefly the dynamic shear testing apparatus as well as recent results of harmonic and earthquake loadings on single rock joints.

2.1. Dynamic Shear Testing Apparatus

The servo-controlled direct shear test apparatus used to conduct the dynamic tests [12,13] was described in an earlier chapter, which addressed the mechanical-hydraulic behavior of rock joints under direct shearing. The mechanical portion of the apparatus and corresponding instrumentation discussed in in this previous chapter is the same as that used to conduct the dynamic direct shear tests discussed in this chapter. For reference, the basic dynamic direct shear test apparatus and instrumentation channels are shown again in Figure 1. The change in the instrumentation plan for the dynamic tests was that accelerometers were mounted to the rock specimen near the joint interface. The horizontal actuator was operated in a displacement-control mode for all the dynamic tests. To conduct the dynamic shear tests under higher normal loads of up to 5 MPa, a horizontal hydraulic actuator of higher capacity than that used for the pseudostatic tests was required.

2.2. Harmonic and Earthquake Input

For the harmonic direct shear tests, the prescribed horizontal shear displacement inputs were sine wave drive signals. The total duration for all the harmonic tests was 30 sec, and the sampling rate was 800 points/sec. The high sampling rate was intended to capture high-frequency responses of joints during harmonic tests. Frequency and amplitude of the input displacement signal varied for the different tests, with the frequency ranging from 1.4 to 3.5 Hz and amplitude from 6.35 to 25.4 mm. This frequency range was considered to be commensurate with typical earthquake displacement histories. Measurements taken during the harmonic tests included normal and shear loads, joint normal and shear displacements, acceleration responses of the top rock block, and displacement response between the horizontal load cell and the top shear box.

The displacement drive signal used for the joint shear tests under earthquake loads was derived from the acceleration response signal recorded from the Guerrero accelerograph array for the 8.1 Richter scale magnitude earthquake of September 19, 1985, in Mexico. The acceleration response signal measured along the south axis was used to generate a displacement drive signal for the planned joint shear tests subjected to earthquake loads. The initial input displacement drive signal was obtained by double-integration of the windowed acceleration data in the frequency domain (Fourier spectra). Before the double-integration, a bandpass filter was applied to the acceleration Fourier spectra. This filter was defined by low- (0.5 Hz) and high-frequency (15 Hz) values. The high pass filtering was intended to eliminate the possibility of developing extremely large-amplitude, low-frequency offset in the data during integration. Low pass filtering of the points was intended to eliminate aliasing of the data due to its limited sampling rate. Figure 2 shows the resulting calculated displacement time history. From spectra analysis, this displacement time history contains a major frequency range from 0 to 2.0 Hz with a dominant frequency at 0.5 Hz. For the various shear tests, two displacement drive signals were used, one with a peak drive signal of 25.4 mm, and 50.8 mm for the other. These two signals were obtained by scaling the displacement signal in Figure 2.

2.3. Experimental Results of Single Rock Joint Dynamic Behavior

Figure 3 shows the measured shear displacement time history and the corresponding shear stress response for a typical test with earthquake load. Only the test results between the 15th and 20th second are presented in the figure for clarity. Results obtained from other earthquake and harmonic tests are similar. There is a phase difference between the shear displacement and shear stress time histories, with the shear displacement lagging. These phenomena were

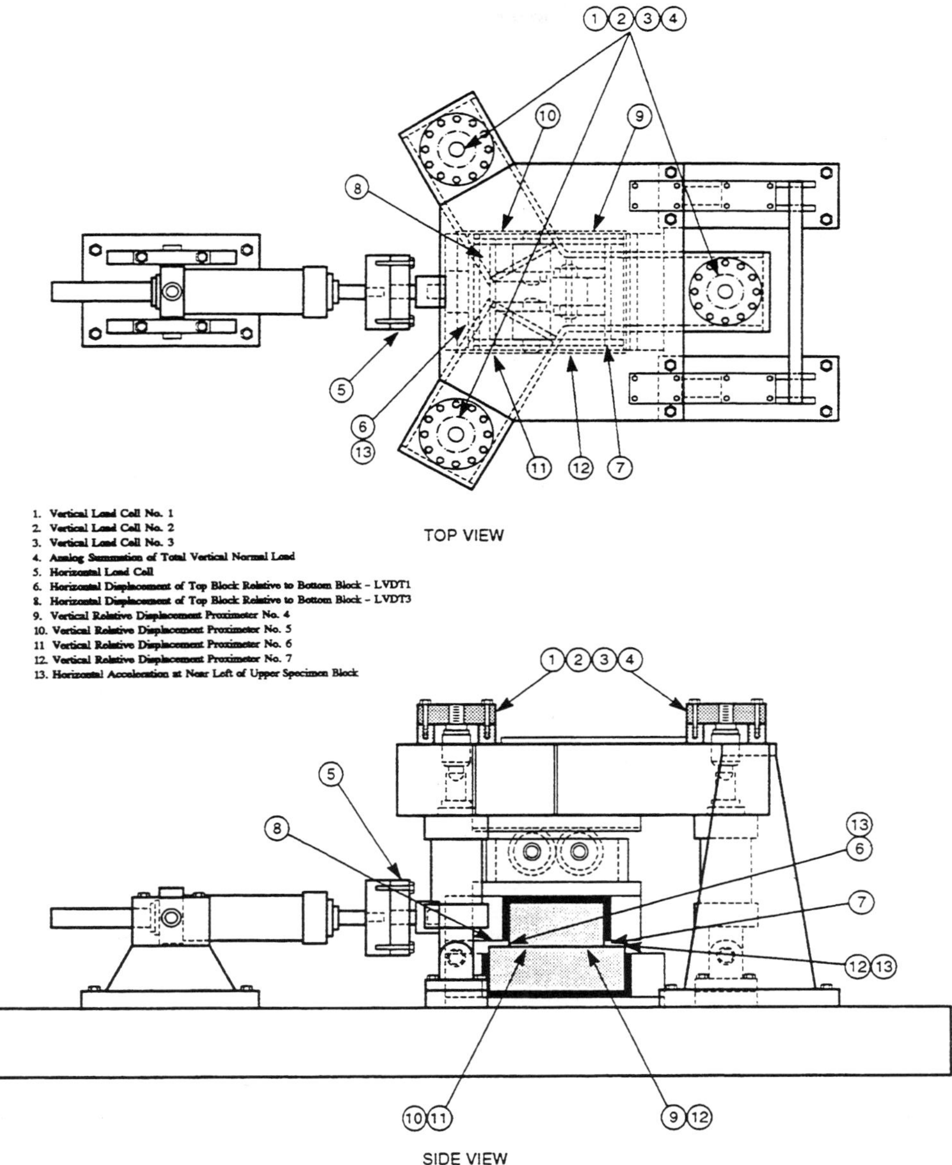

Figure 1. Mechanical apparatus for dynamic direct shear testing of rock joints.

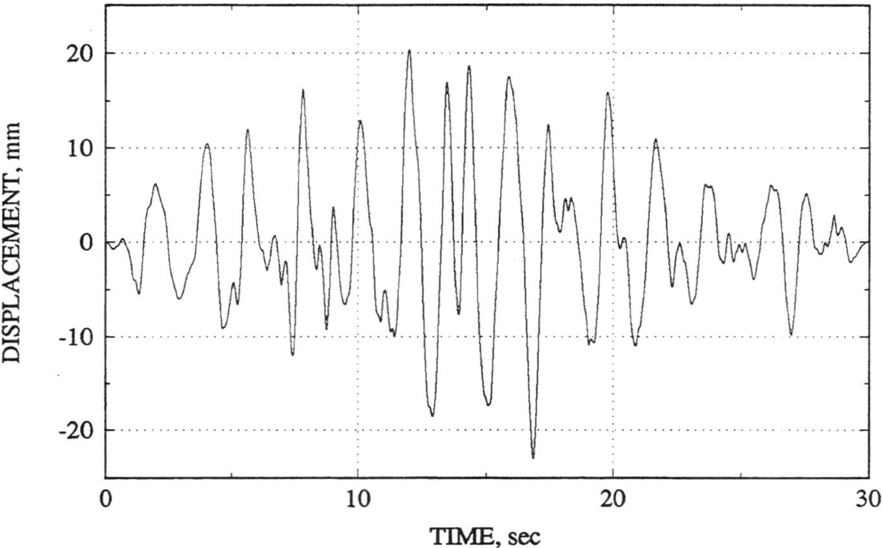

Figure 2. Displacement time history generated from the acceleration time history obtained from the 1985 Mexico City earthquake.

observed for all harmonic and earthquake tests. The source of this phase shift was determined not to be related to the experimental setup. This decision was verified with additional instrumentation. It may be concluded that the phase shift results from shear stress buildup to a level required to initiate joint shear.

As shown in Figure 3, the shear stress magnitude increases during forward shearing. The peak shear stress is reached after a certain amount of shear displacement, and the shear stress decreases afterward. However, the shear stress seems to experience a higher frequency component subsequently, which correlates well with stepwise shear displacements observed in the displacement time history. This higher frequency component appears to be associated with rapid stick-slip (i.e., chatter) of the interface and may be enhanced by the presence of natural vibrational modes in the apparatus, as well. The same behavior is also observed during reverse shearing. This chatter behavior is believed to be excited by the waviness of the joint surface as well as by pieces of rock fragments broken from the joint surface. Visual inspection at the conclusion of each test revealed many rock fragments. When the movement of the top rock block is restrained by asperities on the joint surface, it tends to stop until the shear stress is built up sufficiently to overcome this additional resistance by either breaking or riding over the asperity. Depending upon the size and strength of these asperities, the high-frequency responses will vary in amplitude. After overcoming the obstacles, the shear stress drops sharply. There is a period when the shear stress decreases due to cycling. During this period, the shear stress is actually smaller than the shear resistance. Consequently, the top rock block stops until reverse shearing starts when the negative shear stress begins to increase. This behavior is evident by the flattening of the relative shear displacement near its peak values.

Unlike the pseudostatic direct shear tests, the normal stress for the dynamic tests could not be maintained as constant during the course of the tests as planned. This failure was due to the

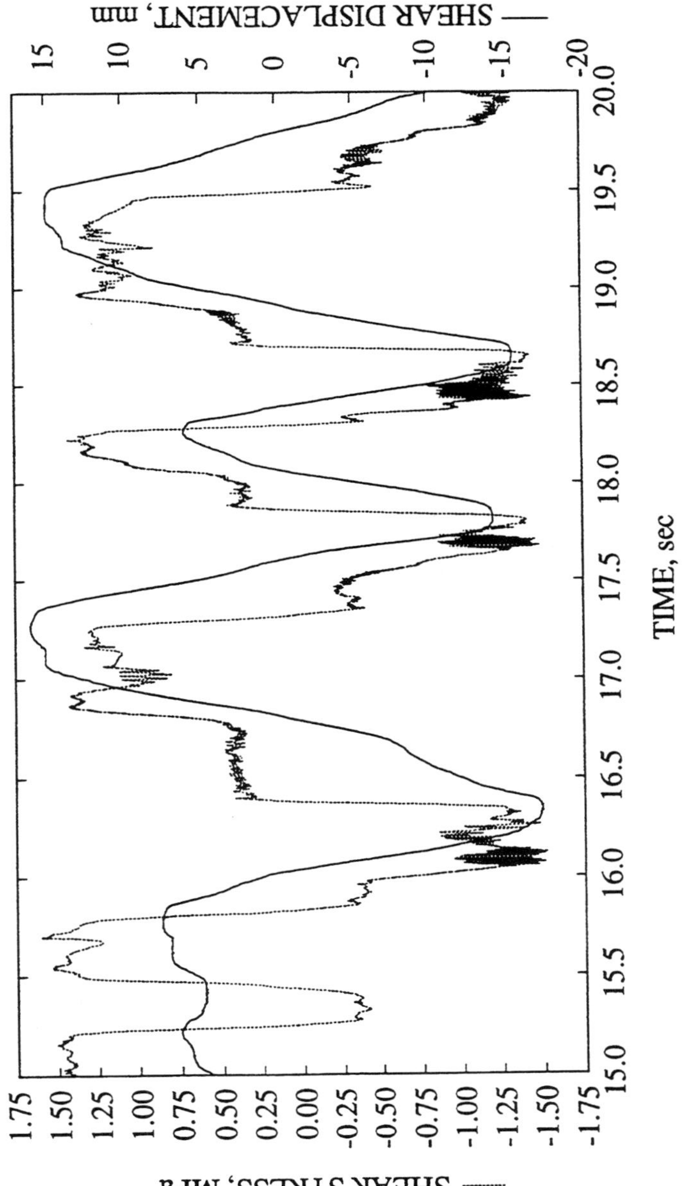

Figure 3. Shear stress and shear displacement history between the 15th and 20th seconds of Test No. 25 using the earthquake signal with maximum nominal input displacement amplitude of 25.4 mm under 1-MPa normal stress.

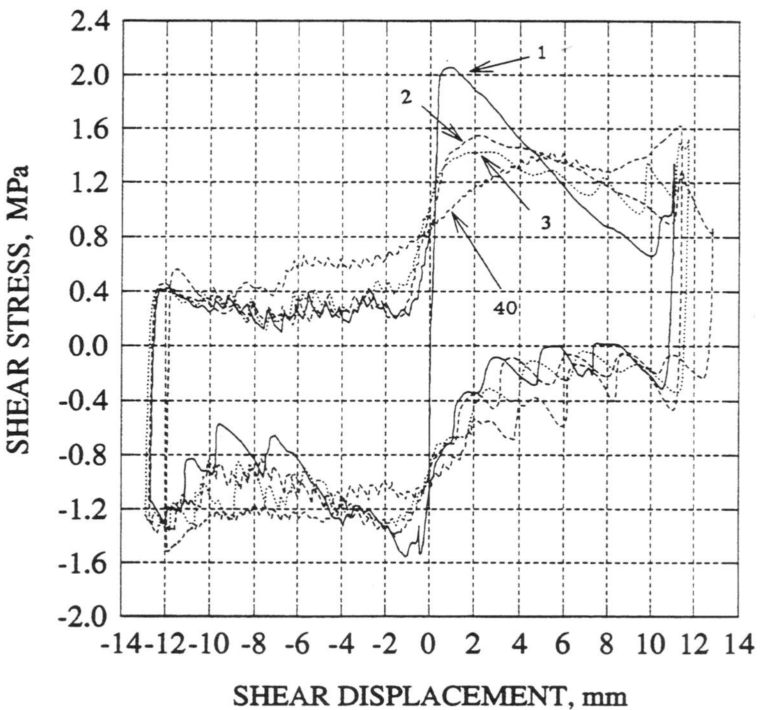

Figure 4. Shear stress versus shear displacement curve for the first phase of Test No. 14 under a harmonic load with 1.4-Hz input frequency and 12.7-mm input displacement amplitude (first three cycles and 40th cycle).

inability of the servo-controlled valve for the vertical actuators to adjust quickly to the sudden changes in the normal stress in response to changes in asperities during shearing. For the harmonic tests conducted under an applied normal stress of 1.0 MPa, the normal stress fluctuated as much as 0.3 MPa. It is reasonable to conclude that, as the normal stress increases, its normalized variation (presented as percentage of change) decreases, and so does the potential impact of normal stress variation for the dynamic tests. Also, the vibration mode of the normal stress is not exactly synchronized with that of the normal displacement, and the variation of the normal displacement at high frequency is quite small. Therefore, it may be concluded that the effect of normal stress variation is likely to be small. The extent of the effect of normal stress variation on the shear response at low normal stress level is difficult to judge. However, given the transient nature of the variation, its impact should also be small.

Figure 4 shows the characteristic hysteresis from Test Number 14 of the joint shear stress versus joint shear displacement from dynamic shear testing using a harmonic input motion with a 1.4-Hz input frequency and an amplitude of 12.7 mm. The test was conducted under a constant normal stress of 1.0 MPa for a duration of 30 sec. Only the response from the first three cycles as well as the 40th cycle are plotted. For these tests, the experiment started with the shearing of the top rock block from its original centered position (represented as 0 shear displacement in the figure) toward one end of the bottom rock block until a predetermined maximum value of shear displacement (positive to the right) was reached. The corresponding shear stress versus shear displacement characteristic curve with this portion of shearing is shown in the first quadrant of Figure 4 (from 0 to +12 mm displacement). After the maximum shear displacement in the first quadrant was reached, the top rock block began to move backward (from +12 to 0 mm displacement) and eventually past its original position (from 0 to -12 mm displacement). The corresponding shear stress versus shear displacement characteristic curves are presented in the fourth and third quadrants of the figures, respectively. After the maximum shear displacement in the third quadrant was reached, the top rock block moved again back to its original position (from -12 to 0 mm displacement) to complete a cycle of shear motion. The associated shear stress versus shear displacement characteristic curve is presented in the second quadrant of the figure. This process was repeated for a number of cycles. Again, these tests were set up to have the top block mated in the middle of the bottom block to allow the top block a maximum allowable travel of about 50.8 mm on either side of the original position along the direction of shearing. As shown in the figure, the shear stress is assigned to be positive when the shearing is along one direction and becomes negative when the shearing follows the opposite direction. Consequently, the sign for the shear stress denotes the direction of the shear instead of the magnitude of the shear stress. Similarly, Figure 5 shows the characteristic hysteresis of the joint shear stress versus joint shear displacement for the first phase of Test Number 30 under earthquake loading.

As was observed in pseudostatic tests, a peak joint shear stress (peak joint shear resistance) was observed for the first cycle for both the harmonic and earthquake shear tests provided the jointed specimens used for the tests had never been shear tested previously or showed signs of past shearing before sample collection. The phenomena of wear of the joint are also clear, shown in Figures 4 and 5, as the shear stress (joint shear resistance) decreases with the number of cycles. Throughout this chapter, the term forward shearing is used to indicate that the top rock block moves away from its original position, while reverse shearing denotes that the top rock block moves toward its original position.

One distinct feature of the shear stress versus shear displacement characteristics in Figures 4 and 5 is the smaller shear resistance upon reverse shearing as compared to that of forward shearing (the first quadrant versus the fourth quadrant, the third quadrant versus the second quadrant). This same behavior is similarly observed on pseudostatic tests on the Apache Leap natural tuff joints and reported by other researchers for rock replicas under pseudostatic loads [14-16]. Both the forward and reverse shearing are likely important phenomena for a rock joint when it is subjected to earthquake loads, whereas only forward shearing is of concern under static loading. The low shear resistance associated with the reverse shearing process may play a key role in determining the stability of an underground opening if the condition is unfavorable. Therefore, a better understanding of the cause of this observation is important to the design of a stable underground excavation.

Jing [14] implied that, on a larger scale, a rock joint surface contains dominant wavelengths called "primary asperities." There also exist, on the joint surface, "higher order asperities" that have much smaller sizes as compared to the primary asperities [Figure 6(a)]. Profiles taken from the Apache Leap tuff joints confirm the existence of the primary and higher order asperities. Three factors have been suggested to affect joint shear behavior [14]: higher order asperities, amplitude to wavelength ratio of the joint surface curvature, and basic friction angle of the rock. The higher order asperities and the basic friction angle of rock provide the fundamental joint resistance to shear, while the amplitude to wavelength ratio determines the magnitude of the tangential component of the normal vertical stress along the

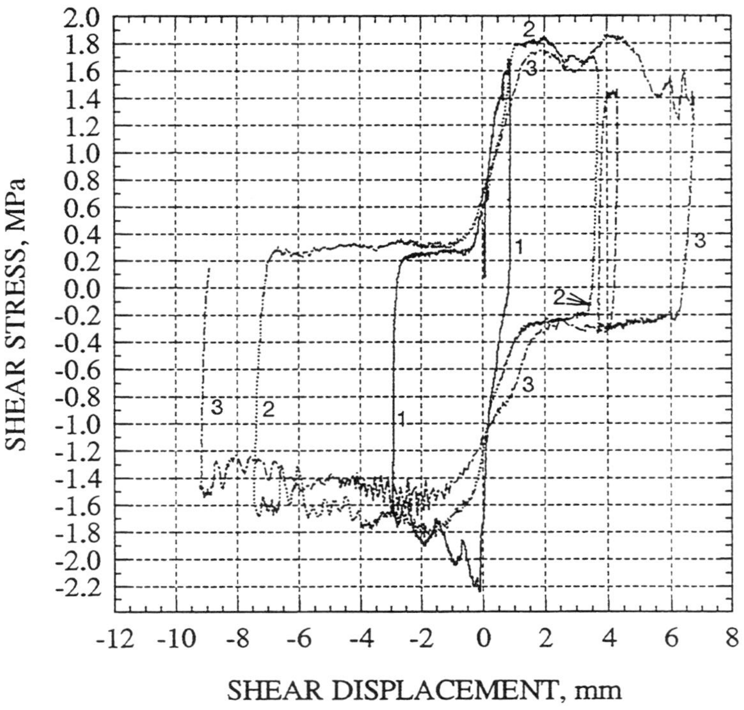

Figure 5. Shear stress versus shear displacement curve for the first phase of Test No. 25 under an earthquake load with a maximum input displacement amplitude of 25.4 mm.

curved surface. Depending upon the direction of shear, this tangential component could either increase or decrease joint shear resistance. As shown in Figure 6, when the top rock block is moving upslope, the "local" direction of shear is opposite to the direction of the tangential component. Consequently, more shear is needed to overcome this tangential component. When the top block moves downslope, the local direction of shear is the same as that of the tangential component of the normal stress. As a result, relatively smaller shear stress is required to overcome the mobilized friction. This concept explains quite well the phenomenon observed in Figure 4. Another important factor not included in the hypothesis proposed by Jing [14] that may also contribute to the difference in shear resistance between forward and reverse shearing is the normal component of the system shear stress applied to the curved surface. When the top block is climbing upward along a primary asperity, a portion of the applied shear stress (horizontal stress) actually becomes localized normal stress, τ_{sn}, as shown in Figure 6(b), which tends to resist shear. As a result, the actual localized shear stress becomes smaller than the system shear stress, τ_s. Consequently, more system shear stress is required to overcome this additional normal stress in order to mobilize the joint. On the other hand, if the top rock block is moving downslope, this normal component of the system shear stress tends to offset the applied local normal stress [Figure 6(c)]. As a result, a smaller shear stress is required to mobilize the joint. It should be noted that the above argument holds if the length of the primary asperity is larger than one-half the shear displacement cycle amplitude,

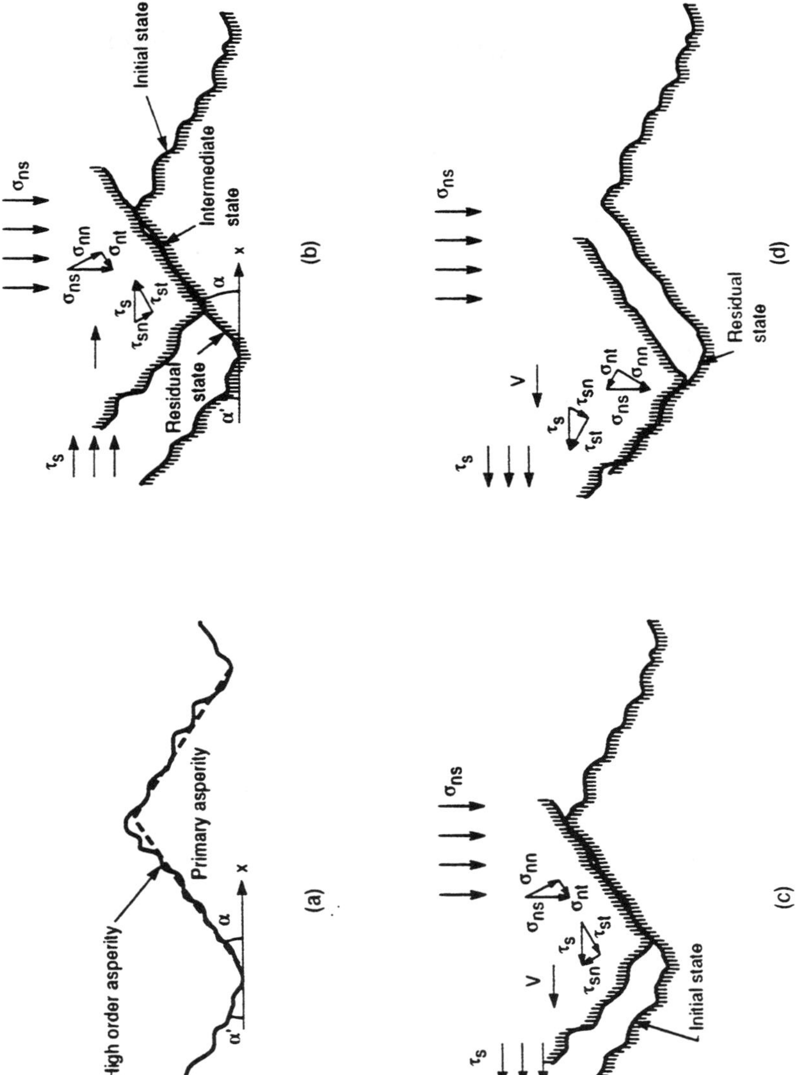

Figure 6. Hypothesis of joint shear behavior during forward and reverse stages.

which was the case for the range of shear displacements (i.e., <5 cm) and particular welded tuff joint samples tested. Otherwise, the top block could go upslope and downslope many times in the same direction (forward or reverse.)

In many cases, the shear stress continued to increase and decrease about the mean value during the course of harmonic or earthquake loading. These fluctuations in shear stress are a result of asperities or rock fragments encountered during the shearing process, commonly referred to as stick-slip behavior. Based on the experimental results, these fluctuations in shear stress or chatter become increasingly pronounced with increase in input frequency, although the applied normal stress was 1 MPa. It is also interesting to note that, in some cases of the dynamic tests, the chatter behavior continued even after a number of cycles of shearing.

Since natural rock joints were used for the direct shear tests under harmonic and earthquake loading conditions, a different joint specimen was needed for each test. The fact that each jointed specimen had its own characteristic roughness made it difficult to evaluate directly the dynamic effect on the joint shear resistance as was done in studies by other researchers using saw-cut joints or replicas. However, based on the present dynamic studies, it was established that the difference between joint shear resistance during reverse shearing and peak shear resistance will be larger for joints with rougher surfaces. In other words, for such joints the shear resistance during reverse shearing will be much smaller compared to the shear resistance during forward shearing.

Figure 7 shows the joint normal displacement versus shear displacement characteristic curve corresponding to Figure 4 for Test No. 14 under harmonic load. Likewise, Figure 8 shows a similar plot obtained under earthquake loading, corresponding to Figure 5. Again, from Figures 7 and 8, the effect of the continuing wear of the joint surfaces is evident. The maximum joint normal displacement continues to decrease through the cycles of shearing. It is interesting to note that joint dilation (positive normal displacement) tends to decrease constantly during reverse shearing and may retain a small amount of dilation as the top rock block returns to its original position. This phenomenon can be explained quite well using the conceptual model shown in Figure 6. Dilation reduces when the top rock block goes downslope, which is always the case during reverse shearing if the top and bottom rock blocks are closely matched before the test. However, for the harmonic tests, small-scale stick-slip oscillations were observed to continue for many cycles. This observation gives an indication of the potential impact of the input frequencies on joint dilation, which may be related to the existence of small-size rock fragments created in the process of shearing. The hysteresis between the normal displacements during forward and reverse shearing of the first cycle for the harmonic and earthquake tests was determined to be smaller than that observed for the pseudostatic tests.

3. DYNAMIC BEHAVIOR OF MULTIPLE ROCK JOINTS LABORATORY-SCALE MODEL EXPERIMENT

3.1. Description of Scale Model Experiment

As a follow-on to the laboratory dynamic studies on single, naturally jointed, welded tuff fractures, a laboratory-scale model experiment was developed and tested to determine the seismic response near a circular opening in a jointed rock mass [17,18]. The 1/15-scale model consisted of an aggregate of simulated rock material blocks that were used to study the earthquake response of a larger segment of the proposed rock mass (welded tuff). A detailed discussion of the derivation of the scaling parameters for the scale model material,

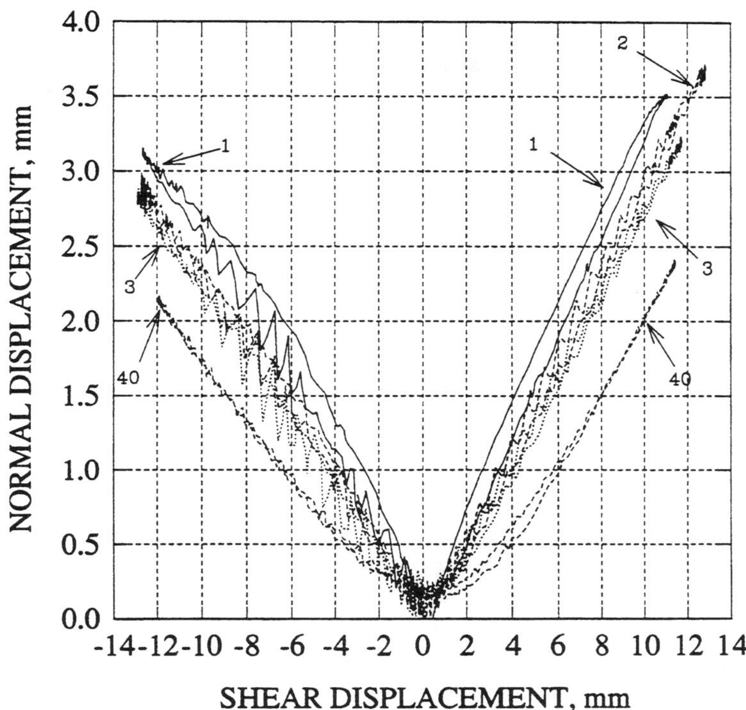

Figure 7. Joint normal displacement (dilation) versus shear displacement for the first phase of Test No. 14 under a harmonic load with 1.4-Hz input frequency and 12.7-mm input displacement amplitude (first three cycles and 40th cycle).

dimensions, and loading conditions is given by Hsiung [17] and Kana [18]. Results of numerical simulation of the experiment using the distinct element approach is given by Hsiung [19].

The final physical design and associated values for various parameters are given in Figure 9. The model consisted of an aggregate of many rock simulant ingots, each 61 cm long, with the interfaces oriented at a 45 degree angle to the horizontal. The ingot cross-sections varied from 5×5-cm square for basic ingots, to half-section ingots at the boundaries, to curved-section ingots around the center circular opening. This opening was 15.2 cm in diameter. The four boundaries of the stack were interfaced with 6.4-mm thick rubber, which is bonded to the rock on the inside and lubricated with silicone at the interface with the confining box boundaries. These boundaries were a very stiff construction of welded aluminum plates and 10.2-cm I-beam frames. The proper pressure, σ_n, was maintained on the system by eight vertical cables and eight horizontal cables. The two end structures were hinged to the bottom support structure at the baseplate and were held against top rollers at each upper corner. Therefore, the end structures can pivot laterally, while the top structure can pivot and float up and down as necessary to follow the confined rock motion.

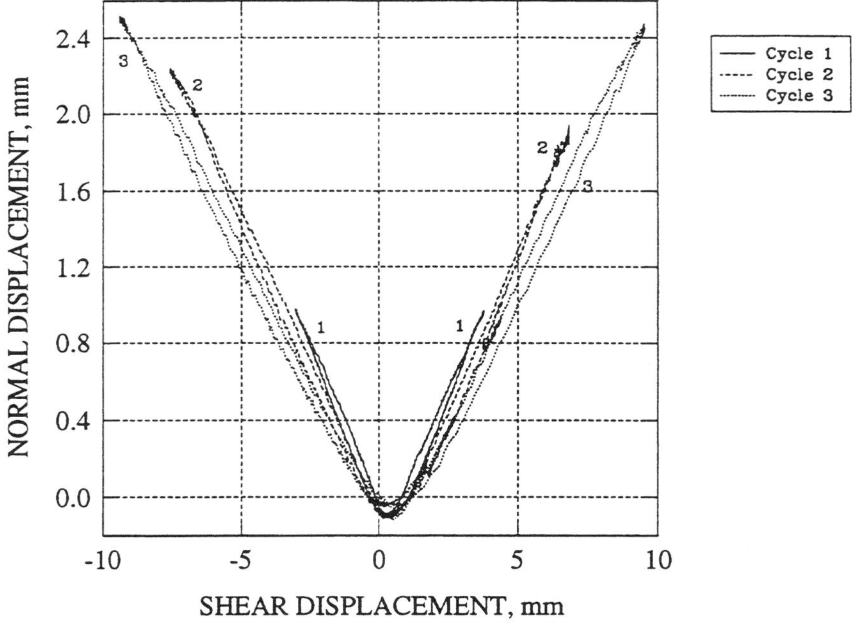

Figure 8. Joint normal displacement (dilation) versus shear displacement of the first phase of Test No. 25 under an earthquake load with a maximum input displacement amplitude of 25.4 mm.

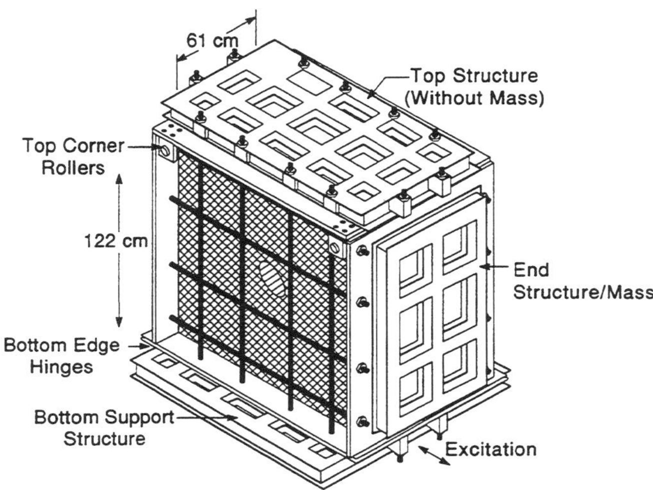

Figure 9. Physical design of scale model rock mass with a circular opening.

3.2. Development of the Rock Simulant

It was recognized that exact modeling of welded tuff behavior was probably neither possible nor was actually necessary for a successful verification of analytical models. Therefore, the technical approach adopted for development of a suitable rock simulant consisted of following the similitude guidelines as much as practical, but allowing deviations as long as they could be quantified. Initial development of the rock simulant was based on repeated trials of various constituent mixtures and testing of material properties of cylindrical specimens cast from these mixtures. The specimens were cast as 5.0-cm (2-in.) diameter by 10.0-cm (4 in.) long that were instrumented with strain gages and tested in uniaxial compression. Table 1 lists the ingredients that were ultimately found to provide a material having the appropriate properties.

Table 1
Properties of rock simulant specimen

Material Constituents (Percent by Weight)	
Type I Portland Cement	25.2
Barite	45.9
Water	25.2
Bentonite	3.4
DARACEM-100 (Plasticizer)	0.3
Vinsol Resin (Air Entrainment)	8.6×10^{-3}
Ivory Liquid Soap	4.6×10^{-2}
Uniaxial Compressive Strength	10.35–13.79 MPa (1,500–2,000 lb/in.^{2})
Material Density	1,682 kg/m^3 (105 lb/ft^3)
Roughness Data	
Average Peak:	±0.2 mm (0.008 in.)
Average Wave Length:	6.4 mm (0.25 in.)

Having developed the above described material, it now became appropriate to develop a scale model rough surface. The intent was to cast specimens of the same size as originally used for the single jointed welded tuff blocks and to perform combined normal and shear tests in the same apparatus to obtain data analogous to Figure 4 or 5. However, it was recognized that scale model conditions also must be considered for these tests. A material was found that produced a random roughness with the average peaks approximately 1/15 geometric scale to those observed for typical welded tuff specimens. However, it was recognized that with each of the two surfaces being independently random, no significant interlocking of surfaces would occur, as was typical for naturally-welded tuff joints. Therefore, the effects of these differences would need to be quantified by shear tests. A series of rock simulant specimens was subjected to both pseudostatic and dynamic shear tests. Figure 10 shows the results for scaled harmonic tests for a typical rock simulant specimen. By comparing Figure 10 with 4, it is obvious that, indeed, no offset in hysteresis occurs for the rock simulant. Nevertheless, corresponding friction properties could still be approximated.

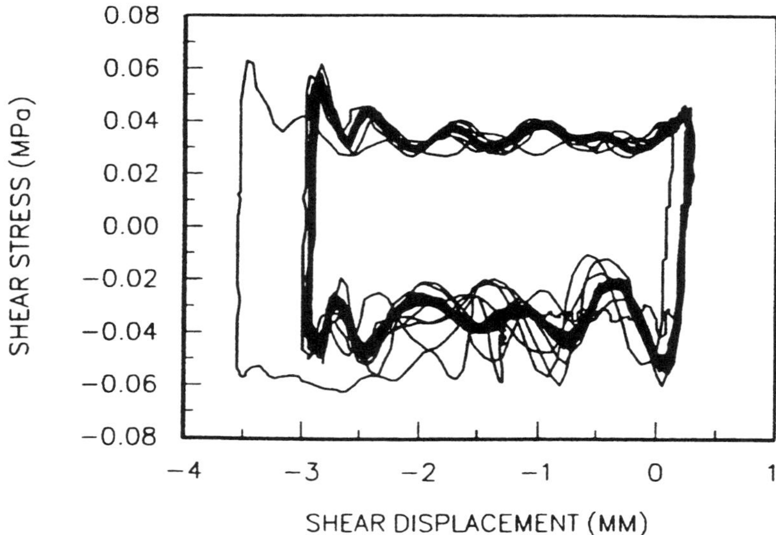

Figure 10. Hysteresis for rock simulant under 0.065 MPa normal stress and 5.4 Hz sinusoidal shear.

3.3. Instrumentation and Data Acquisition

It was recognized that rock interface relative normal and shear displacements and overall rock mass motions were of interest and that the transducers used for such measurements should offer negligible resistance to rock interface motion. Therefore, several types of transducers were selected for measurement of these responses. These transducers included accelerometers, strain gages, specially designed cantilever beam shear displacement measurement devices, Bentley proximeters, and linear variable differential transducers. A photograph of some of the instrumentation on and near the tunnel opening is shown in Figure 11. It was also recognized that relatively large displacements might be expected for the blocks around the opening due to repetitive shaking. The transducers mentioned earlier were not expected to function under large displacements. Consequently, one video camera was mounted at each side of the scale model apparatus along the axis of the tunnel to capture large displacements.

A block diagram of the 50-channel instrumentation system is shown in Figure 12. Data rates were dictated by the capacity of the 486 (66-MHz) digital computer with a 1-gigabyte hard drive, and its associated data acquisition cards. The total duration for each test was 10 seconds. It was determined that 2,800 samples/sec was the fastest data rate feasible for each of 50 data channels sampled sequentially. Hence, for each run a total of 1.4 million samples of data were acquired.

3.4. Test Procedures and Experimental Results

The test runs were started at a very low peak excitation displacement level, and this amplitude was incrementally increased as the runs progressed. Both videotape and digital data were acquired for each run. At the end of each run, all data were converted to engineering units, and a preliminary review of the data was performed visually on the monitor. In some

Figure 11. Instrumentation on near side of tunnel opening.

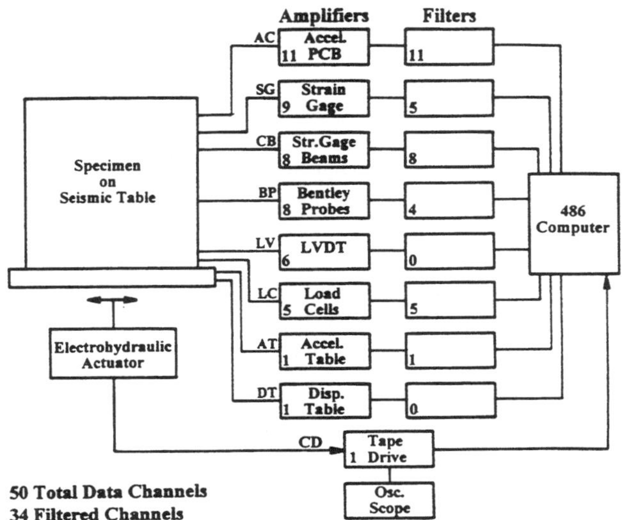

Figure 12. Block diagram of data acquisition system.

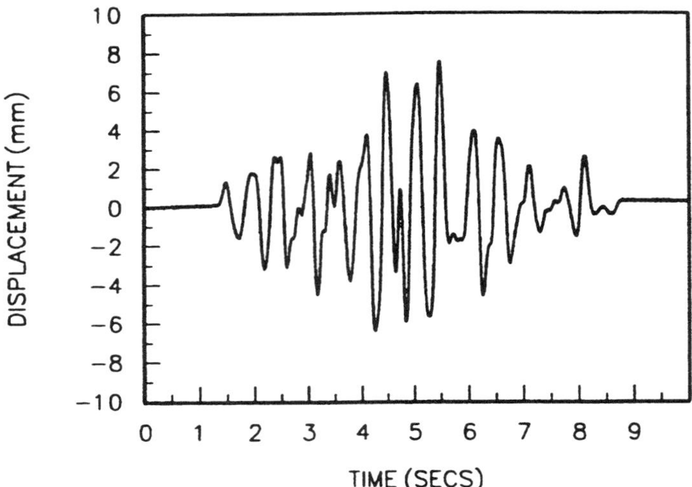

Figure 13. Horizontal table displacement for scale model earthquake data set 14: 1.0-cm displacement amplitude.

infrequent cases, transducer or other component malfunction occurred, and adjustments were performed prior to the next run. Furthermore, some shifting of filter channels and/or transducer locations was performed as response information was acquired.

The horizontal excitation displacement applied to the seismic shaking table (i.e., base of the scale model rock mass) for the intermediate level run is shown in Figure 13. This waveform was again derived from the accelerogram measured at the Guerrero array for the September 1985, Mexico City earthquake. However, unlike the time history used in the single rock joint dynamic tests, the displacement time history waveform in Figure 13 represents a 1/15-scale displacement history (i.e., both the time duration and peak displacement amplitude were scaled down). Corresponding to this excitation, permanent shifts in rock ingot positions are evident in the cantilever beam response of Figure 14 and the Bentley proximeter response of Figure 15. Based on these data, the near side upper right ingot at the opening has shifted downward relative to its adjacent ingots. Further analysis of data showed that both upper ingots, as well as those immediately below them, migrated downward radially into the opening, as would be expected under the influence of gravity. This behavior continued for each test run. Figure 16 shows the condition on the far side opening face after the test series was completed. It is obvious that very significant joint displacements near the opening have occurred. It was noted that the displacements at the near face were not so pronounced. In fact, a fracture of the one top ingot had occurred near the center along its length, so that the marked differences could accumulate. This behavior probably resulted from nonuniformity in the ingots or loading conditions.

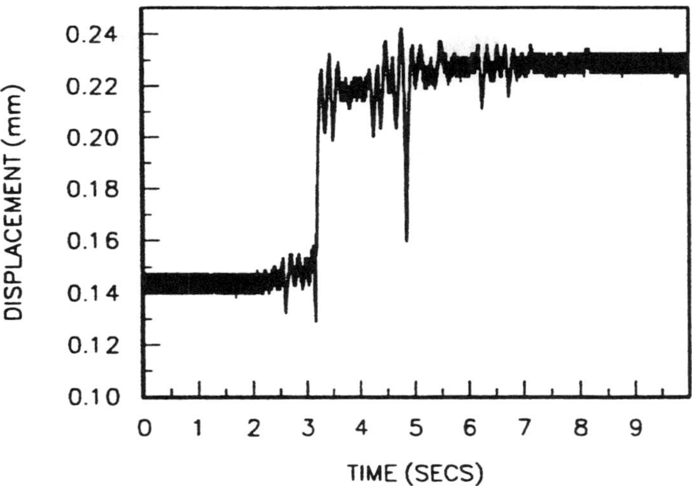

Figure 14. Near side tunnel block displacement: CB 4 data set 14, 1.0-cm displacement amplitude.

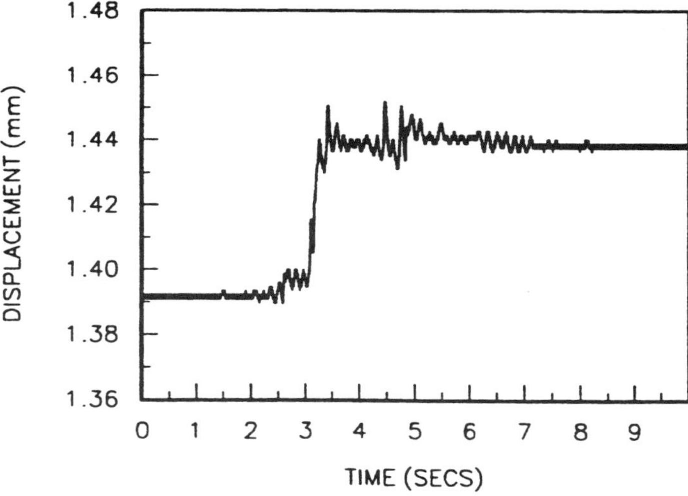

Figure 15. Near side left tunnel wall relative displacement; BP 2 data set 14, 1.0-cm displacement amplitude.

Figure 16. Accumulated permanent shift of rock ingots on far side of opening at end of test series.

4. DYNAMIC BEHAVIOR OF AN UNDERGROUND TUNNEL IN A FRACTURED ROCK MASS (CASE STUDY)

A case study was also conducted to investigate, through field instrumentation, the performance of an underground excavation subjected to repetitive episodes of mining-induced seismic activity. The study was conducted at the Lucky Friday Mine, Mullan, Idaho [3]. The primary purpose was to determine if the progressive accumulation of joint deformation resulting from episodes of dynamic loading has any potentially adverse effect on the performance of underground excavations.

4.1. Site Description and Instrumentation Layout

The Lucky Friday Mine is located in the Coeur d'Alene Mining District in the Idaho panhandle region. The mining depth is approximately 1,615 m below ground surface. The ore-bearing stratum (called the Lucky Friday vein) generally strike north-east and are nearly vertical with about 457 m of mineable strike length. This vein is bound on its north and south extent by faults and is cut by several major faults (Figure 17). The surrounding hanging wall and footwall rock is composed of interbedded units of vitreous quartzite, sericitic quartzite, and greenish siltite-argillite. The bedding planes are sometimes continuous with roughly planar surfaces, which often show evidence of past shearing. The beds dip, in general, south-east with an angle of approximately 70° from the horizontal and strike conformably with the vein. *In situ* stress measurements indicate that the maximum horizontal stress is about 1.35 times the minimum horizontal stress and vertical stress [20]. The maximum horizontal

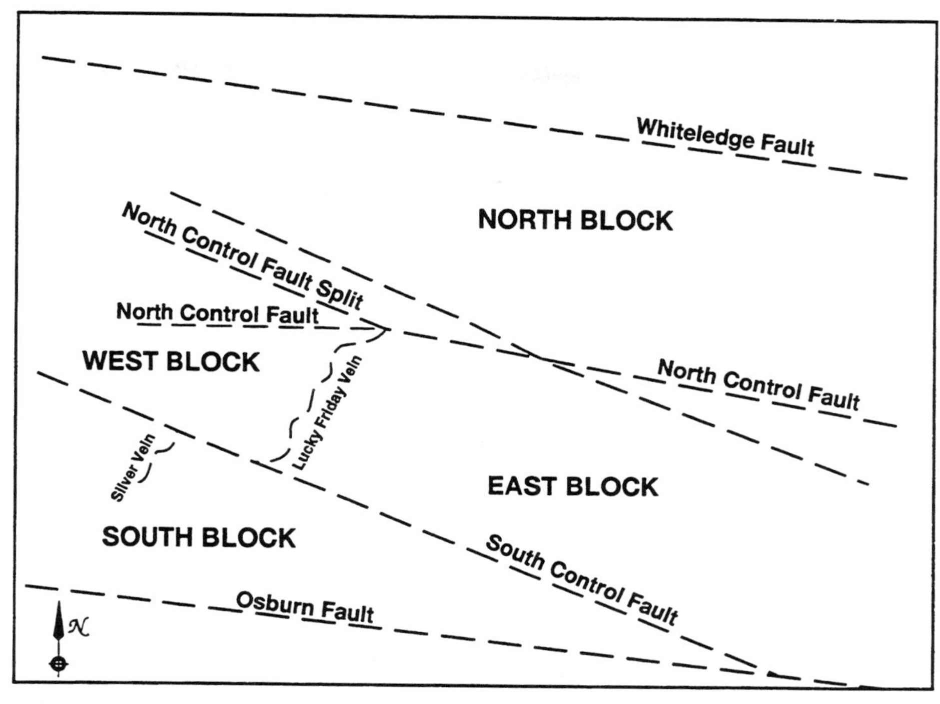

Figure 17. Plan view of the Lucky Friday orebody showing fault structures.

stress in the vicinity of the Lucky Friday Mine is oriented N45°W and perpendicular to the striking direction of the vein.

The mine uses the underhand cut-and-fill mining method. The general advance of the mining is downward. Four stopes are used on each level, which advance downward in a series of 3.05 m high cuts. The haulage development for each stope is a spiralling ramp in the footwall. Most of the seismic events in the mine occurred in the footwall where bedding planes dip toward the orebody.

Two sites near the bottom of the ramp systems under development were selected for instrumentation to monitor rock mass responses around underground excavation under repeated seismic events. Since these two sites were below the stope, they were relatively undisturbed by stoping in the early stages of monitoring. One site (LFM95-C1) was about 1,591 m, and the other (LFM95-C2) about 1,598 m, below ground level. Figure 18 shows a portion of the ramp system where the two instrumentation sites were located. The LFM95-C2 site was about 30.5 m from the orebody and the LFM95-C1 site was about 76.2 m away. The bedding planes intersect the instrumentation cross section of the LFM95-C1 site at an approximately 50° angle and at a 15° angle for the LFM95-C2 site. Both sites were supported by 1.8 m resin-grouted rebars and chain-link wire mesh. Fiber-reinforced shotcrete, 3.81 to 5.1 cm thick, was also used at the LFM95-C2 site. During the period of the monitoring, mining was conducted in the area indicated in Figure 18 from a depth of 1,579 to 1,606 m below ground surface.

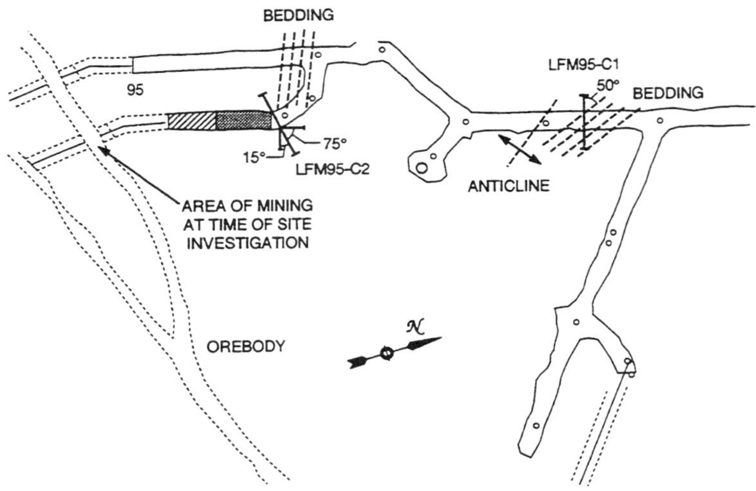

Figure 18. Locations of instrumentation at the 5210 level of Lucky Friday Mine.

Figure 19 shows a cross section of the excavation and location of the instruments for both sites. These cross sections depict the view facing the orebody. Five 5-anchor rod extensometers were installed per site. The extensometer holes were 7.6 cm in diameter. The deep anchor was at approximately 7.9 m down hole, with anchors at roughly 1.5-m intervals. Hydraulically inflated anchors were used to ensure a nonsliping grip. The rod displacement is sensed by linear potentiometers in the extensometer head, which has a range of 5 cm. The circled numbers in Figure 19 denote the extensometer hole numbers for each site.

The layouts for the extensometers are essentially the same for both sites except for the extensometer hole No. 5 at the LFM95-C2 site. This hole was drilled at $15°$ downward relative to the horizontal axis to avoid a potential interference with an up ramp excavation nearby. Point anchors were installed for the measurement of vertical and horizontal closures at both sites. A tape extensometer was used to take closure measurements. Also at each site, one triaxial velocity gauge was grouted into a horizontal borehole about 0.3-m deep for monitoring mining-induced seismic signals.

The extensometer readings were taken automatically via a data acquisition system at an interval of 2.25 hr. This system consisted of two primary components: two underground dataloggers and a surface personal computer. Seismic signals at the site were monitored through a mine-wide macro-seismic monitoring system developed by the U.S. Bureau of Mines, Spokane Research Center. Closure of the excavation was measured manually every 2 weeks.

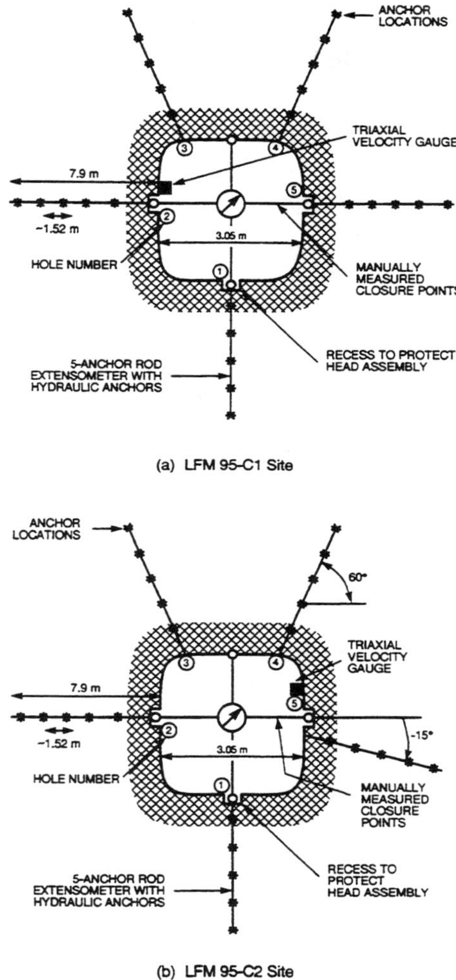

Figure 19. Instrumentation array of cross sections of the 5210 level.

4.2. Results of Field Monitoring

More than 50 seismic events with a magnitude greater than 1 on the Richter scale were recorded during the period of the study. The maximum magnitude experienced was about 3.5. In general, for all the seismic events that were observed, the durations of vibration were relatively short. Most events were over within 0.5 sec. It was therefore not possible to evaluate the potential impact of event durations on mechanical response. The source locations of these events were estimated through a trial-and-error process, with an approximate error of 7.62 m,

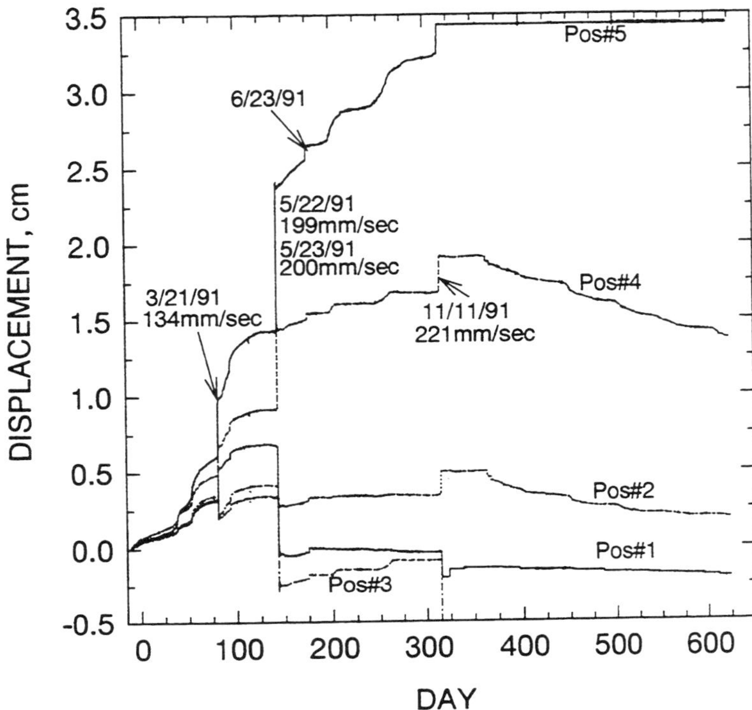

Figure 20. Displacement measurement for EXT No. 1 at LFM95-C2 site.

using measured first-arrival times of seismic signals at different monitoring locations in the mine.

Figure 20 shows a set of typical results from extensometer (EXT) measurements. Position (Pos) Nos. 1 through 5 in each of the figures indicate the anchor positions for an individual extensometer with Pos No. 1 closest to and Pos No. 5 farthest from the excavation. Displacements shown in the figures were measured relative to the assembly head of the extensometer, which is at the collar of the borehole. The recorded displacements were the results of mining-related activities and mining-induced seismic events. Positive values indicate that an anchor and its corresponding assembly head moved away from each other while negative values mean that the two move toward each other.

General observation of Figure 20 indicates that the anchor movements were of two types. The first type showed a gradual increase in displacement. This increase is believed to be a result of mining, which induces stress redistribution, and perhaps time-dependent behavior of the rock mass (creeping). The second type of displacement exhibits a distinct pattern of step increase or decrease in displacements. This type of behavior may be attributed to slip of a joint or a fracture located between an anchor and the assembly head. This joint slip is triggered by stress changes in the region. The stress changes may be either gradual or sudden and induced

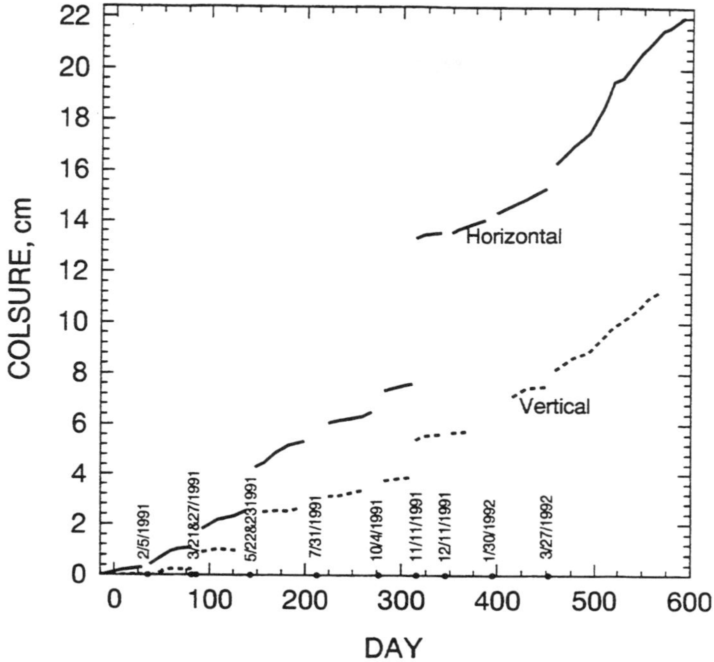

Figure 21. Excavation closure measurement at the LFM95-C2 site.

by mining, mining-induced seismicity, rock mass time-dependency, and other mining-related activities.

Figure 21 shows the closure of the cross section of the excavation for the LFM95-C2 site. More than 220 mm of horizontal closure and 113 mm of vertical closure were observed at the site. The closure curves are broken into line segments to show the effects of seismic events. Both horizontal and vertical closures for the LFM95-C2 site gave a clear sign of seismic effects. Large closures were observed after the March 21, May 22 and 23, and November 11, 1991 events. This observation corresponds well with the extensometer measurements. The closure readings also indicate the effects of the July 31, 1991 and March 27, 1992 events (199 mm/sec, magnitude 2.0), neither of which had any impact on extensometer readings. It is interesting to note that the November 11 event (221 mm/sec, magnitude 2.5) induced more than 50 mm of horizontal closure, which is substantially greater than the horizontal closure induced by the combined effects of the May 22 and 23 events (199 and 200 mm/sec, magnitude 2.5).

Based on the available data, it has been seen that the rock mass at the LFM95-C2 site responded to seismic events of similar or even smaller peak particle velocities with considerably higher displacements than that at the LFM95-C1 site. This phenomenon may be related to differences in the state of stresses at the two sites. As stated earlier, the LFM95-C2 site was about 30.5 m and the LFM95-C1 site was about 76.2 m away from the portion of the orebody where mining occurred during this study. Considerably higher stresses would be expected around the former site relative to the latter site as a result of the mining. The fact that greater closures occurred at the LFM95-C2 site tends to confirm the hypothesized difference in the states of stresses, assuming the rock mass and its behavior is similar at both sites. Normally, a rock mass is relatively weaker when under a higher stress condition, partly because the rock strength is known to be time-dependent [21, 22]. Also, it may require a relatively smaller amount of additional stresses to induce instability in a rock mass that is originally subjected to a higher state of stress. It is therefore logical to conclude, and has been demonstrated by field observation, that the LFM95-C2 site should be more vulnerable to seismic motions.

5. CONCLUSIONS

Experimental studies on both the dynamic behavior of single jointed rock specimens in the laboratory as well as multiple jointed rock masses in both the laboratory and field are presented. The direct shear tests on the Apache Leap welded tuff blocks containing a single, natural fracture showed that the shearing response, namely the shear resistance, could be markedly different in the forward and reverse directions depending on the joint roughness, with the shear resistance in the reverse direction being smaller. As discussed in this chapter, this is a direct consequence of the irregular roughness and interlocking nature of the mated joint surfaces. As a result, constitutive models used in dynamic studies for underground tunnel design purposes should take into account the changes in joint shear resistance with cyclic loading of the joint. The joint dilation response during either harmonic or earthquake loading showed very little hysteresis between the forward and reverse shearing directions as compared to similar dilation measurements taken during pseudostatic direct shear tests. For the most part, the dilation that takes place during forward shearing is fully recovered during shear reversal, with perhaps a small offset due to buildup of gouge within the joint.

Both the laboratory-scale model experiment and the Lucky Friday Mine field experiment to assess the dynamic behavior of a jointed rock mass containing a tunnel showed that the primary mode of deformation of the rock mass was due to stick-slip behavior along the joints. This stick-slip behavior seems to explain quite well the phenomenon of the excavations responding to some seismic events but being unresponsive to others. If the incoming seismic wave cannot provide sufficient energy to increase the shear stress to a level that exceeds the residual shear strength, it is not likely to cause joint slip. It should be noted, that the shear strength may also be reduced with a temporary reduction in normal stress across the joint from seismic wave reflections. The degree to which the seismic motion, or other mining related activities, influences this stick-slip type joint behavior depends on how close the joint is to its residual strength envelope. The joint stick-slip behavior forms a basis for the progressive accumulation of joint permanent deformation and, consequently, rock mass fatigue. Materials are normally weaker under fatigue conditions. This weakened condition implies that similar, or even more, damage to an excavation may be realized through a number of seismic events with relatively smaller magnitudes, as opposed to the damage due to a single seismic event with a strong motion (in terms of peak particle velocity).

6. REFERENCES

1　D.D. Kana, B.H.G. Brady, B.W. Vanzant, and P.K. Nair, Critical Assessment of Seismic and Geomechanics Literature Related to a High-Level Nuclear Waste Underground Repository, NUREG/CR-5440, Washington, DC, U.S. Nuclear Regulatory Commission (1991).

2　E.T. Brown and J.A. Hudson, 1974, Fatigue failure characteristics of some models of jointed rock, Earthquake Eng. and Struct. Dyn. Vol. 2 (1974) 379-386.

3　S.M. Hsiung, A.H. Chowdhury, W. Blake, M.P. Ahola, and A. Ghosh, Field Site Investigation: Effect of Mine Seismicity on a Jointed Rock Mass, CNWRA 92-012, San Antonio, TX, Center for Nuclear Waste Regulatory Analyses (1992).

4　D.R. Gillette, S. Sture, H.K. Ko, M. Gould, and G. Scott, Dynamic behavior of rock joints, Proceedings of the 24th U.S. Symposium on Rock Mechanics (1983) 163-179.

5　Crawford, A.M., and J.H. Curran, The influence of shear velocity on the frictional resistance of rock discontinuities, Int. J. Rock Mech. Min. Sci. & Geomech. Abstr. Vol. 18, New York, NY, Pergamon Press (1981) 505-515.

6　K. Bakhtar and N. Barton, Large scale static and dynamic friction experiments, Proceedings of the 25th U.S. Symposium on Rock Mechanics, Dowding and Singh (eds), Baltimore Maryland, Port City Press, (1984) 457-466.

7　N. Barton and V. Choubey, The shear strength of rock joints in theory and practice, Rock Mechanics, Vol. 10 (1977) 1-54.

8　G. Barla, M. Barbero, C. Scavia, and A. Zaninetti, Direct shear testing of single joints under dynamic loading, Rock Joints. Barton & Stephansson (eds), Rotterdam, Netherlands, A.A. Balkema (1990) 447-454.

9　B.E. Hobbs, A. Ord, and C. Marone, Dynamic behavior of rock joints. Rock Joints, Barton & Stephansson (eds), Rotterdam, Netherlands, A.A. Balkema (1990) 435-445.

10　S.M. Hsiung, D.D. Kana, M.P. Ahola, A.H. Chowdhury, and A. Ghosh, Laboratory Characterization of Rock Joints, NUREG/CR-6178, Washington, DC, U. S. Nuclear Regulatory Commission (1994).

11　S.M. Hsiung, A. Ghosh, A.H. Chowdhury, and M.P. Ahola, Laboratory investigation of rock joint dynamic behavior, Proceedings of the 35th U.S. Symposium on Rock Mechanics, Rotterdam, Netherlands, A.A. Balkema (In publication) (1995).

12　D.D. Kana, D.C. Scheidt, B.H.G. Brady, A.H. Chowdhury, S.M., Hsiung, and B.W. Vanzant, Development of a Rock Joint Dynamic Shear Test Apparatus, CNWRA 90-005, San Antonio, TX, Center for Nuclear Waste Regulatory Analyses (1990).

13　D.D. Kana, A.H. Chowdhury, S.M. Hsiung, M.P. Ahola, B.H.G. Brady, and J. Philip, Experimental techniques for dynamic testing of natural rock joints, Proceedings of the 7th International Congress on Rock Mechanics (1992) 519–525.

14　L. Jing, E. Nordlund, and O. Stephasson, An experimental study on the anisotropic and stress-dependency of the strength and deformability of rock joints, Int. J. Rock Mech. Min. Sci. & Geomech. Abstr. Vol. 29, No. 6, New York, NY, Pergamon Press (1992) 535-542.

15　J.T. Wibowo, B. Amadei, S. Sture, and A.B. Robertson, Shear response of a rock joint under different boundary conditions: an experimental study, Conference on Fractured and Jointed Rock Masses. June 3–5, Lake Tahoe, CA (1992).

16 X. Huang, B.C. Haimson, M.E. Plesha, and X. Qiu, An investigation of the mechanics of rock joints - Part I. Laboratory investigation. Int. J. Rock Mech. Min. Sci. & Geomech. Abstr. Vol. 30, No. 3 (1993) 257-269.

17 S.M. Hsiung, M.P. Ahola, D.D. Kana, A.H. Chowdhury, and S. Mohanty, NRC High-Level Radioactive Waste Research at CNWRA July-December 1994 - Chapter 2: Rock Mechanics, CNWRA 94-02S. San Antonio, TX: Center for Nuclear Waste Regulatory Analyses (1995).

18 D.D. Kana, S.M. Hsuing, and A.H. Chowdhury, A scale model study of seismic response of an underground opening in jointed rock. Proceedings of the 35th U.S. Symposium on Rock Mechanics, Rotterdam, Netherlands, A.A. Balkema (In publication) (1995).

19 S.M. Hsiung, M.P. Ahola, D.D. Kana, A.H. Chowdhury, and S. Mohanty, NRC High-Level Radioactive Waste Research at CNWRA January-June 1994 - Chapter 2: Rock Mechanics, CNWRA 94-01S, San Antonio, TX, Center for Nuclear Waste Regulatory Analyses (1994).

20 M.P. Board and Beus, M.J, In situ measurements and preliminary design analysis for deep mine shafts in highly stressed rock, Report of Investigations, RI 9231, Washington, DC: U.S. Bureau of Mines (1989).

21 C.H. Scholz, Mechanism of creep in brittle rock. J. Geophys. Res. 73 (1968) 3295–3302.

22 W.R. Wawersik, Time-Dependent Rock Behavior in Uniaxial Compression, Hardy, H.R., Stefanko, R. (eds) Proc. 14th Symposium Rock Mechanics. Penn State University, New York, American Society of Civil Engineers (1973) 85–106.

O. Stephansson, L. Jing and C.-F. Tsang (Editors)
Coupled Thermo-Hydro-Mechanical Processes of Fractured Media
Developments in Geotechnical Engineering, vol. 79

Lessons Learned from DECOVALEX

L. Dewiere[a], F. Plas [a] and C.F. Tsang[b]

[a]Agence Nationale pour la gestion des Déchets Radioactifs (ANDRA), Parc de la Croix
  Blanche 1/7 Rue Jean Monnet, 92298 CHATENAY-MALABRY CEDEX, France

[b]Earth Sciences Division, Ernest Orlando Lawrence Berkeley National Laboratory,
  University of California, Berkeley, CA 94720, USA

Abstract

The overall results and lessons learned from the international cooperative project,
DECOVALEX, on coupled thermo-hydro-mechanical processes in fractured media are
presented. A discussion is given on the role of coupled processes in nuclear waste disposal
concept, in the context of performance and safety assessments.

1. BACKGROUND

The goal of disposal of high level and long-lived radioactive wastes in deep geological
formation is the isolation of the wastes and the protection of future generations from their
potentially harmful effects. This isolation keeps the waste far away from human activities
and from the consequences of climatic changes and erosion that affect the ground surface.
Thus, the geological medium must provide a disposal environment that is physically and
chemically stable for as long as is needed for radionuclide decay.

The main factors to be studied in the assessment of the confinement performance of the
geological formation are the water flow, the radionuclides transport, the geochemistry of
water and the speciation of the radionuclides. All these factors will be influenced directly or
indirectly by mechanical, hydraulic and thermal processes at different instants from
construction through the life of the disposal facility. These processes are mainly linked to
the excavation of underground structures (mechanical and hydraulic discharge, introduction
of air into an initially reducing environment) and to the deposition of exothermic wastes that
cause a temporary temperature rise in the geological medium. There is, therefore, a need to
carry out an assessment of the coupled thermo-hydro-mechanical processes of fractured
geological media induced by the disposal facility, especially in terms of its impact on the
confinement performance of geological formations

The international DECOVALEX project is an exercise of cooperative research over 1991–
1994 on the assessment of our ability to model the THM (thermo-hydro-mechanical)
behavior of fractured rocks. It involved seven countries and fourteen research teams. The
work focused on three benchmark test cases (BMTs) and six laboratory or field test cases
(TCs). Details may be found in Chapter 2 in this book and Jing et al (1993, 1994 and 1996).

2. CONCEPTUAL APPROACHES AND RANGE OF PROCESSES STUDIED IN DECOVALEX

Although the DECOVALEX project made no attempt to be exhaustive or complete in its study of coupled THM processes, different conceptual approaches to the fractured media and the related fundamental models with computer codes for modeling thermo-hydro-mechanical (T-H-M) behavior were used in the project.The fourteen research teams used different tools (approaches, numerical methods, constitutive laws, computer codes), simulation strategies with fundamental differences in assumptions and conceptions (continuum vs. discontinuum), alternative representation strategies (fracture network simplification vs. homogenization), and different computer capacities. The project therefore provided a unique opportunity for exchange of knowledge among different schools. As a whole the effort represents an investigation of the state-of-the-art of the mathematical models and computer codes for coupled THM processes in continua and fractured media.

The range of processes and scales studied in DECOVALEX is summarized in Table 1, which presents some characteristics of the three benchmark tests (BMT) and the six test cases (TC). Details can be found in the various chapters in this book.

As a result of these studies, DECOVALEX project has addressed, to various degrees, several specific points of the state of the art in the assessment of coupled T-H-M behavior, mainly:

- Constitutive laws governing (T)HM behavior of an isolated fracture (e.g., TC1, 3, 4 and 5);
- Conceptualization of the THM behavior of a fracture network (e.g., BMT1 and BMT3):
    - Equivalent continuous porous media approaches,
    - Discrete medium approaches,
    - Explicit, discrete fracture network approaches;
- Comparisons of computer codes and algorithms;
- In-situ acquisition of (T)HM parameters and their interpretation (e.g., TC6);
- Application of codes to practical questions for nuclear waste repository (e.g., BMT1, BMT3).

3. OVERALL RESULTS

In view of assessing the T-H-M behavior for a nuclear waste disposal project, the DECOVALEX project leads to the following conclusions for each of the physical variables.

3.1 Temperature

The temperature fields obtained by the different approaches are very consistent and reproduce quite well the observed data from experimental test cases. This is to be expected, since the thermal behavior model is fundamentally the same in all the codes and it is not sensitive to the different fractured medium conceptual models, since fractures represent only a small fraction of the rock volume.

The minor differences observed are generally due to either differences in meshes that overestimates or underestimates the thermal field, or due to the choice whether or not thermal convection is taken into account. The latter is not expected to be significant for fractured rock systems except in the very close vicinity of the heat source.

Table 1
Ranges of processes and scales studied in DECOVALEX.

Problem	Brief Description	Scale	Processes	Remarks
BMT1	Fractured rock with 2 orthogonal sets of persistent fractures and a heat-releasing waste repository.	3000×1000 m	THM	2D far field problem
BMT2	Fractured rock with 4 discrete fractures and a heat source of finite length.	0.75×0.5 m	THM	2D near-field problem
BMT3	Fractured rock with a fracture network of 6580 fractures from Stripa site data.	50×50 m	THM	2D near-field problem
TC1	Laboratory shear-flow test on a rock core sample with a single joint.	8 cm	HM	Fundamental law of HM behavior of an isolated fracture
TC2	Field experiment in fractured rock at Fanay-Augères, France	$12 \times 10 \times 5$ m	THM	3D problem; unsaturated rock; mainly TM
TC3	Large-scale laboratory experiment in engineering buffer material (Big Ben).	5 m high × 6 m diameter	THM	3D radial; unsaturated medium; bentenite blocks.
TC4	Laboratory triaxial stress-flow experiment on rock cores with fracture.	10 cm high × 8.6 cm diameter	(T)HM	Fundamental law of HM behavior of rock joint under triaxial loading. Possible to do experiment in different temperature baths.
TC5	Laboratory shear-flow experiment of a rock block with a single joint.	20×10 cm	HM	To study HM in unsaturated to saturated media.
TC6	Field experiment of injection hydraulic test on fractures at 81 m depth and deeper in fractured rocks.	Radius of influence ~15 m	HM	Comparative study of flow results with constant and with pulse injection using packers.

3.2 Stress/Displacement (from isothermal cases to thermal loading cases)

Stress/strain behaviors (including thermal stress induction and volume expression) are quite consistent among the codes. Qualitatively, there appears to be good agreement also with data from experimental test cases and with major trends observed. Quantitatively, agreement is better for displacement than for stress, and the agreement improves as the distance from the thermal source and/or excavation to the observation point increases. Near the sources, the discrete approach tends to produce larger displacements than the continuum approach, due to its ability to allow larger movement along discontinuities.

3.3 Hydraulic Field (isothermal cases and thermal loading cases)

Unlike thermal and mechanical results, there are significant discrepancies in hydraulic results, in pressure heads as well as in flows, from one conceptual approach to another and sometimes even between models (codes) using the same conceptual approach. The discrepancies are quantitative if not qualitative, both in time and in space.

The origin of these discrepancies in flow of the hydromechanical system is quite fundamental, being related to the unsatisfactory state of science of constitutive laws of rock discontinuities, as well as the designs of the conceptual models used. These models are of necessity simplifications of the rock mass with its many fractures of various lengths and apertures, since it is impossible to measure all these features deterministically. Even if all the information is available, it will be too detailed for a reasonable amount of calculations on even the best computers. One area that nearly all of the conceptual models have difficulties to reproduce is the connected permeability of the fractures which is critical in determining the distribution and magnitudes of flow through the rock mass. Because of the finite lengths and variable aperture of these fractures, fracture connectivity varies greatly over the rock mass. Conceptual models using homogenization methods to obtain equivalent porous medium assume complete connectivity at all points. The discrete element or discrete block models on the other hand assume the identification of rock blocks, which requires a modification of the fracture sets in order to define the blocks. This modification includes the elimination of certain "minor" fractures and lengthening of others. As a consequence the connectivity and flow topology are also artificially modified.

Also most of the models of fractured rocks assume each of the fractures to have constant aperture and thus obey the so-called cubic law. This is also assumed in the homogenization methods. However, it is well known now that the aperture is strongly varying, giving rise to the so-called flow channeling effect. This is a strong effect precisely because of the sensitivity of flow to aperture values. Under changes in mechanical stress, the fracture aperture will change in response, with amplified flow changes.

A number of alternative mechanical stress deformation laws were used in DECOVALEX, including elastic or elastoplastic behaviors, with or without residual openings. These alternatives result in very different aperture and flow distributions. Many of the simplification or homogenization of the fracture systems were designed by considering hydraulic process alone. There is yet no proper homogenization approach that adequately addresses the mechanical process and the detailed interactions involved in the hydromechanical coupling. For example, it may be noted that the equivalent continuum porous medium obtained through an homogenization procedure display only a weak coupled hydromechanical behavior, because its hydraulic property is not very sensitive to porosity changes. However, this result is due to the simplification in the model conceptualization and is not expected for realistic systems.

Thus, on the whole, the geometric simplification procedure needs to be re-evaluated. Often it is difficult to preserve all the physical conservation laws in the simplification. Also, the appropriate conditions at the interface between simplified and unsimplified areas are still an open question.

The above discussions point out the immature state of our ability to model the hydraulic effects from hydromechanical behavior of a realistic volume of a fractured rock mass. The results of the DECOVALEX project have brought this into focus for further international research efforts.

4. LESSONS LEARNED FOR THE EVALUATION OF PERFORMANCE OF AN UNDERGROUND DISPOSAL FACILITY

Within the DECOVALEX project, the benchmark tests and test cases cover a wide range of scales and processes (see Table 1) and the research teams studied them with a variety of codes and modeling strategies, making the project diverse, interesting and significant. Some of these test cases might not have "over-determined" data sets and were not really validation exercises. Rather they were excellent cases for obtaining advanced understanding of coupled THM processes and of the tools for modeling such processes.

By studying the discrepancies between different modeling approaches and evaluating the difficulties of interpreting data of the test cases, one comes to realize that modeling results are the outcome of five different steps, each of which needs to be investigated concerning its correctness and accuracy. These five steps are:

(1)   The conceptual model
(2)   Selected representation of the medium
(3)   Constitutive laws and their parameters
(4)   The mathematical algorithms (numerical codes)
(5)   The calculations (computer)

Absolute evaluation of each of these steps is often not possible; however, sensitivity studies of modeling results on these five items are often instructive and useful in assessing the significance of coupled THM processes in the performance of an underground waste repository.

In addition to the above, other lessons learned are:

(1)   Thermal process in fractured and continuum media is well understood and well represented by the thermal equations in the models and codes in the DECOVALEX project. They can be used, with confidence, as a tool for predicting temperature distribution within and around the repositories for nuclear wastes. For a low-permeability fractured rock and within the studied temperature range (from room temperature to a temperature below boiling), thermal loading was found to have only the effect of mechanical deformation and in this way induce hydromechanical changes. Thus it is not fully coupled to the HM coupling. In other words, we may consider the problem to be partially decoupled as far as the thermal part is concerned. The key is then that we must first understand and model the hydromechanical behavior.

(2)   The mechanical process in the fractured rocks, especially stress distribution, is relatively well understood and may be used for prediction of rock strength around the repositories. However, the reliability of the calculated displacement fields in the neighborhood of excavations depends upon the extent of knowledge about the

geometrical distributions and mechanical behaviors of discontinuities around the excavations for both continuum and discrete approaches and this may or may not be available when an analysis is required.

(3) The hydraulic process is still a not well understood process, especially for fractured rocks, due mainly to the difficulties in adequate representations of fractures, their connectivity, aperture changes due to external loads (stress changes and thermal gradients), and lack of proper homogenization schemes for continuum analysis. Caution should therefore be taken when predictive modeling for hydraulic behavior is required. The differences in hydraulic results from the different models may represent the range of uncertainties in our capability in predicting fluid flow under hydromechanical processes. More research is needed in this area.

(4) Many of the basic parameters associated with the fundamental laws and with geometric features of the fracture medium are not accessible for direct measurements. New measurement methods, especially in-situ ones, are needed. On the other hand appropriate models (appropriate in the sense of the purpose of the model in application) should be developed and used with a view of potentially being able to measure the model parameters. These model parameters may be an averaging or combination of basic parameters.

Given the current state of technology, the gap between the limited potential for characterizing the fractured medium (i.e., fracture network and its THM properties) and the need of input parameters for existing models can perhaps be addressed by limiting the use of "input-intensive" discrete models to small volumes of rock in the near field and performing sensitivity studies on the parameters that cannot be easily measured.

(5) Based on the results and experience from DECOVALEX, we now should put our effort in learning how to study coupled processes at a real site. In this regard, the needed level of characterization and data acquisition, as well as the choice of appropriate modeling approaches for particular regions or problems has to be carefully investigated. For example, the level of characterization needs not be uniform over space. One could reasonably expect that the medium's mechanical behavior is not much affected at large distances from the source of a disturbance, such as heating or excavation. Characterization of the medium at the far field would therefore be mainly on the hydraulic properties. The characterization would become more comprehensive as one approaches the near field, where all THM parameters are involved. However, for a sustained thermal input, the thermal process may cause mechanical effect even in the far field, in which case some degree of characterization of the mechanical properties in the far field is also required, in particular the mechanical properties of major fracture zones.

The appropriate choice of models is also an important issue. For example, both discrete and discontinuous approaches appear to be suitable for near field problems, but an equivalent continuum approach may have great difficulties. This is the region with potentially the largest coupled THM effects and fortunately is also the region more accessible to relatively detailed characterization.

(6) Future scientific work that is necessary to build up our experience and capability on coupled THM processes for performance assessment of an underground waste repository was alluded to several times in the foregoing discussions. Some of the scientific topics to be addressed in continuing work are:

- Fundamental or constitutive laws of hydromechanical behavior of a single fracture, especially under shear stresses. The constitutive laws for predicting mechanical behavior in general, and the changes of conductive aperture during complex deformation paths in particular, are not satisfactory at present.

- Representation of the far-field hydraulic and mechanical behavior of fracture networks in homogenization. It is not yet clear how to represent hydromechanical coupling behavior in equivalent continuum media.

- Characterization strategy of measuring hydraulic and mechanical parameters as a function of distances from thermal or mechanical sources of disturbance. The appropriate strategy is also dependent on the geometric features and fracture distributions at the site.

- Strategy for the choice of models or codes for particular performance assessment objectives for the near field region and far field region respectively.

These scientific topics will be best studied in the context of a major field study of a realistic site.

5. TOWARDS A DEMONSTRATION OF A DISPOSAL CONCEPT — PROPOSED APPROACH

Strictly from the point of view of an organization in charge of high level radioactive waste management and a deep geological repository project, the observations of discrepancies in hydraulic behavior, giving the uncertainty ranges of flow predictions, might be considered to be a major outcome of the DECOVALEX project.

However, the conclusions is only valid in relation to the benchmark tests and test cases in the framework of DECOVALEX. The reality is more complex. Also, the existence of discrepancies, i.e., uncertainties, does not mean that they are inconsistent with demonstrating disposal feasibility. Indeed, the disposal concept must take uncertainties into account. Through an iterative process involving both performance assessment and safety assessment, the disposal concept will be defined and should involve the use of technological solutions that take into account irreducible uncertainties in decision making.

5.1 Use of THM Models and Codes for the Near Field

In the near field, often it is possible to obtain (T)HM data for studying the behavior of THM processes. Then it is realistic to perform local adjustment or calibration of the model parameters to the specific site. This calibration stage can be followed by an experimental stage to test all or part of the thermal, hydraulic and mechanical behaviors and their coupling. The model representativeness of the real site and the possibility of in-situ experimental validation can be envisaged. The extrapolation in time and in the limited space of the near field can then be calculated within the range of validity defined in the experiment.

5.2 Use of THM Models and Codes for the Far Field

In the far field, the practical difficulty of characterizing the geological medium is added to the lack of information on the constitutive laws applicable to the site. Also, in studying the site, there is usually no possibility of thermal phenomena being examined on a large scale because of the extended time needed. Only the study of the mechanical-hydraulic coupling is potentially possible on a large scale during the time when the underground structures are being built.

At the moment, we do not know how to perform time and space extrapolations from intrinsic laboratory parameters to the far field of a specific site. Therefore an important objective of an underground laboratory will be to demonstrate our capacity to take values that are measured or carefully calibrated in a particular field location and extend them in space.

The real disturbance on the scale of a repository is the construction of the access to the underground medium. It is during this stage that most of the mechanical changes have an influence on the hydraulic behavior of the rock mass. According to current knowledge, as demonstrated in DECOVALEX, small variations in mechanical conditions (e.g., stresses and displacements) can cause considerable hydraulic changes. This information and data provide an opportunity to calibrate one or more models and obtain the values for model parameters.

The next stage consists of using this set of calibrated parameters in a new configuration, e.g., in a future horizontal gallery, and to assess the correctness of the predictions. The analysis of the differences between model prediction and observation in the new gallery will yield the degree of uncertainty of the predictions and give some indications of the range of validity of the spatial extrapolation. If the differences between predictions and reality are acceptable and if furthermore the site can be assured to be "homogeneous," then the decision can be made to extend these adjusted parameters to the whole site for the hydro-mechanical part of performance assessment of the repository.

The coupling with thermal loading remains to be assessed. However, from DECOVALEX we see that the codes are satisfactory in their thermal and thermo-mechanical calculations. The main uncertainty is really the hydro-mechanical coupling and in particular the hydromechanical response of large fracture zones.

5.3 The Disposal Concept: The Step Between "Performance Assessment" and "Safety Assessment"

The study of the THM behavior is to gain the specific scientific knowledge bases that allow the nuclear waste managers to integrate the relevant processes and to propose a feasible deep geological disposal concept. The acceptability of the proposed concept must be evaluated by the regulators. Figure 1 illustrates the interactions among the scientific community, the nuclear waste managers and the regulators. Feasibility should be based on scientific and technological knowledge. On the other hand, judgment of acceptability is based on the scientific state of art and the associated risks, including the risks due to uncertainties or lack of knowledge. The overlap area represents a common ground of reasonable assurance of high level of safety. This overlap hopefully will be as large as possible.

The DECOVALEX project demonstrated that predictions of THM behavior of fractured media entail considerable uncertainties, mainly in hydraulic results. The experiences gained during this project enable us to identify the sources of these uncertainties more clearly.

The question that must be answered is what is the impact of the coupled THM processes on the confinement performance of the geological barrier. The models and codes that have been tested in the DECOVALEX project enable us to distinguish various couplings and to give anestimation of the range of variation or uncertainties of results, especially for calculations

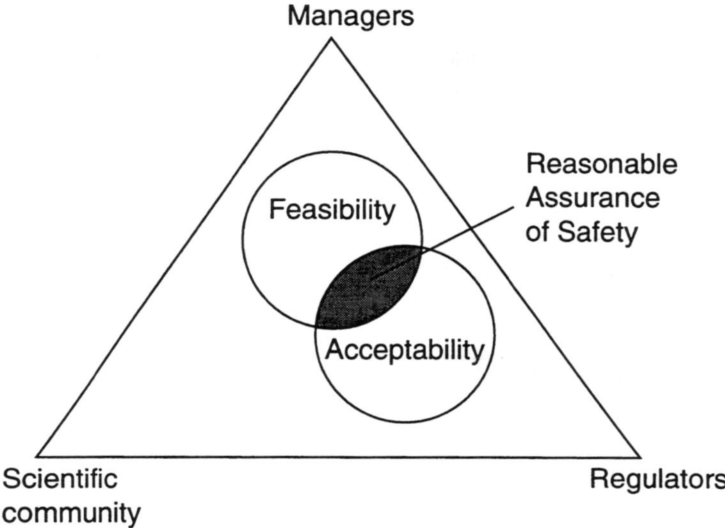

Figure 1. Interactions between the waste managing agencies, scientific community and the regulatory bodies.

of water flow in the geological barrier. These uncertainties must be incorporated in the safety assessment calculations, in order to estimate their level of acceptability from the safety point of view.

There are two possible scenarios:

(1) The safety is not affected by this uncertainty; i.e., the entire THM processes do not have a predominant influence on the safety of a disposal facility. In this case, existing tools are sufficient. However, we need to mtaintain progress in understanding these phenomena. If necessary, the assessment should be repeated.

(2) The uncertainties concerning the geological medium and its THM behavior are unacceptable for safety. Then these uncertainties must be reduced.

To reduce uncertainties, we can:

- modify the design parameters of the disposal concept (geometry, thermal loads, etc.);
- improve the characterization of the medium;
- acquire further data on the intrinsic parameters associated with constitutive laws or models used;
- improve the tools for evaluating THM effects: fundamental laws, computer codes and visualization.

These four ways of reducing uncertainties can be carried out with the aid of performance assessment, which is used to evaluate the roles of various components involved. A new safety assessment must be made each time there is a major modification of the disposal system. This iterative-recursive approach is illustrated in Figure 2. In this diagram, the role of safety assessment is to determine the area of acceptability, while performance assessment is to determine the limits of possibilities and range of uncertainties in predictions. The final product is a disposal concept involving a series of safeguards that mutually reinforce each other until the system reaches a level of risk that is called "As Low As Reasonably Possible" (also referred to as the ALARA principle).

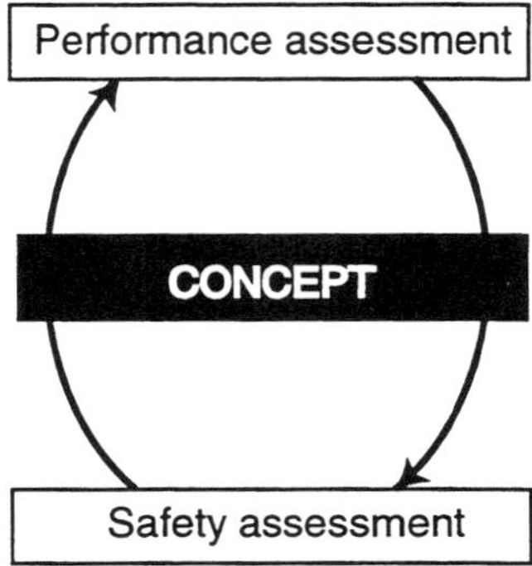

Figure 2. The iterative-recursive approach for performance and safety assessments.

ACKNOWLEDGMENTS

The authors thank all the participants in the DECOVALEX project for many stimulating discussions, and the Commission of the European Community for its financial support. The conclusions, interpretations and viewpoints presented in this paper are those of the authors and are not necessarily shared by the participants in the DECOVALEX project.

REFERENCES

1. L. Jing, J. Rutqvist, O. Stephansson, C-F. Tsang and F. Kautsky, DECOVALEX - mathematical models of coupled T-H-M processes for nuclear waste repositories, Report of Phase I, SKI Technical Report 93:31, Swedish Nuclear Power Inspectorate (SKI), Stockholm, Sweden, 1993.
2. L. Jing, J. Rutqvist, O. Stephansson, C-F. Tsang and F. Kautsky, DECOVALEX - mathematical models of coupled T-H-M processes for nuclear waste repositories, Report of Phase II, SKI Technical Report 94:16, Swedish Nuclear Power Inspectorate (SKI), Stockholm, Sweden, 1994.
3. L. Jing, J. Rutqvist, O. Stephansson, C-F. Tsang and F. Kautsky, DECOVALEX - mathematical models of coupled T-H-M processes for nuclear waste repositories, Report of Phase III, Swedish Nuclear Power Inspectorate (SKI), Stockholm, Sweden, 1996 (in press).

Short Descriptions of Computer Codes applied in DECOVALEX

Code

VIPLEF: S-M. Tijani
FLAC: J. Israelsson
UDEC & 3DEC: J. Israelsson
NAPSAC: P. Wilcock
FRACON: S. Nguyen
THAMES: Y. Ohnishi & A. Kobayashi
ROCMAS: Jahan Noorishad and Cin-Fu Tsang
CASTEM 2000 & TRIO-EF: A. Millard
ABAQUS: L. Börgesson

O. Stephansson, L. Jing and C.-F. Tsang (Editors)
Coupled Thermo-Hydro-Mechanical Processes of Fractured Media
Developments in Geotechnical Engineering, vol. 79
© 1996 Elsevier Science B.V. All rights reserved.

507

Short description of VIPLEF code

S-M. Tijani

Ecole Nationale Supérieure des Mines de Paris
Centre de Géotechnique et d'Exploitation du Sous-sol
35, Rue Saint Honoré
77305 Fontainebleau CEDEX (France)

1. INTRODUCTION

Numerical codes have always been needed for the geomechanical activities of the Centre de Géotechnique et d'Exploitation du Sous-sol (CGES) of the Ecole des Mines de Paris. Since its creation (1968), the CGES's strategy has been to develop its own codes in order that the mathematical formulation of the physical and mechanical phenomena derived from the experiments carried out in its laboratory or reported should not be influenced by the limitations of the existing codes. As a consequence of this strategy in all CGES Finite Element Codes the algorithms are chosen in such a way as to allow, simply and rapidly, the implementation of any governing equation or constitutive law [1]. For instance, a specialist of VIPLEF (like the one who has written the code!) is able to introduce, within one day , any new rheological model (even a complex one) including validity tests. The numerical VIPLEF Fortran code, was written in 1978 [2]. Its final "Computing Structure" was revised in 1985 [3] (Bidimensional Thermo-Hydro-Mechanical Finite Element Codes: CHEF, HYDREF and VIPLEF). A three-dimensional version was developed in 1993 (Codes: CHALEF3D, HYDREF3D and VIPLEF3D).

2. GOVERNING EQUATIONS

VIPLEF is frequently modified, principally to take into account the new industrial and scientific requirements and secondarily to take advantage of any new numerical technique provided it does not restrict the scientific aim of the code. The governing equations described hereafter correspond to the 1995 version of the code (which was used for the DECOVALEX benchmark tests). The various simplifications of the governing equations existing in this version of VIPLEF and indicated hereafter are not imposed by the numerical processes but have been chosen by the scientific experts.

2.1 Balance laws

The mass balance law (fluid mass flow throughout the skeleton of a porous continuum and inside the hydraulic joints) is simplified assuming that the fluid density has small variations in space so that the continuity equation is equivalent to the fluid volume balance law. In the momentum balance laws, static approach is used (inertia forces are neglected) and the classical Darcy's law governs the fluid flow. In the energy balance law only thermal energy is taken into account, neglecting energies associated with irreversible dissipation, thermal dilatancy and fluid transport.

2.2 Constitutive laws

Each of the three following problems may be solved separately using the relevant code (either CHALEF or HYDREF or VIPLEF).

2.2.1 Thermal conduction

Linear or non-linear transient or steady-state thermal conduction (Fourier's Law) problems can be solved with CHEF (or CHALEF3D) which allows:

- initial temperature field,

- complex time dependent boundary conditions as prescribed thermal flux or temperature or relationships between these two physical variables (for instance black-body radiation transfer condition of an external heat source defined by the equation: $\Phi = K(T^4 - T_{ext}^4))$,

- heterogeneity and anisotropy,

- dependence of the thermal parameters upon the temperature using any given governing equation, possibly non-linear,

- any given time dependent right hand side of the energy balance equation (in order to take into account not only volumetric heat generation, but also other energies usually neglected),

- variability of the geometry of the solid where the thermal conduction occurs (this is useful in the case of finite deformation and when the geological solid is modified by excavations, lining ...),

- joints with any given relationship between the thermal flux and the discontinuity of the temperature.

2.2.2 Hydraulic diffusion

Linear or non-linear transient or steady-state hydraulic diffusion (Darcy's Law) problems can be solved with HYDREF (or HYDREF3D) which allows:

- initial pore pressure field,

- complex time dependent boundary conditions (prescribed pore pressure or fluid flow rate or hydraulic head or relationship between these two physical variables),

- heterogeneity and anisotropy,

- dependence of the hydraulic parameters upon the pressure (or upon specified mechanical or thermal variables) using any given governing equation, even non-linear relationships.

- any given time dependent right hand side of the fluid mass balance equation (in order to take into account not only distributed hydraulic sources but also the variation of the connected porosity),

- variability of the geometry of the porous solid where the hydraulic diffusion occurs (this is useful in the case of finite deformation and when the geological solid is modified by excavations, lining ...),

- joints with a permeability which may be any given function of the pore pressure and mechanical and thermal external variables (for instance, in the well known cubic law, the permeability depends on the aperture of the joint which is defined by the displacement of its opposite sides, it depends also on the temperature which has an influence on the viscosity of the fluid).

2.2.3 Mechanical deformation

Linear or non-linear static finite (Jauman's objective stress rate) or infinitesimal deformation can be modeled with VIPLEF (or VIPLEF3D) which allows:

- initial stresses and internal rheological fields,

- complex time dependent boundary conditions (prescribed loads or displacement or relationship between these two physical variables),

- heterogeneity and anisotropy,

- time dependent temperature (thermal dilatancy) and pore pressure (Biot's law) fields,

- any rheological model (non-linear elasticity, linear or non-linear viscoelasticity, elastoplasticity, elastoviscoplasticity, even with instantaneous plasticity ...),

- joints with any given constitutive law (for instance, the well known normal hyperbolic law associated with friction and dilatancy when sliding occurs).

- any prescribed governing equations for the influence of the temperature on the mechanical parameters of solids and joints.

3. NUMERICAL METHODS

3.1 Discretisation

All continuous time dependent fields are replaced by discret sets of their values at the nodes of the mesh and at a finite number of given times. The space approximation (interpolation) is based on the Finite Element Method and the three codes have a library of all classical types of elements ranging from the 2 node linear segments to the 27 node quadratic Lagrangian hexaedral elements. All the second-order elements are isoparametric (the edges may be curvilinear). The time interpolation is linear between two successive time steps.

3.2 Transient conduction and diffusion
The time integration is based on the implicit Euler's method with a prescribed or automatically controlled time step (CHALEF3D and HYDREF3D).

3.3 Plasticity
In VIPLEF, all rheological non-linearities similar to plasticity are treated with an iterative process based on the Initial Stress Method. But the actual constitutive laws are used without any linearisation (i.e. the rheological equations are fully integrated by hands before the programming of the model) [2].

3.4 Viscoplasticity
In the case of viscosity, the Initial Stress Method is used with a step by step semi-implicit integration technique which needs only the response of the rheological model to a creep test (given constant stresses and temperature during the step), starting from any given initial state (Multi-Step Creep Technique) [1].

3.5 Finite deformation
The Updated Lagrangian Method is used. The actual deformation is divided into infinitesimal deformations at the end of which the geometry, the stresses (Jauman's rate) and the material tensors are updated.

3.6 Coupled t.h.m. processes
An external automatic "code" performs an iterative loop of calls to each of the three codes (CHALEF -> HYDREF -> VIPLEF -> CHALEF ...) until the convergence is achieved. The straightforwardness of this iterative technique is required in order to ensure the easiness of introduction of any new theoretical formulation based on the actual observations and measurements.

4. COMPUTING ASPECTS

Written in Fortran the three codes are fully autonomous (i.e. no special mathematical or graphic external library is needed) :

  - Pre-processors help the user to construct the mesh (semi-automatic and automatic griding codes).

  - All needed mathematical operations are programmed in the codes even the linear algebraic systems solver which uses the Active Column Direct Gauss Method associated with Sky Line Storage Technique (Block decomposition and file storage when necessary) and an internal automatic Profile Reduction.

- In order to be easily installed in any computer environment the codes have their own graphic library, written in Fortran initially in 1978 and frequently modified since that time, to take into account the new graphic devices (X11, Post-script, HPGL ...).

5. REFERENCES

1. S-M. Tijani, Résolution numérique des problèmes d'élastoviscoplasticité - Application aux cavités de stockage de gaz en couches salines profondes, Thèse de Docteur-Ingénieur. P. & M. Curie. Paris VI (1978).

2. F. De Grenier and S-M. Tijani, Logiciel VIPLEF - Résolution des problèmes d'élasto-viscoplasticité, Méthodes Numériques dans les Sciences de l'Ingénieur. Congrès International. DUNOD. Paris (1979).

3. B. Plischke, A Survey of Computer Programs in Rock Mechanics - Research and Engineering Practice, Int. J. Rock Mech. & Min. Sci., 25(4) (1988) 194.

O. Stephansson, L. Jing and C.-F. Tsang (Editors)
Coupled Thermo-Hydro-Mechanical Processes of Fractured Media
Developments in Geotechnical Engineering, vol. 79
© 1996 Elsevier Science B.V. All rights reserved.

513

Short Description of FLAC Version 3.2 [*]

Jan I. Israelsson

Itasca Geomekanik AB, Box 17, 781 21 Borlänge, Sweden.

1. INTRODUCTION

FLAC (*F*ast *L*agrangian *A*nalysis of *C*ontinua) [1] is a two-dimensional explicit finite-difference continuum code that simulates the behavior of structures built of soil, rock or other materials which may undergo plastic flow when their yield limit are reached. It is developed specifically to perform analyses on IBM-compatible microcomputers or SUN SPARCStations.

2. DISTINGUISHING FEATURES OF FLAC

Included among the code's capabilities are the following:

- large-strain simulation of continua, with optional interfaces
- explicit solution scheme, giving a stable solution to unstable physical processes
- library of material models (e.g., elastic, Mohr-Coulomb plasticity, ubiquitous joint, double-yield, viscous and strain-softening)
- several grid generators for common shapes
- statistical distribution of any property
- convenient specification of general boundary conditions
- groundwater flow, with full coupling to mechanical calculation (including negative pore pressure for saturated soil)
- structural elements (including grouted cables), with general coupling to the grid
- optional thermal and creep calculations
- a built-in language to add user-defined features (e.g., new constitutive models, new variables or new commands)
- sliding and opening interfaces between objects (continuum and structural)
- coupling to infinite elastic boundary
- dynamic option, with seismic input, free-field calculation and pore-pressure coupling

Problems in geotechnical engineering encompass a wide range of physical processes. The power of *FLAC* is its ability to simulate these processes either individually or in combination. Mechanical, fluid flow and thermal analyses can be performed as separate or coupled calculations. For example, saturated and unsaturated groundwater flow modeling can be

---

[*] Adapted from the *FLAC* version 3.2 manual [1].

performed independently or coupled to mechanical-stress calculations. Similarly, heat-transfer calculations can be run alone or coupled to thermal-stress calculations. Models can be run to a static equilibrium solution and then subjected to dynamic excitations. Structural element calculations can be run independently or coupled to the *FLAC* grid. The degree to which the analyses are integrated is the prerogative of the user, and the couplings can be turned on and off at the user's discretion.

Types of processes that can be treated in *FLAC*, as well as typical problem applications, include:

- mechanical loading capacity and deformations — in slope stability and foundation design
- evolution of progressive failure and collapse — in hard rock mine and tunnel design
- time-dependent creep behavior of viscous materials — in salt and potash mine design
- restraint provided by structural support on geologic materials — in tunnel lining, rock bolting, tiebacks and soil nailing
- saturated and unsaturated fluid flow and pore pressure build-up and dissipation for undrained and drained loading — in groundwater flow and consolidation studies of earth-retaining structures and earthen slopes, and in reservoir engineering
- coupled mechanical-fluid flow interaction — in depletion of reservoirs
- dynamic loading on slip-prone geologic features — in earthquake engineering and mine rockburst studies
- dynamic effects of explosive loading and vibrations — in tunnel driving or in mining operations
- seismic excitation of structures such as dams and the coupled effect of fluid on time-dependent pore-pressure change — in liquefaction phenomena in foundations, dams
- deformation and mechanical instability resulting from thermal-induced loads — in performance assessment of underground repositories of high-level radioactive waste

Materials are represented by quadrilateral elements, or zones, that form a grid that is adjusted by the user to fit the shape of the object to be modeled. Internally, FLAC subdivides each element into two overlayed sets of constant-strain triangular elements (Figure 1). The four triangular sub-elements are termed a, b, c and d.

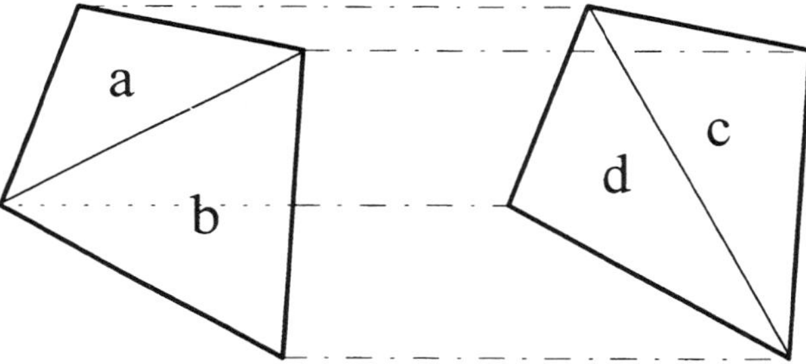

Figure 1. Overlayed quadrilateral elements used in *FLAC*.

Each element behaves according to a prescribed linear or non-linear stress/strain law in response to the applied forces or boundary restraints. FLAC is based upon a "Lagrangian" calculation scheme that is well suited for modeling large distortions. Furthermore, FLAC has several built-in constitutive models which permit the simulation of highly non-linear, irreversible response representative of geologic, or similar, materials.

The mechanical material models are the transversely-isotropic model, double-yield (cap) plasticity model, isotropic elastic model, Mohr-Coulomb plasticity model, strain-softening plasticity model, and ubiquitous joint model. The creep (visco-elastic) models are the two-component creep power law, classical viscosity, and the WIPP reference creep formulation. The thermal models are anisotropic heat conduction, isotropic heat conduction (with thermal conductivity of the form $k(T) = k1 + k2T^n$) and isotropic heat conduction.

The explicit finite-difference formulation of the code makes it ideally suited for modeling geomechanical problems that consist of several stages, such as sequential excavation, backfilling and loading. Stable solutions can be given to unstable physical processes.

FLAC is a command-driven (rather than menu-driven) computer program which offers several advantages when applied for engineering studies. The built-in programming language (FISH) enables the user to define new variables and functions. For example, new variables may be plotted or printed, special grid generators may be implemented, servo-control may be applied to a numerical test, unusual distributions of properties may be specified, and parameter studies may be automated.

3. GOVERNING EQUATIONS

The general calculation sequence embodied in FLAC (see Figure 2) first invokes the equation of motion to derive new velocities and displacements from stresses and forces. Then, strain rates are derived from velocities, and new stresses from strain rates.

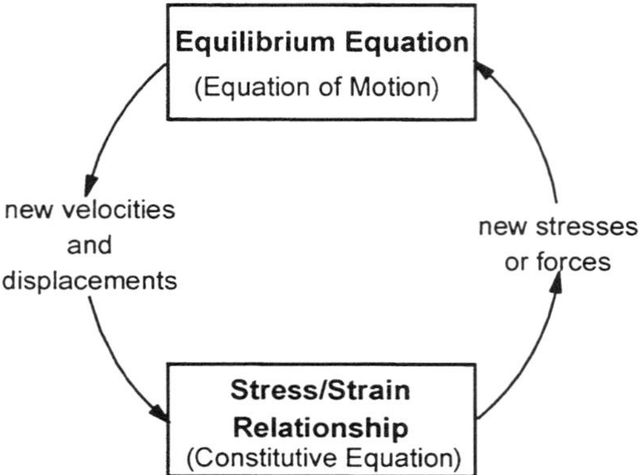

Figure 2. Basic explicit calculation cycle used in *FLAC*.

For every cycle around the loop, one timestep is taken. With a timestep so small that information cannot physically pass from one element to another in that interval, all grid variables will be updated from known values. The central concept is that the computational "wave speed" always keeps ahead of the physical wave speed. Explicit methods are best for ill-behaved systems (e.g., non-linear, large strain, physical instability), although it takes a large number of steps due to the small timesteps required for numerical stability.

The difference equations for a triangle are derived from the generalized form of Gauss' divergence theorem (e.g., [2]):

$$\int_S n_i f \, ds = \int_A \frac{\partial f}{\partial x_i} \, dA \tag{1}$$

where $\int_S$ is the integral around the boundary of a closed surface,
  n_i is the unit normal to the surface, s,
  f is a scalar, vector or tensor,
  x_i are position vectors,
  ds is an incremental arc length, and
  $\int_A$ is the integral over the surface area, A.

Defining the average value of the gradient of f over the area A as

$$\left\langle \frac{\partial f}{\partial x_i} \right\rangle = \frac{1}{A} \int_A \frac{\partial f}{\partial x_i} \, dA \tag{2}$$

one obtains, by substitution into Eq. (1),

$$\left\langle \frac{\partial f}{\partial x_i} \right\rangle = \frac{1}{A} \int_s n_i \, f \, ds \tag{3}$$

For a triangular sub-element, the finite-difference form of Eq. (3) becomes

$$\left\langle \frac{\partial f}{\partial x_i} \right\rangle = \frac{1}{A} \sum_s \langle f \rangle \, n_i \, \Delta s \tag{4}$$

where Δs is the length of a side of the triangle, and the summation occurs over the three sides of the triangle. The value of $\langle f \rangle$ is taken to be the average over the side.

If the body is at equilibrium, or in steady-state flow (e.g., plastic flow), the net nodal force vector (F_i, on the node) will be zero; otherwise, the node will be accelerated according to the finite-difference form of Newton's second law of motion:

$$\dot{u}_i^{(t+\Delta t/2)} = \dot{u}_i^{(t-\Delta t/2)} + \sum F_i^{(t)} \frac{\Delta t}{m} \tag{5}$$

where the superscripts denote the time at which the corresponding variables are evaluated.

For large-strain problems, Eq. (5) is integrated again to determine the new coordinate of the gridpoint:

$$x_i^{(t+\Delta t)} = x_i^{(t)} + \dot{u}_i^{(t+\Delta t/2)} \Delta t \tag{6}$$

Equation (4) enables strain rates, $\dot{e}_{ij}$, to be written in terms of nodal velocities for a zone by substituting the average velocity vector of each side for f:

$$\frac{\partial \dot{u}_i}{\partial x_j} \cong \frac{1}{2A} \sum_s (\dot{u}_i^{(a)} + \dot{u}_i^{(b)}) n_j \, \Delta s \tag{7}$$

$$e_{ij} = \frac{1}{2}\left[\frac{\partial \dot{u}_i}{\partial x_j} + \frac{\partial \dot{u}_j}{\partial x_i} \right] \tag{8}$$

where (a) and (b) are two consecutive nodes on the triangle boundary.

Note that the averaging equation (7) is identical to that derived by exact integration, if there is a linear variation in velocity between nodes.

The volumetric strain is averaged over each pair of triangles, while the deviatoric strains remain unchanged. The strain rates in triangles a and b of Figure 1 are adjusted in the following way, where subscript m denotes "mean" and subscript d denotes "deviatoric:"

$$\dot{e}_m = \frac{\dot{e}_{11}^a + \dot{e}_{22}^a + \dot{e}_{11}^b + \dot{e}_{22}^b}{2} \tag{9}$$

$$\dot{e}_d^{\,a} = \dot{e}_{11}^{\,a} - \dot{e}_{22}^{\,a}$$

$$\dot{e}_d^{\,b} = \dot{e}_{11}^{\,b} - \dot{e}_{22}^{\,b} \tag{10}$$

$$\dot{e}_{11}^{\,a} = \frac{\dot{e}_m + \dot{e}_d^{\,a}}{2}$$

$$\dot{e}_{11}^{\,b} = \frac{\dot{e}_m + \dot{e}_d^{\,b}}{2}$$

$$\dot{e}_{22}^{\,a} = \frac{\dot{e}_m + \dot{e}_d^{\,a}}{2}$$

$$\dot{e}_{22}^{\,b} = \frac{\dot{e}_m + \dot{e}_d^{\,b}}{2} \tag{11}$$

Similar adjustments are made for triangles c and d. The component $\dot{e}_{12}$ is unchanged.

In general, non-linear constitutive laws are written in incremental form because there is no unique relation between stress and strain. The simplest example of a constitutive law is that of isotropic elasticity. Equation (12) provides a new estimate for the stress tensor in isotropic elasticity, given the old stress tensor and the strain rate:

$$\sigma_{ij} := \sigma_{ij} + \left\{ \delta_{ij} \left(K - \frac{2}{3} G \right) \dot{e}_{kk} + 2G \dot{e}_{ij} \right\} \Delta t \tag{12}$$

where $:=$ means "replaced by",
        δ_{ij} is the Kronecker delta,
        Δt is timestep, and
        $G,\ K$ are the shear and bulk moduli, respectively.

There is another contribution to the stress tensor, due to the finite rotation of a zone during one timestep: the stress components referred to the fixed frame of reference change as follows:

$$\sigma_{ij} := \sigma_{ij} + (\omega_{ik}\sigma_{kj} - \sigma_{ik}\omega_{kj})\Delta t \tag{13}$$

where

$$\omega_{ij} = \frac{1}{2}\left\{ \frac{\partial \dot{u}_i}{\partial x_j} - \frac{\partial \dot{u}_j}{\partial x_i} \right\} \tag{14}$$

The adjustment of Eq. (13) is only done in large-strain mode and is, in fact, applied before Eq. (12). Stress adjustments due to other strain components are not made.

The constitutive law, which is presented for an isotropic elasticity in Eq. (12) and rotation adjustment, Eq. (13), are then used to derive a new stress tensor from the strain-rate tensor. Mixed discretization is invoked again — but on the stresses, in order to equalize isotropic stress between the two triangles in a pair, using area weighting:

$$\sigma_0^{(a)} = \sigma_0^{(b)} := \left[\frac{\sigma_0^{(a)} A^{(a)} + \sigma_0^{(b)} A^{(b)}}{A^{(a)} + A^{(b)}} \right] \tag{15}$$

where $\sigma_0^{(a)}$ is the isotropic stress in triangle (a), and $A^{(a)}$ is the area of triangle (a).

A form of damping is used in *FLAC* in which the damping force is such that energy is always dissipated. Equation (5) is replaced by the following equation, which incorporates the new damping scheme:

$$\dot{u}_i^{(t+\Delta t/2)} = \dot{u}_i^{t-\Delta t/2)} + \left\{ \sum F_i^{(t)} - \left\{ \alpha \left| \sum F_i^{(t)} \right| sgn(\dot{u}_i^{(t-\Delta t/2)}) \right\} \right\} \frac{\Delta t}{m_n} \tag{16}$$

where α is a constant (set to 0.8 in *FLAC*), and m_n is a fictitious nodal mass.

Either stress or displacement may be applied at the boundary of a solid body in *FLAC*.

4. GROUNDWATER FLOW

The groundwater equations in FLAC are expressed in terms of pressure rather than head, although the latter is more common in soil mechanics. It seems more consistent to work in terms of pressure, since FLAC models the coupling between solid and fluid — the interaction between the two phases is critically dependent on the relative magnitude of pressure and stress, not of head and stress.

Darcy's law for an anisotropic porous medium is

$$V_i = K_{ij} \frac{\partial P}{\partial x_j} \tag{17}$$

where V_i is the specific discharge vector (units of velocity),
       P is the pressure, and
       K_{ij} is a "permeability" tensor with the units $[L^3 T/M]$

Each quadrilateral element is divided into triangles in two different ways. The specific discharge vector can be derived for a generic triangle using the equation (3):

$$V_i \approx \frac{K_{ij}}{A} \sum P n_j s \tag{18}$$

where $\sum$ is the summation over three sides of the triangle.

The specific discharge vector is then converted to scalar volumetric rates at the nodes by making dot products with the normals to the three sides of the triangle. The general expression is

$$Q = V_i n_i s \tag{19}$$

The "stiffness" matrix [M] of the whole quadrilateral element is defined in terms of the relation between the pressures at the four nodes and the four nodal flow rates.

$$\{Q\} = [M]\{P\} \tag{20}$$

The flow imbalance, $\sum Q$, at a node causes a change in pore pressure at a saturated node as follows:

$$\frac{\partial P}{\partial t} = -\frac{K_w}{nV} \sum Q \tag{21}$$

where nV is the pore volume associated with the nodes (n is the porosity, and V is the total volume.)

In finite-difference form, Eq. (21) becomes

$$P := P - \frac{K_w \left(\sum Q \Delta t + \Delta V_{mech} \right)}{nV} \tag{22}$$

where ΔV_{mech} is the equivalent nodal-volume increase arising from mechanical deformations of the grid. If the nodal pore pressure computed by Eq. (22) were to result in a pressure more negative than the tension limit, then the pressure is set to zero, and the nodal outflow used to reduce the saturation, s, is as follows:

$$s := s - \frac{\sum Q \Delta t + \Delta V_{mech}}{nV} \tag{23}$$

In *FLAC*, an implicit solution scheme for confined flow calculations is also provided. If the implicit option is selected, an iterative method is used to obtain the solution within each timestep.

Permeabilities and porosities may be applied as functions of accumulated volume strain, and particle tracking may be done by introducing particles into the flow at any time and place.

5. THERMAL-MECHANICAL LOGIC

This logic in FLAC allows simulation of transient heat conduction in materials and the development of thermally induced displacements and stresses. There are three material models for the thermal behavior of the material: isotropic conduction, anisotropic conduction, and temperature-dependent conduction. Different zones may have different models and properties. Any of the mechanical models may be used with any of the thermal models, and several different thermal boundary conditions may be prescribed. Heat sources may be inserted into the material either as line sources or as volume sources. These sources may be made to decay exponentially with time. Both explicit and implicit solution algorithms are available.

The basic equation of conductive heat transfer is Fourier's law, which can be written as

$$Q_i = -k_{ij}\frac{\partial T}{\partial x_j} \tag{24}$$

where Q_i = flux in the i-direction (W/m^2),

$\quad\quad\quad k_{ij}$ = thermal conductivity tensor (W/mK), and

$\quad\quad\quad T$ = temperature.

Also for any mass, the change in temperature can be written as

$$\frac{\partial T}{\partial t} = \frac{Q_{net}}{C_p M} \tag{25}$$

where Q_{net} = net heat flow into the mass (W),

$\quad\quad\quad C_p$ = specific heat (J/kg K), and

$\quad\quad\quad M$ = mass(kg).

The diffusion equation for two-dimensional heat transfer will then be;

$$\frac{\partial T}{\partial t} = \frac{1}{C_p\rho}\frac{\partial}{\partial x}\left[k_x\frac{\partial T}{\partial x}\right] + \frac{\partial}{\partial y}\left[k_y\frac{\partial T}{\partial y}\right] = \frac{1}{C_p\rho}\left[k_x\frac{\partial^2 T}{\partial x^2} + k_y\frac{\partial^2 T}{\partial y^2}\right] \tag{26}$$

if k_x and k_y are constant.

Temperature changes cause stress changes according to the equation

$$\Delta\sigma_{ij} = -\delta_{ij}3K\alpha\Delta T \tag{27}$$

where $\Delta\sigma_{ij}$ = change in stress ij,

δ_{ij} = Kronecker delta (δ_{ij} = 1 for $i = j$ and 0 for $i \neq j$),
K = bulk modulus,
α = linear thermal expansion coefficient, and
ΔT = temperature change.

The coupling occurs in one direction only (i.e., the temperature may result in stress changes, but mechanical changes resulting from force application do not induce temperature change).

6. REFERENCES

1 Itasca Consulting Group, Inc. (1993) **FLAC Version 3.2**. Minneapolis, Minnesota: ICG.

2 Malvern, L. E. (1969) "Introduction," in **Introduction to the Mechanics of a Continuous Medium**. Engelwood Cliffs, New Jersey: Prentice Hall, Inc.

O. Stephansson, L. Jing and C.-F. Tsang (Editors)
Coupled Thermo-Hydro-Mechanical Processes of Fractured Media
Developments in Geotechnical Engineering, vol. 79
© 1996 Elsevier Science B.V. All rights reserved.

Short Descriptions of UDEC[*] and 3DEC[†]

Jan I. Israelsson

Itasca Geomekanik AB, Box 17, 781 21 Borlänge, Sweden

1. UDEC VERSION 2.0

1.1 Introduction

The Universal Distinct Element Code (*UDEC*), [1] is a two-dimensional numerical program based on the distinct-element method for discontinuum modeling. *UDEC* simulates the response of a discontinuous medium (such as jointed rock mass), represented as an assemblage of discrete blocks, subjected to either static or dynamic loading. The discontinuities are treated as boundary conditions between blocks; large displacements along discontinuities and rotations of blocks are allowed. Individual blocks behave as either rigid or deformable material. Deformable blocks are subdivided into a mesh of finite-difference elements, and each element responds according to prescribed linear or non-linear force-displacement relations for movement in both the normal and shear directions. The relative motion of the discontinuities is also governed by linear or non-linear force-displacement relations for movement in both the normal and shear directions.

UDEC is primarily intended for analysis in rock-engineering projects ranging from studies of the progressive failure of rock slopes to evaluations of the influence of rock joints, faults, bedding planes, etc. on underground excavations and rock foundations. *UDEC* is ideally suited to study potential modes of failure directly related to the presence of discontinuous features.

1.2 Distinguishing Features of *UDEC*

The distinguishing features of *UDEC* are described below.

* Simulation of large displacements (slip and opening) along distinct surfaces in a discontinuous medium (e.g., jointed rock masses)

* Discontinuous medium treated as an assemblage of discrete (convex or concave) polygonal blocks with rounded corners

* Explicit solution scheme, giving a stable solution to unstable physical processes

* Rigid or deformable blocks (can be mixed)

---

[*] Adapted from UDEC version 2.0 manual [1].
[†] Adapted from 3DEC version 1.5 manual [3].

- Library of material models for deformable blocks (e.g., elastic, Mohr-Coulomb plasticity, ubiquitous joint, double-yield and strain-softening)

- Library of material models for discontinuities (e.g., Coulomb slip, continuously-yielding, and Barton-Bandis [optional])

- Full dynamic capability, with absorbing boundaries and wave input

- "Null" blocks for excavation and backfill simulation

- Coupled fluid flow in joints and pressure in cavities

- Boundary-element coupling for "infinite domain" problems

- Structural elements (including non-linear cables), with general coupling to continuum blocks (spatially extensive) or discontinuities (local reinforcement)

- Tunnel generator and statistically based joint-set generator

1.3 Governing Equations

Equations governing the mechanical behavior of rigid and fully deformable blocks and the thermal and hydraulic flow processes in *UDEC* are described in Chapter 7, [2].

1.4 Coupling processes

Coupling processes in *UDEC* are described in Chapter 7, [2].

2. 3DEC VERSION 1.5

2.1 Introduction

3DEC (3-Dimensional Distinct Element Code) [3] is a three-dimensional numerical program based on the distinct-element method (DEM) for discontinuum modeling. The basis for this code is the extensively tested numerical formulation used by *UDEC*, previously described in Chapter 7, [2].

The distinguishing features of *3DEC* are described below.

- Simulation of large displacements (slip and opening) along distinct surfaces in a discontinuous medium (e.g., jointed rock masses)

- Discontinuous medium treated as an assemblage of discrete (convex or concave) polyhedra

- Discontinuities treated as boundary conditions between blocks; six modes of contact automatically recognized

- Relative motion along discontinuities governed by linear and non-linear force-displacement relations for movement in both the normal and shear directions

- Explicit solutions to unstable physical processes

- Rigid or deformable blocks (cannot be mixed)

- Two types of material models for deformable blocks (elastic or Mohr-Coulomb plasticity)

- Two types of material models for discontinuities (Coulomb-slip or continuously-yielding)

- Full dynamic capability, with absorbing boundaries and wave input

- "Null" blocks for excavation and backfill simulation

- Preprocessor program available to read AutoCAD files of solid objects and output *3DEC* data files to generate polyhedra

- Joint structure viewed separately from block structure; vectors and contours plotted on joint plane

- Interactive manipulation of screen images (shaded perspective views, cross-sections, wire frames, vectors, tensors, contours, etc.)

- Inner/outer region coupling and automatic, radially graded mesh generation within polyhedra for modeling "infinite domain" problems

- Structural elements (including grouted cables), with general coupling to continuum blocks (spatially extensive) or discontinuities (local reinforcement) and triangular plate elements to model concrete or shotcrete linings

- Tunnel generator and statistically based joint-set generator

2.2 Governing Equations

Equations governing the mechanical behavior of rigid and fully deformable blocks in *3DEC* are basically the same as for the two-dimensional DEM code *UDEC*, which are described in Chapter 7, [2].

2.3 Block Contact Detection

In *3DEC*, a "common plane" is used for contact detection between neighboring blocks. The "common plane" is analogous to a plate that is held loosely between the two blocks (see Fig. 1). If the blocks are held tightly and brought together slowly, the plate will be deflected by the blocks and will become trapped at some particular angle when the blocks finally come into contact. As the blocks are brought together, the plate will take up a position mid-way between them, at a maximum distance from both.

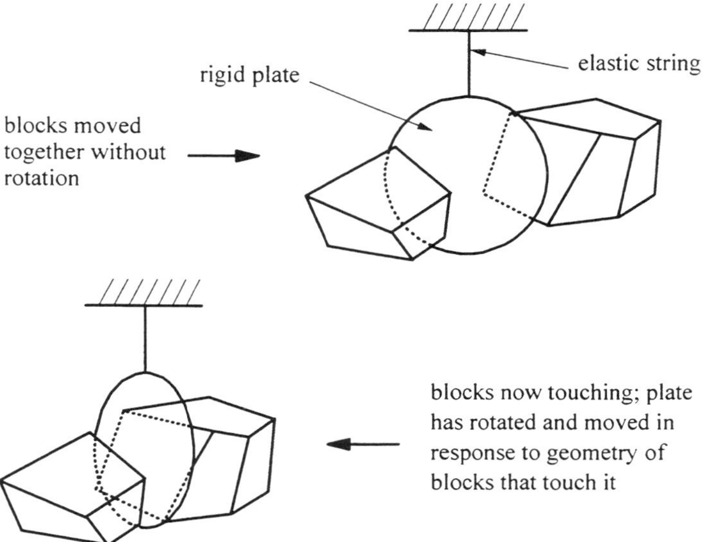

Figure 1. Visualization for positioning of common-plane in response to block geometry.

2.4 Thermal Analysis

3DEC is coupled to a simplified three-dimensional thermal-analysis model for simulation of internal heat sources specifically oriented to solving design problems associated with nuclear waste disposal. The model simulates transient conduction of heat in material and the subsequent development of thermally induced stresses. Heat sources can be added and can be made to decay exponentially with time.

The advantages of this method of calculating the temperatures are that:

- the infinite thermal boundary is automatically incorporated;

- the calculation procedure is extremely quick, compared with the finite-element or other numerical methods;

- the mechanical boundary conditions can be applied correctly, because the standard mechanical logic in *3DEC* is used for the mechanical part of the calculation; and

- inhomogeneous and anisotropic mechanical properties can be used.

The limitations are that:

- the material is assumed to be thermally homogeneous and isotropic with constant properties; and

- only a few restricted thermal boundary conditions can be applied — namely, adiabatic and isothermal planes.

The temperature at a point due to any spatial and temporal distribution of heat sources can be found by considering the distribution as a sum of point sources. In the limit as the number of sources becomes larger, and the strength of each is reduced so that the total heat output is constant, the sum becomes an integral over time, space (a line, surface or volume), or both.

For example, the temperature due to a time-dependent heat source of strength Q(t) can be written as

$$T = \frac{1}{8(\pi K)^{3/2} \rho C_p} \int_0^t \frac{Q(t')e^{-r^2/4K(t-t')}}{(t-t')^{3/2}} dt' \tag{1}$$

where Q = quantity of heat in source,
  C_p = specific heat,
  ρ = density,
  K = thermal diffusivity = $k/\rho\, C_p$
  r = distance between source and observation point,
  t = time after source released, and
  k = thermal conductivity.

For the case that the heat source is constant (i.e., $Q(t) = H(t)$, where $H(t)$ is the Heaviside step function), Eq. (1) becomes, [4]:

$$T = \frac{1}{4\pi Kr}\, erfc \frac{r}{(4Kt)^{1/2}} \tag{2}$$

where $erfc(x)$ = complimentary error function.

The temperature due to an exponentially decaying source of the form $Q(t) = Q_e\, exp(-At)$ is given by [5]:

$$T = \frac{Q_e}{4\pi Kr}\, exp\left(\frac{-r^2}{4Kt}\right) \cdot Re\left[\omega(At)^{1/2} + \frac{ir}{(4Kt)^{1/2}}\right] \tag{3}$$

where $\omega(z)$ $= \dfrac{2iz}{\pi} \int_0^{inf} \dfrac{e^{-t^2}}{z^2 - t^2} dt$, $Im\{z\}>0$,
     $Re\{z\}$ = real part of z,
     $Im\{z\}$ = imaginary part of z, and
     i $= (-1)^{1/2}$.

Heat source characteristics such as number of components, decay constants and proportion of each component are input by the user. In addition, the location and initial strengths of arrays of point sources are input. Symmetric and isothermal planes may be specified, as well as the initial temperature and the time at which the solution is desired. Once all these quantities have been input, the temperatures are calculated at all gridpoints. Stresses are not calculated at this stage but, when mechanical cycling is performed, the average temperature change for each zone is used to calculate a stress change in each zone. The stress state thus produced will not be in equilibrium but is treated as an initial condition for the mechanical calculation. The standard *3DEC* mechanical calculation is used to reach equilibrium.

3. REFERENCES

1   Itasca Consulting Group, Inc., (1993) "UDEC Universal Distinct Element Code, Version 2.0: Volume I: User's Manual; Volume II: Verification Problems and Example Applications," Minneapolis, Minnesota: ICG.

2   Ahola, M. P., A. Thoraval and A. Chowdhury. "Distinct Element Models for the Coupled T-H-M processes: Theory and Implementation," In this volume.

3   Itasca Consulting Group, Inc., (1994) "3DEC 3-Dimensional Distinct Element Code, Version 1.5: Volume I: User's Manual; Volume II: Appendices and Example Problems," Minneapolis, Minnesota: ICG.

4   Carslaw, H. S., and J. C. Jaeger. (1959) "Conduction of Heat in Solids," 2nd Ed. Oxford: Clarendon Press.

5   Crouch, S. L., and W. C. McClain. (1978) "Interim Report on Development of a Semi-Empirical Numerical Model for Simulating the Deformational Behavior of a High-Level Radioactive Waste Repository in Bedded Salt," University of Minnesota, Report to Oak Ridge National Laboratory, ORNL/TM-5462.

O. Stephansson, L. Jing and C.-F. Tsang (Editors)
Coupled Thermo-Hydro-Mechanical Processes of Fractured Media
Developments in Geotechnical Engineering, vol. 79
© 1996 Elsevier Science B.V. All rights reserved.

The NAPSAC fracture network code.

P. Wilcock

AEA Technology, 424.4 Harwell, Didcot, Oxfordshire, OX11 0RA, United Kingdom

1. INTRODUCTION

The NAPSAC fracture network code (see [1], [2] and [3]) was developed, by AEA Technology, to simulate steady-state and transient flow and tracer transport through fractured rock. It is assumed that fluid flow through such rock is restricted to a connected network of fracture planes and that there is no flow through the intervening rock matrix. The flow field is calculated as a function of the pressure which NAPSAC calculates at points along the intersections between fracture planes. NAPSAC works with the non-hydrostatic component of pressure, P, which is given in terms of the total pressure, P_{TOTAL}, by

$$P = P_{TOTAL} + \rho g z \tag{1}$$

where ρ is the density of the groundwater, g is the acceleration due to gravity and z is the vertical distance above some reference level. P will subsequently be referred to as the pressure.

Realistic fracture networks typically incorporate thousands of fractures. In order to represent the flow accurately, a fine discretisation must be used on each fracture. This might mean that NAPSAC requires several million finite elements to solve the flow problem through a fracture network. To solve this computationally intensive problem, an approach based on the finite-element method is adopted which uses an efficient numerical technique on each individual fracture. A NAPSAC calculation consists of three main phases. First, the problem to be solved is defined. Second, a solution for the groundwater flow is obtained. Third, the results are output.

2. THE NAPSAC MODEL GENERATION PHASE

The first phase in a NAPSAC calculation is the generation of a network of rectangular fracture planes. The model is defined within a domain formed from the union of a number of (possibly irregular) hexahedra, which will be referred to as the flow domain. An individual fracture is completely defined by the location of its centre, three orientation angles, the lengths of each side (or in the case of square fractures, a single length) and an effective aperture. Fractures can either be 'known' in which case the above properties are specified explicitly, or 'random' in which case the fracture properties are sampled from a wide range of statistical distributions. NAPSAC can also represent random variations of aperture within a fracture.

Figure 1: Fracture network generated by NAPSAC.

Up to six separately parameterised sets of random fractures can be defined at one time. The locations of the centres of the fractures are distributed uniformly within a cuboidal region whose boundaries are set by the user. This region should be sufficiently larger than the flow domain, bearing in mind the expected size of the fractures, that there is no reduced density of fractures near the edge of the flow domain. For each of the other fracture properties, the user specifies the distribution type and its parameters. Constant, uniform, normal, lognormal and two parameter exponential distributions are supported in Release 3.0. Given this information, fractures are generated randomly up to a user prescribed density. A typical NAPSAC generated fracture network is shown in Figure 1.

Intersections between the planes and the boundaries of the flow domain are calculated, and part or all of a fracture falling outside the flow domain is discarded. Boundary conditions are set on the boundaries of the flow domain. By default, NAPSAC treats any boundary for which no condition is set as impermeable. For a permeable boundary, there are three ways of prescribing the pressure distribution on the surface: a constant value may be set over the whole surface, a linear pressure variation may be specified, or a lattice of points on the surface may have a specified pressure, read in from a file, with the pressures at intermediate points obtained by bilinear interpolation. In addition to the boundary conditions set on the flow domain boundaries, the user may specify the pressure or flux on individual boreholes.

All fracture intersections are located by checking suitable pairs of fractures. Intersections with the flow domain boundaries and boreholes are also located. The flow field is discretised by assigning a number of nodes, referred to as the global flow nodes, to each intersection.

2 THE NAPSAC SOLUTION PHASE

2.1. The NAPSAC flow model.

The groundwater flow is assumed to be occur entirely within the fracture network, and the groundwater is assumed to be incompressible so that mass conservation implies

$$\nabla \cdot \mathbf{q} = 0, \tag{2}$$

where $\mathbf{q}$ is the groundwater volume flux. The pressure at each global flow node is calculated by solving equation 2 subject to the prescribed boundary conditions. The finite-element approach is used. The weak form of the problem is obtained by multiplying equation 2 by a suitable test function and then integrating over the flow domain, Ω. The pressure field is approximated by a function of the pressure values at the global nodes, P_I, with piecewise linear basis functions defined along each intersection. These basis functions are defined over the fracture planes, away from the intersections in the following way.

Noting that the groundwater flux is linearly dependent on the gradient of the pressure, and using the Galerkin finite-element formulation, the problem of solving the mass conservation equation becomes one of solving

$$\sum_J \hat{T} \int_\Omega \Psi_I \nabla^2 \Psi_J P_J = 0, \tag{3}$$

for suitable basis functions Ψ_I, where $\hat{T}$ is a constant related to the transmissivity T by $T = \rho g \hat{T}$. Defining

$$F_{IJ} = \hat{T} \int_\Omega \Psi_I \nabla^2 \Psi_J \tag{4}$$

so that F_{IJ} is the net flux due to unit pressure at node J and zero pressure at all other global flow nodes, integrated over the basis function for node I along the intersection containing node I (mass conservation on the fracture plane requires that $\nabla^2 \Psi_J$ is zero elsewhere in the solution domain). Both the fractures that include the intersection containing global flow node I will make a contribution, $f_{IJ}^{(k)}$, to F_{IJ}.

The next stage of the NAPSAC groundwater flow calculation is to evaluate $f_{IJ}^{(k)}$ for each fracture plane k and each pair of global flow nodes I, J. The finite-element method is used to solve the mass conservation equation (2) on the fracture plane. Assuming laminar viscous flow between two parallel plates with a separation equal to the fracture aperture, the groundwater volume flux $\mathbf{q}$ over the surface of the fracture plane is

$$\mathbf{q} = -\frac{e^3}{12\mu} \nabla P \tag{5}$$

where e is the fracture aperture and μ is the viscosity of the groundwater. The transmissivity of the fracture plane is given by

$$T = \frac{\rho g e^3}{12\mu}, \tag{6}$$

the parallel-plate law of transmissivity.

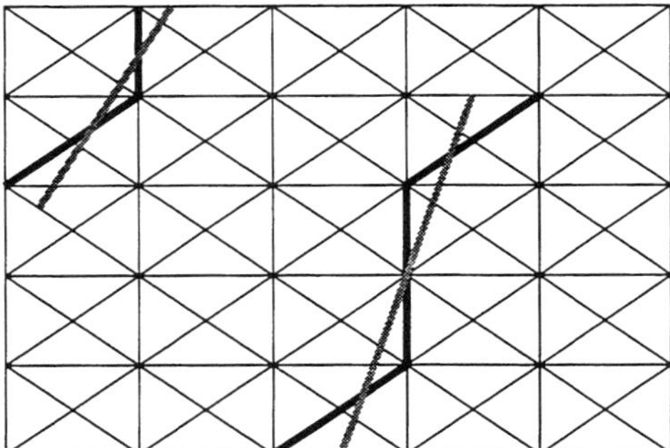

Figure 2: Discretisation of fracture plane. Fracture intersections shown in grey with the best-fit line of finite-element edges.

Each fracture plane is discretised into a user-specified number of linear triangular finite-elements as shown in Figure 2. The line of a fracture intersection is approximated so as to coincide with the edges of the triangular elements generated by the fracture discretisation. An example of this is also shown in Figure 2. The boundary conditions for calculating Ψ_J on plane k are specified pressures at all the finite-element nodes along the intersections, the values of which are interpolated from the condition of unit pressure at global flow node J and zero pressure at all other global flow nodes on the plane. The solution of these equations for the different boundary conditions corresponding to the different choice of global flow node, J, on the plane provide a complete set of groundwater volume flux responses which are assembled into a global matrix.

The final stage of the groundwater flow calculation is to evaluate the pressure field in the fracture network. The matrix equation in (3) is assembled from the individual $f_{IJ}^{(k)}$. The global boundary conditions are imposed, and a direct frontal solver is used to calculate the pressure values at the global flow nodes from the matrix equation. The pressure distribution across a fracture plane can be recovered by the superposition of the individual basis functions calculated earlier, weighted by the pressure solution at the global flow nodes.

2.2. Transient flow modelling

As the field experiments from which the data to generate fracture networks are derived usually involve transient flows, a transient flow modelling capability has been developed in NAPSAC. An approach consistent with the steady-state approach is adopted. This ensures that the code is applicable to the complex networks that the steady-state code is able to handle. The equation describing transient flow through a fracture network is

$$\frac{S}{\rho g}\frac{\partial P}{\partial t} = \frac{T}{\rho g}\nabla^2 P \tag{7}$$

where S is the fracture storativity which is dependent both on fluid and rock compressibility. The choice of a suitable model for fracture storativity is important for an accurate transient flow model. Three models for fracture storativity are available in NAPSAC:

$$S = \rho gAe,$$
$$S = \rho g \left[1/RKN + eC_f\right], \tag{8}$$
$$S = \alpha T^\beta,$$

where A, RKN, C_f, α and β are constants that can be specified by the user.

The fracture network is discretised in the way described previously and a forward difference is used to approximate the time derivative. The finite-element equation to be solved for the pressure values at the global flow nodes becomes

$$\sum_J \left(\frac{\hat{S}}{\Delta t} \int_\Omega \Psi_I \Psi_J P_J^{n+1} - \hat{T} \int_\Omega \Psi_I \nabla^2 \Psi_J P_J^{n+1} \right) = \sum_J \frac{\hat{S}}{\Delta t} \int_\Omega \Psi_I \Psi_J P_J^n, \tag{9}$$

where $\hat{S}$ is related to the storativity by $S = \rho g \hat{S}$. This equation is solved iteratively for a fixed timestep Δt to give P^n the pressure solution at time $t = n\Delta t$. The second term on the left-hand side of this equation is simply the flux term,

$$\sum_J F_{IJ} P_J^{n+1}, \tag{10}$$

that appears in the steady-state equation (3), and the first term on the left-hand side of this equation will be referred to as the storativity term

$$\sum_J S_{IJ} P_J^{n+1}. \tag{11}$$

The second step of a transient groundwater flow calculation is analogous to that in the steady-state calculation. The contributions from the individual planes to the global matrix are calculated. As before, the contributions to the flux term from each plane are evaluated by solving the mass conservation equation on each plane subject to a number of different boundary conditions. In addition, the contributions to the storage term are evaluated for each fracture plane.

These contributions are then assembled into the global matrix, ready to start timestepping. In order to simplify generating the right-hand sides of equation 8 for each timestep, two global matrices are assembled, one containing the flux term, F_{IJ}, and the other containing the storativity term, S_{IJ}. It is assumed that the boundary conditions are fixed with respect to time, and therefore can be imposed by deleting terms from the storativity matrix and changing the flux matrix in the same way as for the steady-state model. A boundary condition vector is constructed at this point. The two global matrices are added together, component by component, to form one global matrix. The resulting global matrix and the boundary condition vector are unchanged for all timesteps of the same size, since they depend on time only through the timestep size Δt.

The next stage is to solve the matrix equations for each timestep. The right-hand side of equation 8, for the first timestep, is evaluated by multiplying the storativity matrix by an initial solution which is prescribed by the user. This initial solution may be a constant pressure, a linearly varying pressure or an existing solution previously found by NAPSAC. The contribution from the boundary conditions is added to this right-hand side and the matrix equation in (8) is solved using the same direct frontal solver as for a steady-state calculation to produce the pressure solution at the first timestep. This solution is used to evaluate the next right-hand side and the timestepping loop is repeated producing a sequence of solutions P^{n+1} which define the flow field at the $(n+1)^{th}$ timestep.

As with steady-state problems, this method allows quite coarse meshes to be used on very large systems. However more detailed refinement might be required for smaller networks, or near sources and sinks. To deal with this the transient model permits optional local refinement for significant fractures. This option involves solving the transient mass conservation equation on the fracture and then adding this contribution to the refined matrix. At each timestep the local pressure solution is saved on the finite-element mesh of the refined plane and this solution is used to calculate the next solution to the transient mass conservation equation on the refined plane.

2.3. Boreholes

Since field experiments in fractured rock usually involve boreholes, a submodel is provided to facilitate their incorporation in simulations. Close to a point where a borehole intersects a fracture, the pressure field behaves like that within a parallel plane having a single sink:

$$P = P_W - \frac{\rho g Q}{2\pi T} \ln\left(\frac{r}{r_W}\right) \tag{12}$$

where P_W is the borehole pressure, r is the distance from the borehole, r_W is the borehole radius and Q is the volumetric flow rate from the borehole into the fracture.

This logarithmic behaviour is poorly approximated by a regular linear finite-element discretisation, and so a correction is applied to the pressure calculated by the finite-element method at the borehole, P_{FE}, using the analytical solution in (11), to obtain the borehole pressure, P_W:

$$P_W = P_{FE} + \frac{Q}{\Gamma} \tag{13}$$

where Γ is a productivity index and is dependent on the size of the finite element mesh, the borehole radius and the transmissivity of the fracture. Although this model is based on a steady-state analytical solution, tests have shown that the correction factor to be reasonable when modelling transient flow.

2.4. Two-dimensional Networks

The complicated geometry that NAPSAC is able to represent in three-dimensions puts restrictions on how well NAPSAC can model the physical factors affecting the flow. As better models of the physical factors affecting flow through fractured rock are required, it becomes necessary to approximate a three-dimensional network by a two-dimensional one. A two-dimensional version of NAPSAC was developed so as to use much of the existing three-dimensional code. The flow model and the approach used to solve the equations are analogous to the three-dimensional version. The network can be considered as a slice through a three-dimensional network in the x-z coordinate plane, with constant flow in the y-direction, and mass conservation governing flow through the fracture. The fractures are represented as straight line segments, and are defined by an orientation angle, a length and an effective aperture. As in three-dimensional simulations, the parameters describing each fracture may be sampled from statistical distributions. A more physical way of generating a two-dimensional network, and the one usually adopted, is to generate a three-dimensional fracture network and to map traces onto a plane.

The flow solution in the two-dimensions is less complicated than in three-dimensions. The flow field is discretised by assigning one global flow node to each intersection between fractures. Linear basis functions, which are zero outside the fractures, are defined at each node J by

$$\begin{aligned} \Psi_J &= 1 \qquad \text{at node } J \\ \Psi_J &= 0 \qquad \text{at node } I \neq J, \end{aligned} \tag{14}$$

and are defined by linear interpolation along fractures between nodes. The contributions to the matrix equation ((3) or (8)) for each plane are calculated directly without needing to calculate the response functions on each plane. Boundary conditions are imposed and the resulting matrix equation is solved using the direct frontal solver.

2.5. Modelling the effect of stress on the fracture network

NAPSAC is designed to deal with complex fracture networks, and as a consequence it is only practical to incorporate relatively simple models for stress. Two types of problem can be tackled. In the first, the effect on the flow of a change in stress caused by disturbing the surrounding fracture network, for example by drilling a repository tunnel, can be calculated. In the second, the effect on the flow of a change in stress caused by imposing a pressure field can be modelled. In both cases the network is assumed initially to be in hydro-mechanical equilibrium, with the apertures of the fractures being those that apply to the in-situ network under the specified equilibrium stress field.

NAPSAC does not calculate the stress field discretely. In the first case, an analytical solution may be used to determine the changed stress field, or the results of field experiments can be used to obtain an empirical specification of the changed stress field. Thus the normal stress acting at any point on a fracture may be calculated. In the second case, the changed normal stress, σ_n, is calculated from the in-situ normal stress acting on the fracture, σ_{n0}, and the change in the pressure, P, of the fluid flowing through it by $\sigma_n = \sigma_{n0} - P$.

Having calculated the change in normal stress, a stress-aperture coupling is used to change the fracture aperture. In three-dimensions, there are two levels of discretisation available, a coarse discretisation computes the change in aperture from the values of fluid pressure and normal stress at the centre of the fracture plane, but for more detailed simulations the effective stress, and hence the change in aperture, is calculated for each finite element on the plane. In two-dimensions, the change in normal stress acting on the fracture is calculated at the centre point of each section of fracture between intersections.

There are several stress-aperture couplings available. Three that have been used in NAPSAC modelling are

$$\begin{aligned} &T = \text{constant}, \\ &T/T_0 = (\sigma_n/\sigma_{n0})^{-\alpha}, \\ &e - e_0 = \max\left(-(\sigma_n - \sigma_{n0})/RKN, e_{min}\right). \end{aligned} \tag{15}$$

In the first coupling, the transmissivity, T, of each fracture is assumed to be unchanged by the excavation of the tunnel. NAPSAC directly converts transmissivities into apertures using the parallel-plate law, as defined by equation 5, and so the apertures remain constant. The second coupling is a compliance law relating the change in fracture transmissivity, T, to changes in normal stress, σ_n, through a power law with exponent α. The value of α is obtained from laboratory tests carried out on the rock. For fractured rock α typically has values between 0.1 and 1.0. The third law relates the change in aperture, e-e_0, directly to the change in stress and the fracture normal stiffness, RKN. It is necessary to define a minimum aperture, e_{min} to ensure that all apertures remain positive. Again, the results of laboratory tests are used to determine the value of RKN.

2.6. Tracer transport

The tracer transport option in NAPSAC is designed to calculate the migration and dispersal of tracer through a discrete fracture network. Within the groundwater, it is assumed that tracer transport is dominated by advection, so that molecular diffusion can be ignored, and the major cause of dispersion is due to the existence of a number of different paths through the fracture network. It is also assumed that the fracture apertures are small enough that the tracer diffuses quickly over the cross-section.

The transport calculations are based upon a particle-tracking algorithm. The problem is split up into the calculation of single fracture responses followed by the calculation of the transport of a particle swarm through the network. For each fracture plane a representative number of pathlines between the intersections on the plane are calculated. Intersections are discretised by transport nodes and pathlines are calculated from each transport node. There are two algorithms available for calculating these pathlines.

The first algorithm provides 'exact particle tracking'. For each fracture, the flow field is discretised in terms of linear triangular finite-elements. The flow is determined by the pressure field, and since the pressure varies linearly over each triangle, the groundwater velocity,

$$\mathbf{v} = \frac{\mathbf{q}}{e} = -\frac{e^2}{12\mu}\nabla P, \tag{16}$$

is constant on each element. The pathlines are calculated on each fracture by stepping the path across the mesh, one element at a time. On reaching a fracture intersection, the path is complete. Once the pathlines from the transport nodes on each fracture plane have been calculated, the possible connections for that node are determined. A list of possible destinations, travel times, distances and relative probabilities for a particle leaving each node are calculated. In this way, a library of paths is created for every transport node in the network. The model relies on the calculation of a very accurate flow solution. If a low accuracy solution is used, then problems with local flow sinks on fractures may occur, resulting in the loss of a significant fraction of the particle swarm.

The second algorithm provides 'approximate particle tracking'. NAPSAC is able to create a database that records the net flux between all the intersections for a flow solution. This network of flux connections links the centre of every intersection on a given fracture with every other intersection centre on the fracture. A transport option has been developed that is based on this flux database in which particles are transported between intersections. This is a more robust method as it does not need a highly accurate flow solution, just a good flow balance at the network intersections and is more computationally efficient. One disadvantage is that this model cannot accurately model dispersion on a single fracture. However, where dispersion is dominated by the different paths through the network rather than dispersion on an individual fracture plane, these inaccuracies may be small and so this method is most appropriate for large networks.

The next step in the transport calculation consists of following a large swarm of particles across the network. Particles can be started on any surface of the fracture network region where there is an inflow or from any number of boreholes. Particles are tracked through the network from node to node, building up the path taken by each of the particles using the information calculated in the first step.

3. NAPSAC OUTPUT

There are many NAPSAC output options, giving a wide variety of ways of displaying and analysing the results of the first two stages. Most of the options produce graphical output via an internal graphics package which can be interfaced to a wide range of commercial graphics packages, such as postscript.

There are several options that allow detailed inspection of the fracture network. A simple plot of the planes may be requested, where the user can specify the perspective of the diagram and choose whether to view the whole network or a subset of the planes (see Figure 1). The user can specify a planar cross-section through the network and view the intersecting planes or the network can be probed by dropping line segments resembling boreholes and collecting data on the intersecting planes. If the variable aperture option is used, then the user can request a colour plot of the apertures on individual fractures.

To display the flow solution, there are many options available. The user may request a plot of the fracture network showing pressure contours or flux vectors, represented by arrows, on the fractures. Plots of pressure on individual fractures may also be requested. Pressure profiles in which graphs of pressure against distance along a line segment through the network are plotted can be drawn for steady-state and transient flow options. Plots of pressure at a point against time are also available. When locally-varying apertures are specified, histograms of fluxes across a line within a fracture can be plotted.

Several types of diagram are available for displaying the results of tracer transport simulations. To look at the results of a large swarm of particles sent through the network, graphs showing the proportion of particles leaving the network as a function of time or path length are available. A three-dimensional view of the network showing the swarm of particles at user specified times or a diagram of a boundary surface showing the arrival of particles can also be selected. Individual particle tracks can be investigated by producing a three-dimensional plot showing the tracks or by plotting graphs showing aperture against time or distance for individual particle tracks.

4. VERIFICATION AND VALIDATION

A comprehensive test library for NAPSAC is in existence which provides a suite of cases against which releases of NAPSAC can be verified. The basic NAPSAC flow model has been extensively verified. For instance, as part of the International Stripa project, a cross-verification [4] study was carried out in which it was demonstrated that NAPSAC (version 2c) could accurately perform the tasks of generating a deterministic fracture network for a given set of parameters and solve the steady-state groundwater flow equation.

Release 3.0 of NAPSAC includes the ability to simulate steady-state and transient flow through both two-dimensional and three-dimensional fracture networks. Verification of the transient flow algorithm has been performed during the development stage using test cases for which analytical solutions exist. The results of NAPSAC simulations are compared with these solutions to ensure that the NAPSAC transient flow model is an accurate representation of flow through a fracture network.

The validity of using fracture network models and the use of NAPSAC to represent flow and transport through fractured rock has been proved by a number of validation exercises. For instance, as part of the Stripa project, three phases of predictive modelling were carried out ([5-7]). NAPSAC has also been used to predict the outcome of experiments carried out at a field site in Cornwall, UK. These exercises have shown NAPSAC to be a powerful tool in the modelling of groundwater flow and tracer transport in fractured rocks.

5. ACKNOWLEDGEMENT

Funding for the production of this chapter was provided under a contract from UK Nirex Ltd. to AEA Technology.

6. REFERENCES

1   Grindrod, P., A.W. Herbert, D.L. Roberts and P.C. Robinson, NAPSAC
     Technical Document. AEA D&R 0270, AEA Technology, Harwell, 1992.
2   Herbert, A. W., NAPSAC (Release 3.0) Summary Document.
     AEA D&R 0271 Release 3.0, AEA Technology, Harwell, 1994.
3   Herbert, A. W. and G. W. Lanyon, NAPSAC (Release 3.0) Command Reference
     Manual. AEA D&R 0273 Release 3.0, AEA Technology, Harwell, 1994.
4   Schwartz, F.W., and G. Lee, Cross-verification Testing of Fracture Flow
     and Mass Transport Codes. Stripa Project Technical Report 91-29,
     SKB, Stockholm, 1991.
5   Herbert, A.W. and B.A. Splawski, Prediction of inflow into the D-Holes
     at Stripa Mine. AEA D&R 0023, AEA Technology, Harwell, 1990.
6   Herbert, A.W., J.E. Gale, G.W. Lanyon and R. MacLeod, Prediction of flow
     through fractured rock at Stripa. AEA D&R 0276, AEA Technology, Harwell, 1992.
7   Herbert, A.W. and G.W. Lanyon, Modelling tracer transport through
     fractured rock at Stripa. AEA D&R 0277, AEA, Harwell, 1992.

O. Stephansson, L. Jing and C.-F. Tsang (Editors)
Coupled Thermo-Hydro-Mechanical Processes of Fractured Media
Developments in Geotechnical Engineering, vol. 79
© 1996 Elsevier Science B.V. All rights reserved.

539

Description of the computer code FRACON

T.S. Nguyen

Atomic Energy Control Board, Wastes and Impact Division, 280 Slater Street, Ottawa, Canada,
K1P 5S9

Abstract

The finite element computer code FRACON solves the equations of thermal consolidation, derived from a generalization of Biot's theory of consolidation. Plane strain and axisymmetric problems of coupled T-H-M processes in sparsely fractured rock masses can be simulated with the code. The unfractured rock is represented by eight-noded isoparametric elements while discrete joints are represented by special six-noded joint elements. The unfractured rock is assumed to be linearly elastic. An elastoplastic constitutive relationship, with strain-softening is adopted for the joint behaviour.

1. GOVERNING EQUATIONS

The governing equations in FRACON were derived from a generalization of Biot's theory of consolidation to include thermal effects [1]:

$$\frac{\partial}{\partial x_i}\left(\kappa_{ij}\frac{\partial T}{\partial x_j}\right) + q = \rho C \frac{\partial T}{\partial t} \tag{1a}$$

$$G\frac{\partial^2 u_i}{\partial x_j \partial x_j} + (G+\lambda)\frac{\partial^2 u_j}{\partial x_i \partial x_j} + \alpha\frac{\partial p}{\partial x_i} - \beta K_D \frac{\partial T}{\partial x_i} + F_i = 0 \tag{1b}$$

$$\frac{\partial}{\partial x_i}\left(\frac{k_{ij}}{\mu}\left(\frac{\partial p}{\partial x_j} + \rho_f g_j\right)\right) - c_e\frac{\partial p}{\partial t} + \alpha\frac{\partial}{\partial t}\frac{\partial u_i}{\partial x_i} + \beta_e\frac{\partial T}{\partial t} = 0 \tag{1c}$$

where:

$$c_e = n\left(\frac{1}{K_f} - \frac{1}{K_s}\right) + \frac{\alpha}{K_s} \tag{2a}$$

$$\beta_e = (1-\alpha)\beta - (1-n)\beta_s - n\beta_f$$

(2b)

where:
the unknowns are the displacement u_i [m], temperature T [°C] and pore pressure p [Pa]

κ_{ij} is the thermal conductivity tensor of the bulk medium [W/m/°C]
p is the density of the bulk medium [kg/m^3]
C is the specific heat per unit mass of the bulk medium[J/kg/°C]

G and λ are the Lame's constants [Pa]

$\alpha = 1 - K_D/K_s$
K_D, K_s and K_f are respectively the bulk moduli of the drained material, the solid phase and the fluid phase [Pa]
F_i is the volumetric body force [N/m^3]
β, β_s and β_f are respectively the coefficient of thermal expansion of the drained material, the solid phase and the fluid phase [°C^{-1}]
n is the porosity of the medium [dimensonless]
ρ_f is the density of the fluid [kg/m^3]
μ is the viscosity of the fluid [kg/m/s]
k_{ij} is the intrinsic permeability tensor [m^2]
g_i is the ith component corresponding to the acceleration due to gravity [m/s^2]

In the above equations, the Cartesian tensor notation, with Einstein's summation convention on repeated indices is adopted. In addition, the sign convention for stress and fluid pressure is considered positive for tension fields. In developing these equations, we invoke the basic principles of continuum mechanics (namely conservation of mass, momentum and energy). These principles are universally applicable, independent of the nature of the medium being considered. Also, in order to arrive at a set of equations in which the number of unknowns equals the number of equations, the following additional assumptions are necessary:

i) Darcy's law governing pore fluid flow is considered valid. Darcy's law is applicable with reasonable accuracy to almost all types of geological materials [2] including soils and rocks, provided that the hydraulic gradients are within the laminar flow range, and above a threshold gradient within which the pore fluid is virtually immobile.

ii) A generalized principle of effective stress is adopted. Several forms of this principle exist [3-4]. We adopt the form of generalized principle of effective stress formulated by Zienkiewicz et al. [4]. In contrast to Terzaghi's [5] principle of effective stress, this generalized principle takes into consideration the compressibility of the pore fluid and the solid phases . Omission of the compressibility of the pore fluid and the solid phase could lead to an overprediction of pore pressures in competent rocks [1].

iii)Hooke's law for linear isotropic elastic behaviour is adopted for the unfractured rock. An elastoplastic model was derived for the joint behaviour. This model is based on the formulation given by Plesha [6]. It allows for strain softening due to asperity degradation and also for the reduction of the joint permeability due to gouge production [7].

iv) Only heat conduction is considered and heat convection is neglected.

2. FINITE ELEMENT FORMULATION

The governing equations are approximated by matrix equations via a standard Galerkin finite element procedure [8] . Let us consider a domain R with boundary B where the above equations apply. Considering standard finite element procedures, the domain R is discretized into N_e elements. Two types of elements are considered:

i) <u>Plane isoparametric elements</u>: This element (Figure 1) is used to represent the unfractured rock mass. Displacements within the element are interpolated as functions of the displacements at all 8 nodes, while the pore pressure and temperature are interpolated as functions of the same values at only the four corner nodes i, k, m and o. A detailed description of this element is given , for example, by Smith and Griffiths [9].

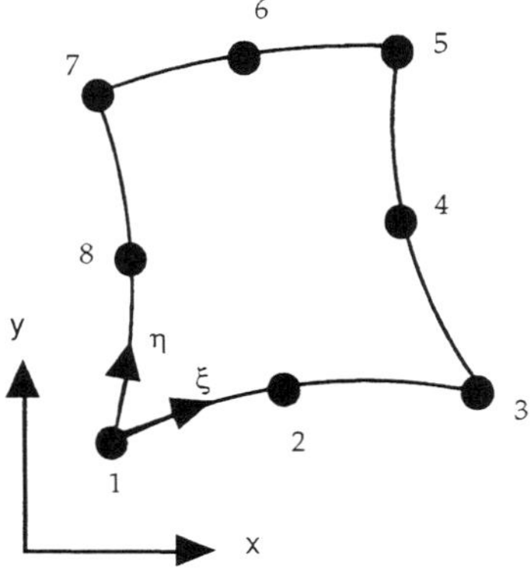

Figure 1. Eight-noded isoparametric element

ii) <u>Joint Element</u>: This element (Figure 2) is used to simulate discontinuities in the rock mass such as joints, fracture zones, and fault zones. It is a very thin element, characterized by a thickness b and length L. Nodal displacements are obtained at all six nodes (Figure 2) while nodal pore pressures and temperatures are obtained only at the corner nodes i, k, l and n. The mechanical behaviour of the element is dictated by its shear stiffness D_s , normal stiffnesses D_n and off-diagonal terms D_{sn} and D_{ns} (allowing for shear dilation). The joint's hydraulic and thermal behaviour are governed by the transverse and longitudinal permeabilities $k_{y'y'}$ and $k_{x'x'}$, and the transverse and longitudinal thermal conductivities $\kappa_{y'y'}$ and $\kappa_{x'x'}$, respectively .

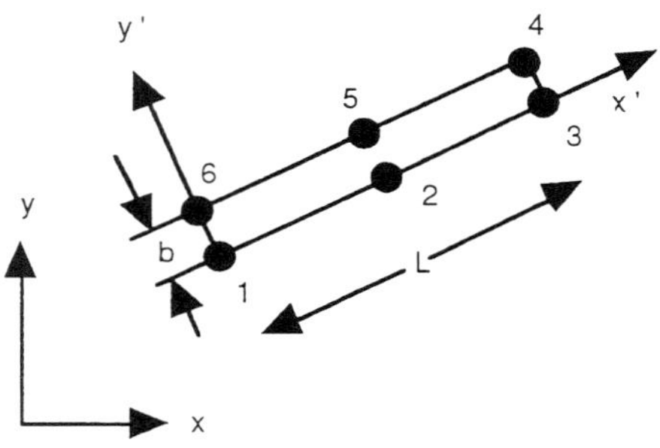

Figure 2. Joint element

With the above two types of elements being fully defined, a Galerkin procedure is applied to the differential equations (1a to c) of thermal consolidation. One then obtains the matrix equations of the form:

$$[\gamma[KH] + \rho C[CM]]\{T\}^1 = \{FH\}^\gamma + \{FQ\}^\gamma + [(\gamma-1)[KH]+\rho C/Dt[CM]]\{T\}^0$$

$$\begin{bmatrix} [K] & \alpha[CP] \\ \alpha[CP]^T & \gamma Dt[KP] - c_e[CM] \end{bmatrix} \begin{Bmatrix} \{r\}^1 \\ \{p\}^1 \end{Bmatrix} =$$

$$\begin{bmatrix} [K] & \alpha[CP] \\ \alpha[CP]^T & (1-\gamma)Dt[KP] - c_e[CM] \end{bmatrix} \begin{Bmatrix} \{r\}^0 \\ \{p\}^0 \end{Bmatrix} +$$

$$\begin{Bmatrix} \{F_a\}^1 - \{F_a\}^0 + \{F_b\}^1 - \{F_b\}^0 \\ \{F_Q\}^\gamma Dt + \{F_g\}^\gamma Dt \end{Bmatrix} + \begin{bmatrix} \beta K_D[CP] & [0] \\ [0] & -\beta_e[CM] \end{bmatrix} \begin{Bmatrix} \{T\}^1 - \{T\}^0 \\ \{T\}^1 - \{T\}^0 \end{Bmatrix}$$

$$(3)$$

where:

the unknown are the nodal displacements $\{r\}^1$, the nodal temperatures $\{T\}^1$ and the nodal pore pressures $\{p\}^1$ at the current time step,

$\{r\}^0$, $\{T\}^0$ and $\{p\}^0$ are the nodal displacements , nodal temperatures, and nodal pore pressures at the previous time step,

$\{f\}$ is the "force" vector ,$\{FQ\}$ and $\{FH\}$ are heat flux vectors, γ is a time integration constant All the other matrices , $[K]$, $[CP]$, etc.   are assembled from element matrices , which are dependent on thermal, mechanical and hydrological properties of the individual elements and the interpolation functions used.

The time integration constant γ varies between 0 and 1. The thermally induced pore pressure varies very fast. For these problems, a value of γ equal to 1 is recommended.

3. CODE VERIFICATION AND COMPARISON WITH EXPERIMENTS

For isothermal cases, analytical solutions for one-dimensional consolidation are given by Terzaghi [5], for two-dimensional plane strain or axisymmetric isothermal consolidation results are given by McNamee and Gibson [10] .To our knowledge, the only analytical solution under transient conditions and including thermal effects is that given by Booker and Savvidou [11] for the consolidation of an infinite porous medium due to a volumetric heat source. The FRACON code was verified with the above analytical solutions.

The code was compared against experimental results from a coupled T-H-M laboratory experiment in artificial rock, DECOVALEX TC6 (borehole injection tests) and a series of laboratory experiments on joints mechanical and hydraulic-mechanical behaviour [7].

4. REFERENCES

1 Selvadurai A.P.S. and Nguyen T. S.   Finite element modelling of consolidation of fractured porous media. *Proc. of the 1993 Canadian Geotechnical Conference*, 79-88 (1993).
2 Freeze R.A. and Cherry J.A. <u>Groundwater</u>. Prentice Hall (1979).
3 Rice J.R. and Cleary M.P. Some basic stress diffusion solutions for fluid-saturated elastic porous media with compressible constituents. *Rev. Geophys. Space Phys.*, **14,** 227-241 (1976).
4 Zienkiewicz O.C. , Humpheson C. and Lewis R.W. A unified approach to soil mechanics problems (including plasticity and viscoplasticity). In <u>Finite Element in Geomechanics,</u> edited by G. Gudehus, 151-178, John Wiley & Sons (1977) .
5 Terzaghi K. Die Berechnung der Durchlassigkeitsziffer des Tones aus dem Verlauf der hydrodynamischen Spannungserscheinungen'. *Ak. der Wissenschaften in Wien, Sitzungsberichte mathematisch-naturwissenschaftliche Klasse*, part IIa, 132(3/4), 125-38, (1923).
6 M. E. Plesha , ' Constitutive models for rock discontinuities with dilatancy and surface degradation', Int. Journal for Num. and Analy. Methods in Geomech.,**11**, 345-362 (1987).
7 Nguyen T.S. and Selvadurai A.P.S.   A model for coupled mechanical and hydraulic behaviour of a rock joint, submitted to the International Journal of Numerical and Analytical methods in Geomechanics.

8 Huyakorn P.S. and Pinder G.F. <u>Computational methods in subsurface flow</u> . Academic Press (1983).
9 Smith I.M. and Griffiths D.V. '<u>Programming the finite element method</u>. John Wiley & Sons (1988).
10 McNamee J. and Gibson R.E. Plane strain and axially symmetric problems of the consolidation of a semi-infinite clay stratum. *Quart. J. Mech. and Applied Math.*, **XIII**, Pt. 2 (1960).
11 Booker J.R. and Savvidou C. Consolidation around a point heat source. *Int. J. Num. and Analytical Methods in Geomechanics*, **9**, 173-184 (1985).

O. Stephansson, L. Jing and C.-F. Tsang (Editors)
Coupled Thermo-Hydro-Mechanical Processes of Fractured Media
Developments in Geotechnical Engineering, vol. 79

THAMES

Y. Ohnishi[a] and A. Kobayashi[b]

[a]School of Civil Engineering, Kyoto University, Yoshida-honmachi, Sakyo-ku, Kyoto. Japan

[b]Department of Agricultural Engineering, Iwate University, 3-18-1, Ueda, Morioka, Iwate, Japan

1. INTRODUCTION

THAMES (Thermal, Hydraulic And MEchanical System analysis) is a three dimensional finite element code designed to analyze fully coupled thermal, hydraulic and mechanical behaviors of a saturated-unsaturated geologic medium. Such an analysis is essential in performance of assessment for deep geologic isolation of nuclear wastes.

THAMES code takes into account: (a) heat transferred by conduction and convection; (b) fluid flow in the saturated-unsaturated medium; (c) buoyancy induced by temperature raise; (d) thermal stress; (e) mechanical work done by temperature change; and (f) coupled deformation.

Major assumptions used in THAMES are; (a) isotropic and poro-elasticity of the medium; (b) validity of Darcy's law; (c) no change of water phase; and (d) validity of Fourier's law.

The mathematical formulation for the model utilizes the Biot's theory with Duhamel-Neuman form of Hooke's law and energy balance equation. The governing equations were derived with the fully coupled thermal, hydraulic and mechanical relationships. The three coupled equations are solved simultaneously.

Solutions to steady-state and transient problems may be obtained. Initial condition may vary with spatial position. Source and sink strength of fluid flow and thermal flow, and fixed displacement condition may vary with time.

Nonlinear parameter of rocks; heat conductivity, specific heat and thermal expansivity may be used in calculation. Nonlinear parameter of water; heat conductivity, specific heat and dynamic viscosity may be also used.

In order to simulate fractured media, a joint element is introduced for a discrete analysis. Major discontinuity is represented by joint elements. A large numbers of discontinuities are simulated as the equivalent continuum by using crack tensor proposed by Oda[1]. Stress dependency of permeability of fractures may be considered according to the equation proposed by Iwai[2]. For the analysis of soils, permeability as a function of void ratio may also be used.

THAMES does have a number of limitation which must be recognized. Perhaps most serious limitation is the fact that presently the code accounts for only mechanically elastic medium. Considerable investigation is necessary to introduce the elasto-plastic theory under the condition which allows the temperature change. Other limitation of the code include its inability to simulate phase change of water induced by temperature change.

2. DESCRIPTION OF "THAMES" CODE

2.1. Assumptions
The governing equations are derived under the following assumptions:
   1) The continuum medium is isotropic and poro-elastic.
   2) Darcy's law is valid for flow of water in the saturated-unsaturated medium.
   3) Energy flow occurs only in the solid and liquid phase. Phase change of water between liquid and gas are not considered.
   4) Heat transferred among three phases (solid, liquid and gas) are neglected.
   5) Fourier's law holds for heat flux.
   6) Water density varies depending upon the temperature and pressure of water.
   7) Deformation based on the infinitesimal strain is used.

2.2. Equilibrium Equation
Equation of motion for the medium in a static case is known as an equilibrium equation. It is derived, based on the fundamental continuum mechanics theory.

Temperature effects can be implemented in the constitutive law for a solid medium. For an isotropic linear elastic material, Duhamel-Neuman relationship can be used to derive constitutive law.

2.3. Energy Conservation Law
The ground in the earth consists of the materials with three phases; solid, liquid and gas. It is not so easy to analyze the behavior of heat transfer since the thermal process is different in each phase. It was assumed that the pore in the porous medium is filled by the material in liquid phase. It means that ground water does not change its phase from liquid to gas or vise versa. Faust and Mercer [3] proposed that the movement of water through porous media is sufficiently slow and the surface areas of all phases are sufficiently large that it is reasonable to assume that local thermal equilibrium among phases is achieved instantaneously.

With the above assumptions, the energy conservation law was derived without the effects of viscous dissipation for groundwater, based upon the one proposed by Bear and Corapcioglu [4]. It takes into account the existence of unsaturated zone.

The energy conservation law takes into account the stress-deformation and groundwater flow, the dependency of energy and the energy change due to heat convection. It also express the energy change due to heat conduction, pore water pressure change and reversible energy change caused by solid deformation.

3. NUMERICAL TECHNIQUES

3.1. General
The Galerkin type finite element technique is employed to formulate the finite element discretization. Linear quadrilateral isoparametric elements are used to represent the behavior of total head h and temperature T. Quadratic quadrilateral isoparametric elements are used to express the displacement . For joint elements, linear quadrilateral ones for h and T and 6 nodes quadrilateral ones for displacement are employed in the code. In order to integrate time derivatives, time weighting factor is introduced and so any type of finite difference scheme may be applied.

The solution procedure is repeated until the maximum difference of nodal values between successive iterations become smaller than a prescribed allowance for the particular time step. At this stage, a solved matrix is updated using the information of the previous iteration step. The computer program then proceeds to the next iteration step. This solution procedure is performed for each time step until the final time level is reached.

3.2. Discontinuities

To deal with fractured media fractures as well as porous matrix blocks must be taken into account. Both the discrete fractures approach and equivalent continuum approach are applied in this code. Major discontinuities whose geometry are known are represented by joint elements. Minor and a large numbers of fractures in rock masses are converted into the equivalent continuum model by using the crack tensor of Oda.

The material properties of fractures have been the subject of detailed laboratory investigations. In this code, the stress-shear strain relation is assumed to be elastic-perfectly plastic. Failure is specified according to the linearized Mohr-Coulomb yield criterion. Fracture is assumed to have no strength in tension. The permeability of fracture may be specified according to the cubic law or the function of void ratio. The heat conductivity is calculated by assuming that the heat is conducted through liquid phase. The heat conductivity is obtained from the empirical formula when the permeability is the function of void ratio.

3.3. Nonlinearity

To solve the governing equation with the finite element method, the following parameters are needed in this code.

a) Young's Modulus	b) Poisson's ratio
c) Thermal expansivity of solid	d) Thermal expansivity of fluid
e) Density of solid	f) Permeability
g) Initial porosity	h) Compressibility of fluid
i) Reference density of fluid and water contents	j) Relation between permeability
k) Heat capacity of solid	l) Heat capacity of fluid
m) Thermal conductivity of solid	n) Thermal conductivity of fluid

Some of the above parameters of water and granite rock mass show strong non-linearity. In this code, thermal conductivity of water and granite rocks, heat capacity of granite rocks and thermal expansivity of rocks may be changed depending on the temperature change. Viscosity of water and heat capacity of water may be changed depending on the temperature change.

The dependency of permeability of granite rock mass including many fractures on the stress change have been investigated by many researchers. The stress in the rock mass which change the permeability in it is the results of not only loading forces exerted on the medium but thermal stress induced by the temperature change. In this code, Iwai's experimental equation are introduced as the equation appearing the intrinsic permeability change of granite rock mass induced by the stress change. Therefore, the use of this dependency of permeability of granite rock mass on the stress change is limited to the fractured medium in which fluid flow is mainly influenced by the fracture in it because that the permeability in the matrix block mass in it is not considered in this equation.

3.4. Solution Procedure

Once the coefficient matrix and force vector are formed, the equation can be solved for the nodal values of the displacements, water pressure and temperature at any time value using the nodal values at the previous time t. It is important to recognize that the response at time of the system is purely elastic because no dissipation of the fluid and no transfer of heat has taken place. To capture this initial response, the use of very small time t is recommended for the first time step. For the subsequent time steps larger values can be used.

The flow chart for the calculation process of THAMES is shown in the figure to explain the structure of the computer code.

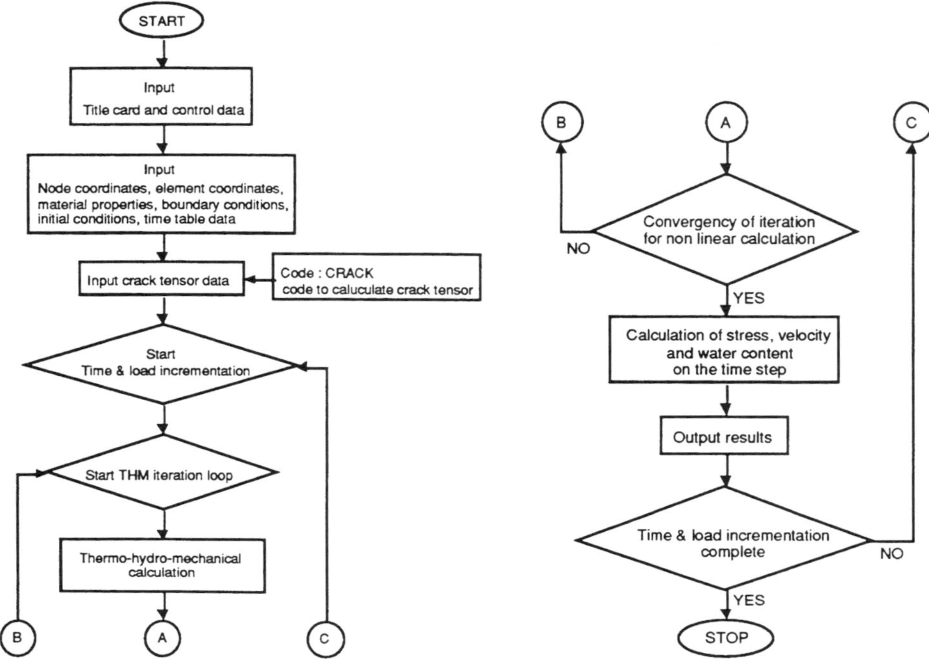

Flow chart of THAMES

4. VERIFICATION

Function of this code has to be verified by the comparison with analytical or experimental solutions of fully coupled thermal, hydraulic and mechanical behavior. To test the numerical solution techniques and their coding and to demonstrate various capabilities of the code, five test problems are provided.

The problems considered in the verification are:
a) One-dimensional consolidation problem.
   Analytical solution of this problem can be found in Terzerghi[5].
b) Stress-strain problem of thick walled circular cylinder subjected to a uniform internal and external pressure.
   Analytical solution of this problem can be found in Timoshenko and Goodier [6].
c) Thermal stress problem of thick walled circular cylinder subjected to a uniform temperature gradient.
   Analytical solution of this problem can be found in Boley and Weiner [7].
d) One-dimensional heat conduction problem.
   Analytical solution of this problem can be found in Smith [8].
e) Coupled thermal and hydraulic problem. No analytical solution can be found in literature, experimental one, however, can be found in Sato [9]. He experimented hot water seepage in saturated sand sample.

5. REFERENCES

1 Oda, M. (1986) :"An Equivalent Model for Coupled Stress and Fluid Flow Analysis in Jointed Rock Masses", Water Res. Research, 22, pp.1945-1956
2 Iwai, K. (1976): "Fundamental Studies of Fluid Flow through a Single Fracture", Ph.D. dissertation, Univ. of Calif., Berkeley
3 Faust, C.R. and J.W. Mercer (1979):"Geothermal Reservoir Simulation 1. Mathematical Models for Liquid- and Vapor-dominated Hydrothermal System, Water Res. Research, Vol.15, pp.23-30
4 Bear, J and M.Y. Carapcioglu (1981):"A Mathematical Model for Consolidation in a Thermoelastic Aquifer due to Hot Water Injection or Pumping", Water Res. Research, Vol.17, No.3, pp.723-736
5 Terzaghi, K. (1925):"Settlement and Consolidation of Clay", Engineering News Records, 26
6 Timoshenko, S. and J.N. Goodier (1951):"Theory of Elasticity, McGraw- Hill, NY
7 Boley, B.A. and J.H. Weiner (1960):"Theory of Thermal Stresses", Wiley, pp.288-291
8 Smith, G.D. (1965):"Numerical Solution of Partial Differential Equation", The Clarendon Press, Oxford
9 Sato, K. (1982):"Experimental Determination of Transfer Parameters of Heat Flow through Porous Media by means of A New-designed Apparatus in Laboratory", Transaction of JSCE, No.320, pp.57-65 (in Japanese)

O. Stephansson, L. Jing and C.-F. Tsang (Editors)
Coupled Thermo-Hydro-Mechanical Processes of Fractured Media
Developments in Geotechnical Engineering, vol. 79

551

ROCMAS Simulator; A Thermohydromechanical Computer Code

Jahan Noorishad and Chin-Fu Tsang

Earth Sciences Division, Ernest Orlando Lawrence Berkeley National Laboratory, One Cyclotron Road, Berkeley, California 94720

Abstract

A brief description of the capabilities of the THM code ROCMAS along with the code's flow chart is presented.

1. INTRODUCTION

This simulator is designed to address the coupled occurrences of heat flow, fluid flow, and deformation in geologic media. The theoretical foundation of the code is explained in Chapter 3.

2. COMPUTER IMPLEMENTATION

The code uses an incremental Galerkin finite element algorithm that is also explained in Chapter 3. A general Newton-Raphson (N-R)scheme, with options for mixed and modified methods, is used for linearization purposes within a sub incremental context. In addition, a direct iteration approach is also available. Time integration is performed using a predictor corrector scheme. The system of equations are solved by LU decomposition and back substitution method. Isoparmetric quadrilateral and cubical elements are employed for continuum representation, while isoparmetric line and plane are deployed for joints, in 2D and 3D applications respectively.

3. MATERIAL MODELS

The library of material models include
Mechanical models:
- Linear elastic solid
- Associated and non-associated strain softening/hardening elastoplastic continuum
- Sandler/DeMaggio cap plasticity
- Oriented plasticity
- Compressible, dilating and strain softening elastoplastic joint
- No tension continuum

Hydrologic
- Linear continuum
- Nonlinear continuum
- Nonlinear joint

Thermal
- Linear continuum

4. CODE VERIFICATION

Because of the complexity of the coupled THM phenomena, there are only limited number of analytic solutions available for verification purposes. However, by verifying various single phenomenon performances of the code on one hand, code to code verifications of coupled phenomena, on the other, and by engineering judgmental of results of some of the code solutions, the soundness and robustness of the ROCMAS code has been verified. In the following accounts some of these efforts are provided.
Analytical verifications:
Hydrologic:
- Radial flow to a well in an infinite confined aquifer (Theis, 1935)

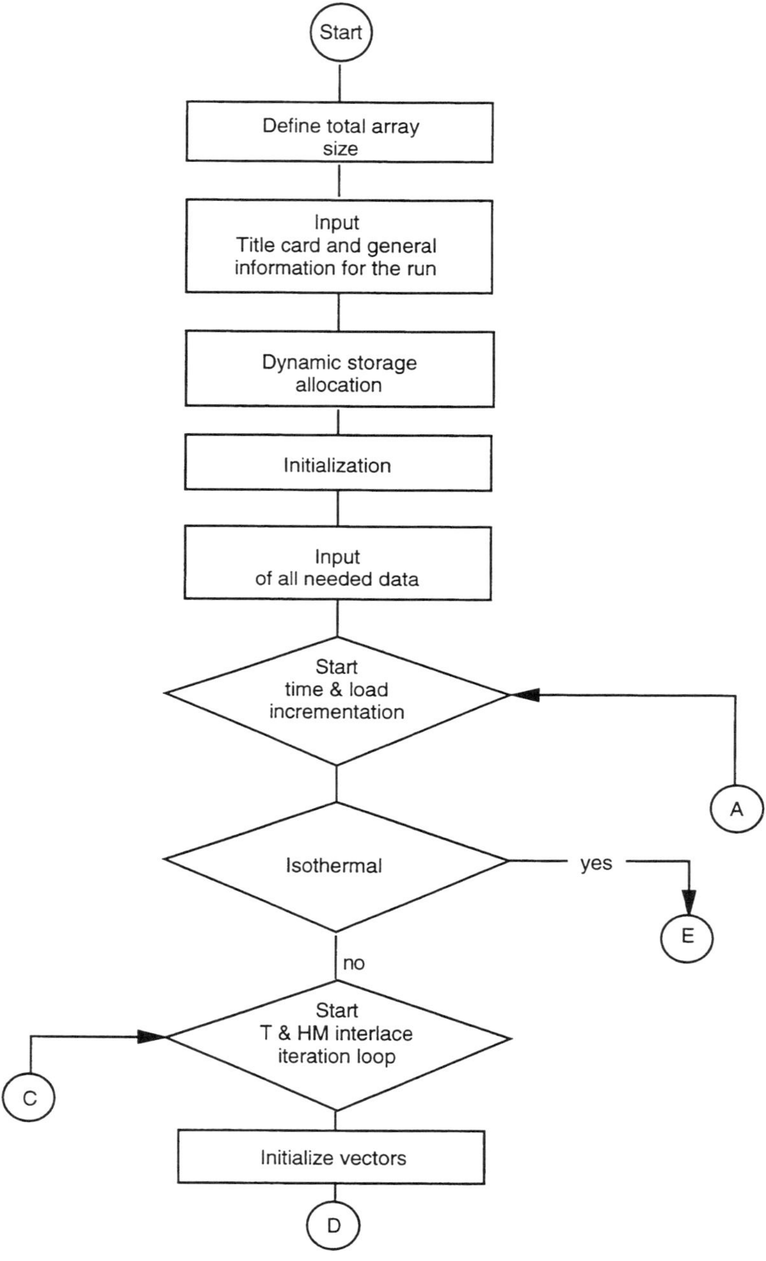

Figure 1a. ROCMAS CODE FLOW CHART

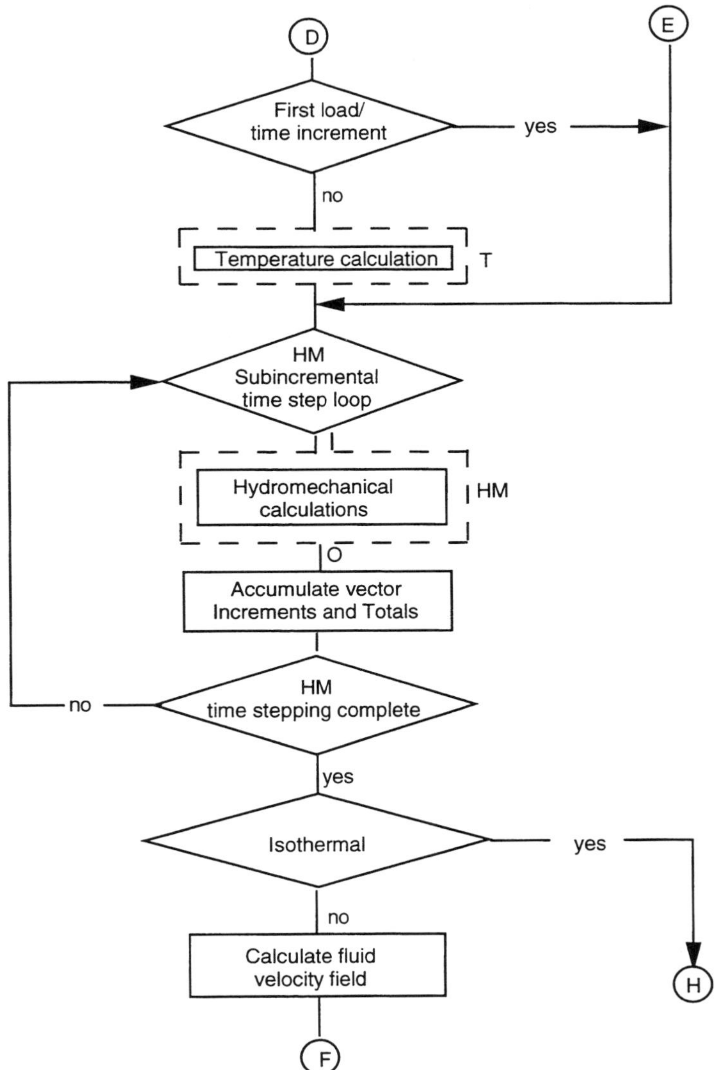

Figure 1b. ROCMAS CODE FLOW CHART

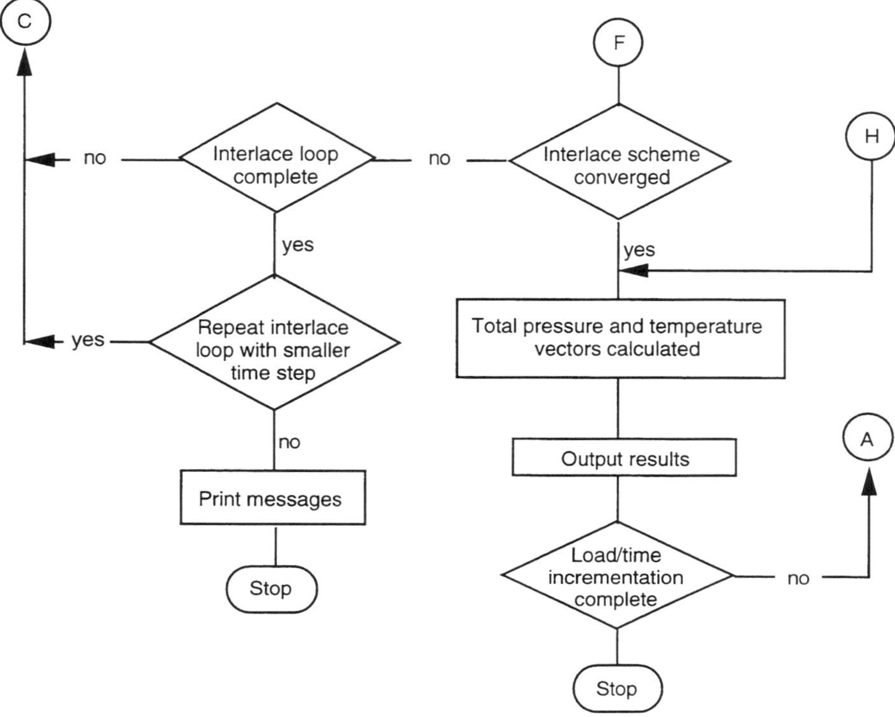

Figure 1c. ROCMAS CODE FLOW CHART

- A zero radius well in a finite vertical fracture in a reservoir (Raghavan et al., 1936)
- Radial flow from a horizontal fracture to a well in a reservoir (Gringarten and Ramy, 1974)

Mechanical (M):
- Infinite plate with a circular hole (Kirsch solution), (Hoek and Brown, 1980)
- Uniformly pressurized thick wall cylinder (Hoek and Brown, 1980)
- Circular tunnel in an elastoplastic rock (Hoek and Brown, 1980)

Thermal (T):
- Finite line source in an infinite medium (Carslaw and Jeager, 1962)
- Semi-infinite space subject to constant temperature boundary (Carslaw and Jeager, 1962)

Coupled Hydroelastic (HM)
- Radial flow from a horizontal fracture to a well in reservoir (Wijesinghe, 1986)

Numerical verifications (various levels of coupling):

(M)	-	Internally pressurized tunnel in strain-softening rocks (Yurtzine et al., 1982)
(THM)	-	Thermohydroelastic soil consolidation (Aboustit et al., 1982)
(HM)	-	Tunnel excavation in fractured rocks (Noorishad et al., 1992)
(M)	-	Fracture constitutive modelings (Goodman, 1975)
(HT)	-	Cold water injection and hot water production (Lippmann et al., 1977)
(HM)	-	Tunnel excavation in elastoplastic rocks (Noorishad and Tsang, 1994)
(HM)	-	Tunnel excavation in ubiquitously jointed rocks (Noorishad and Tsang, 1994)
(HM)	-	Non-associated drained and undrained behavior of an internally pressurized thick walled cylinder (Small et al., 1979)
(HM)	-	Non-associated drained and undrained behavior of a uniformly strip loaded soil foundation (Small et al., 1979)
(TH)	-	Solution of a number of convection dominated problems (Noorishad et al., 1994)

5. FLOW CHART

A general logical flow chart of the ROCMAS code is presented in Figures 1a, b, c, and Figures 2a and b. This chart explains the interlaced scheme operation of the code. Each round of calculation of heat flow (T) for a thermal time step is followed by a number of smaller steps (with their sum equal to the thermal step) of HM calculations. There is an option for iterating between HM and T calculations. Because of complexity of the HM part another general logical flow chart for HM is presented. Due to the simplicity of thermal finite element approach, thermal flow chart is not offered. The treatment of convection part in this code follows the explanation offered in chapter three.

6. REFERENCES

Carslaw, H.S. and Jaeger, J.C., <u>Conduction of Heat in Solids,</u> Oxford Press, London, 1962.

Carter, J.P., Booker, J.R., and Small, J.C. "The analysis of finite elasto-plastic consolidation," *Int. J. for Numer. and Analy. Meth. in Geomech.* 3, 107–129 (1979).

Hoek, E. and Brown, E.T. (1980). <u>Underground Excavation in Rock</u>, The Institution of Mining and Metallurgy, London.

Lippmann, M.J., Tsang, C.F., and Witherspoon, P.A.: "Analysis of the Response of Geothermal Reservoirs Under Injecti on and Production Procedures," paper SPE 6537 presented at the 1977 SPE California Regional Meeting, Bakersfield, April 13–15.

Noorishad, J. , Tsang, C.F., Perrochet, P., and Musy, A. "A perspective on the numerical solution of convection-dominated transport problems: A price to pay for the easy way out," *Water Resources Research,* 28(2), 551–561, 1992.

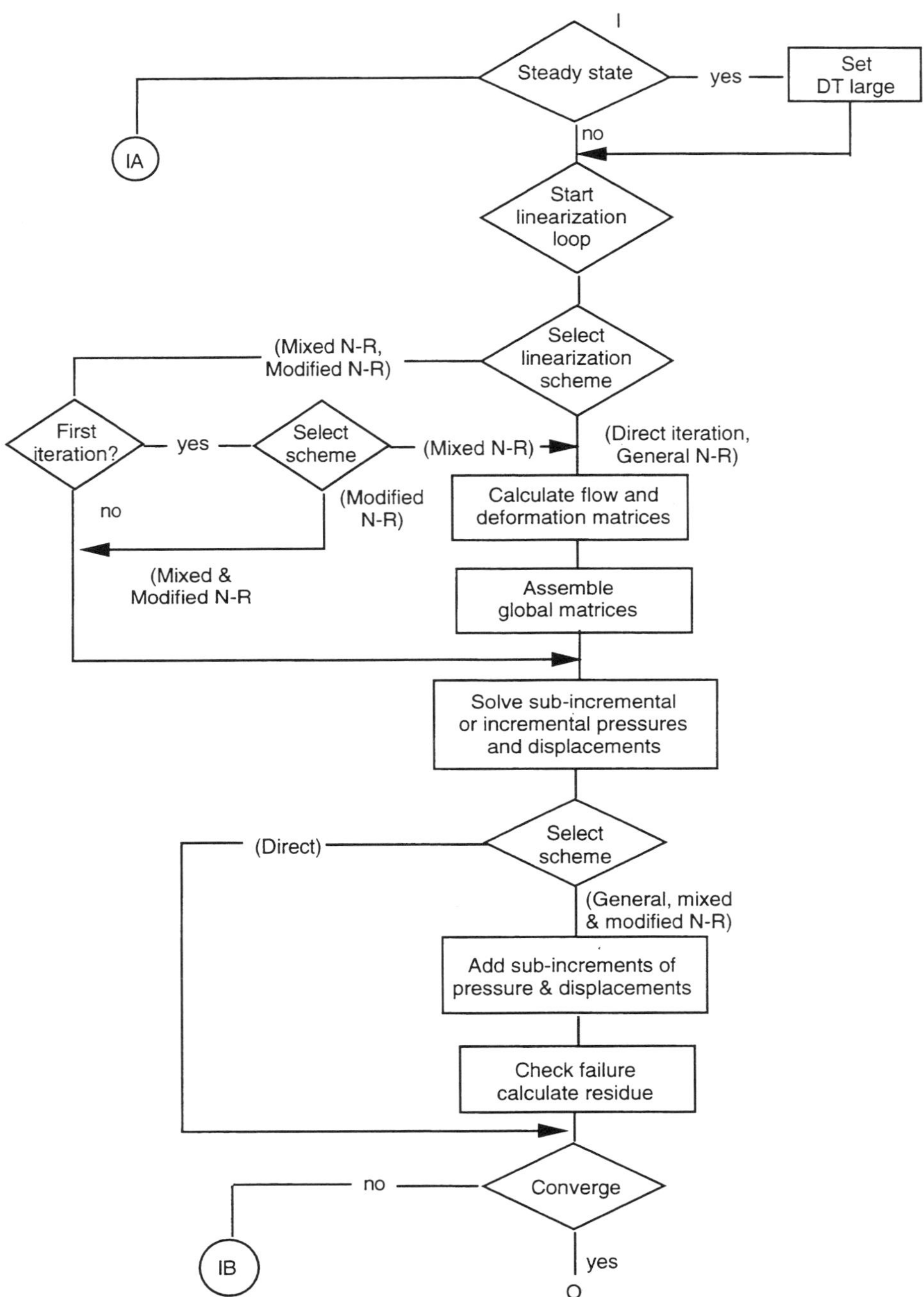

Figure 2a. ROCMAS SUB-FLOW CHART HM

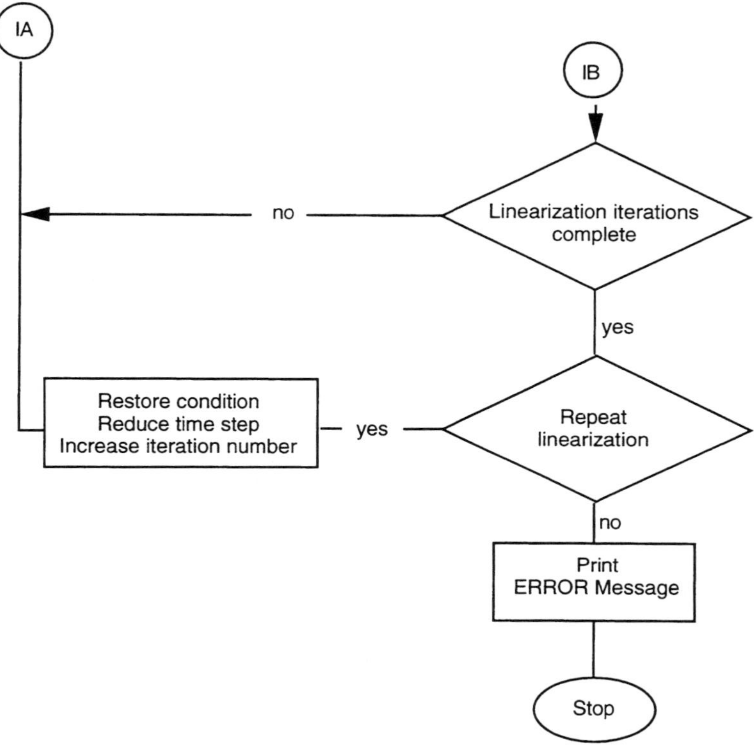

Figure 2b. ROCMAS SUB-FLOW CHART HM

Noorishad, J. and Tsang, C.F., Hydromechanical Modeling of the SCV Block Tunneling -
- use of ubiquitous joint material models, Swedish Nuclear Power Inspectorate,
SKI#TR Report, Feb. 1994.
Noorishad, J. and Tsang, C.F., Hydromechanical Modeling of the SCV Block Testing,
Swedish Nuclear Power Inspectorate, *SKI#TR Report,* June 1994.
Theis, C.V., "The relationship between the lowering of piezometric surface and the rate
and duration of discharge using ground-water storage," <u>Trans. AGU</u>, 519, 1935.
Wijesinghe, M.A., "A similarity solution for coupled deformation and fluid flow in
discrete fractures. *Second Int. Radioactive Waste Management Conf.*, Winnipeg,
Canada (1986).

O. Stephansson, L. Jing and C.-F. Tsang (Editors)
Coupled Thermo-Hydro-Mechanical Processes of Fractured Media
Developments in Geotechnical Engineering, vol. 79
© 1996 Elsevier Science B.V. All rights reserved.

Short description of CASTEM 2000 and TRIO-EF

A. Millard

CEA/DMT/SEMT, CEN Saclay
91191 GIF/YVETTE Cedex FRANCE

1. GENERAL PRESENTATION OF CASTEM 2000 AND TRIO-EF

Since the beginning of simulations of mechanical systems with computers, many general purpose finite element codes have been developed. They have become more and more complex and unfortunately more and more difficult to use, to develop and to maintain. At the same time, the problems to be solved are becoming more and more coupled, as occurs in the case of DECOVALEX (coupling between hydraulics, mechanics, and heat transfer).

These general considerations have led to the development of a new generation of structural mechanics and thermohydraulics codes, CASTEM 2000 and TRIO-EF, at CEA/DMT [1]. The starting point was to consider that a given PDE-type problem can be solved by a sequence of elementary operations. Therefore, the computer program consists in a collection of such operators. The generic syntax of an operation is as follows:

Result = OPERATOR (Input_data);

Any operator can work provided that the relevant information ("Input_data") is supplied from the database. When an operation is performed, the information produced ("result") is added to the database. Subsequent access to this information is achieved through a name, given by the user.

There is neither priority, nor dependency between the operations. In order to solve his or her own problem, the user can decide how to organize the sequence of operations. Some specialized operations like "do loops" and "tests", as well as algebraic calculations operations, are available as an aid for writing algorithms. Moreover, the user can write some meta-operations, called "procedures", which can be used as standard CASTEM 2000 operators [2]. It must be emphasized that all this is done in user's language, without any need for compilation of subroutines.

There are now more than 300 elementary operators in CASTEM 2000, which are used for pre-processing tasks, heat transfer analyses, mechanical and structural analyses, magnetostatic analyses, and post processing tasks.

CASTEM 2000 is what is usually called a general purpose, object oriented, finite element program. It can be used for static or dynamic analysis, linear or non linear, thermal or mechanical problems, steady state or transient evolutions [3]. A large variety of unidimensional (truss and beam elements), two dimensional and three dimensional solid or plate and shell elements can be used. Various material models are available for linear as well as non linear studies on steel, concrete, soils and rocks. Geometrical non linearities and stability problems can be analysed, as well as fluid-structure interaction problems.

The program is widely used in France and in many universities and research centers abroad. It has been intensively validated by comparison to analytical solutions, experimental results, international comparative benchmarks, etc.

2. THM CALCULATIONS

As far as THM calculations are concerned, one major interest of CASTEM 2000 is to make available thermal and hydromechanical analysis utilities (operators) in the same computer program, thus enabling coupled calculations when necessary.

The basic partial differential equations, which were solved for the DECOVALEX benchmark tests are briefly summarized hereafter.

Notations:

B^{ij} : Biot's tensor,
M : Biot's modulus,
D^{ijkl} : Hooke's law for the solid skeleton, i.e. drained mixture,
k^{ij} : Intrinsic permeability tensor,
η : Shear viscosity of the fluid,
ρ_f : fluid density,
σ^{ij} : stress tensor,
f^j : body forces,
ξ : fluid mass production,
w : Darcy's velocity,
p : pressure,
ε_{ij} : strain tensor,
u_i : displacement.

Momentum conservation:

$$\partial_i \, \sigma^{ij} + f^j = 0 \tag{1}$$

Fluid mass conservation:

$$\frac{\partial \xi}{\partial t} + \operatorname{div} w = 0 \tag{2}$$

State laws (here in the particular case of poroelasticity):

$$\begin{cases} \sigma^{ij} = D^{ijkl}\,\varepsilon_{kl} - B^{ij}\,.\,p \\ p = M\left[-B^{ij}\,\varepsilon_{ij} + \xi\right] \end{cases}$$

$$(3)$$
$$(4)$$

Darcy's law:

$$w^i = -\frac{k^{ij}}{\eta}\left[\frac{\partial p}{\partial x_j} + \rho_f\,g_j\right]$$

$$(5)$$

Kinematic equations:

$$\varepsilon_{ij} = \frac{1}{2}\left(\frac{\partial u_i}{\partial x_j} + \frac{\partial u_j}{\partial x_i}\right)$$

$$(6)$$

For finite element discretization, the displacement field u and the pressure p are chosen as primal variables. The six node triangle and the eight node quadrangle have been selected. The following shape functions are assumed for the unknowns:

$$u(x,y) = N(x,y)\,.\,U$$

$$(7)$$

$$p(x,y) = \overline{N}(x,y)\,.\,P$$

$$(8)$$

where U and P are the vectors of nodal unknown. The N functions are chosen quadratic, whereas the $\overline{N}$ functions are chosen linear.

For the poro-elastic mixture, the variational principle used is the principle of virtual work:

$$\int_\Omega \delta\varepsilon^t\,.\,\sigma\,d\Omega = \int_\Omega f^t\,.\,\delta u\,d\Omega + \int_{\partial\Omega_1}\phi^t\,.\,\delta u\,dS$$

$$(9)$$

where the ϕ functions are surface loads prescribed on a part $\partial\Omega_1$ of the boundary.

For the fluid, on the other hand, we use a classical Galerkin approach leading to the following functional:

$$\int_\Omega \nabla q^t\,.\,\frac{k}{\eta}\left(\nabla p - \rho_f\,g\right)d\Omega + \int_\Omega q\left(\frac{1}{M}\frac{\partial p}{\partial t} + B\frac{\partial\varepsilon}{\partial t}\right)d\Omega + \int_{\partial\Omega_2}q\,\phi_f\,.\,dS = 0$$

$$(10)$$

where ϕ_f stands for a prescribed fluid flux through a part $\partial\Omega_2$ of the boundary.

Inserting the above mentioned shape functions, the following time differential system is then obtained:

$$KU - LP = F$$

$$(11)$$

$$L^t \frac{\partial U}{\partial t} + S \frac{\partial P}{\partial t} + HP = F_f \tag{12}$$

For the time integration of equation (12), the classical θ method is used, that is:

$$L^t \left[\frac{U_{t+\Delta t} - U_t}{\Delta t} \right] + S \left[\frac{P_{t+\Delta t} - P_t}{\Delta t} \right] + H \left[(1-\theta) P_t + \theta P_{t+\Delta t} \right] = (1-\theta) F_{f_t} + \theta F_{f_{t+\Delta t}} \tag{13}$$

Finally, using the incremental notation ΔU, ΔP, the discretized system can be written in the form:

$$\left. \begin{aligned} K \cdot \Delta U - L \cdot \Delta P &= \Delta F \\ L^t \cdot \Delta U + (S + \theta \cdot \Delta t \cdot H) \cdot \Delta P &= \Delta t \left[(1-\theta) F_{f_t} + \theta F_{f_{t+\Delta t}} \right] - \Delta t \cdot HP_t \end{aligned} \right\} \tag{14}$$

These equations are solved for each time-step in CASTEM 2000. For the DECOVALEX calculations, a value of $\theta = 0.55$ has been used.

In the case of an elastoplastic behaviour, the strains ε are partitioned on one hand in elastic strains, and on the other hand in plastic strains:

$$\varepsilon = \varepsilon^e + \varepsilon^p \tag{15}$$

The elastic strains are known by the compliances matrix C:

$$\varepsilon^e = C\sigma \tag{16}$$

The plastic strains comply with the plastic flow rule:

$$\dot{\varepsilon}^p = \frac{\partial f}{\partial \sigma} \dot{\lambda} \tag{17}$$

where f is the yield criterion and λ the plastic multiplier.

The system of equilibrium non linear algebraic equations is solved through a Newton-Raphson's iterative technique.

$$[R_T] \{q^{(n)}\} \cdot \delta q^{(n+1)} = F_{ext} - \int_{\Omega^t} B \, \sigma\big(q^{(n)}\big) \, d\Omega \tag{18}$$

$$q^{(n+1)} = q^{(n)} + \delta q^{(n+1)} \tag{19}$$

where n is the iteration index and $[R_T]$ the tangent stiffness matrix.

Thermics

In case of temperature variations, we assume that the temperature field can be determined independently of the hydromechanical response of the rock-mass. Therefore, a thermal analysis is first performed and the history of the temperature field is used as input for the coupled hydromechanical analysis.

The heat equation derived from the conservation of energy and Fourier's law is as follows:

$$\rho C \frac{\partial T}{\partial t} = \lambda \, \Delta t + q \tag{20}$$

where:

λ is the thermal conductivity,
C   the specific heat capacity,
ρ the density,
q   the volumetric heat source.

Using a Galerkin method, the following set of differential equations is obtained:

$$M \frac{dT}{dt} + DT = Q \tag{21}$$

where:

T   is the set of unknown nodal temperatures,
M   is the heat capacity matrix,
D   is the heat diffusion matrix,
Q   is the vector of nodal heat source values.

A previously, a θ method is used for the time integration, leading to the set of algebraic equations:

$$M \cdot \left[\frac{T_{t+\Delta t} - T_t}{\Delta t} \right] + D \left[(1-\theta) T_t + T_{t+\Delta t} \right] = (1-\theta) Q_t + \theta Q_{t+\Delta t} \tag{22}$$

And finally, the updated temperatures $T_{t+\Delta t}$ are obtained through a direct inversion of the matrix resulting from the above equations.

Hydraulics:

The pure hydraulics equation (12) are solved with TRIO-EF. First we solve:

$$\text{div} \left(k \cdot \text{grad } h \right) = s = -\frac{\partial \theta}{\partial t} \tag{23}$$

where θ is the porosity.

The discretization leads to a matrix system:

$$L \cdot H = S \tag{24}$$

where:
L   is the permeability matrix (similar to diffusion matrix),
S   the vector of source terms,
H   the vector of hydraulic heads.

The velocity field is defined at each node of the mesh. It is calculated in two steps :
- First, we calculate:

$$M \cdot V = {}^{t}G \cdot H \tag{25}$$

with the following notations:
M   is the mass matrix,
V   the vector of velocities,
${}^{t}G$ the matrix of gradient operator,
$[G]$ the matrix of divergence operator.

The so obtained field does not generally satisfy the conservation equation : $\mathrm{div}\ \vec{V} + \dfrac{\partial \theta}{\partial t} = 0$

(particularly if there is a strong permeability variation).
- For the second step, we use a penalization method.
The problem is to find V^* such as:

$$M\left(V^* - V\right) + {}^{t}G\ \lambda = 0 \quad \text{and} \quad G\ V^* - S = 0, \tag{26}$$

where λ are Lagrange's multipliers. Let's set:

$$\lambda = w \cdot \left(G \cdot \left\{v^*\right\} - \left\{s\right\}\right) \tag{27}$$

Thus we get:
$$M\left(V^* - V\right) + w \cdot {}^{t}G \cdot \left(G\ V^* - S\right) = 0 \tag{28}$$

A choice of w large enough ($10^6 < w < 10^{10}$) ensures correctly the constraint (26).

3.   REFERENCES

1   P. Verpeaux, A. Millard, T. Charras, "Castem 2000, une approche moderne du calcul des structures", in Calcul des structures et intelligence artificielle, Ed. Pluralis, 1989.
2   L. Ebersolt, A. Combescure, A. Millard, P. Verpeaux, "Non linear algorithms solved with the help of the Gibiane macro language", Prof. of SMIRT 9 Conference, Lausanne, Ed. Balkema, 1987.
3   P. Verpeaux, A. Millard, T. Charras, A. Combescure, "A modern approach of large computer codes for structural analysis", Proc. of SMIRT 10 Conference, Los Angeles, Ed. Hadjian, 1989.

ABAQUS

L. Börgesson

Clay Technology AB, Ideon, S-223 70 Lund, Sweden

1. INTRODUCTION

ABAQUS/Standard is a general purpose finite element program designed specifically for advanced structural and heat transfer analysis. It is designed for both nonlinear and linear stress analysis of both very small and extremely large structures.

The element library provides a complete geometric modeling capability. Solids in one, two, and three dimensions as well as shells, beams, pipes and pipe bends with deforming sections, cables etc. can be modelled using first, second or third order interpolation. The multilevel substructuring capability is another useful facility.

The material library contains several different constitutive models, e.g. linear and non-linear elasticity, rubber, plasticity, concrete, sand, soils, acoustic etc.

ABAQUS has a built-in automatic and adaptive choice of time incrementation. This approach provides uniform accuracy throughout the solution history and is in most cases significantly more efficient and practical than user controlled fixed time incrementation.

The built-in postprocessor utilizes state-of-the-art interactive color graphics to present the results as displaced shape, color fringe plots, x-y plots etc.

A wide variety of problems may be addressed with the modeling tools available within the program. These are described below.

ABAQUS is thoroughly documented with the following manual set:

- **User's Manual**
- **Theory Manual**
- **Example Problems Manual**
- **Verification Manual**

The latest release is version 5.4

Besides ABAQUS/Standard the following modules are included in the ABAQUS software suite:

ABAQUS/Explicit

It provides a complete explicit transient dynamics capability in ABAQUS that is suited for defence applications, vehicle crash simulation, nuclear safety, high-speed metal forming etc. The module is fully vectorized, to make best use of supercomputers.

ABAQUS/Aqua

This is a set of wave loading, drag and buoyancy calculation capabilities that may be addressed to the ABAQUS/Standard module to provide modelling of offshore piping or cable systems.

ABAQUS/USA

Interface that may be added to ABAQUS/Standard to allow Lockheed's Underwater Shock Analysis to be used with ABAQUS.

ABAQUS/Pre

Interactive pre-processor for generation of complicated large 3D-geometries, properties, loads and boundary conditions for ABAQUS.

ABAQUS is written, supported and steadily upgraded by HKS (Hibbitt, Karlsson and Sorensen) with headquarters address:

HKS Inc.
1808 Main Street
Pawntucket, RI 02860-4847
USA

2. GEOMETRY AND ELEMENTS

Structures and continua can be modeled. One-, two- and three-dimensional continuum elements are provided, as well as beams, membranes, shells and infinite elements etc. ABAQUS/Standard is a modular code: any combination of elements, each with any appropriate material model, can be used in the same analysis.

The program includes planar, axisymmetric, shell and 3-D elements which have both displacement and linear scalar field degrees of freedom. The scalar field can be temperature, pore pressure or electric potential.

These elements are used for coupled temperature/stress problems; for effective stress/groundwater flow problems in soil mechanics; and for piezoelectric analysis.

Any number of linear or nonlinear user defined element types may be introduced into a model. Stiffness or mass matrices can be specified directly for linear user elements. Alternatively, a user subroutine may be used to define linear or nonlinear behavior for such elements.

3. CONTACT AND INTERACTIONS

ABAQUS/Standard contains an extensive set of tools for modeling contact and interface problems for stress analysis, heat transfer analysis, coupled stress-heat transfer cases, coupled pore fluid-stress analysis, and coupled acoustic pressure-structural response analysis.

Contact is typically modeled by identifying surfaces which may interact and pairing them by name. Interactions between deforming bodies or between a deforming body and a rigid body are allowed. Both small and finite sliding may be modeled in either two or three dimensions. A Coulomb friction model may be used for shear interaction or, for more complex response, a user subroutine may be used to define the frictional behavior.

In certain cases, contact elements, rather than contact surfaces are used to define the interaction between bodies.

4. FRACTURE MECHANICS

A contour integral capability is provided for calculating fracture mechanics parameters. The domain integral method is used to evaluate these quantities. Multiple contours around the crack tip are extracted automatically to verify path independence of the relevant quantity. The direction of crack growth may be defined by specifying the crack extension direction directly, or by specifying a normal to the crack plane. If the normal is given, the crack extension direction is determined automatically from the normal and the tangent to the crack front. There is a crack growth procedure to model material debonding.

5. MATERIAL DEFINITION

Examples of different material models and capabilities:

Temperature dependence: Most material constant definitions allow temperature dependence of the data. Most material parameters can also be made to depend on any number of predefined field variables, such as the density of a particular phase in a multi-phase material.

Orientation option: This option is provided so that a local coordinate system may be defined at each point for material property input and stress/strain component output. This is particularly convenient for laminated composite shell analysis.

Elasticity: Several different definitions of elastic behavior are provided. For linear elasticity the elastic moduli, including coefficients of thermal expansion, may be isotropic, orthotropic or anisotropic. Hypoelasticity allows the moduli to be dependent on strain. A voided material model is provided, in which the elastic part of the volume change depends on the logarithm of the pressure stress. For large strain elasticity two hyperelastic models are included: a general polynomial strain energy function (includes Mooney-Rivlin) and Ogden's model, both of which can be used for fully incompressible or almost incompressible response. Fully incompressible behavior is allowed through the use of hybrid (mixed displacement and pressure) elements.

Material constants for all large strain elasticity models may be specified directly or are calculated by ABAQUS/Standard from user supplied test data.

Viscoelasticity: Both time domain and frequency domain viscoelastic models are provided. The time domain model is available for small and finite strain and uses a Prony series representation. The Prony series constant are calculated by ABAQUS/Standard based on user specified test data. Viscoelastic materials can be specified as thermo-rheologically simple (TRS); the Williams-Landel-Ferry relationship is provided, and other TRS relationships can be specified via a user subroutine.

Metal plasticity: Isotropic with von-Mises yield criterion; anisotropic with Hill's anisotropic yield criterion. The flow rule is associated (normal) flow and the hardening rules are isotropic, kinematic and ORNL theory with perfect plasticity default. Gurson's porous plasticity model for porous metals is available. This includes both void nucleation and strain hardening of the matrix material.

Creep: Isotropic or anisotropic with time or strain hardening laws. Special creep laws can be defined with a user subroutine. ABAQUS/Standard automatically switches from explicit to implicit time stepping when the explicit time step is restricted by numerical stability considerations, thus providing for efficient solution of long time creep problems.

Volumetric swelling: Isotropic or anisotropic volume change with time as a function of field variables; tabular or user subroutine input.

Cam-clay model: Critical state plasticity, for clay-like soils. A generalized Cam-clay model that includes third invariant dependence in the yield definition and a "cap" on the yield surface with respect to pressure stress, with the Cam-clay strain hardening/softening rule.

Extended Drucker-Prager model: Provides a pressure dependent yield surface and includes strain hardening/softening, rate dependence, and with non-associated flow, for granular materials such as sand, and for material with different yield in tension and compression, such as polymers and cast iron.

Capped Drucker-Prager model: Combines the ingredients of the critical state model and the extended Drucker-Prager model. It is suitable for many geotechnical applications.

Jointed material model: For analysis of jointed or faulted materials. Frictional sliding along the joints, joint opening/closing and bulk material failure are included in the model.

Strain rate dependent plasticity: Hardening in all of the metal plasticity models as well as the extended Drucker-Prager and crushable foam models may be rate independent or rate dependent (visco-plastic). Rate dependent hardening data, derived from experiments, may be defined in a simple tabular fashion and will then be interpolated directly.

Concrete: Elastic-plastic-damage theory for concrete, including tension cracking, compression crushing, concrete-rebar interaction (via tension stiffening) and post-crack response using damaged elasticity concepts.

Permeability: Isotropic, orthotropic or fully anisotropic permeability with voids ratio and saturation dependence. May be velocity dependent.

Acoustic medium properties: May be frequency for temperature dependent.

Thermal conductivity: Isotropic, orthotropic or fully anisotropic, temperature dependent.

Specific heat: Temperature dependent.

User materials: User subroutine UMAT allows any material model to be implemented. ABAQUS provides for an arbitrary number of solution dependent state variables to be stored at each material calculation point, as well as for any number of material constants to be read as data, for use in this subroutine. This capability is very popular with many groups working on advanced material behavior.

6. ANALYSIS PROCEDURES

In ABAQUS/Standard, the analysis procedures can be mixed in any reasonable fashion. For example, a single simulation may include a nonlinear static analysis followed by nonlinear dynamics, in which case the final static solution provides the initial conditions for the dynamic response. Another example is natural frequency extraction which may include preloading effects due to an initial nonlinear static analysis.

In nonlinear analysis, the initial condition for each step is the state of the model at the end of the previous step. This provides a method for following complex loading histories.

A procedure is specified within each step. Available procedures are:

- **Linear and nonlinear stress/displacement analysis.**
- **Dynamic stress/displacement analysis for linear and nonlinear problems.**
- **Creep and swelling analysis.**
- **Transient and steady-state heat transfer analysis.**
- **Natural frequency extraction.**
- **Eigenvalue buckling estimates.**
- **Sequentially coupled temperature and thermal stress analysis.**
- **Fully coupled, transient or steady state, temperature-displacement analysis.**
- **Fully coupled acoustic-structural vibration analysis.**
- **Consolidation.**
- **Partially saturated flow.**
- **Mass diffusion analysis.**

7. PROCEDURES AND MODELS MADE BY CLAY TECHNOLOGY AND FEM-TECH

Besides the standard procedures and models included in the commercial version of ABAQUS adaptations of these to the behavior of buffer materials have been made as well as own procedures and models.

Thermally driven water vapor flow in water unsaturated soil is a process that has been modelled, coded and completely coupled to the other ABAQUS processes. The process has been modelled as a diffusion process driven by the temperature gradient.

Modified cap model. A material model for water saturated bentonite-rich buffer materials has been formulated and coded. The model is based on the capped Drucker-Prager model but has a curved failure envelope, a curved critical state line, and a variable Poisson's ratio.

Subject Index

Author Index